Advances in Number Theory

Advances in Number Theory

The Proceedings of the Third Conference of the
Canadian Number Theory Association
August 18–24, 1991
The Queen's University at Kingston

Edited by

FERNANDO Q. GOUVÊA
Department of Mathematics and Computer Science, Colby College
and
NORIKO YUI
Department of Mathematics and Statistics, Queen's University

in consultation with
A. Granville, R. Gupta, E. Kani, H. Kisilevsky,
R. Mollin, and C. Stewart

CLARENDON PRESS · OXFORD
1993

MATH-STAT.

Oxford University Press, Walton Street, Oxford OX2 6DP

Oxford New York Toronto
Delhi Bombay Calcutta Madras Karachi
Kuala Lumpur Singapore Hong Kong Tokyo
Nairobi Dar es Salaam Cape Town
Melbourne Auckland Madrid

and associated companies in
Berlin Ibadan

Oxford is a trade mark of Oxford University Press

Published in the United States by
Oxford University Press Inc., New York

A catalogue record for this book is available from the British Library

Library of Congress Cataloging in Publication Data
Canadian Number Theory Association. Conference (3rd : 1991 : Queen's University)
Advances in number theory : the proceedings of the Third
Conference of the Canadian Number Theory Association, August 18–24,
1991, the Queen's University at Kingston / edited by Fernando Q.
Gouvêa and Noriko Yui in consultation with A. Granville . . . [et al.].
1. Number theory—Congresses.   I. Gouvêa, Fernando Q. (Fernando
Quadros)   II. Yui, Noriko.   III. Title.
QA241.C275   1991   512'.7—dc20   92-42788
ISBN 0-19-853668-2

Printed in Great Britain by
Bookcraft (Bath) Ltd
Midsomer Norton, Avon

Table of Contents

Chapter I: Plenary Addresses

Chapter II: Analytic Number Theory

Invited Addresses

Contributed Talks

Chapter III: Arithmetical Algebraic Geometry

Invited Addresses

Contributed Talks

Chapter IV: Diophantine Approximation

Invited Addresses

Contributed Talks

Chapter V: Session in Honour of Paulo Ribenboim

Plenary Address

Invited Addresses

Speeches at the Banquet in Honour of P. Ribenboim

Preface

The Canadian Number Theory Association was founded in 1987 with the purpose of enhancing scientific communication and exchange among number theorists in Canada. One of the means to that goal is the organization of conferences that bring together many number-theorists from all over the world. The first such conference was held at the Banff Conference Center in 1988, and the second at the University of British Columbia in 1989.

The third conference of the CNTA was hosted by Queen's University in August 18–24, 1991, as part of the celebration of Queen's sesquicentennial anniversary. The conference was partially dedicated to Paulo Ribenboim on the occasion of his retirement from Queen's. There were 177 registered participants: 50 from Canada and 127 from abroad.

The conference met in plenary sessions in the mornings, and was divided into sessions focusing on specific areas in the afternoon. These areas were

  i. Analytic Number Theory,

  ii. Arithmetical Algebraic Geometry, and

  iii. Diophantine Approximation.

A special session was organized on the afternoon of the 24th in honour of Paulo Ribenboim, who has been a central figure among Canadian number-theorists and who will be retiring from Queen's University in the near future.

There were 33 invited addresses: 10 were presented at plenary sessions and 23 were given at the various specific sessions. A full list of these speakers is included in this volume. In addition, 37 contributed papers were presented at the conference.

The conference activities were overseen by the Scientific and Organizing Committee, whose members were Andrew Granville (Georgia), Fernando Gouvêa (Queen's and Colby), Rajiv Gupta (UBC), Ernst Kani (Queen's), Hershy Kisilevsky (Concordia), Richard Mollin (Calgary), Cameron Stewart (Waterloo) and Noriko Yui (Queen's).

The conference was supported in part by a conference grant from the Natural Sciences and Engineering Research Council of Canada (NSERC). It also received financial support for young participants from the National

Science Foundation and the National Security Agency of the United States. Invited speakers for the special session in honor of P. Ribenboim received partial support from Queen's University. We thank all of these organisations for their generosity. Thanks are also due to the supporting members of the organizing committee, M. Lambert and C. Burns, for their help in organizing the conference.

This book contains the Proceedings of the Third Conference of the Canadian Number Theory Association, edited by Fernando Gouvêa and Noriko Yui in consultation with the Scientific Committee.

The editors were not able to persuade all the lecturers to write up their talks for these proceedings. Nevertheless, the size of this volume (which contains 38 articles) is a testimony to the variety and interest of the material presented at the conference. For the most part, the articles in this volume reflect the contents of the talks given at the conference, though of course they often go beyond those limits and report also on later developments and/or related material. We hope the result is not only a record of the conference activities, but also of intrinsic mathematical interest.

The editors wish to express warm appreciation to all the contributors for their effort in preparing manuscripts. We would also like to offer the warmest thanks to the many referees who helped to evaluate the papers submitted for the Proceedings.

That this volume exists at all in the present form is a tribute to the power and flexibility of $\TeX$. In the course of its preparation, we made use of most of the popular flavors of $\TeX$ (including "plain" $\TeX$, $\LaTeX$, $\mathcal{A}\mathcal{M}\mathcal{S}$-$\TeX$, $\mathcal{A}\mathcal{M}\mathcal{S}$-$\LaTeX$, and even $\LaTeX\mathcal{A}\mathcal{M}\mathcal{S}$-$\TeX$) and of implementations of the basic $\TeX$ program for UNIX, OS/2, and Macintosh operating systems. We would like to express our indebtedness to Donald E. Knuth, author of $\TeX$, to the authors of the various macro packages, and to the authors of emTeX (for PC's running DOS or OS/2) and of DirectTeX (for the Macintosh). Finally, we would offer exhausted thanks to those authors that followed our instructions in the formatting of their papers.

The next conference of the Canadian Number Theory Association will be held at Dalhousie University in 1994. We hope it will be as successful as the third conference. See you there!

Fernando Q. Gouvêa
Noriko Yui
August 1992

List of Invited Addresses

Takayuki Oda, *Galois Action on the Pro-ℓ Fundamental Group of an Algebraic Curve*

Lucien Szpiro, *On the Size of Solutions of Certain Algebraic Equations*

Yuri G. Zarhin, *Abelian Varieties of K3 Type and Ordinary K3 Surfaces over Finite Fields*

JOINT SPECIAL SESSION IN ARITHMETICAL ALGEBRAIC GEOMETRY AND DIOPHANTINE APPROXIMATION

Paul Vojta, *Rational Points on Subvarieties of Abelian Varieties*

SPECIAL SESSION IN DIOPHANTINE APPROXIMATION

David Boyd, *Linear Recurrence Relations for some Generalized Pisot Sequences*

W. Dale Brownawell, *How Fast Can Polynomials Get Small?*

Jan-Hendrik Evertse, *Lower Bounds of Discriminants of Binary Forms*

Julia Mueller, *Trinomial Thue's Equations over Function Fields*

Jeff Vaaler, *On the Number of Irreducible Factors of a Polynomial*

SPECIAL SESSION IN HONOUR OF PAULO RIBENBOIM

Karl Dilcher, *A New Criterion for the First Case of Fermat's Last Theorem*

Michael Filaseta, *Short Interval Results for k-Free Values of Irreducible Polynomials*

Ján Mináč, *Demuškin Groups as Galois Groups*

Richard Mollin, *On Real Quadratic Fields, Continued Fractions, Reduced Ideals, Prime-Producing Quadratic Polynomials, and Quadratic Residue Covers*

List of Contributed Talks

S. Zhang, *Positive Line Bundles in Arithmetic Surfaces*

T. Crespo, *Construction of $\tilde{S}_4$-Fields and Modular Forms of Weight 1*

J. Morales, *Invariants for Pairs of Quadratic Forms*

H. Cohn, *Projection from an Algebraic Quadratic Form to Rational Quadratic Forms*

A. Laing, *Shimura Reciprocity for Modular Functions with Rational Fourier Coefficients*

P. Feit, *On the Transition from Local to Global*

A. Escassut and M.-C. Sarmant, *Conditions for Injectivity of Analytic Elements*

P.-J. Cahen, *Parties Pleines d'un Anneau Noetherien*

Sessions in Diophantine Approximation

Deanna Caveny, *Transcendence and Algebraic Independence Results for Well-Approximated Numbers*

Marie J. Bertin *K–Nombres de Pisot et de Salem*

A. Van der Poorten, *Continued fractions of formal power series*

P.-G. Becker, *An Invariance Principle for the Transcendency of Power Series with Finitely Many Different Coefficients*

M. Langevin, *On the Generalized Problems of Favard and Lehmer*

N. Hirata-Kohno, *Points Entiers sur les Courbes Algébriques de Genres ≥ 1*

D. Roy, *On the v-adic Independence of Algebraic Numbers*

H. Faure, *Discrepance et Diaphonie en Dimension Un*

R. Cook, *Cubic Inequalities of Additive type*

Unscheduled Talks

Y. Tschinkel, *Rational Points on Varieties*

U. Jannsen, *Motives, numerical equivalence and semi-simplicity*

T. Viswanathan, *Two Algorithms for Completing Latin Squares and the Theorems of Ryser and Evans*

Conference Participants

S. Arno	Supercomputing Research Centre
M. Bean	University of Waterloo
P.-G. Becker	Pennsylvania State University
R. A. Bell	City College, CUNY
M. Bennett	University of British Columbia
G. Berger	Columbia University
M. J. Bertin	Université Paris VI
F. Beukers	University of Utrecht
E. Bombieri	Institute for Advanced Study
P. Borwein	Dalhousie University
D. Boyd	University of British Columbia
D Bradley	University of Illinois
W. D. Brownawell	Pennsylvania State University
A. Brumer	Fordham University
E. Burger	University of Waterloo
D. M. Busey	
P. J. Cahen	Université d'Aix-Marseille III
J. W. S. Cassels	University of Cambridge
D. M. Caveny	College of Charlston, South Carolina
J.-L. Chabert	
C. E. Chace	Williams College
W. K. Chan	University of Michigan
D. Clark	McGill University
J. H. Coates	University of Cambridge
H. Cohn	City College, CUNY
J. B. Conrey	Oklahoma State University
R. J. Cook	University of Sheffield
T. Crespo	University of Barcelona
C. David	McGill University
L. Davidson	Laurentian University
J.-M. DeKoninck	Laval University
K. Dilcher	Dalhousie University
E. Dubois	Université de Caen
D. S. Dummit	University of Vermont

H. M. Edgar	San Jose State University
A. Escassut	
J. H. Evertse	University of Leiden
D. Farmer	Oklahoma State University
H. Faure	Université de Provence
P. Feit	Princeton University
M. Filaseta	University of South Carolina
G. Fractman	Princeton University
G. Frei	ETH-Zürich
J. B. Friedlander	University of Toronto
C. Friesen	Erindale College, University of Toronto
F. Garvan	University of Waterloo
G. Giordano	Ryerson Polytechnical Institute
B. Gold	Ohio State University
D. Goldfeld	Columbia University
D. Goldston	San Jose State University
Y. Goto	Queen's University
F. Gouvêa	Queen's University
A. Granville	University of Georgia
R. Gupta	University of British Columbia
S. Gurak	University of California at San Diego
F. Hajir	Massachusetts Institute of Technology
R. Heath-Brown	Magdalen College, University of Oxford
D. Hedi	Université Paris Sud
N. Hirata-Kohno	Tokyo Institute of Technology
J. Hoffstein	Brown University
D. Holland	McMaster University
J. Holt	University of Texas
W. L. Hoyt	Rutgers University
J. G. Huard	Carleton University
H. Iwaniec	Rutgers University
U. Jannsen	Universität Köln
K. Johnson	Dalhousie University
M. Jutila	University of Turku, Finland
S. Kanemitsu	Kinki University, Japan
E. Kani	Queen's University
P. Kaplan	Université de Nancy I
Y. Karamatsu	Utsunomiya University
Y. Kato	Chuo University, Japan

M. A. Kenku	University of Lagos, Nigeria
H. Kisilevsky	Concordia University
M. Klassen	University of Arizona
A. Knopfmacher	University of the Witwatersrand
T. Kodama	Kyushu University, Japan
M. Kolster	McMaster University
A. Kuribayashi	Tokoha Galuen Hamamatsu University
M. Kuwata	McGill University
J. Labute	McGill University
J. Lagarias	AT&T Bell Laboratories
A. J. Laing	University of Maryland
E. Lamprecht	Universität des Saarlandes
W. E. Lang	Brigham Young University
M. Langevin	Institut Henri Poincaré
J.-C. Lario	Universitat Politècnica de Catalunya
O. Lecacheux	Université Paris VI
C. Levesque	Laval University
D. J. Lewis	University of Michigan
P. Liardet	Université de Provence
Q. Liu	Université de Bordeaux 1
L. Mai	University of Montreal
M. Marshall	University of Saskatchewan
T. Matala-Aho	University of Oulu, Finland
M. Matignon	Université de Bordeaux
W. McDaniels	University of Missouri
J. McKay	Concordia University
J. Meng	Columbia University
J. Mináč	University of Western Ontario
R. Mollin	University of Calgary
H. L. Montgomery	University of Michigan
J. Morales	Louisiana State University
C. J. Mozzochi	Institute for Advanced Study
J. Mueller	Fordham University
E. Murray	University of Waterloo
R. Murty	McGill University
V. K. Murty	University of Toronto
K. Nagasaka	College of Engineering, Hosei University
T. Nakahara	Saga University, Japan
J. Neukirch	Universität Regensburg

J.-L. Nicolas	Université de C. Bernard
N. O. Nygaard	University of Chicago
T. Oda	RIMS, Kyoto University
J. Oesterlé	Institut Henri Poincaré and Université Paris VI
K. Oskolkov	Steklov Institute, Moscow
E. N. Ozgener	Oklahoma State University
C. A. Ozgener	Oklahoma State University
F. Pappalardi	McGill University
J. Pihko	University of Helsinki
J. Pila	Columbia University
C. G. Pinner	Univeristy of Texas
N. Pitt	Rutgers University
R. Ramakrishna	Princeton University
G. Rhin	Université de Metz
F. Rhin	Université de Metz
P. Ribenboim	Queen's University
L. Roberts	University of British Columbia
M. Robinson	Supercomputing Research Center
F. Rodriguez Villegas	Princeton University
D. Roy	Laval University
J. W. Sands	University of Vermont
I. Satake	Chuo University, Japan
A. L. Schmidt	University of Copenhagen
E. J. Scourfield	University of London
R. D. Silverman	MITRE Corp.
C. Simons	McGill University
L. D. Simons	St. Michael's College
C. J. Smyth	University of Edinburgh
V. Snaith	McMaster University
A. Sofer	Ohio State University
L. Somer	
C. Spiro	
H. Stark	University of California at San Diego
T. Stefanicki	McGill University
C. Stewart	University of Waterloo
O. Such	
L. Szpiro	Université Paris VII
W. Tautz	Universität GHS Essen

N. Q. Thang	University of Waterloo
P. E. Thomas	University of California at Berkeley
J. Thunder	University of Michigan
J. Top	Erasmus University, Rotterdam,
Y. Tschinkel	Massachusetts Institute of Technology
J. Vaaler	University of Texas
A. van der Poorten	Macquarie University, Australia
P. van Mulbregt	Wellesley College
M. van Rossum-Wijsmuller	LaSalle University
T. M. Viswanathan	Uviversity of North Carolina
P. Vojta	University of California at Berkeley
P. Voutier	University of Colorado
S. S. Wagstaff, Jr.	Purdue University
M. Waldschmidt	Institut Henri Poincaré
	and Université Paris VI
G. Walsh	University of Waterloo
L. Washington	University of Maryland
K. Williams	Carleton University
S. Wong	Massachusetts Institute of Technology
R. Xiao	University of Waterloo
H. Yokota	Hiroshima Institute of Technology
J. T. Yu	University of Notre Dame
N. Yui	Queen's University
D. Zagier	Max Planck Insitut, Bonn
	and University of Utrecht
Y. Zarhin	Academy of Science, USSR
B. Zhang	Columbia University
S. Zhang	Institute for Advanced Study
Y. Zhang	McGill University

List of Contributors

Invited Addresses:

F. BEUKERS, Department of Mathematics, University of Utrecht, P.O. Box 80.010, 3508 TA Utrecht, the Netherlands

D. BOYD, Department of Mathematics, University of British Columbia Vancouver, British Columbia V6T 1Z2, Canada

W. D. BROWNAWELL, Department of Mathematics, Pennsylvania State University, University Park, PA 16802, USA

J.-H. EVERTSE, Department of Mathematics and Computer Science, University, of Leiden, P.O. Box 9512, 2300 RA Leiden, the Netherlands

J. B. FRIEDLANDER, Scarborough College, University of Toronto, Toronto, Ontario M5S 1A1, Canada

A. GRANVILLE, Department of Mathematics, University of Georgia, Athens, GA 30602, USA

R. HEATH-BROWN, Magdalen College, University of Oxford, Oxford OX1 3LB, England

R. IWANIEC, Department of Mathematics, Rutgers University, New Brunswick, NJ 08903, USA

M. JUTILA, Department of Mathematics, University of Turku, SF-20500, Turku, Finland

H. W. LENSTRA JR., Department of Mathematics, University of California at Berkeley, Berkeley, CA 94720, USA

S. LOUBOUTIN, Mathematics Department, University of Caen, F-14032 Caen Cedex, France

J. MINÁČ, Department of Mathematics, University of Western Ontario, London, Ontario N6A 5B7, Canada

R. MOLLIN, Department of Mathematics, University of Calgary, Calgary, Alberta T2N 1N4, Canada

H.L. MONTGOMERY, Department of Mathematics, University of Michigan, Ann Arbor, MI 48019-1003, USA

J. MUELLER, Department of Mathematics, Fordham University, New York, NY 10458, USA
Current address: Department of Mathematics, Columbia University, New York, NY 10027, USA

V. K. MURTY, Department of Mathematics, University of Toronto, Toronto, Ontario M5S 1A1, Canada

M. R. MURTY, Department of Mathematics, McGill University, Montreal, Québec H3A 2K6, Canada

T. ODA, Research Institute for Mathematical Sciences, Kyoto University, Kyoto 606, Japan

F. RODRIGUEZ VILLEGAS, Department of Mathematics, Princeton University, Princeton, NJ 08544, USA

P. VOJTA, Department of Mathematics, University of California at Berkeley, Berkeley, CA 94720, USA

M. WALDSCHMIDT, Université P. et M. Curie (Paris VI), Institut Henri Poincaré, 11, rue P. et M. Curie, F-75231 Paris Cedex 05, France

H.C. WILLIAMS, Department of Computer Science, University of Manitoba, Winnipeg, Manitoba R3T 2N2, Canada

D. B. ZAGIER, Max-Planck-Institut für Mathematik, Gottfried-Claren Strasse 26, 5300 Bonn 3, Germany, and University of Utrecht, P.O. Box 80.010, 3508 TA Utrecht, the Netherlands

YURI ZARHIN, Institute for Mathematical Problems in Biology, Russian Academy of Sciences, Pushchino, Moscow Region 142292, Russia
Current address: Department of Mathematics, Pennsylvania State University, University Park, PA 16802, USA

Contributed Talks:

M.J. BERTIN, Université P. et M. Curie (Paris VI), Mathématiques, 4 Place Jussieu, F-75005 Paris, France

J. BRÜDERN, Mathematisches Institut, Universität Göttingen 3400 Göttingen, Germany

H. COHN, Mathematics Department, City College (CUNY), New York, NY 10031, USA

R.J. COOK, Department of Pure Mathematics, University of Sheffield, Sheffield S3 7RH, England

T. CRESPO, Departament d'Àlgebra i Geometria, Universitat de Barcelona, Gran Via de les Corts Catalanes 585 – 08007 Barcelona, Spain

H. FAURE, UFR–MIM et URA 225, Université de Provence, 3 Place Victor-Hugo, F-13331 Marseilles Cedex 3, France

P. FEIT, Department of Mathematics, University of Toledo, Toledo, OH 43606, USA
Current address: University of Texas/Permian Basin, 4901 E. University Blvd., Odessa, TX 79762, USA

N. HIRATA-KOHNO, Department of Mathematics, Tokyo Institute of Technology, Oh-Okayama, Meguro Tokyo 152, Japan

A. KNOPFMACHER, Department of Computational and Applied Mathematics, University of the Witwatersrand, Johannesburg, South Africa

J. KNOPFMACHER, Department of Mathematics, University of the Witwatersrand, Johannesburg, South Africa

A.J. LAING, Department of Mathematics, University of Maryland, College Park, MD 20742, USA

M. LANGEVIN, UFR de Mathématiques, Université de Lille, F-59655 Villeneuve d'Ascq Cedex, France

J.-C. LARIO, Departament de Matemàtica Applica II. UPC. Barcelona, Spain

O. LECACHEUX, Université Paris VI, Place Jussieu F-75252 Paris Cedex 06, France

T. NAKAHARA, Department of Mathematics, Faculty of Science and Engineering, Saga University, Saga 840, Japan

J. PIHKO, Department of Mathematics, University of Helsinki, Hallituskatu 15, SF-00100 Helsinki, Finland

A.J. VAN DER POORTEN, School of Mathematics, Computing and Electronics, Macquarie University, NSW 2109, Australia

D. ROY, Départment de Mathématiques, Université Laval, Ste-Foy, P. Québec G1K 7P4, Canada

J. TOP, Erasmus University Rotterdam, Vakgroep Wiskunde, P.O. Box 1738, 3000 DR Rotterdam, the Netherlands
Current address: Department of Mathematics, University of Groningen, P. O. Box 800, 9700 AV Groningen, the Netherlands

Y. TSCHINKEL, Department of Mathematics, Massachussets Institute of Technology, Cambridge, MA 02139, USA
Current address: Department of Mathematics, Harvard University, Cambridge, MA 02138, USA

T. M. VISWANATHAN, Mathematical Sciences, University of North Carolina at Wilmington, Wilmington, NC 28403, USA

R. WARLIMONT, Fachbereich Mathematik, Universität Regensburg, Universitätsstrasse 31, 8400 Regensburg, Germany

Chapter I

Plenary Addresses

Exotic values of G-functions

F. Beukers

University of Utrecht

1 Introduction

Let $f(z)$ be a power series of the form

$$f(z) = \sum_{n=0}^{\infty} f_n z^n \tag{1.1}$$

where the numbers f_n belong to an algebraic number field K ($[K : \mathbb{Q}] < \infty$). For the moment we shall assume K embedded in $\mathbb{C}$. Suppose that the radius of convergence ρ of $f(z)$ is positive. Suppose also that $f(z)$ satisfies a linear differential equation of order m,

$$p_m(z)f^{(m)} + p_{m-1}(z)f^{(m-1)} + \cdots + p_1(z)f' + p_0(z)f = 0 \tag{1.2}$$

with $p_i(z) \in K[z]$ for all i. The question we are interested in is the following,

Question: *Let $\xi \in \bar{\mathbb{Q}}$ be an algebraic number with $|\xi| < \rho$. Is $f(\xi)$ rational, irrational or transcendental ?*

It turns out that with the conditions imposed so far one cannot say very much in general. However, there are two special cases, both initiated by C. L. Siegel, to which transcendence methods can be applied. The first case is the case of *E-functions*. The series $f(z)$ is called an E-function if the coefficients f_n can be written in the form $b_n/n!$, where

i. $\overline{|b_n|} = O(n^{\epsilon n})$ for all $n > 0$ and any $\epsilon > 0$.

ii. (common denominator of $b_0, \ldots, b_n$) $= O(n^{\epsilon n})$ for all $n > 0$ and any $\epsilon > 0$.

Here $\overline{|\xi|}$ denotes the maximum of the absolute values of all algebraic conjugates of ξ. The best known examples are e^z (hence the name E-functions) and the Bessel function

$$J_0(z) = \sum_{n=0}^{\infty} \frac{(-z^2/4)^n}{(n!)^2} = \sum \binom{2n}{n} \left(-\frac{1}{4}\right)^n \frac{z^{2n}}{(2n)!}.$$

3

Transcendence results for values of E-functions are well-established and it is known that if $f(z)$ is an E-function and not a polynomial then $f(\xi)$ is transcendental for any algebraic $\xi \notin \{0, \mathcal{S}$ where $\mathcal{S}$ is the set of singularities of the differential equation (see [S], [Sh]).

The second class is the class of *G-functions*. The series $f(z)$ is called a G-function if

i. $\exists c_1$ such that $\overline{|f_n|} = O(c_1^n)$

ii. $\exists c_2$ such that (common denominator of $f_0, \ldots, f_n$) $= O(c_2^n)$.

The best known examples are of course the geometric series (hence the name G-functions), the algebraic functions, the polylogarithms

$$L_k(z) = \sum_{n=1}^{\infty} \frac{z^n}{n^k}$$

and the hypergeometric functions

$$F(a, b, c|z) = \sum_{n=0}^{\infty} \frac{(a)_n (b)_n}{(c)_n n!} z^n$$

with $a, b, c \in \mathbb{Q}$. Before we discuss the arithmetic nature of values of G-functions we like to draw attention to another interesting problem. Notice that the G-function condition (ii) entails that the denominators of the coefficients of f increase at most exponentially in n. However, if one writes down an arbitrary linear differential equation then the denominators of the coefficients of a power series solution tend to increase with a power of $n!$. As an example consider

$$z(z^2 - 1)f'' + (3z^2 - 1)f' + (z + B)f = 0 \tag{1.3}$$

where B is an arbitrary rational number. The coefficients f_n of a power series solution $f(z)$ satisfy the recurrence relation

$$n^2 f_n = B f_{n-1} + (n-1)^2 f_{n-2} \qquad f_0 = 1, \ f_1 = B.$$

So, when $B \in \mathbb{Z}$, we see from the recurrence that the denominators of f_n grow a priori with $(n!)^2$. Therefore a small miracle is required in order to have exponential growth of the denominators. This happens for example when $B = 0$. Then $f_{2n+1} = 0, f_{2n} = \binom{2n}{n}^2 / 16^n$. Nothing is known in this respect for other values of B, but we expect that $B = 0$ is the only value of B for which Eq(1.3) has a G-function solution. This automatically raises the following question,

Question: *What are G-functions?*

A conjecture of E. Bombieri and B. Dwork states that differential equations having G-function solutions 'arise' from Gauss-Manin systems corresponding to 1-parameter families of algebraic varieties. For a more extended formulation and a discussion we refer to [A, Chapters 1, 2].

To return to our original question, in the case of G-functions relatively little is known about the arithmetic nature of function values at algebraic points. For example, is $\zeta(5) = L_5(1)$ (ir)rational, transcendental? For the few things that are known we refer to [CC] or [Bo]. A few years ago J. Wolfart [W] discovered that there exist transcendental G-functions which assume algebraic values in a topologically dense set of algebraic points. The best known example is $F(\frac{1}{12}, \frac{5}{12}, \frac{1}{2}|z)$ and as example of a peculiar evaluation we quote from [BW],

$$F\left(\frac{1}{12}, \frac{5}{12}, \frac{1}{2} \Big| \frac{1323}{1331}\right) = \frac{3}{4}\sqrt[4]{11}. \tag{1.4}$$

One may notice that the series on the left hand side of Eq(1.4) also converges 7-adically. The 7-adic radius of convergence of the hypergeometric function is 1 and the 7-adic valuation of $1323/1331$ is less than 1. Numerical experiment convinced me that

$$F\left(\frac{1}{12}, \frac{5}{12}, \frac{1}{2} \Big| \frac{1323}{1331}\right)_7 = \frac{1}{4}\sqrt[4]{11}.. \tag{1.5}$$

and recently this was proved, along with many other such examples, in [Be]. One may note the striking similarity between Eq(1.4) and Eq(1.5). Evaluations such as these will be the topic of this paper. In section 2 we shall give some background on v-adic radii of convergence and related subjects. Then, in Section 3 we discuss families of elliptic curves and the Picard-Fuchs equations that arise from them. Finally, in Section 4 we give an overview of the results on function values together with examples. For full proofs of the theorems we have to refer to the forthcoming paper [Be].

2 Radii of convergence

In the introduction we assumed that the number field K was embedded in $\mathbb{C}$ and we found a corresponding radius of convergence of f. Denoting the embedding by ϕ we observe that this automatically gives us an absolute value or valuation $||.||$ on K via $||a|| = |\phi(a)|$ where $a \in K$ and $|.|$ is the ordinary absolute value on $\mathbb{C}$. However there may exist different embeddings $K \hookrightarrow \mathbb{C}$ and corresponding valuations, each of which gives rise to a radius of convergence of f. We would now like to take all valuations of K into consideration, including the finite ones.

Let v be a valuation on K and let K_v be the completion with respect to v. We shall assume the valuations of K normalized as follows,

i. $|x|_v = |x|^{e_v}$ for infinite v. Here $|.|$ is the ordinary absolute value on $\mathbb{C}$ and $e_v = 1$ if v is real, $e_v = 2$ if v is complex.

ii. $|p|_v = p^{-[K_v:\mathbb{Q}_p]}$ for finite v and where p is the unique prime with v-adic norm less than 1.

With these normalisations we have the following rule,

Theorem 2.1. (Product rule) *For any $x \in K$, $x \neq 0$ we have*

$$\prod_v |x|_v = 1$$

where the product is taken over all normalized valuations of K.

Using valuations the G-function condition can be stated more elegantly. Namely, $f(z)$ is a G-function if $\exists c > 0$ such that

$$\sum_v \log(\max(|f_0|_v, |f_1|_v, \ldots, |f_n|_v)) < c \cdot n \qquad (2.1)$$

for all $n \in \mathbb{N}$ and where the sum runs over all normalized valuations v of K.

Now consider our power series solution $f(z) \in K[[z]]$ of a linear differential equation again. For any valuation v of K we can define the v-adic radius of convergence ρ_v of f by

$$\rho_v = \liminf_{n \to \infty} |f_n|_v^{-1/n}.$$

One easily sees that $f(\xi)$ converges v-adically whenever $|\xi| < \rho_v$. It is also very easy to see that if $\rho_v < 1$ for a finite valuation v then there must be an exponentially increasing contribution of primes p, corresponding with v, to the denominators of f_n. In fact, we can be more precise in this respect via a theorem of Dwork and Robba [DR, Thm 4.3] of which we state a slightly weakened version.

Theorem 2.2. *Let v be a finite valuation of K. Suppose that 0 is not a singularity of Eq(1.2) and that there exists a basis of solutions in $K[[z]]$ of Eq(1.2) having common v-adic radius of convergence ρ_v. Suppose also that the Wronskian of Eq(1.2) does not vanish in the v-adic disk $\{x \mid |x|_v < \rho_v\}$. Let $f \in K[[z]]$ be a solution of Eq(1.2) with Taylor coefficients f_n. Then,*

$$|f_n|_v \leq |\rho|_v^{-n} |\text{lcm}(1, \ldots, n)|_v^{-(m-1)} \max(|f_0|_v, \ldots, |f_{m-1}|_v)$$

where $\text{lcm}(1, \ldots, n)$ denotes the lowest common multiple of $1, \ldots, n$.

Remark 2.3. *As a consequence of the prime number theorem one can show that $\log(\text{lcm}(1, \ldots, n))/n \to 1$ as $n \to \infty$, i.e $\text{lcm}(1, \ldots, n)$ is bounded exponentially in n.*

We have the following corollary of Theorem 2.2 which is of interest to us.

Corollary 2.4. *Suppose that zero is not a singularity of Eq(1.2). Let $u_1, \ldots, u_n$ be a basis of solutions of Eq(1.2) in $K[[z]]$. For each valuation v let ρ_v be their common radius of v-adic convergence. Suppose that $\rho_v > 0$ for all infinite v and $\rho_v = 1$ for almost all finite v ('almost' means 'with finitely many exceptions'). Then $u_1, \ldots, u_n$ are G-functions.*

Proof. Let S be the set consisting of all infinite valuations, all finite valuations with $\rho_v < 1$ and all finite valuations such that the wronskian of Eq(1.2) has a zero with v-adic valuation less than 1. Notice that the set S is finite. For each $v \in S$ we can use the rough estimate $\log(|f_n|_v) < n(1 - \log(\rho_v)) + O(1)$. For each $v \notin S$ we can use Theorem 2.2 to find $\log(|f_n|_v) \leq -(m-1)\log(|\mathrm{lcm}(1, \ldots, n)|_v)$. Hence

$$\sum_v \log(\max(|f_0|_v, |f_1|_v, \ldots, |f_n|_v))$$

is less than

$$\sum_{v \notin S} -(m-1)\log(|\mathrm{lcm}(1, \ldots, n)|_v) + O(n).$$

Using the product formula we see that the latter sum equals

$$\sum_{v \in S} (m-1)\log(|\mathrm{lcm}(1, \ldots, n)|_v).$$

Using the fact that finite valuations of an integer are at most 1, the latter sum can be bounded above by

$$\sum_{v \text{ infinite}} (m-1)\log(|\mathrm{lcm}(1, \ldots, n)|_v) = (m-1)[K : \mathbb{Q}]\log(\mathrm{lcm}(1, \ldots, n)).$$

According to Remark 2.3 the latter term is $O(n)$. $\qquad\square$

A question, which seems to be very difficult, is whether the converse of Corollary 2.4 holds. So, suppose we have a basis of G-function solutions of Eq(1.2). Is it true then that $\rho_v = 1$ for almost all v?

Let us finish this section by noting we how unpredictably function values can behave depending on the kind of convergence we impose. Consider for example the power series

$$-\log(1 - z) = \sum_{n=1}^{\infty} \frac{z^n}{n}.$$

Let ζ be a p^r-th root of unity, where p is a prime and $r \in \mathbb{N}$ arbitrary. Consider a complex valuation such that $|1-\zeta| < 1$. Then, in the completion of $\mathbb{Q}(\zeta)$ with respect to this valuation (so, $\mathbb{C}$) we have,

$$-\log(\zeta) = -\log(1 - (1 - \zeta)) = 2\pi i m/p^r$$

where $m \in \mathbb{Z}$ depends on the valuation we have chosen. For any p-adic number u with $|u|_p < 1$ we have the rule $\log(1-(1-(1-u)^{p^r})) = p^r \log(1-u)$ for any $r \in \mathbb{N}$. Let v be a finite valuation on $\mathbb{Q}(\zeta)$ which extends the p-adic valuation on $\mathbb{Q}$, then $|1-\zeta|_v < 1$. Apply the above rule with $u = 1-\zeta$ to discover that

$$-p^r \log(\zeta) = -p^r \log(1 - (1 - \zeta)) = \log(1 - (1 - \zeta^{p^r})) = 0.$$

In particular, $\log(1 - z)$ has infinitely many algebraic zeros in the p-adic domain of convergence, which is utterly unlike the complex case. On the other hand it might also be that complex and p-adic evaluations are very similar, as suggested by the example in the introduction. The theorems from section 4 will contain examples of both behaviours.

3 Picard–Fuchs equations

As usual we let K be an algebraic number field and consider the family of elliptic curves $\mathcal{E}_t$ given by

$$y^2 = x^3 + A(t)x + B(t), \qquad A(t), B(t) \in K[t].$$

We shall assume that we have good reduction at $t = 0$ that is, $\Delta(0) \neq 0$, where $\Delta(t) = \Delta(\mathcal{E}_t) = 4A(t)^3 + 27B(t)^2$ and we assume that $j(t) = j(\mathcal{E}_t)$ is non-constant. Consider the differential forms

$$\Omega(t) = \frac{dx}{2y}, \qquad N(t) = x\frac{dx}{2y} + \gamma\frac{dx}{2y}$$

on $\mathcal{E}_t$, where $\gamma \in K$ is to be specified later if necessary. There is a natural connection on the De Rham cohomology of $\mathcal{E}_t$ which is known as the Gauss–Manin connection. In a naive way one can describe this as follows. Differentiate $\Omega(t), N(t)$ with respect to t, assuming that x is horizontal, that is, $dx/dt = 0$. Then the results modulo exact forms will be linear combinations of $\Omega(t), N(t)$ with coefficients in $K(t)$. Hence there exists $\mathcal{A} \in M_2(K(t))$ such that

$$\frac{d}{dt}\begin{pmatrix} \Omega(t) \\ N(t) \end{pmatrix} \equiv \mathcal{A}\begin{pmatrix} \Omega(t) \\ N(t) \end{pmatrix} \quad \text{(Modulo exact forms)}.$$

Explicit calculation with $\gamma = 0$ gives us

$$12\Delta(t)\frac{d}{dt}\begin{pmatrix}\Omega(t)\\N(t)\end{pmatrix} = \begin{pmatrix}-\Delta'(t) & 18\delta(t)\\6A(t)\delta(t) & \Delta'(t)\end{pmatrix}\begin{pmatrix}\Omega(t)\\N(t)\end{pmatrix}$$

$$+d\left(\frac{1}{y}\begin{pmatrix}12A(t)\delta(t) - \Delta'(t)x + 18\delta(t)x^2\\-18B(t) - 6A(t)\delta(t)x - \Delta'(t)x^2\end{pmatrix}\right)$$

where $\delta(t) = -2A(t)B'(t) + 3B(t)A'(t)$. The two by two system of first order linear differential equations given by

$$\frac{d}{dt}\begin{pmatrix}y_1\\y_2\end{pmatrix} = A\begin{pmatrix}y_1\\y_2\end{pmatrix} \tag{3.1}$$

is called the *Picard-Fuchs equation* corresponding to $\mathcal{E}_t, \Omega(t), N(t)$. Because $\Delta(0) \neq 0$, A has no pole at $t = 0$, hence there exists a unique $M(t) \in M_2(K[[t]])$ such that

$$\frac{d}{dt}M(t) = A(t)M(t) \text{ and } M(0) = \begin{pmatrix}1 & 0\\0 & 1\end{pmatrix}$$

Notice that $\frac{d}{dt}(\det M) = \text{Tr}A \cdot \det M$. Since $\text{Tr}A = 0$ we have $\det M(t) = 1$. We call $M(t)$ the *fundamental solution matrix* of Eq(3.1) at $t = 0$.

Remark 3.1. *When we consider $H^1_{DR}(\mathcal{E}_t)$ over $K[[t]]$ one easily checks that the components of $M(t)^{-1}\begin{pmatrix}\Omega(t)\\N(t)\end{pmatrix}$ form a basis of H^1_{DR} which is horizontal with respect to d/dt.*

Remark 3.2. *From differential Galois theory it easily follows that the transcendence degree of the field generated by the elements of $M(t)$ over $\mathbb{C}(t)$ is three.*

Remark 3.3. *Each component of a solution of Eq(3.1) satisfies a second order linear differential equation as can be verified via the following general rule. Let y_1, y_2 satisfy*

$$\frac{d}{dt}\begin{pmatrix}y_1\\y_2\end{pmatrix} = \frac{1}{\Delta}\begin{pmatrix}p & q\\r & s\end{pmatrix}\begin{pmatrix}y_1\\y_2\end{pmatrix}.$$

Then

$$q\Delta^2 y_1'' + \Delta(q\Delta' - q'\Delta - q(p+s))y_1' + ((pq' - p'q)\Delta + q(ps - qr))y_1 = 0$$

and

$$r\Delta^2 y_2'' + \Delta(r\Delta' - r'\Delta - r(p+s))y_2' + ((sr' - s'r)\Delta + r(ps - qr))y_2 = 0.$$

Using the explicit formulas given above, it is straightforward to verify the following two examples.

Example 3.4. Let $\mathcal{E}_t$: $y^2 = x^3 - x - t$ and let $\Omega(t) = dx/2y$, $N(t) = xdx/2y$. Then $\Delta(t) = 27t^2 - 4$,

$$A(t) = \frac{1}{2(27t^2 - 4)} \begin{pmatrix} -9t & -6 \\ 2 & 9t \end{pmatrix}$$

and

$$M(t) = \begin{pmatrix} F(\frac{1}{12}, \frac{5}{12}, \frac{1}{2} | \frac{27}{4} t^2) & \frac{3}{4} t F(\frac{7}{12}, \frac{11}{12}, \frac{3}{2} | \frac{27}{4} t^2) \\ -\frac{1}{4} t F(\frac{5}{12}, \frac{13}{12}, \frac{3}{2} | \frac{27}{4} t^2) & F(-\frac{1}{12}, \frac{7}{12}, \frac{1}{2} | \frac{27}{4} t^2) \end{pmatrix}.$$

Example 3.5. Let $\mathcal{E}_t$: $y^2 = x^3 + tx + 1$ and let $\Omega(t) = dx/2y$, $N(t) = xdx/2y$. Then $\Delta(t) = 4t^3 + 27$,

$$A(t) = \frac{1}{2(4t^3 + 27)} \begin{pmatrix} -2t^2 & 9 \\ 3t & 2t^2 \end{pmatrix}$$

and

$$M(t) = \begin{pmatrix} F(\frac{1}{12}, \frac{7}{12}, \frac{2}{3} | -\frac{4}{27} t^3) & \frac{1}{6} t F(\frac{5}{12}, \frac{11}{12}, \frac{4}{3} | -\frac{4}{27} t^3) \\ \frac{1}{36} t^2 F(\frac{7}{12}, \frac{13}{12}, \frac{5}{3} | -\frac{4}{27} t^3) & F(-\frac{1}{12}, \frac{5}{12}, \frac{1}{3} | -\frac{4}{27} t^3) \end{pmatrix}.$$

Denote for any valuation v of K the v-adic radius of convergence of the elements of $M(t)$ by ρ_v. We have the following result,

Theorem 3.6. *When v is infinite, $\rho_v = \min\{|\tau|_v| \mid \Delta(\tau) = 0\}$. Suppose v is finite and that $|A(0)|_v, |B(0)|_v \leq 1, |6\Delta(0)|_v = 1$. Then $\rho_v \geq 1$.*

For the proof we refer to [Be]. In combination with Corollary 2.4 we see that the elements of $M(t)$ are G-functions.

4   Theorems and examples

In this Section we state our results on values of the G-functions that occur in our matrix $M(t)$. For a proof of these theorems we refer to [Be], here we content ourselves by giving a sketch of the method. The examples are all obtained by application of the theorems to the family of elliptic curves in Example 3.4 In what follows we adhere to the notation introduced in Section 3. We denote the member of the family $\mathcal{E}_t$ with parameter value $t = a$ by E_a. The corresponding specialisations of $\Omega(t), N(t)$ are denoted by ω_a, η_a.

Theorems 4.1,4.5 and Corollary 4.2 all derive from the fact that the elliptic curve E_0 is isogenous to E_a for some a which lies in the v-adic region of convergence of $M(t)$ for some valuation v. Notice that the resulting relations we will find are all linear.

Roughly speaking, an isogeny $\phi : E \to E'$ between two elliptic curves E and E' is a non-constant rational map ϕ which sends the zero of the group law on E to the zero of the group law on E'. It is automatically a morphism and a homomorphism of groups . Isogenies of an elliptic curve E to itself are called endomorphisms. Using the group law on elliptic curves, one easily sees that the endomorphisms of E form a ring, denoted by $\mathrm{End}(E)$. When E is a curve defined over a field of characteristic zero it is known that either $\mathrm{End}(E) = \mathbb{Z}$ (trivial case) or $\mathrm{End}(E)$ is an order in an imaginary quadratic field. In the latter case we say that E admits *complex multiplication* or E has C.M. A standard example is $y^2 = x^3 - x$ which admits $(x, y) \mapsto (-x, \sqrt{-1}y)$ as endomorphism. We have $\mathrm{End}(E) = \mathbb{Z}[\sqrt{-1}]$ in this case. Notice also that this curve is precisely the curve E_0 of 3.4. For much more on elliptic curves see for example Silverman's book [Si].

Theorem 4.1. *Suppose E_0 is isogenous over K to E_a via $\phi : E_0 \to E_a$. Let $N = \deg \phi$. Let $\alpha, \beta \in K$ be defined by*

$$\phi^* \omega_a = \alpha \omega_0 \qquad \phi^* \eta_a = \frac{N}{\alpha} \eta_0 + \beta \omega_0.$$

Let v be a valuation of K such that $|a|_v < \max(1, \rho_v)$. When v is finite, also suppose that $|A(0)|_v, |B(0)|_v \le 1$ and $|6N\Delta(0)|_v = 1$, $|-2A(0)B'(0) + 3B(0)A'(0)|_v = 1$. Then,

$$\mathrm{Tr}\left(\begin{pmatrix} N/\alpha & 0 \\ -\beta & \alpha \end{pmatrix} M(a)_v \right) \in \mathbb{Z} \qquad (4.1)$$

where $M(a)_v$ denotes the v-adic evaluation of $M(a)$. Moreover, when v is finite, the trace on the left hand side of Eq(4.1) is $\le 2\sqrt{N}$.

When v is infinite, we use integrals of ω_b, η_b along closed loops on E_b, known as *periods*. Our theorem is proved fairly directly by using period relations that exist between the isogenous curves E_0, E_a. In the case when v is finite it is not clear what periods of an elliptic curve are, the idea of integration along a loop does not exist p-adically. In that case we use a mechanism of lifting the isogeny $E_0 \to E_a$ to an isogeny of the whole family $\mathcal{E}_t$. Remarkably enough this mechanism works in the case of finite v only.

In the following Corollary we shall assume that E_0 is a curve with complex multiplication (C.M.). We shall also choose γ such that for every $e \in \mathrm{End}(E_0)$ there exists $\epsilon \in \mathbb{C}$ such that $e^* \omega_0 = \epsilon \omega_0$, $e^* \eta_0 = \bar{\epsilon} \eta_0$. In such a case we shall say that $\mathrm{End}(E_0)$ *acts diagonally* on the basis ω_0, η_0 of $H^1_{DR}(E_0)$.

Corollary 4.2. *Suppose in addition to the assumptions of Theorem 3.1 that E_0 is a C.M.curve. Choose γ such that $\mathrm{End}(E_0)$ acts diagonally on ω_0, η_0. Let $-d = \mathrm{dicsr}(\mathrm{End}(E_0))$ and write*

$$M(t) = \begin{pmatrix} m_{11}(t) & m_{12}(t) \\ m_{21}(t) & m_{22}(t) \end{pmatrix}.$$

Then,

$$\frac{N}{\alpha} m_{11}(a)_v, \ \alpha m_{22}(a)_v - \beta m_{12}(a)_v \in \frac{1}{\sqrt{-d}} \mathbb{Z} \left[\frac{d + \sqrt{-d}}{2} \right].$$

Proof. Let $e \in \mathrm{End}(E_0)$ and apply Theorem 3.1 to $\phi \circ e$ to obtain

$$\mathrm{Tr}\left(\begin{pmatrix} \bar{\epsilon} & 0 \\ 0 & \epsilon \end{pmatrix} \begin{pmatrix} N/\alpha & 0 \\ -\beta & \alpha \end{pmatrix} M(a)_v \right) \in \mathbb{Z}.$$

Written out explicitly,

$$\bar{\epsilon} \frac{N}{\alpha} m_{11}(a)_v + \epsilon \left(\alpha m_{22}(a)_v - \beta m_{12}(a)_v \right) \in \mathbb{Z}.$$

This holds in particular for $\epsilon = 1$ and $\epsilon = (d + \sqrt{-d})/2$. Hence our assertion follows. $\square$

We give two applications.

Example 4.3. *Take $a = 11^{3/2}/14$. Then $j(E_a) = 66^3$ and there is an isogeny of degree 2 from E_0 to E_a. Let*

$$f(z) = 2 \cdot 11^{-1/4} F\left(\frac{1}{12}, \frac{5}{12}, \frac{1}{2} | z \right)$$

then it follows from Corollary 4.2 that

$$f\left(\frac{1323}{1331} \right)_\infty = \frac{3}{2} \qquad f\left(\frac{1323}{1331} \right)_7 = \frac{1}{2}.$$

Moreover, from Theorem 4.1,

$$\frac{21}{242} F\left(\frac{7}{12}, \frac{11}{12}, \frac{3}{2} \Big| \frac{1323}{1331} \right)_\infty + F\left(-\frac{1}{12}, \frac{7}{12}, \frac{1}{2} \Big| \frac{1323}{1331} \right)_\infty = \frac{3}{2} \cdot 11^{-1/4}$$

$$\frac{21}{242} F\left(\frac{7}{12}, \frac{11}{12}, \frac{3}{2} \Big| \frac{1323}{1331} \right)_7 + F\left(-\frac{1}{12}, \frac{7}{12}, \frac{1}{2} \Big| \frac{1323}{1331} \right)_7 = \frac{1}{2} \cdot 11^{-1/4}.$$

Example 4.4. *Take a such that* $j(E_a) = (3(724 + 513\sqrt{2}))^3$. *Then* E_a *is isogenous to* E_0 *of degree 4. Let*

$$f(z) = 4 \cdot (91 + 60\sqrt{2})^{-1/4} F\left(\frac{1}{12}, \frac{5}{12}, \frac{1}{2}|z\right)$$

and

$$\xi = 3^3 \cdot 7^2 \cdot 11^2 (3 - 2\sqrt{2})(5 + \sqrt{2})^3 (7 - \sqrt{2})^3 / (23 \cdot 47)^3.$$

Then it follows from Corollary 4.2 that

$$f(\xi)_\infty = \frac{5}{2} \qquad f(\xi)_{3+\sqrt{2}} = \frac{3}{2} \qquad f(\xi)_{3-\sqrt{2}} = \frac{1+i}{2} \qquad f(\xi)_{11} = \frac{1+2i}{2}.$$

In the following theorem we denote by $\overline{E_0}$ the curve E_0 reduced modulo the prime ideal corresponding to a finite valuation v. We say that $\overline{E_0}$ is *ordinary* if $\mathrm{End}(\overline{E_0})$ is a quadratic order, we say it is *supersingular* if $\mathrm{End}(\overline{E_0})$ is a quaternion algebra.

Theorem 4.5. *Suppose* E_0 *is a C.M.curve and suppose* γ *is chosen such that the action of* $\mathrm{End}(E_0)$ *on* ω_0, η_0 *is diagonal. Let* v *be a finite place of* K *which satisfies* $|A(0)|_v$, $|B(0)|_v \leq 1$ *and* $|6\Delta(0)|_v = |-2A(0)B'(0) + 3B(0)A'(0)|_v = 1$. *Suppose* E_0 *is isogenous over* K *to* E_a *via* $\phi : E_0 \to E_a$ *and suppose that* $\phi(\mathrm{mod}\ v)$ *is a power of the Frobenius map. Let* $N = \deg \phi$. *Let* $\alpha, \beta \in K$ *be defined by*

$$\phi^* \omega_a = \alpha \omega_0 \qquad \phi^* \eta_a = \frac{N}{\alpha} \eta_0 + \beta \omega_0.$$

Suppose that $\overline{E_0}$ *is ordinary and* $|a|_v < 1$. *Let* $-d = \mathrm{dicsr}(\mathrm{End}(\overline{E_0}))$. *Then there exists* $\epsilon \in \mathbb{Z}[(d + \sqrt{-d})/2]$ *with* $\epsilon\bar{\epsilon} = N$ *such that*

$$M(a)_v = \frac{1}{N}\begin{pmatrix} \alpha & 0 \\ \beta & N/\alpha \end{pmatrix}\begin{pmatrix} \bar{\epsilon} & 0 \\ 0 & \epsilon \end{pmatrix}.$$

Example 4.6. *Let a be such that* $j(E_a) = (6(2927 + 1323\sqrt{5}))^3$. *Then* E_0 *is 5-isogenous to* E_a. *Moreover,* $\overline{E_0}$ *is ordinary. Let*

$$\xi = 2 \cdot 3^3 \cdot \cdot 7^2 \sqrt{5}(47 - 20\sqrt{5})^2(673 + 357\sqrt{5})^3/(11 \cdot 59 \cdot 71)^3.$$

Then it follows from Theorem 4.5 that

$$F\left(\frac{1}{12}, \frac{5}{12}, \frac{1}{2}|\xi\right)_5 = -\frac{1}{2+i}(161 + 120\sqrt{5})^{1/4}$$

$$F\left(\frac{7}{12}, \frac{11}{12}, \frac{1}{2}|\xi\right)_5 = 0$$

$$F\left(\frac{5}{12}, \frac{13}{12}, \frac{3}{2}|\xi\right)_5 = -\frac{(1229 - 515\sqrt{5})(161 + 120\sqrt{5})^{1/4}}{7(47 - 20\sqrt{5})(2+i)}$$

$$F\left(-\frac{1}{12}, \frac{7}{12}, \frac{1}{2}|\xi\right)_5 = -(2+i)(161 + 120\sqrt{5})^{-1/4}.$$

The next theorem shows that if both E_0 and E_a are curves with complex multiplication and a lies in the region of convergence of $M(t)$ for some suitable v, then we obtain quadratic relations between the elements of $M(a)$.

Theorem 4.7. *Suppose E_0 is C.M.curve and γ is chosen such that the action of $\mathrm{End}(E_0)$ on ω_0, η_0 is diagonal. Let E_a, $a \in K$ be another C.M.curve and suppose that there exists a valuation v on K such that $|a|_v < \max(1, \rho_v)$. When v is finite we also assume that $|A(0)|_v, |B(0)|_v \leq 1$ and $|6\Delta(0)|_v = |-2A(0)B'(0) + 3B(0)A'(0)|_v = 1$. Let δ be such that the action of $\mathrm{End}(E_a)$ on $\omega_a, \eta_a + \delta\omega_a$ is diagonal. Then,*

$$\frac{m_{11}(a)_v(m_{22}(a)_v + \delta m_{12}(a)_v)}{m_{12}(a)_v(m_{21}(a)_v + \delta m_{11}(a)_v)} \in \mathbb{Q}(\mathrm{End}(E_0), \mathrm{End}(E_a)).$$

Example 4.8. *Let a be such that $j(E_a) = 20^3$. Then $\mathrm{End}(E_a) = \mathbb{Z}[\sqrt{-2}]$ and $\delta = 1/\sqrt{30}$. Denote*

$$f(z) = F(\tfrac{1}{12}, \tfrac{5}{12}, \tfrac{1}{2}|z) \qquad g(z) = F(\tfrac{7}{12}, \tfrac{11}{12}, \tfrac{3}{2}|z)$$

$$h(z) = F(\tfrac{5}{12}, \tfrac{13}{12}, \tfrac{3}{2}|z) \qquad k(z) = F(-\tfrac{1}{12}, \tfrac{7}{12}, \tfrac{1}{2}|z)$$

and

$$Q(z) = \frac{f(z)(2250k(z) + 105g(z))}{g(z)(-49h(z) + 105f(z))}.$$

Then it follows from Theorem 4.7 that

$$Q\left(\frac{98}{125}\right)_\infty = (\sqrt{2} + 1)^4 \qquad Q\left(\frac{98}{125}\right)_7 = -\frac{(2\sqrt{2} - 1)^2}{7}.$$

References

[A] Y. André, *G-functions and Geometry*, Aspects of Math. 13, Vieweg, Bonn 1989.

[Be] Algebraic values of G-functions, preprint 675, University of Utrecht 1991. Submitted to J. für Reine Angew. Math.

[Bo] E. Bombieri, On G-functions, Recent progress in analytic number theory, vol. 2, 1–67, Durham 1979, Academic Press, London, 1981.

[BW] F. Beukers, J. Wolfart, Algebraic values of hypergeometric functions, 68–81 in *New advances in transcendence theory* (A. Baker ed.), Cambridge University Press 1988.

[CC] D. V. Chudnovsky, G. V. Chudnovsky, Applications of Padé- approximations to diophantine inequalities of values of G-functions, Lecture Notes in Math. 1135, 9–51, Springer 1985.

[CD] G. Christol, B. Dwork, Effective p-adic bounds at regular singular points, Duke Math. J. 62(1991), 689–719.

[DR] B. Dwork, P. Robba, Effective p-adic bounds for solutions of homogeneous linear differential equations, Trans. Am. Math. Soc. 259(1980), 171–204.

[Iy] S. Iyanaga, *The Theory of Numbers*, North-Holland 1975.

[La] S. Lang, *Algebraic number theory*, Springer 1986.

[S] C. L. Siegel, *Transcendental Numbers*, Annals of Math. Studies 16(1949), Princeton University Press.

[Sh] A. B. Shidlovskii, *Transcendental Numbers*, Studies in Math. 12, W. de Gruyter, New York, Berlin 1989.

[Si] J. Silverman, *The Arithmetic of Elliptic Curves*, Springer 1986.

[W] J. Wolfart, Werte Hypergeometrische Funktionen, Inv. Math. 92 (1988), 187-216.

Frits Beukers
Department of Mathematics
University of Utrecht
P. O. Box 80.010, 3508 TA Utrecht
The Netherlands
e-mail beukers@math.ruu.nl

Irregularities in the Distribution of Primes

John B. Friedlander

University of Toronto

In this paper, we survey a body of results, stemming from an idea of Maier [M], which has demonstrated that the distribution of primes in certain basic integer sequences is not as uniform as it was once believed to be. We also make some speculations, both as to further problems that might be amenable to these methods, and as to the nature of some (still apparently) reasonable conjectures to modify those which have been found no longer tenable.

1. Primes in Integer Sequences

Let x be real and let $\pi(x)$ denote the number of primes not exceeding x. The logarithmic integral

$$\ell i\, x \doteq \int_2^x \frac{dt}{\log t} = \frac{x}{\log x} + O\left(\frac{x}{\log^2 x}\right)$$

provides a good approximation to $\pi(x)$ and, in particular, the Prime Number Theorem (cf [D]) gives the asymptotic formula

(1)
$$\pi(x) \sim \ell i\, x.$$

In other words, letting $E(x) = \pi(x) - \ell i\, x$, we have

(2)
$$E(x) = o\left(x/\log x\right).$$

It is an important problem in analytic number theory to obtain analogous asymptotic formulae for the prime counting functions of a variety of integer sequences which are of natural interest.

Thus, for example, consider the sequence of integers of the form $p - 2$ where p runs through primes. The integers in this sequence which are themselves prime are just those which are the first members in a pair of 'twin primes' and it is a famous old problem to provide an asymptotic formula for these (or even to prove that there are infinitely many such).

Armed with only the knowledge of (1), it is still possible to make a reasonable guess as to the shape of such an asymptotic formula. As a first guess one might interpret (1) as a statement that the 'probability' that the 'random' integer n be prime is $1/\log n$, hence that the chances of n and

$n-2$ both being prime is $(1/\log n)(1/\log(n-2))$. From this it would follow that the number of prime pairs $(p, p-2)$ with $p \leq x$ is $\sim x/\log^2 x$. Such a rationale would, however, be too simplistic since it presupposes that the events n and $n-2$ being prime are independent. That this is not the case is most easily seen by noting that the information that p is prime (and ≥ 3) implies that p is odd. Thus $p-2$ is odd, thereby 'doubling' the chances that it is prime. An only slightly more complicated argument applies to the residue class of $p-2$ modulo $3, 5, 7$, and the other primes. Following this argument one is led to conjecture that the number of such prime pairs $\leq x$ is $\sim cx/\log^2 x$ where

$$c = 2 \prod_{p>2} \left(1 - \frac{1}{(p-1)^2} \right).$$

Another well known example concerns the sequence $f(n)$, $n = 1, 2, 3, \ldots$ of values taken on by the polynomial f with integer coefficients. Here, in order that there be infinitely many primes $f(n)$ it is clearly necessary that f be irreducible and, in that case, heuristic reasoning similar to the above leads to the conjecture that the number $\pi_f(x)$ of $n \leq x$ for which $f(n)$ is prime satisfies

(3) $$\pi_f(x) \sim Cx/\log|f(x)|$$

where

$$C = \prod_p \left(1 - \frac{\omega(p)}{p} \right) \Big/ \left(1 - \frac{1}{p} \right),$$

$\omega(p)$ denoting the number of roots of the congruence $f(n) \equiv 0 \pmod{p}$.

This conjecture implies that in most cases there are infinitely many such primes, but not invariably so, since we may have $C = 0$. Thus, for example, $n^2 + n + 2$, although irreducible, is always even and so $\omega(2) = 2$.

In spite of our general lack of success in providing asymptotic formulae in the above problem there are two important classes of sequences for which considerable progress has been achieved; these pertain to the special case of the latter example when the polynomial is linear.

2. SHORT INTERVALS, ASYMPTOTIC RESULTS

The first class of interest is that where we take the sequence to be the integers in an interval $(x-y, x]$. Here y is taken to be a function of x and we should like to be able to provide a simple approximation for the number $\pi(x) - \pi(x-y)$ of primes with the function y, certainly small compared to x (we call such an interval a short interval), and hopefully as small a function as possible. The simplest possible behaviour for which one could reasonably hope is that one has the asymptotic formula

$$(4) \qquad \pi(x) - \pi(x - y) \sim \int_{x-y}^{x} \frac{dt}{\log t}$$

which is at least consistent with Prime Number Theorem and which, in case y is not very short (for example $y = \epsilon x$), may be deduced from it.

A natural goal then is the proof of the asymptotic formula (4) with y as small as possible and also the determination of accurate estimates for the threshold size of y at which it begins to break down. (By the Prime Number Theorem it must break down, at $y = \frac{1}{2} \log x$ for example.) The fundamental achievement in the first aspect of this problem has been the theorem of Hoheisel that there exists a constant $\theta_0 < 1$ such that the asymptotic formula (4) holds provided that $y = x^\theta$, $\theta > \theta_0$. A great deal of cumulative effort has, in the intervening years, succeeded in lowering the admissible value of θ to 7/12, [H, HB]. It has long been known that, on the assumption of the Riemann Hypothesis, one may take $\theta_0 = \frac{1}{2}$, and it is widely believed that in fact $\theta_0 = 0$ is admissible.

The problem of obtaining lower bound estimates on the size of the functions for which the interval $(x - y, x]$ can sometimes (for arbitrarily large x) contain no primes at all, has also received much attention but, in spite of several recent inventive ideas [MP], the estimate of Erdös-Rankin that, for arbitrarily large x, there do exist such intervals with $y > c(\log x)(\log_2 x)(\log_4 x)/(\log_3 x)^2$ has over many years received no improvement other than in the value of the constant c. (Here, and throughout, $\log_k$ denotes the k-th iterate of the logarithm.) The problem of determining whether there are larger functions y for which there might exist primes, but in insufficient number for the asymptotic formula to hold, for a long time received no attention and it was fairly generally believed that the asymptotic formula (4) actually held for $y = (\log x)^N$ for any fixed $N > 2$. Such an expectation was a consequence of certain heuristic considerations of a probabilistic nature, largely initiated by Cramér [C]. That this probabilistic model was too simplistic was brought forcefully home when Maier [M] showed that, for every N, the asymptotic formula (4) was doomed to fail with $y = (\log x)^N$. Maier's basic idea, albeit a simple one, has sparked a great deal of interest and further work.

3. Arithmetic Progressions, Asymptotic Results

The second basic class of sequences for which one has made significant progress in counting the primes is that of arithmetic progressions. Let a and q denote relatively prime integers with $q \geq 1$ and let $\pi(x; q, a)$ count the number of primes $p \leq x$ which are in the progression a modulo q. The expected answer that

(5)
$$\pi(x; q, a) \sim \frac{\ell i \, x}{\varphi(q)}$$

(where $\varphi(q)$, Euler's function, is the number of those residue classes modulo q which consist of integers relatively prime to q) follows (cf [D]) from a combination of the ideas of Dirichlet which proved the infinitude of primes in the progression with the ideas of Hadamard and of de la Vallée-Poussin which proved the Prime Number Theorem itself.

We define $E(x; q, a) = \pi(x; q, a) - \ell i \, x / \varphi(q)$ and so (5) becomes

(6)
$$E(x; q, a) = o \left(\frac{1}{\varphi(q)} \frac{x}{\log x} \right),$$

the result holding with a and q fixed and $x \to \infty$.

We now change the problem somewhat and consider not a single arithmetic progression but rather a family of them. We regard q (and perhaps also a) as a function of x and ask whether (6) still holds if $q \to \infty$ as $x \to \infty$. Such a viewpoint is fruitful for very many applications, but to discuss these here would carry us too far afield.

Viewed in this light the problem becomes in some respects rather similar to the earlier problem of the estimation of primes in short intervals, with the number of integers in the sequence being approximately x/q, the latter corresponding to the length y of the short interval which approximated the number of integers therein. There are however some important differences (for example, 'exceptional' zeros of L-functions) which render this problem somewhat more difficult and indeed there has been no proof given of the natural analogue of Hoheisel's theorem, that there exist a constant $\alpha_0 > 0$, ($\alpha_0 = 1 - \theta_0$), such that the asymptotic formula (6) hold under the restriction $q = x^\alpha$ for every $\alpha < \alpha_0$. Although a fundamental theorem of Linnik (cf [B]) ensures the existence of some primes under such an assumption, the strongest known version of (6) requires that q satisfy the far more severe 'Siegel-Walfisz' condition

(7)
$$q < (\log x)^N \qquad \text{for some fixed } N > 0.$$

In greater accord with the analogy to the short interval problem one does know that the Generalized Riemann Hypothesis implies (6) under the condition $q < x^{\frac{1}{2} - \delta}$ (for any $\delta > 0$), it is widely believed that $q < x^{1-\delta}$ suffices, and, until recently seemed quite likely (as follows from the Cramér probabilistic model) that $q < x/(\log x)^N$ sufficed for any fixed $N > 2$. However the analogy in question could also be shown to carry over to the argument that Maier had employed to deduce his above referenced result and in this way it was shown in [FG₁] that, for every N, the asymptotic formula (5) cannot hold uniformly in the range $q < x/(\log x)^N$.

4. THE BASIC ARGUMENT

In this section we briefly sketch Maier's idea in a slightly more general context. We consider an arithmetic function $f(n) \geq 0$, integers $(a, q) = 1$, and the error

$$(8) \qquad \Delta(x; q, a) = \sum_{\substack{n \leq x \\ n \equiv a(q)}} f(n) - \frac{1}{\varphi(q)} \sum_{\substack{n \leq x \\ (n,q)=1}} f(n).$$

We expect that Δ is not too large but wish to show, that for some large q, it may also be not too small. We also expect and intend to use the knowledge that, when the modulus q is small, then Δ is indeed small (for example, in the range (7)) and can be proven to be so.

We consider integers n of the form $n = rP + sq$ where r, s run through all integers in certain intervals $R < r \leq 2R$, $S < s \leq 2S$. Here R, S, P are parameters to be chosen (with P an integer) and q is an integer for which we are attempting to show that there is, for our given f and some x and a, a value of $|\Delta|$ that is relatively large.

We note that, for each fixed r, as s varies n will run through the integers in a segment of the arithmetic progression $rP(\bmod\ q)$. Thus, we have $a = rP$ and x will have size about $RP + Sq$. Here q will be large, nearly as large as x (except that in the case of short intervals we simply take $q = 1$), and S will be small by comparison to x and q. To show that, for some r, the sum $\sum_{n \equiv rP(q)} f(n)$ is, say, larger than expected, it suffices to show that the sum is larger than expected in aggregate over r. To show this (which can happen only under a proper choice of the parameters) we interchange the order of summation and consider the contribution with fixed s and varying r. The integer n will now run through a segment of the arithmetic progression $sq(\bmod\ P)$. We shall thus want the modulus P small compared to x so that we can sum $f(n)$ in this progression fairly precisely.

In the case at issue f is the characteristic function on the primes and we have the primes being divided in roughly equal numbers into those progressions $(\bmod\ P)$ for which $(sq, P) = 1$. We shall choose the parameter P so that $(q, P) = 1$ and we are thus required to estimate φ, the number of s with $S < s \leq 2S$ and $(s, P) = 1$.

If P is taken to be any integer which has an unusually large number of prime factors, this sifting function φ will exhibit (see [FGHM, Theorem B2]) an oscillatory behaviour about its expected mean $S \prod_{p|P}(1 - 1/p)$, depending on the choices of the parameters P and S. In the particular case where one makes the simple choice of P as the product of all primes less than z (other than those which are divisors of q, which are forbidden by the method) then this oscillatory behaviour is particularly large and has been studied rather extensively. One may choose S and z so to arrive above (or below) the mean and the result follows.

Although the argument sketched above has so far apparently been applied only to the case of primes, there appears to be the possibility of using it to study other arithmetic functions, for example the divisor functions, and the square-free and k-free numbers. In these cases the support of the function f involved does not split entirely into the reduced residue classes mod q. This will lead to the study of the oscillatory behaviour of slightly more complicated functions φ.

5. Error Terms

In studying the distribution of primes, and other arithmetic functions, one is interested in obtaining not only asymptotic results but, where possible, bounds for the error terms that are stronger than those implied by the accompanying asymptotic formulae. The replacement of (2) by the stronger estimate

$$(9) \qquad\qquad E(x) \ll x^{\frac{1}{2}+\epsilon},$$

for every fixed $\epsilon > 0$, has long been known to be equivalent to the Riemann Hypothesis. (Here $A \ll B$ means that A/B is bounded for sufficiently large x.) Indeed, even the existence of any $\delta > 0$ for which $E(x) \ll x^{1-\delta}$, which would immediately imply Hoheisel's theorem (and which would be equivalent to a 'quasi-Riemann Hypothesis') seems very far from being proven. Nevertheless, such results have been expected to be true not only, as in (9) for the set of all primes, but also for the corresponding problems for short intervals and for arithmetic progressions. Thus one might expect, and the probabilistic model predicts that, for $y = (\log x)^N$, $N > 2$, the error term in the short interval problem,

$$(10) \qquad E(x, y) \doteq \pi(x) - \pi(x - y) - \int_{x-y}^{x} \frac{dt}{\log t}$$

satisfies a bound of the type

$$(11) \qquad\qquad E(x, y) \ll y^{\frac{1}{2}+\epsilon}$$

with an implied constant depending on ϵ, and for the case of the arithmetic progressions, again satisfies

$$(12) \qquad\qquad E(x; q, a) \ll y^{\frac{1}{2}+\epsilon},$$

where in this latter case, we define $y = x/q$.

That these cannot hold in case y is bounded by a fixed power of $\log x$ follows already from [M] and [FG$_1$] since even the asymptotic formulae fail in these ranges. However, although the method sketched in the previous section appears to be unable to destroy the asymptotic formula when y grows more quickly than any fixed power of $\log x$, it nevertheless can be adapted to contradict the assumption that the asymptotic formula hold with 'good' error term. This has been carried out in [HM] for the short interval problem and then in [FGHM] for arithmetic progressions. The main difficulty in adapting the method comes from the analysis of the oscillation of the sifting functions φ which is much smaller and harder to detect in the ranges of the parameters corresponding to these larger values of y.

In this argument the range of y to which the method is applicable varies in accordance with the strength of the error term that one wishes to contradict. Thus, whereas the contradiction of the weak error term equivalent to the asymptotic formula has been demonstrated only in the limited range $y < (\log x)^N$ where N is fixed, the assumption of the strong probabilistic error term given in (12), can be contradicted in the much wider range

$$(13) \qquad y < \exp\left((\log x)^{\frac{1}{2}-\epsilon}\right),$$

for every $\epsilon > 0$. Similarly, more general results, of which the above are special cases, are given in [HM] and [FGHM]. As even the statements of these are rather technical we do not reproduce them here. Essentially they cover the whole range from $y = (\log x)^N$ to that given by (13). Actually the method works for y even larger than that given by (13) but is of less interest then because the error terms it contradicts are so small that they can be contradicted more directly.

6. ARITHMETIC PROGRESSIONS, STATISTICAL RESULTS

One aspect of the study of primes in arithmetic progressions is the question of producing results on the error term $E(x; q, a)$ which hold for most q, or for most $a \bmod q$, or for most a and most q. Similar results are not entirely absent from the study of primes in short intervals. A beautiful theorem of Selberg [S] asserts that, on the Riemann Hypothesis, the asymptotic formula (4) for primes in short intervals holds for most $x \leq X$, i.e. outside a set of measure $o(X)$, as long as y is sufficiently large that $y/\log^2 X \to \infty$.

Nevertheless it is probably fair to say that this statistical aspect of the theory is much more important in the case of arithmetic progressions, perhaps in part because of the absence here of any "Hoheisel" theorem, but for the most part because of the many applications that ensue. It is because of its importance for applications that the central theorem of this type is the Bombieri-Vinogradov Theorem (cf [B,D])

(14)
$$\sum_{\substack{q<Q \\ (q,a)=1}} |E(x;q,a)| \ll \frac{x}{(\log x)^A}$$

valid for any $A > 0$, with $Q = x^{\frac{1}{2}}/(\log x)^{B(A)}$ for some $B(A)$, and where the implied constant depends only on A. The result saves an arbitrary power of $\log x$ as opposed to trivial estimates; this saving is sufficient for most applications, and in any case would seem difficult to improve, since any significant improvement would have implications for the problem of exceptional zeros.

The Bombieri-Vinogradov Theorem implies that in some sense the asymptotic formula 'usually' holds for $q < Q$. Its importance derives from the large size of the moduli with which it deals, the same size that can be treated in an individual non-statistical sense only under the assumption of the Generalized Riemann Hypothesis.

For the same reason it is of great importance to prove results of this type with larger values of Q; such theorems lead to quantitative (and, less frequently, qualitative) sharpenings in applications. One would like to know what is the best that might be hoped for in this direction; the famous Elliott-Halberstam conjecture [EH] asserts that it should hold, certainly for $Q = x^{1-\epsilon}$, and perhaps a good deal further.

The method described in §4 has the feature that when the parameters are chosen so as to produce unusually large (or small) values of $\pi(x;q,a)$ this property persists for other moduli q having nearly the same size. This made it possible for the argument to be adapted in [FG$_1$] to disprove the strongest plausible forms of the Elliott-Halberstam conjecture, which had been predicted by the probabilistic model, that (14) might hold in the range $Q = x/(\log x)^{B(A)}$ with some $B(A)$. This result of [FG$_1$] was strengthened in [FGHM] to give the following theorem; as in the case of the results mentioned in §5, it is the special case of a more general statement.

Theorem. *Let $\epsilon > 0$, $A > 1$. For every $x > x_0(\epsilon, A)$, there exists an integer a such that, for all Q with*

(15)
$$Q \geq x \exp\left(-\frac{(1-\epsilon)(A-1)(\log_2 x)^2}{\log_3 x}\right)$$

we have

(16)
$$\sum_{\substack{q<Q \\ (q,a)=1}} |E(x;q,a)| > \frac{x}{(\log x)^A}.$$

A theorem only slightly weaker follows from the method described in §4 but in fact the result given here was proven in [FGHM] using a somewhat different method which, rather curiously, works better when one considers sums $\sum_q |E(x; q, a)|$ but does not apply to the individual error terms $E(x; q, a)$. In this method we first note that it suffices to bound below the smaller sum $|\sum_q E(x; q, a)|$ where we have moved the absolute value signs outside the sum. This allows us to work with the sum $\sum_q \pi(x; q, a)$, the sum over the main terms being easy. This new sum may be transformed, using the equation $p - a = qr$ to define r, into a sum $\sum_r \pi(x; r, a)$. Since we have $p \leq x$ and q nearly as large as x the complementary factor r is rather small and so the new transformed sum deals with primes in progressions where the small size of the modulus allows for accurate estimations. The ensuing sums of main terms can be shown to exhibit the requisite oscillatory behaviour provided that the integer a is chosen to have many prime factors as was P chosen in §4.

7. ARITHMETIC PROGRESSIONS, THE DEPENDENCE ON THE RESIDUE CLASS

In connection with the problem of obtaining uniform upper bounds for the size of the error term $E(x; q, a)$ the size of the variable q has historically been the major issue; the role of the variable a has been relatively insignificant. Thus in the two main upper bounds for $|E(x; q, a)|$, the Siegel-Walfisz estimate (that (6) hold subject to (7)) and the Bombieri-Vinogradov estimate (14), the results hold with complete uniformity in a (maintaining of course the trivial requirement that $(a, q) = 1$). Nevertheless, in some of the more recent investigations, the methods used are such that the quality of the result exhibits a dependence on the integer a. Moreover it can be shown, at least in some circumstances, that this dependence is inevitable.

As was indicated in the previous section, it is an important goal, both theoretically and for the sake of applications, to extend to range of validity of the estimate (14) to values of Q larger than that given in the Bombieri-Vinogradov Theorem. In the strictest sense this has not yet been accomplished but, if one allows a little bending of the rules, (14) has been improved in a number of ways. In all of these there has been the requirement that the integer a be fixed (or at least not too large); in other words the implied constant in (14) is now permitted to depend on a as well as A. Subject to this weakening and no other, the estimate (14) was shown in [BFI$_2$] to hold for any $A < 3$ with Q slightly larger than $x^{\frac{1}{2}}$.

An additional bending of the rules occurs when one introduces weights λ_q and replaces (14) by the estimate

$$(17) \qquad \sum_{\substack{q < Q \\ (q,a)=1}} \lambda_q E(x; q, a) \ll \frac{x}{(\log x)^A},$$

the implied constant being permitted to depend on a and A.

For many interesting weights λ_q one can treat (17) with Q significantly larger than $x^{\frac{1}{2}}$. Of greatest importance for applications are the "well-factorable" weights introduced by Iwaniec [I]. For a great many applications the estimate (17) with these weights is equally as good as the estimate (14) with the same Q. However, with these weights the estimate (17) can be proven with a value $Q = x^\alpha$, $\alpha > \frac{1}{2}$. This was first accomplished by Fouvry and Iwaniec in [Fo-I]; the largest admissible value found to date is $\alpha = \frac{4}{7} - \epsilon$ for any $\epsilon > 0$, given in [BFI$_1$].

This residue class dependence in the recent work has occurred not only in the upper bounds like (17) but also in the lower bounds that limit the range of uniformity for which such upper bounds can hold. Thus, while upper bounds like (17) have proved more tractable for large Q when a is taken fixed, it has also been proved more difficult to limit the range of Q for which such results may hold. The arguments described in §4 and §6 both specifically require that the integer a be taken as a parameter which grows to infinity.

In more recent work [FG$_3$] it has been shown that the argument described in §6 can be modified to yield the result that, even for fixed a, the asymptotic formula (5) fails to hold in the range $q < x/(\log x)^N$. In this newer argument one replaces the condition that a have many prime factors (so that a may now be kept fixed) by the condition that some other early integer in the progression have many prime factors. This however places a restriction on q so that the resulting estimates do not apply to sufficiently many moduli q to allow one to confound the average estimate (17), even in the range $Q < x/(\log x)^N$. Indeed, this is not entirely a fault in the method since it can be shown for certain weights λ_q, and is probably true for most weights λ_q, that the maximum allowable value for Q in (17) actually depends on the integer a. Thus in the case where we take λ_q to be identically equal to one it was shown in [FG$_{2.5}$] that, if we start with an integer a, $0 < |a| < x$, for which the number of distinct prime factors ν satisfies $\nu(a) < 2\log_2 x$ (as is the case for most integers a), then one has for each A, some $B(A)$, and for every $Q < x/(\log x)^B$

$$(18) \qquad \left| \sum_{\substack{q<Q \\ (q,a)=1}} E(x;q,a) \right| < \frac{x}{(\log x)^A}$$

for all sufficiently large x. On the other hand, if one weakens the restriction on ν to $\nu(a) < (\log_2 x)^{6/5+\epsilon}$ (or, if the Riemann Hypothesis is assumed, to $\nu(a) < (\log_2 x)^{1+\epsilon}$) then by [FG$_{2.5}$] there exist integers a for which the conclusion (18) is no longer true.

This dependence of the largest admissible Q on the integer a can be demonstrated in the case that λ_q is identically equal to one because, in that simple case, the upper bound (17) (which is here (18)) can be proven with unusual precision. It seems likely that this dependence is also the case for more general weights λ_q but, since we are so far from optimal results in the general case, this will likely be difficult to prove.

8. POLYNOMIAL AND OTHER SEQUENCES

It has already been mentioned in §1 that, although the problem of getting asymptotic formulae for the prime counting functions of more general integer sequences seems to be very difficult indeed, there are in many cases reasonable conjectures which at least predict what these asymptotic formulae should be; see [BH], and the references therein. In comparison to the methods which are used to obtain the asymptotic formulae in the case of linear polynomials, the methods described above for the converse problem do generalize more readily.

Thus, in the case of polynomials of higher degree it was shown in [FG$_4$] that, for arbitrary degree $d \geq 1$, there exist families of irreducible polynomials f in $\mathbb{Z}[X]$ of degree d for which the expected asymptotic formula (3) fails in the range $x \asymp (\log|f|)^N$, for any $N > 0$, where here $|f|$, which plays the role of y in the earlier cases, is the product of the absolute values of the non-zero coefficients of f. In other words $\log|f|$ is the height of the polynomial. The method applied to prove this is rather similar to that described in §4; let us just mention that the polynomials considered are of the form

$$(19) \qquad f_r(n) = Pn^d + (rP+1)n + 1$$

where P is an integer with many prime factors that produces the oscillation necessary to prove the result for some r in a range $R < r < 2R$. It may be shown that for $R > 1$, $P > 1$, all of the polynomials f_r are irreducible.

There are certainly possibilities of applying these methods to count primes in other integer sequences. For example one might try to extend the last result to simultaneous primality of several polynomials, or to prove that there is a non-trivial limitation to the range of uniformity for which the 'prime k-tuple' conjecture can hold, or perhaps a non-trivial limitation to the error term in the conjectured asymptotic formula in the Goldbach problem.

9. SOME SPECULATIONS

We consider two questions which naturally arise in connection with the results described above. The first of these is the question of whether the phenomenon detected by these methods is the only one responsible for the production of error terms larger than those predicted by the probabilistic model. In this case we might expect that the lower bounds provided are not

far from the truth and so the ranges of y, q, Q given above are essentially
best possible. Alternatively, it may be that there are still unknown reasons
for large deviations from asymptotic behaviour which will prevent the expected uniformity in ranges larger than those already found. Following the
time-honoured scientific tradition that the simpler explanation is the more
likely, the former alternative is chosen here.

The authors in [FGHM] framed the following general hypothesis which
would imply that the results given there are essentially best possible.

Hypothesis. *Let $\epsilon > 0$. For all q, a, and x, with $q \geq q_0(\epsilon)$, $(a, q) = 1$, and
$x \geq q(\log q)^{1+\epsilon}$, we have*

$$(20) \qquad\qquad E(x; q, a) \ll y^{1-(1-\epsilon)\delta(x,y)}$$

where here $y = x/\varphi(q)$ and

$$\delta(x, y) = \min\left(\frac{1}{2}, \frac{\log(\log y/\log_2 x)}{\log_2 x}\right),$$

*(and the same bound holds for $E(x, y)$, the error term in the short interval
problem).*

We thus expect the asymptotic formula to hold, but with very weak
error term, essentially as soon as $y/(\log x)^N \to \infty$ for every $N > 0$, for
the error to gradually improve as y gets larger until, after the restriction
$y > \exp(\sqrt{\log x})$, we have the 'expected' error $E \ll y^{\frac{1}{2}+\epsilon}$ for all $\epsilon > 0$.
Perhaps, in both of the above extremes for y, some iterate of the logarithm
is at issue.

A second question that arises concerns the ranges $y < (\log x)^N$ in which
the expected asymptotic formula fails. What are we to replace it with?
Consider for example the interval $(x - y, x]$ with $y = (\log x)^N$ for some
fixed $N > 1$.

For the shorter intervals with $y = \lambda \log x$ (λ constant) one has long
known the distribution to be non-uniform. Moreover this distribution is, at
least in some sense (see [G]), related to the location of the interval relative
to divisibility criteria modulo each (small) prime in much the same way
that the constant c in the twin prime problem (primes in the ultimately
short interval) discussed in §1, could not be taken to be equal to one as a
naive first guess might have suggested.

We believe that this phenomenon persists to the (only slightly) longer
intervals of length $(\log x)^N$, $N > 1$. Strong evidence is provided by the
argument of Maier itself (take $q = 1$ in §4) which produces short intervals
with many (or few) primes; these intervals are in close proximity to an
integer which is a multiple of very many small primes.

One might hope perhaps for an 'asymptotic' formula of the type

$$(21) \qquad \pi(x) - \pi(x - y) = F(x, y) \int_{x-y}^{x} \frac{dt}{\log t} + O\left(y^{\frac{1}{2} + \epsilon}\right)$$

where F can be written as a product over primes

$$(22) \qquad F(x, y) = \prod_p F_p(x, y),$$

where F satisfies the bound,

$$(23) \qquad F(x, y) - 1 \ll y^{-(1-\epsilon)\delta_0}$$

$$\text{with} \quad \delta_0 = \delta_0(x, y) = \frac{\log(\log y / \log_2 x)}{\log_2 x},$$

and where the local factors $F_p(x, y)$ are fairly smooth having partial derivatives of modest size and having a form that is not too complicated.

ACKNOWLEDGEMENTS. The author's research in irregularities of distribution has been done in collaboration, throughout with A. Granville, and to a large extent with A. Hildebrand and H. Maier. Beyond the "Hypothesis" quoted from [FGHM], all of these colleagues are blameless for the speculations in this article.

This paper was written while the author was enjoying a Research Professorship at MSRI, Berkeley. The excellent working atmosphere was greatly appreciated.

REFERENCES

[BH]   P.T. Bateman and R.A. Horn, *Primes represented by irreducible polynomials in one variable*, Proc. Symp. Pure Math. 8, AMS (Providence) 1965, 119–132.

[B]   E.Bombieri, *La Grand Crible dans la Théorie Analytique des Nombres*, Astérisque Vol. 18 Soc. Math. France, 2nd ed. (Paris) 1987.

[BFI$_1$]   E. Bombieri, J.B. Friedlander and H. Iwaniec, *Primes in arithmetic progressions to large moduli*, Acta Math. **156** (1986), 203–251.

[BFI$_2$]   _____ , *II*, Math. Ann. **277** (1987), 361-393; III, J. Amer. Math. Soc. **2** (1989), 215–224.

[C]   H. Cramér, *On the order of magnitude of the difference between consecutive prime numbers*, Acta Arith. **2** (1936), 23–46.

[D]   H. Davenport, *Multiplicative Number Theory, 2nd. Ed.*, Springer (New York) 1980.

[EH]   P.D.T.A. Elliott and H. Halberstam, *A conjecture in prime number theory*, Symp. Math. **4**; Rome (1968–69), 59–72.

[Fo-I]   E. Fouvry and H. Iwaniec, *Primes in arithmetic progressions*, Acta Arith. **42** (1983), 197–218.

[FG$_1$]   J. Friedlander and A. Granville, *Limitations to the equi-distribution of primes I*, Ann. Math **129** (1989), 363–382.

[FG₃]   ———— , *III*, Comp. Math. **81** (1992), 19–32.

[FG₄]   ———— , *IV*, Proc. Roy. Soc. Lond. Ser. A, **435** (1991), 197–204.

[FG₂.₅]   ———— , *Relevance of the residue class to the abundance of primes*, Amalfi Conf. Proc., to appear.

[FGHM] J. Friedlander, A. Granville, A. Hildebrand, and H. Maier, *Oscillation theorems for primes in arithmetic progressions and for sifting functions*, J. Amer. Math. Soc. **4** (1991 pages 25–86).

[G]   P.X. Gallagher, *On the distribution of primes in short intervals*, Mathematika **23** (1976), 4–9.

[HB]   D.R. Heath-Brown, *The number of primes in a short interval*, J. Reine Angew. Math. **389** (1988), 22–63.

[HM]   A. Hildebrand and H. Maier, *Irregularities in the distribution of primes in short intervals*, J. Reine Angew. Math. **397** (1989), 162–193.

[H]   M. Huxley, *On the difference between consecutive primes*, Invent. Math. **15** (1972), 164–170.

[I]   H. Iwaniec, *A new form of the error term in the linear sieve*, Acta Arith. **37** (1980), 307–320.

[M]   H. Maier, *Primes in short intervals*, Michigan Math. J. **32** (1985), 221–225.

[MP]   H. Maier and C. Pomerance, *Unusually large gaps between consecutive primes*, Trans. Amer. Math. Soc. **322** (1990), 201–237.

[S]   A. Selberg, *On the normal density of primes in small intervals and the difference between consecutive primes*, Arch. Math. Naturvid. **47** (1943), 87–105.

John B. Friedlander
Department of Mathematics
Scarborough College
University of Toronto
Toronto, Ontario M5S 1A1
e-mail frdlndr@math.toronto.edu

The Dirichlet Divisor Problem

D. R. Heath-Brown

Magdalen College, Oxford

The Dirichlet divisor problem concerns the error term $\Delta(x)$ in the asymptotic formula

$$\sum_{n \leq x} d(n) = x(\log x + 2\gamma - 1) + \Delta(x).$$

It was shown by Dirichlet that $\Delta(x) \ll x^{1/2}$, and numerous authors have improved this, the best published result being the bound $\Delta(x) \ll x^{7/22+\varepsilon}$ of Iwaniec and Mozzochi [2]. However there is evidence from a variety of sources in support of the conjecture that $\Delta(x) \ll x^{1/4+\varepsilon}$. Thus, for example the approximate Voronoï summation formula states that

$$\Delta(x) = \frac{x^{1/4}}{\pi\sqrt{2}} \sum_{n \leq x} \frac{d(n)}{n^{3/4}} \cos(4\pi\sqrt{nx} - \frac{\pi}{4}) + O(x^{\varepsilon}),$$

and it is natural to expect the oscillating series to total $O(x^{\varepsilon})$ only.

Here we shall be interested in the distribution of $\Delta(x)$. We have

Theorem 1. *The function $x^{-1/4}\Delta(x)$ has a distribution function $f(\alpha)$ in the sense that, for any interval I we have*

$$X^{-1} mes\{x \in [1, X] : x^{-1/4}\Delta(x) \in I\} \rightarrow \int_I f(\alpha)d\alpha$$

as $X \rightarrow \infty$. The function $f(\alpha)$ and its derivatives satisfy the growth condition

$$\frac{d^k}{d\alpha^k} f(\alpha) \ll_{A,k} (1 + |\alpha|)^{-A}$$

for $k = 1, 2, \ldots$ and any constant A. Moreover $f(\alpha)$ extends to an entire function on the complex plane.

Clearly f is non-negative, and since it extends to an entire function on the complex plane its zeros form a discrete set. Thus

$$\int_I f(\alpha)d\alpha > 0$$

for any interval I of positive length. It follows that $x^{-1/4}\Delta(x)$ spends a positive proportion of its time in any such interval.

Although all moments of $f(\alpha)$ are finite it does not follow that all moments of $\Delta(x)$ converge. However we can use Theorem 1 to prove results for a suitable range of exponents.

Theorem 2. *For any exponent $k \in [0, 9]$ the mean value*

$$X^{-1-k/4} \int_0^X |\Delta(x)|^k dx$$

converges to a finite limit as X tends to infinity. Moreover the same is true for the odd moments

$$X^{-1-k/4} \int_0^X \Delta(x)^k dx$$

for $k = 1, 3, 5, 7$ and 9. The limits one obtains are the corresponding moments of $f(\alpha)$.

It is natural to ask for more information about $f(\alpha)$. A theorem of Voronoï shows that

$$\int_0^X \Delta(x)dx = o(X^{5/4}),$$

so that f has mean value zero. Similarly recent work of Tsang [3] yields

$$\int_0^X \Delta(x)^3 dx \sim cX^{7/4},$$

with a positive constant c, so that f has a positive third moment, and hence is asymmetric. Moreover one can be more precise about the rate of decay of $f(\alpha)$ as $|\alpha| \to \infty$. For example, one has

$$f(\alpha) \ll \exp(-|\alpha|^\theta)$$

for any fixed $\theta < 4$.

Detailed proofs of Theorems 1 and 2 may be found in Heath-Brown [1]. The argument applies equally well to the error terms

$$P(x) = \sum_{n \le x} r(n) - \pi x$$

in the circle problem, and

$$E(T) = \int_0^T |\zeta(\frac{1}{2} + it)|^2 dt - T(\log \frac{T}{2\pi} + 2\gamma - 1),$$

for the mean value of the zeta-function, and exactly the same theorems can be obtained. For the generalized divisor problem one can handle

$x^{-1/3}\Delta_3(x)$ in place of $x^{-1/4}\Delta(x)$, but one can only deal with the range $k \in [0, 3)$ in the analogue of Theorem 2. However the method breaks down for $\Delta_k(x)$ with $k \geq 4$.

The proof applies to suitable functions $F(t)$ which can be approximated by an oscillating series in the following way. Let $a_1(t), a_2(t), \ldots$ be continuous real valued functions of period 1, and suppose that $\gamma_1, \gamma_2, \ldots$ are non-zero constants. We shall assume that $F(t)$ satisfies the condition

$$\lim_{N \to \infty} \limsup_{T \to \infty} \frac{1}{T} \int_0^T |F(t) - \sum_{n \leq N} a_n(\gamma_n t)| dt = 0. \tag{0.1}$$

This already suffices to prove that

$$\frac{1}{T} \int_0^T p(F(t)) dt$$

converges as $T \to \infty$, for any continuous, piecewise differentiable function p for which both p and $\dot{p}$ are integrable. We would like to apply this to the function $p(\alpha) = \chi_I(\alpha)$ of course. In order to prove Theorem 1 in this general setting one must assume that the γ_n are linearly independent over the rationals, and that the functions $a_n(t)$ satisfy certain growth conditions. These assumptions suffice to establish the existence of a distribution function f and to establish the properties described. To handle $\Delta(x)$ one would choose $F(t) = t^{-1/2}\Delta(t^2)$ and put

$$a_n(t) = \frac{1}{\pi\sqrt{2}} \sum_{r=1}^{\infty} \frac{d(nr^2)}{(nr^2)^{3/4}} \cos\{2\pi r t - \frac{\pi}{4}\}$$

for n square-free, and $a_n(t) = 0$ otherwise. One can then define $\gamma_n = \sqrt{n}$ for square-free n. It is precisely the fact that the integers must be grouped according to their square-free kernel that makes the distribution function f so difficult to describe.

We conclude with an argument which leads to an expression for f. Let $p(\alpha)$ be a smooth approximation to the characteristic function of the interval I, so that

$$L = \lim_{T \to \infty} T^{-1} \int_0^T p(F(t)) dt$$

is an approximation to

$$\lim_{T \to \infty} T^{-1} mes\{t \in [0, T] : F(t) \in I\}.$$

The Fourier inversion theorem then yields

$$L = \lim_{T \to \infty} T^{-1} \int_0^T \int_{-\infty}^{\infty} \hat{p}(\alpha) e(\alpha F(t)) d\alpha dt,$$

where $e(x) = \exp(2\pi i x)$, as usual. The approximation condition (1) now allows us to replace F by a sum of functions $a_n(\gamma_n t)$, to give

$$L = \lim_{N\to\infty} \lim_{T\to\infty} T^{-1} \int_0^T \int_{-\infty}^{\infty} \hat{p}(\alpha) e(\alpha \sum_{n=1}^N a_n(\gamma_n t)) d\alpha dt.$$

At this point we apply Fubini's theorem, whence

$$L = \lim_{N\to\infty} \lim_{T\to\infty} \int_{-\infty}^{\infty} \hat{p}(\alpha) T^{-1} \int_0^T \prod_{n=1}^N b_{n,\alpha}(\gamma_n t) dt d\alpha, \qquad (0.2)$$

where $b_{n,\alpha}(t) = e(\alpha a_n(t))$. We now call on the following lemma.

Lemma 3. *Let $b_n(t)$ be continuous complex valued functions of period 1, and let γ_n be non-zero real numbers, linearly independent over the rationals. Then*

$$T^{-1} \int_0^T \prod_{n=1}^N b_n(\gamma_n t) dt \to \prod_{n=1}^N \{\int_0^1 b_n(t) dt\}$$

as $T \to \infty$.

By Lebesgue's dominated convergence theorem we may therefore replace (2) by

$$\lim_{N\to\infty} \int_{-\infty}^{\infty} \hat{p}(\alpha) K_N(\alpha) d\alpha,$$

where

$$K_N(\alpha) = \prod_{n=1}^N \{\int_0^1 e(\alpha a_n(t)) dt\}.$$

Since K_N is boundedly convergent as N tends to infinity we obtain

$$L = \int_{-\infty}^{\infty} \hat{p}(\alpha) K(\alpha) d\alpha,$$

with

$$K(\alpha) = \prod_{n=1}^{\infty} \{\int_0^1 e(\alpha a_n(t)) dt\}.$$

Finally, Parseval's identity yields

$$L = \int_{-\infty}^{\infty} \hat{p}(\alpha) K(\alpha) d\alpha = \int_{-\infty}^{\infty} p(\alpha) \hat{K}(\alpha) d\alpha.$$

One therefore sees that the required distribution function $f(\alpha)$ is $\hat{K}(\alpha)$.

References

1. D.R. Heath-Brown, The distribution and moments of the error term in the Dirichlet divisor problem, Acta arithmetica, 60 (1992), 389-414.

2. H. Iwaniec and C.J. Mozzochi, On the divisor and circle problems, J. Number Theory, 29 (1988), 60-93.

3. K.-M. Tsang, Higher moments of $\Delta(x)$, $E(t)$ and $P(x)$, Proc. London Math. Soc. (3), to appear.

Roger Heath-Brown
Magdalen College
University of Oxford
Oxford OX1 4AU
England

A Motivated Introduction
to the Langlands Program

M. Ram Murty
McGill University

Dedicated to the memory of Professor R. Sitaramachandra Rao

§1. Motivation.

This paper is partially expository and is intended as an introduction to the Langlands program for number theorists. The new theorem in the paper is that automorphic induction map for Hecke characters implies both the Artin conjecture and the Langlands reciprocity law. (See below for definitions.) In the last section, we describe some recent work with K. Murty [18] that applies the theory of base change to elliptic curves. The paper is not an exhaustive survey. We have tried to use some classical problems of number theory as motivation for discussion. For instance, we concentrate on GL_n though the functoriality conjecture predicts that this is not a limitation. Nevertheless, our discussion is sufficiently motivated from the number theoretic point of view that a non-specialist in the field can appreciate the depth and profundity of these ideas.

We begin by considering two open problems confronting number theory: Fermat's last theorem and the Sato-Tate conjecture.

First, we must understand the notion of a modular form. Let $SL_2(\mathbb{Z})$ denote the full modular group. That is,

$$SL_2(\mathbb{Z}) = \left\{ \begin{pmatrix} a & b \\ c & d \end{pmatrix} \quad : \quad a, b, c, d \in \mathbb{Z}; \; ad - bc = 1 \right\}.$$

If $\mathfrak{h}$ denotes the upper half-plane, a holomorphic function $f : \mathfrak{h} \to \mathbb{C}$ is called a modular form for $SL_2(\mathbb{Z})$ of weight k if

$$f\left(\frac{az + b}{cz + d}\right) = (cz + d)^k f(z) \quad \forall \begin{pmatrix} a & b \\ c & d \end{pmatrix} \in SL_2(\mathbb{Z})$$

and f is "holomorphic at infinity". Since $f(z + 1) = f(z)$, such a function f has a Fourier expansion and the condition "holomorphic at infinity" can be stated by saying that f has a Fourier expansion of the form

Research partially supported by an NSERC grant.

$$f(z) = \sum_{n=0}^{\infty} a_n e^{2\pi i n z}.$$

More generally, we may consider for each natural number N, the subgroup $\Gamma_0(N)$ defined as the subgroup of matrices

$$\gamma = \begin{pmatrix} a & b \\ c & d \end{pmatrix}$$

of $SL_2(\mathbb{Z})$ satisfying $c \equiv 0 (\bmod N)$. A modular form of weight k for $\Gamma_0(N)$ is defined analogously. (All rational numbers are $SL_2(\mathbb{Z})$-equivalent to $i\infty$. The $\Gamma_0(N)$ equivalence classes of rational numbers are called **cusps** for $\Gamma_0(N)$ and sometimes we refer to a representative of the class as a cusp. We require that the modular form for $\Gamma_0(N)$ be holomorphic at each of these cusps.)

Now let $A, B, C \in \mathbb{Z}$ be coprime integers such that $A + B + C = 0$ with $32|B$ and $4|(A+1)$. Let E be the curve

$$E: \qquad y^2 = x(x - A)(x + B).$$

For each prime p not dividing ABC let N_p be the number of solutions of E mod p. Define

$$a_p = p - N_p.$$

Then a classical theorem of Hasse (conjectured by E. Artin in his doctoral thesis) states that $|a_p| \le 2\sqrt{p}$. Set

$$L_E(s) = F(s) \prod_{p \nmid ABC} \left(1 - \frac{a_p}{p^s} + \frac{1}{p^{2s-1}}\right)^{-1}$$

where

$$F(s) = \prod_{p|A}\left(1 - \left(\frac{B}{p}\right)\frac{1}{p^s}\right)^{-1} \prod_{p|B}\left(1 - \left(\frac{A}{p}\right)\frac{1}{p^s}\right)^{-1} \prod_{p|C}\left(1 - \left(\frac{-B}{p}\right)\frac{1}{p^s}\right)^{-1}$$

By virtue of Hasse's inequality, this infinite product converges for $\mathrm{Re}\,(s) > 3/2$ and so in this half-plane, we can write $L_E(s)$ as a Dirichlet series

$$L_E(s) = \sum_{n=1}^{\infty} \frac{a_n}{n^s}.$$

This defines the a_n (which coincides with a_p when n is prime, so the notation is consistent). Define

$$N = \prod_{p|ABC} p, \qquad f_E(z) = \sum_{n=1}^{\infty} a_n e^{2\pi i n z}.$$

In 1955, Taniyama [32] made the astounding conjecture that $f_E(z)$ is a modular form of weight 2 for $\Gamma_0(N)$. In 1985, Frey [7] noticed that there may be a link between Taniyama's conjecture and Fermat's last theorem. This led Serre [28] to formulate more precise conjectures concerning the ramification of modular Galois representations which culminated in K. Ribet [22] proving in 1989 the following remarkable theorem:

THEOREM 1.   (Ribet, 1989) Taniyama's conjecture implies Fermat's last theorem.

Taniyama's conjecture reduces to the problem of determining when a given sequence of numbers $\{a_n\}_{n=1}^{\infty}$ is the sequence of Fourier coefficients of a modular form. In 1967, A. Weil [34] answered this question in the following way. Let

$$L(s) = \sum_{n=1}^{\infty} \frac{a_n}{n^s}$$

and for each Dirichlet character χ mod c define

$$L(s, \chi) = \sum_{n=1}^{\infty} \frac{a_n \chi(n)}{n^s}.$$

Suppose that the a_n's are of polynomial growth and for each primitive character χ mod c and $(c, N) = 1$, $L(s, \chi)$ extends to an entire function and satisfies the functional equation

$$(c\sqrt{N}/2\pi)^s \Gamma(s) L(s, \chi) = w_\chi (c\sqrt{N}/2\pi)^{k-s} \chi(-N) L(k - s, \bar{\chi})$$

where w_χ is a complex number of absolute value 1. Then

$$f(z) = \sum_{n=1}^{\infty} a_n e^{2\pi i n z}$$

is a modular form of weight k for $\Gamma_0(N)$.

In view of Weil's theorem, Theorem 1 reduces Fermat's last theorem to an assertion about analytic continuation of certain Dirichlet series. This is not the first time that such an event has taken place in number theory. In retrospect, we see that the introduction of the zeta function to solve problems of the distribution of prime numbers or the use of Dirichlet L-functions to describe the behaviour of primes in arithmetic progressions foreshadowed this event.

As we shall see, the Taniyama conjecture is a special case of the Langlands program which seeks to unify representation theory, number theory and arithmetic algebraic geometry. The binding link between all these diverse disciplines is the notion of an L-function of an automorphic representation and the relation between its analytic properties and the underlying algebraic structures. The L-functions of automorphic forms and automorphic representations generalise the classical zeta and L-functions of Riemann, Dirichlet and Hecke.

To further motivate our understanding, I would like to describe the Sato-Tate conjecture. Let E be an elliptic curve defined over $\mathbb{Q}$. By Hasse's inequality, we know $|a_p| \le 2\sqrt{p}$. Let us write

$$a_p = \sqrt{p}(e^{i\theta_p} + e^{-i\theta_p}) = 2\sqrt{p}\cos\theta_p.$$

Sato and Tate independently asked the question how does θ_p vary as p varies and were led to conjecture that if the elliptic curve is not of CM type, then the θ_p's are uniformly distributed with respect to the measure

$$\frac{2}{\pi}\sin^2\theta d\theta.$$

In his McGill lectures given in 1967, Serre [27] reformulated this conjecture as follows. Let $\alpha_p = e^{i\theta_p}$ and $\beta_p = e^{-i\theta_p}$. For each m, define the L-series

$$L_m(s) = \prod_p \prod_{j=0}^{m} \left(1 - \frac{\alpha_p{}^{m-j}\beta_p^j}{p^s}\right)^{-1}.$$

Each $L_m(s)$ converges for $\mathrm{Re}\,(s) > 1$. Suppose that each $L_m(s)$ extends to an entire function for all $s \in \mathbb{C}$ and $L_m(1 + it) \ne 0$ for all real values of t. Then, Serre [27] showed that the θ_p's are uniformly distributed with respect to the (Sato-Tate) measure $2\sin^2\theta/\pi$. In 1979, Kumar Murty [17] showed that analytic continuation of each $L_m(s)$ to $\mathrm{Re}\,(s) = 1$ alone suffices to imply the Sato-Tate conjecture.

Since the a_p's behave like Fourier coefficients of cusp forms of weight 2, it is reasonable to expect the same type of behaviour from Fourier coefficients of cusp forms which are eigenfunctions of Hecke operators. To illustrate, consider the Ramanujan τ function defined by the power series

$$q \prod_{n=1}^{\infty}(1 - q^n)^{24} = \sum_{n=1}^{\infty}\tau(n)q^n.$$

If we set $q = e^{2\pi i z}$, then the series defines a cusp form of weight 12 for the full modular group. In 1916, Ramanujan [20] conjectured that τ satisfies

(1) $\tau(nm) = \tau(n)\tau(m)$ if $(n, m) = 1$ and
(2) $|\tau(p)| \le 2p^{11/2}$ whenever p is prime.

(1) was proved by Mordell in 1928, but he overlooked the depth of the ideas that went into his proof. Hecke [12] saw in it the theory of certain operators acting on the space of cusp forms. But (2) defied many attempts

until Deligne [4] in 1974 proved it as a consequence of his proof of the Weil conjectures. We can therefore write, in analogy with elliptic curves,

$$\tau(p) = p^{11/2}(e^{i\theta_p} + e^{-i\theta_p}) = p^{11/2}(\alpha_p + \beta_p).$$

With this notation, the series $L_m(s)$ can be analogously defined. Serre [27] conjectured that the θ_p's are uniformly distributed with respect to the Sato-Tate measure. In terms of L-functions and Kumar Murty's theorem, this means we must continue each $L_m(s)$ to the line $\text{Re}(s) = 1$. If $m = 1$, $L_1(s)$ is the Mellin transform of a cusp form and so has analytic continuation by the work of Ramanujan and more generally by the work of Hecke. If $m = 2$, Rankin [21] and Selberg [26] (independently) showed in the 1940's that $\zeta(s)L_2(s)$ has an analytic continuation and satisfies a functional equation. Since the ζ-function does not vanish on the line $\text{Re}(s) = 1$, it follows that $L_2(s)$ extends to an entire function to $\text{Re}(s) = 1$. That in fact $L_2(s)$ extends to an entire function for all values of s was proved by Shimura [30] in 1975 by a slight modification of the Rankin-Selberg method. In 1982, Shahidi [29] showed that $L_3(s)$ and $L_4(s)$ have analytic continuation up to $\text{Re}(s) = 1$. In 1985, Garrett [8] has also obtained results which imply these by another method. Their results imply

Theorem. If $L_1(s)$ has no real zeroes in $(1/2, 1)$, then $L_3(s)$ is entire.

In 1952, Gelfand and Fomin [9] showed how a modular form gives rise to a representation of $SL_2(\mathbb{R})$. Langlands' idea is to look at representations of GL_2 of the adele ring of the rational numbers. More generally, he attaches L-functions to representations of adele groups. These L-functions play a central role in the Langlands program.

2. Artin L-series and Hecke's L-series.

One can view the construction of L-series in a purely formal way. This is described by Serre [27] in his McGill lectures alluded to earlier. Let G be a compact group and μ its normalised Haar measure. Let X be the space of conjugacy classes. Suppose S is a finite set of primes and for each prime $p \notin S$, let X_p be a conjugacy class. We can ask the question of how the X_p's are distributed as p varies.

Let ρ be an irreducible representation of G and define

$$L(s, \rho) = \prod_{p \notin S} \det\left(1 - \rho(X_p)p^{-s}\right)^{-1}.$$

This defines an analytic function for $\text{Re}(s) > 1$. Suppose that each $L(s, \rho)$ extends to an analytic function on $\text{Re}(s) = 1$ and does not vanish there. Then, the X_p's are uniformly distributed with respect to the image of the

Haar measure in X. This theorem is proved using the standard Tauberian theory.

Examples.

1. Let $\tau(p) = 2p^{11/2}\cos\theta_p$ or $a_p = 2p^{1/2}\cos\theta_p$. For each prime p, we can associate

$$\begin{pmatrix} e^{i\theta_p} & \\ & e^{-i\theta_p} \end{pmatrix} \in SU(2, \mathbb{C}).$$

Let ρ be the standard representation of $SU(2, \mathbb{C})$. Then, all irreducible representations of $SU(2, \mathbb{C})$ are $\mathrm{Sym}^m(\rho)$ and $L(s, \mathrm{Sym}^m(\rho))$ is $L_m(s)$ defined in the previous lecture. The image of the Haar measure in the space of conjugacy classes of $SU(2, \mathbb{C})$ is $2\sin^2\theta/\pi$.

2. Artin L-series. Let F be an algebraic number field and K/F be a Galois extension. For each prime ideal $\mathfrak{p}$ of F, we have the factorization

$$\mathfrak{p}O_K = \mathcal{P}_1^e \cdots \mathcal{P}_r^e,$$

where $\mathcal{P}_1$, ..., $\mathcal{P}_r$ are prime ideals of K. There is a finite set of primes S (namely the ramified primes of K) such that for $p \notin S$, the decomposition group

$$D_\mathcal{P} = \{\sigma \in G : \mathcal{P}^\sigma = \mathcal{P}\}$$

is cyclic. One can show that this group is canonically isomorphic to the Galois group of the extension of finite fields $O_K/\mathcal{P}$ over $O_F/\mathfrak{p}$ which is generated by $x \mapsto x^{N_{F/\mathbb{Q}}(\mathfrak{p})}$. We denote by $\sigma_\mathcal{P} \in D_\mathcal{P}$ which corresponds to this element. That is,

$$\sigma_\mathcal{P}(x) \equiv x^{N_{F/\mathbb{Q}}(\mathfrak{p})}(\bmod \mathcal{P})$$

for all $x \in O_K$. If $\mathcal{P}|\mathfrak{p}$, then the $\sigma_\mathcal{P}$ are conjugate and we denote by $\sigma_\mathfrak{p}$ the conjugacy class to which the $\sigma_\mathcal{P}$ belong. $\sigma_\mathfrak{p}$ is called the Artin symbol of $\mathfrak{p}$. It is a generalisation of the familiar Legendre symbol which distinguishes quadratic residues from non-residues. For each irreducible complex representation of G, $L(s, \rho)$ is called the Artin L-series attached to ρ.

Artin's conjecture. If ρ is irreducible and $\neq 1$, then $L(s, \rho)$ extends to an entire function of s.

This is one of the major unsolved problems in number theory. If ρ has degree 1, then Artin showed that his conjecture is true by showing that $L(s, \rho) = L(s, \psi)$ where ψ is a Hecke character. Hecke had already shown that such L-series $L(s, \psi)$ arise as Mellin transforms of generalised theta functions and so established their analytic continuation and functional equation. This equality is a deep statement which is called the Artin reciprocity law. In the simplest case when $F = \mathbb{Q}$ and K is a quadratic extension, the identity is equivalent to the law of quadratic reciprocity. In the general case, Brauer showed that each $L(s, \rho)$ has a meromorphic continuation for all $s \in \mathbb{C}$. This was a consequence of his induction theorem which says that every character of a finite group G can be written as an integral linear combination of monomial characters. Thus, every non-Abelian L-series can be written as a quotient of two entire functions each of which is a product of abelian L-series. Since these L-series do not vanish on the line $\mathrm{Re}(s) = 1$, we deduce immediately that the Artin symbols are uniformly distributed in G. This is the famous Chebotarev density theorem, which plays a central role in many problems of number theory. For instance, if $F = \mathbb{Q}$ and $K = \mathbb{Q}(\zeta_k)$, then $\mathrm{Gal}(K/\mathbb{Q}) \simeq (\mathbb{Z}/k\mathbb{Z})^*$ and the Artin symbol of p is $p \mod k$. Thus, we recover Dirichlet's theorem on primes in arithmetic progressions.

Brauer's theorem is essentially group theoretic in nature. By a similar argument, one can show that Artin's conjecture is true for all supersolvable groups where every irreducible character is monomial. But these are group theoretic arguments which have limitations. The natural question is what can we say about representations of degree 2. As we shall see, the Langlands program will make a precise conjecture as to what $L(s, \rho)$ should be in general. In 1975, Langlands made significant progress by proving Artin's conjecture in the "tetrahedral case" by using ideas of representation theory. More precisely, if

$$\rho : G \to GL_2(\mathbb{C}) \overset{nat}{\to} PGL_2(\mathbb{C})$$

is a 2 dimensional representation, then the image of G in $PGL_2(\mathbb{C})$ is one of five possibilities: cyclic, dihedral, A_4, S_4 or A_5. It was the case of A_4 (tetrahedral) that Langlands [14] dealt with after developing the theory of base change for GL_2. Tunnell [33] in 1982 showed the same ideas work for S_4 (the octahedral case). In his doctoral thesis, Buhler [2] showed that Artin's conjecture is true in the icosahedral case for certain cases. The general case is still open.

Artin's conjecture suggests the following question: given a sequence $\{a_n\}_{n=1}^{\infty}$, define

$$L(s) = \sum_{n=1}^{\infty} \frac{a_n}{n^s}.$$

When does $L(s)$ have analytic continuation and functional equation. Hecke obtained such L-series as Mellin transforms of theta functions of several variables. These are L-series attached to grossencharactere. He also considered Mellin transforms of modular forms and showed that these series also enjoyed analytic continuation and functional equations. Let me illustrate with a concrete example. Let

$$\Delta(z) = \sum_{n=1}^{\infty} \tau(n) e^{2\pi i n z}.$$

Let $SL_2(\mathbb{Z})$ be the group of matrices

$$\begin{pmatrix} a & b \\ c & d \end{pmatrix}, \quad a, b, c, d \in \mathbb{Z}, \quad ad - bc = 1.$$

Then,

$$\Delta\left(\frac{az + b}{cz + d}\right) = (cz + d)^{12} \Delta(z)$$

and in particular $\Delta(-1/z) = z^{12} \Delta(z)$. Consider the Mellin transform

$$\int_0^{\infty} \Delta(iy) y^s \frac{dy}{y} = \int_0^{\infty} \sum_{n=1}^{\infty} \tau(n) e^{-2\pi n y} y^s \frac{dy}{y}$$

$$= (2\pi)^{-s} \Gamma(s) \sum_{n=1}^{\infty} \frac{\tau(n)}{n^s}$$

$$= \int_0^1 \Delta(iy) y^s \frac{dy}{y} + \int_1^{\infty} \Delta(iy) y^s \frac{dy}{y}$$

$$= \int_1^{\infty} \Delta(i/y) y^{-s} \frac{dy}{y} + \int_1^{\infty} \Delta(iy) y^s \frac{dy}{y}$$

from which we get analytic continuation and functional equation upon using the modular relation.

To describe Hecke theory in some more detail, let $\mathfrak{h}$ denote the upper half plane. Let $GL_2^+(\mathbb{R})$ denote the non-singular 2×2 matrices with positive determinant. $GL_2^+(\mathbb{R})$ acts on $\mathfrak{h}$. Consider Γ, a congruence subgroup of $SL_2(\mathbb{Z})$. That is, $\Gamma \supseteq \Gamma(N)$ where

$$\Gamma(N) = \left\{ \begin{pmatrix} a & b \\ c & d \end{pmatrix} \equiv I \pmod{N} \right\}$$

for some natural number N. For example, Hecke's group

$$\Gamma_0(N) = \left\{ \begin{pmatrix} a & b \\ c & d \end{pmatrix} \quad c \equiv 0 \pmod{N} \right\}$$

a congruence subgroup. Define $j(g, z) = (cz + d)(\det g)^{-1/2}$ for all $g \in GL_2^+(\mathbb{R})$ and

$$g = \begin{pmatrix} * & * \\ c & d \end{pmatrix}.$$

Set

$$(f|_k \sigma)(z) = j(\sigma, z)^{-k} f(\sigma z)$$

for all positive integers k. A fundamental domain $\mathfrak{F}$ for Γ is any connected set in $\mathfrak{h}$ such that no two interior points of $\mathfrak{F}$ are Γ equivalent and every point in $\mathfrak{h}$ is equivalent to some point in $\mathfrak{F}$. For example, for $\Gamma = SL_2(\mathbb{Z})$, it is easy to see that a fundamental domain is $\mathfrak{F} = \{z = x + iy : \quad -1/2 \leq x \leq 1/2, \quad |z| \geq 1\}$ because $SL_2(\mathbb{Z})$ is generated by

$$\begin{pmatrix} 1 & 1 \\ 0 & 1 \end{pmatrix} \quad \text{and} \quad \begin{pmatrix} 0 & 1 \\ -1 & 0 \end{pmatrix}.$$

For an arbitrary Γ, a fundamental domain is easily derived from this one as follows. Let $\gamma_1, \gamma_2...$ be the left coset representatives of Γ in $SL_2(\mathbb{Z})$. Then a fundamental domain for Γ is $\cup_i \gamma_i^{-1} \mathfrak{F}$.

A **cusp** for Γ is an element of $\mathbb{Q} \cup \{i\infty\}$ which is fixed by an element of trace ± 2. In the case of $SL_2(\mathbb{Z})$, $i\infty$ is the only cusp up to equivalence. Let $\mathfrak{h}^* = \mathfrak{h} \cup$ cusps of Γ. Then, $\Gamma \backslash \mathfrak{h}^*$ has a complex analytic structure of a Riemann surface. We also identify this with a fundamental domain of Γ.

An **automorphic form** for Γ of weight k is a complex valued function f on $\mathfrak{h}$ such that
(1) $f|_k \gamma = f$ for all $\gamma \in \Gamma$,
(2) f is holomorphic on $\mathfrak{h}^*$.

A **cusp form** for Γ is an automorphic form which vanishes at all the cusps of Γ. Let $M_k(\Gamma)$ denote the space of modular forms for Γ and $S_k(\Gamma)$ the space of cusp forms. For $SL_2(\mathbb{Z})$, $\dim M_k(\Gamma) = \dim S_k(\Gamma) + 1$ where the term $+1$ arises from the subspace spanned by the Eisenstein series:

$$E_k(z) = \sum_{(m,n) \neq (0,0)} (mz + n)^{-k}.$$

In the general case for k even and $k > 2$, $\dim M_k(\Gamma)$ - $\dim S_k(\Gamma)$ is equal to the number of inequivalent cusps of Γ so that there is an Eisenstein series attached to each cusp. The space of cusp forms is an inner product space where the inner product is defined by

$$(f, g) = \int \int_{\Gamma \backslash \mathfrak{h}^*} f(z)\overline{g(z)} y^k \frac{dx\,dy}{y^2}.$$

Hecke's First theorem. Let $f \in S_k(\Gamma_0(N))$ and write

$$f(z) = \sum_{n=1}^{\infty} a_n e^{2\pi i n z}$$

for its Fourier expansion at $i\infty$. Then,

$$L(s, f) = \sum_{n=1}^{\infty} \frac{a_n}{n^s}$$

has an analytic continuation and functional equation relating $L(s, f)$ and $L(s, f|w_N)$ where $w_N = \begin{pmatrix} 0 & 1 \\ -N & 0 \end{pmatrix}$.

The Artin conjecture for 2-dimensional Galois representations can now be reformulated and made more precise. Every 2-dimensional Artin L-series is the Mellin transform of a cusp form of weight 1 on $\Gamma_0(N)$ for some natural number N. This can be viewed as the "2-dimensional reciprocity law".

Hecke was led to ask which $L(s, f)$ have Euler product expansions because only these will correspond to Artin L-series. He defined certain linear operators (called Hecke operators) on the space $S_k(\Gamma)$ and showed that if f is an eigenfunction of these operators, then $L(s, f)$ has an Euler product. Such f's are called eigenforms. In 1976, Deligne and Serre showed that for each eigenform of weight one on $\Gamma_0(N)$ of "odd type", there is a 2-dimensional Galois representation ρ_f over $\mathbb{Q}$ such that $L(s, \rho_f) = L(s, f)$.

3. Hecke operators.

Let $\Gamma = \Gamma_0(N)$ and let $S_k(\Gamma)$ be the space of cusp forms of Γ of weight k. Define operators

$$T_p(f) = p^{k-1} f(pz) + \frac{1}{p} \sum_{b \bmod p} f\left(\frac{z+b}{p}\right)$$

where $p \nmid N$ is a prime. Then, T_p is a linear operator on $S_k(\Gamma)$.

Hecke's Second Theorem.   Let

$$f(z) = \sum_{n=1}^{\infty} a_n e^{2\pi i n z}, \qquad a_1 = 1.$$

The a_n's are multiplicative if and only if f is an eigenfunction of all the T_p's. In this case, $L(s, f)$ has an Euler product of the form

$$L(s, f) = \prod_{p \nmid N} \left(1 - \frac{a_p}{p^s} + \frac{1}{p^{2s+1-k}}\right)^{-1} \prod_{p \mid N} \left(1 - \frac{a_p}{p^s}\right)^{-1}.$$

Another way of thinking about the T_p's is via double cosets. Let

$$\Gamma \begin{pmatrix} p & 0 \\ 0 & 1 \end{pmatrix} \Gamma = \cup \Gamma \gamma_i.$$

The γ_i can be taken to be

$$\begin{pmatrix} p & 0 \\ 0 & 1 \end{pmatrix} \quad \begin{pmatrix} 1 & b \\ 0 & p \end{pmatrix} \qquad b \bmod p.$$

It is then easily verified that

$$T_p(f) = \sum_i f|\gamma_i.$$

This interpretation will be useful in later generalizations.

4. Modular forms as representations of $SL_2(\mathbb{R})$.

In 1952, Gelfand and Fomin [9] noticed that cusp form of $S_k(\Gamma)$ define representations of $SL_2(\mathbb{R})$ in the following way. In $SL_2(\mathbb{R})$, let

$$N = \{\begin{pmatrix} 1 & n \\ 0 & 1 \end{pmatrix}\}, \quad A = \{\begin{pmatrix} a & \\ & a^{-1} \end{pmatrix}\},$$

$$K = \{\begin{pmatrix} \cos\theta & -\sin\theta \\ \sin\theta & \cos\theta \end{pmatrix} = R_\theta : 0 \le \theta \le 2\pi\}.$$

Then, $SL_2(\mathbb{R}) = NAK$ (Iwasawa decomposition).

If G is any group acting transitively on a set S, then S can be identified with the coset space G/Γ_x where Γ_x is the stabilizer of any $x \in S$. Since $SL_2(\mathbb{R})$ acts transitively on $\mathfrak{h}$, and $\Gamma_i = K$, we can identify the upper half plane with $SL_2(\mathbb{R})/K$. Thus, elements of $SL_2(\mathbb{R})$ can be thought of as pairs (z, θ) with $z \in \mathfrak{h}$. Let $G = SL_2(\mathbb{R})$. Define for $g \in G$ and $f \in S_k(\Gamma)$,

$$\phi_f(g) = j(g, i)^{-k} f(g \cdot i).$$

This satisfies

(1) $\phi(\gamma g) = \phi(g)$ for all $\gamma \in \Gamma$,
(2) $\phi(g R_\theta) = e^{-ik\theta}\phi(g)$,
(3) $\int_{\Gamma\backslash G} |\phi(g)|^2 dg < \infty$,
(4) for all $\rho \in SL_2(\mathbb{Z})$, and all $g \in G$,

$$\int_{\mathbb{Z}\backslash\mathbb{R}} \phi\left(\rho\begin{pmatrix} 1 & xh \\ 0 & 1 \end{pmatrix} g\right) dx = 0$$

where h is the "width" of the cusp $\rho(i\infty)$.

Define the Laplace operator on G

$$\Delta = -y^2\left(\frac{\partial^2}{\partial x^2} + \frac{\partial^2}{\partial y^2}\right) - y\frac{\partial^2}{\partial x\partial\theta}.$$

Then ϕ_f satisfies

(5') $\Delta\phi_f = -k(k-2)\phi_f/4$.

Theorem. Any function $\phi \in L^2(\Gamma\backslash G)$ satisfying (1) - (4) and (5') is necessarily a ϕ_f for some $f \in S_k(\Gamma)$.

If we replace (5') by

(5) ϕ is an eigenfunction of Δ

we get real analytic forms in addition to holomorphic forms. Such forms were first studied by Maass and are called Maass wave forms. They have a Fourier expansion of the type

$$f(z) = \sum_n a_n e^{2\pi i n x} \sqrt{y} K_{ir}(2\pi |n|y)$$

where $r^2 + 1/4$ is an eigenvalue of

$$-y^2 \left(\frac{\partial^2}{\partial x^2} + \frac{\partial^2}{\partial y^2} \right)$$

and

$$K_v(z) = \int_0^\infty e^{-z \cosh t} \cosh vt \, dt$$

is the Bessel function. Connected with these eigenvalues is the famous conjecture of Selberg that predicts that if $\lambda = r^2 + 1/4$ is an eigenvalue for $\Gamma_0(N)$, then $\lambda \geq 1/4$. This is equivalent to saying that r cannot be purely imaginary. This conjecture is known if $\Gamma = SL_2(\mathbb{Z})$. Selberg [26] showed that $\lambda \geq 3/16$ for $\Gamma_0(N)$. Iwaniec and Selberg have noted that there are arithmetic applications of this conjecture.

Let $L^2(\Gamma \backslash G / K) = V$ be the vector space of square integrable, left-invariant by Γ and right K-invariant functions on G. Define the right regular representation of G by

$$(R(g) \cdot \phi)(h) = \phi(hg).$$

Then, $\Delta R = R \Delta$. Thus, irreducible representations occurring in R coincide with eigenfunctions of Δ. (We are ignoring certain growth conditions.) These ideas can be generalised to $GL_n(\mathbb{R})$.

5. $GL_n(\mathbb{R})$.

Let $K = O_n(\mathbb{R})$ be the orthogonal group, Z the group of scalar matrices. Let $\Gamma = GL_n(\mathbb{Z})$. The analog of the upper half plane is the symmetric space

$$\mathcal{H} = \Gamma \backslash GL_n(\mathbb{R}) / KZ.$$

Every element of G/KZ can be written as a product

$$g = \begin{pmatrix} 1 & x_{12} & \cdots & \cdots & x_{1n} \\ & 1 & x_{23} & \cdots & x_{2n} \\ & & \cdots & \cdots & \cdots \\ & & & & 1 \end{pmatrix} \begin{pmatrix} y_1 y_2 \cdots y_n & & & \\ & y_2 \cdots y_n & & \\ & & \cdots & \\ & & & y_{n-1} \\ & & & & 1 \end{pmatrix}.$$

For example, if $n = 2$,

$$\begin{pmatrix} 1 & x \\ 0 & 1 \end{pmatrix} \begin{pmatrix} y & 0 \\ 0 & 1 \end{pmatrix} = \begin{pmatrix} y & x \\ 0 & 1 \end{pmatrix}$$

corresponds to $x + iy \in \mathfrak{h}$.

We will consider functions satisfying

(1) $\phi(\gamma g) = \phi(g)$ for all $\gamma \in \Gamma$.

(2) ϕ is square integrable,

(3) ϕ should be an eigenfunction of $n - 1$ "Laplacians". More precisely, if $\mathcal{D}$ is algebra of G-invariant differential operators, it is isomorphic to a polynomial ring of rank $n - 1$. We require $D\phi = \rho(D)\phi$ for all $D \in \mathcal{D}$.

If these conditions are satisfied by ϕ in addition to the following growth condition

(4) $|\phi(g)| \le C(y_1 \cdots y_{n-1})^A$

then we say that ϕ is an automorphic form for G. An automorphic form is called **cuspidal** if

(5) $\int \phi \left(\begin{pmatrix} I & X \\ 0 & I \end{pmatrix} g \right) dx = 0$ for all $g \in G$ and X is an $n_1 \times n_2$ matrix with $n_1 + n_2 = n$.

(This definition is due to Gelfand.)

One can equally well develop the theory of Hecke operators. Let

$$\xi(p, i) = \begin{pmatrix} pI_i & \\ & I_{n-i} \end{pmatrix}.$$

Let

$$\Gamma\xi(p, i)\Gamma = \cup\Gamma\gamma_{p,i,v}$$

be a double coset decomposition. Define the Hecke operator $T_{p,i}$ by

$$(T_{p,i}\phi)(g) = \sum_v \phi(g\gamma_{p,i,v}).$$

If ϕ is an eigenfunction for all Hecke operators $T_{p,i}$, we will write

$$T_{p,i}\phi = \lambda_{p,i}p^{i(n-i)}\phi, \quad 1 \le i \le n.$$

Define for such a ϕ,

$$L(s, \phi) = \prod_p \left(1 - \lambda_{p,1}p^{-s} + \lambda_{p,2}p^{-2s} + \cdots \pm \lambda_{p,n}p^{-ns}\right)^{-1}.$$

One has the contragredient form $\tilde{\phi}$ defined by $\tilde{\phi}(g) = \phi({}^t g)$ where

$${}^t g = w^t g^{-1} w$$

and

$$w = \begin{pmatrix} & & & -1 \\ & & -1 & \\ & \cdots & & \\ -1 & & & \end{pmatrix}.$$

Then, it is known that:

Theorem. $L(s, \phi)$ has an analytic continuation and functional equation relating $L(s, \phi)$ to $L(s, \tilde{\phi})$ where $\tilde{\phi}$ is the contragredient automorphic form.

These L-functions are related to the ones described previously in the case $n = 1$ (Hecke L-series attached to grossencharactere) and $n = 2$ (Hecke L-series attached to modular forms or L-series attached to Maass wave forms).

Langlands has formulated a strong reciprocity conjecture: every Artin L-series $L(s, \rho)$ is an $L(s, \phi)$ for some automorphic form ϕ on $GL_r(\mathbb{R})$, where $r = \deg \rho$. The truth of this conjecture implies Artin's conjecture in view of the Theorem above. With reference to the Sato-Tate conjecture, it is also conjectured that $L_m(s)$ should be $L(s, \phi)$ for some automorphic form ϕ on $GL_{m+1}(\mathbb{R})$. Again, such a reciprocity conjecture would imply the Sato-Tate conjecture. One can also show that Selberg's eigenvalue conjecture and Ramanujan conjectures also follow from this reciprocity. Thus, reciprocity is the problem of converse theory (that is generalisation of Weil's theorem) to higher GL_n.

6. Adeles and ideles.

For $x \in \mathbb{Q}$, put $v_p(x) = \operatorname{ord}_p(x)$. Then,
(1) $v_p(x + y) \geq \min(v_p(x), v_p(y))$,
(2) $v_p(xy) = v_p(x) + v_p(y)$,
(3) $v_p(0) = \infty$.

Define a metric on $\mathbb{Q}$ by $|x - y|_p = e^{-v_p(x-y)}$. Denote by $\mathbb{Q}_p$ the completion of $\mathbb{Q}$ with respect to this p-adic metric. We call $\mathbb{Q}_p$ the p-adic number field. Let $\mathbb{Z}_p$ be the subring of p-adic integers. Every p-adic number x can be written uniquely as

$$x = \sum_{i=-N}^{\infty} a_i p^i, \qquad a_i \in \{0, 1, ..., p-1\}.$$

We will say that two metrics are equivalent if they induce the same topology on $\mathbb{Q}$. We have the famous:

Theorem. (Ostrowski) Up to equivalence, the only metrics on $\mathbb{Q}$ are $|\cdot|_p$ and $|\cdot|_\infty$ (the usual absolute value).

A similar result holds for any algebraic number field. Metrics which do not correspond to a finite prime (ideal) are called Archimedean valuations (or infinite primes). Thus, for a number field K, we can consider analogously K_v the field of v-adic numbers and the ring O_v of v-adic integers.

Consider the following problem: what are the characters of the additive group of rational numbers $\mathbb{Q}$?

We can write down some obvious ones: given $\alpha \in \mathbb{R}$, let $\chi_\alpha(x) = e^{2\pi i \alpha x}$. If $\beta \in \mathbb{Q}_p$, then $\beta = \sum_{i=-N}^{\infty} a_i p^i$. Let $\tilde{\beta} = \sum_{i=-N}^{-1} a_i p^i$ be the "fractional part" of β. Define $\chi_\beta(x) = e^{2\pi i \bar{\beta} x}$ is also a character of $\mathbb{Q}$. It turns out that all characters of $\mathbb{Q}$ are finite products of such characters. To state this neatly, we can say every character of $\mathbb{Q}$ is given by a sequence

$$a = (a_\infty, a_2, a_3, a_5, \ldots)$$

where $a_\infty \in \mathbb{R}$ and $a_p \in \mathbb{Q}_p$ and $a_p \in \mathbb{Z}_p$ for all p sufficiently large. Such a sequence is called an **adele**. Defining componentwise addition and multiplication, we see that the set of all adeles is a ring, called the adele ring of $\mathbb{Q}$ and denoted $\mathbb{A}_\mathbb{Q}$ or simply $\mathbb{A}$. Hence, for $a \in \mathbb{A}_\mathbb{Q}$ we set

$$\chi_a(x) = e^{2\pi i (a_\infty x + a_2 x + \cdots)}$$

then every character of $\mathbb{Q}$ is of this form. Is this a one-one correspondence? It turns out to be not. It is a simple exercise to show that two characters χ_a and χ_b are equal if and only if $a - b = (x, x, x, \ldots)$ for some rational number x. Thus, we can view $\mathbb{Q}$ as a subgroup of $\mathbb{A}_\mathbb{Q}$ embedded diagonally in this way. Then we have:

Theorem. The character group of $\mathbb{Q}$ is isomorphic to $\mathbb{A}_\mathbb{Q}/\mathbb{Q}$ as an abstract group.

Since $\mathbb{Q}$ has a natural topology on it, one can ask whether this isomorphism can be made as topological groups. For this purpose, it is necessary to define a topology on $\mathbb{A}_\mathbb{Q}$ so as to have this. To this end, we begin by putting the product topology on

$$\mathbb{A}^0 = \mathbb{R} \times \prod_p \mathbb{Z}_p$$

and declaring that $\mathbb{A}^0$ is an open neighbourhood of 0. Thus, a sequence of adeles $a^{(n)} = (a_\infty^{(n)}, a_2^{(n)}, \cdots)$ tends to zero if and only if $a_\infty^{(n)}$ tends to zero and $a_p \in \mathbb{Z}_p$ for all n sufficiently large. Thus, this is stronger than the product topology. Thus, the adele ring with this topology is the restricted direct product

$$\mathbb{A} = \prod_p (\mathbb{Q}_p : \mathbb{Z}_p).$$

In general, if K is a number field, the adele ring $\mathbb{A}_K$ is the restricted direct product

$$\mathbb{A}_K = \prod_v (K_v : O_v).$$

If G is a linear algebraic group, we can define topological group of its adelic points by the restricted direct product

$$G(\mathbb{A}_K) = \prod_v (G(K_v) : G(O_v)).$$

$\mathbb{A}_K$ becomes locally compact topological group and as such has a Haar measure on it. A similar result holds for $G(\mathbb{A})$. In the course of this paper, we will be interested in $G = GL_n$. If $G = GL_1$, then $G(K)\backslash G(\mathbb{A}_K)$ is called the **idele class group** of K.

7. Automorphic representations of adele groups.

Let Z be the group of scalar adeles and consider the vector space

$$V = L^2(ZG(\mathbb{Q})\backslash G(\mathbb{A}_\mathbb{Q})).$$

We will be interested in irreducible unitary representations of $G(\mathbb{A})$. Every such representation π factors as a restricted tensor product:

$$\pi = \otimes_p \pi_p$$

where π_p is an irreducible unitary representation of $G(\mathbb{Q}_p)$. In case $G = GL_1$, these are Hecke's grossencharactere. In case $G = GL_2$, every irreducible unitary representation arises from either a modular form or a Maass wave form. Our goal will be to attach an L-function to such a representation π. There is a finite set S of primes (namely the ramified primes of π) such that for each prime $p \notin S$, we can associate a conjugacy class A_p in $GL_n(\mathbb{C})$. The exact assignment needs more background (see [15]) which we will not discuss here. Suffice it to say that in case f is a Hecke eigenform, and π_f is the corresponding automorphic representation, then $\pi_{f,p}$ is associated to $\begin{pmatrix} \alpha_p \\ & \beta_p \end{pmatrix}$ with notation as in §1.

$G(\mathbb{A})$ acts on $L^2(ZG(\mathbb{Q})\backslash G(\mathbb{A}))$ by right translation:

$$(R(g) \cdot \phi)(x) = \phi(xg).$$

An automorphic representation is an irreducible constituent of R. (Cuspidal automorphic representations form a subspace corresponding to cusp forms.)

Let K be a number field and π an automorphic representation of $G(\mathbb{A}_K)$. Let A_v be the conjugacy class attached to π_v. Define

$$L(s, \pi_v) = \det(1 - A_v N v^{-s})^{-1}$$

for non-Archimedean valuations. Set

$$L_S(s, \pi) = \prod_{v \notin S} L(s, \pi_v).$$

Theorem. $L_S(s, \pi)$ has an analytic continuation and at the places $v \in S$, $L(s, \pi_v)$ can be defined so that the global L-function $L(s, \pi)$ has a functional equation.

This theorem, for $n = 1$ is the theory of Hecke's L-series attached to grossencharacter and Tate, who formulated it adelically in his doctoral thesis. For $n = 2$, it is due to Hecke, Maass and the final adelic formulation is the work of Jacquet and Langlands [13]. The general case for GL_n is due to Godement and Jacquet. A good write up for the Jacquet-Langlands theory is found in Robert [23].

In this adelic framework, Langlands conjectures that every Artin L-series $L(s, \rho)$ is and $L(s, \pi)$ for some automorphic representation π on $GL_n(\mathbb{A}_K)$ where $n = \deg \rho$. Thus, to each ρ there should be an automorphic representation $\pi(\rho)$. In case $\deg \rho = 1$, this is Artin's reciprocity law since Hecke's grossencharactere are automorphic representations of $GL_1(\mathbb{A})$. We have already discussed the situation if $\deg \rho = 2$.

To make further progress, we must clarify the converse problem and resolve it. Namely, given a Dirichlet series

$$L(s) = \sum_{n=1}^{\infty} \frac{a_n}{n^s}$$

when is $L(s) = L(s, \pi)$ for some automorphic representation π of $GL_n(\mathbb{A})$. Weil's theorem is an incomplete converse theorem for GL_2 in that it applies to only modular forms. Jacquet-Langlands [13] filled this gap to cover Maass forms as well and generalised to an arbitrary number field. Thus the analog of Weil's theorem holds for GL_2. For GL_3, Gelbart, Jacquet and Piatetski-Shapiro proved an analogous result. However, it was already shown by Piatetski-Shapiro [19] that for GL_4, one needs to twist by not just abelian characters but all 2-dimensional representations and establish analytic continuation and functional equations of the right type before one can conclude that $L(s) = L(s, \pi)$ for some automorphic representation π on $GL_4(\mathbb{A})$. In general, one expects to twist by all representations of degree $d \leq m - 2$ before we can conclude that π lives on $GL_m(\mathbb{A})$. To establish continuation and functional equations, one needs the Rankin-Selberg method.

8. The Rankin-Selberg method.

Recall that the problem of the Sato-Tate conjecture was equivalent to the problem of analytic continuation of $L_m(s)$ to the line $\operatorname{Re}(s) = 1$. $L_1(s)$ was treated by Hecke. We have mentioned that $\zeta(s)L_2(s)$ was shown to extend to a meromorphic function with only a simple pole at $s = 1$ by Rankin and Selberg (independently) in the 1940's. Their idea is fundamental in understanding the mechanism of converse theory so we will describe this in some detail. I will illustrate it for the full modular group $SL_2(\mathbb{Z})$.

Suppose that $f(z) = \sum_{n=1}^{\infty} a_n e^{2\pi i n z}$ and $g(z) = \sum_{n=1}^{\infty} b_n e^{2\pi i n z}$ are two cusp forms of weight k for the full modular group. (Weaker assumptions on f and g still lead to the same results but then, our exposition will be bogged down by unnecessary technical details.) Define the Eisenstein series

$$E(z,s) = \sum_{\gamma \in \Gamma_\infty \backslash \Gamma} Im(\gamma z)^s$$

where

$$\Gamma_\infty = \left\{ \begin{pmatrix} 1 & n \\ 0 & 1 \end{pmatrix} : n \in \mathbb{Z} \right\}.$$

Note that

$$y^k f(z)\overline{g(z)}$$

is $SL_2(\mathbb{Z})$-invariant. Consider

$$\int\int_{\Gamma\backslash\mathfrak{h}} y^k f(z)\overline{g(z)} E(z,s) \frac{dxdy}{y^2}.$$

Inserting the definition of $E(z,s)$ as a sum and interchanging the sum and integration we get

$$\sum_{\gamma \in \Gamma_\infty \backslash \Gamma} \int\int_{\Gamma\backslash\mathfrak{h}} (Im(\gamma y))^k f(\gamma z)\overline{g(\gamma z)}(Im(\gamma z))^s \frac{dxdy}{y^2}$$

$$= \sum_{\gamma \in \Gamma_\infty \backslash \Gamma} \int\int_{\Gamma\backslash\mathfrak{h}} y^k f(z)\overline{g(z)} y^s \frac{dxdy}{y^2}$$

$$= \int\int_{\Gamma_\infty\backslash\mathfrak{h}} y^k f(z)\overline{g(z)} y^s \frac{dxdy}{y^2}$$

because

$$\Gamma_\infty\backslash\mathfrak{h} = \cup_{\gamma \in \Gamma_\infty \backslash \Gamma}\gamma(\Gamma\backslash\mathfrak{h})$$

which can be taken to be the region $-1/2 \leq x \leq 1/2$, $y > 0$. Thus, the above is

$$= \int_0^\infty \int_{-1/2}^{1/2} y^k f(z)\overline{g(z)} y^s \frac{dxdy}{y^2}$$

$$= \int_0^\infty y^{k+s} \sum_{n,m} a_n \overline{b_m} e^{-2\pi(n+m)y} \int_{-1/2}^{1/2} e^{2\pi i(n-m)x} dx \frac{dy}{y^2}$$

$$== \int_0^\infty \sum_{n=1}^\infty a_n \overline{b_n} y^{k+s} e^{-4\pi n y} \frac{dy}{y^2}$$

$$= (4\pi)^{k+s+1}\Gamma(k+s+1) \sum_{n=1}^\infty \frac{a_n \overline{b_n}}{n^s}.$$

It is known that

$$E^*(z, s) = \zeta(2s)E(z, s)$$

has an analytic continuation except for a simple pole at $s = 1$ and satisfies a functional equation relating the value at s to the value at $1 - s$. Therefore, from this integral representation, we obtain the analytic continuation of the series

$$\sum_{n=1}^{\infty} \frac{a_n \overline{b_n}}{n^s}.$$

Now suppose that f and g are Hecke eigenforms. That is, a_n and b_n are multiplicative. Our goal is to write the above Dirichlet series as an Euler product. Let us write

$$a_p = 2p^{(k-1)/2}(\alpha_p + \beta_p), \quad b_p = 2p^{(k-1)/2}(\gamma_p + \delta_p).$$

Let $\tilde{a}_n = a_n/n^{(k-1)/2}$ and $\tilde{b}_n = b_n/n^{(k-1)/2}$. It is not difficult to show that

$$\tilde{a}_{p^n} = \frac{\alpha_p^{n+1} - \beta_p^{n+1}}{\alpha_p - \beta_p}$$

and a similar formula holds for $\tilde{b}_{p^n}$ whenever p is prime. Then,

$$\sum_{n=1}^{\infty} \frac{\tilde{a}_n \tilde{b}_n}{n^s} = \prod_p \left(\sum_{n=1}^{\infty} \left(\frac{\alpha_p^{n+1} - \beta_p^{n+1}}{\alpha_p - \beta_p} \right) \left(\frac{\gamma_p^{n+1} - \delta_p^{n+1}}{\gamma_p - \delta_p} \right) p^{-ns} \right).$$

Now we invoke Ramanujan's identity:

$$\sum_{n=1}^{\infty} \left(\frac{\alpha^{n+1} - \beta^{n+1}}{\alpha - \beta} \right) \left(\frac{\gamma^{n+1} - \delta^{n+1}}{\gamma - \delta} \right) T^n =$$

$$= \frac{1 - \alpha\beta\gamma\delta T^2}{(1 - \alpha\gamma T)(1 - \alpha\delta T)(1 - \beta\gamma T)(1 - \beta\delta T)}.$$

Thus, we get the product equal to

$$\zeta(2s)^{-1} \prod_p \left(1 - \frac{\alpha_p \gamma_p}{p^s} \right)^{-1} \left(1 - \frac{\alpha_p \delta_p}{p^s} \right)^{-1} \left(1 - \frac{\beta_p \gamma_p}{p^s} \right)^{-1} \left(1 - \frac{\beta_p \delta_p}{p^s} \right)^{-1}.$$

This is precisely $L(s, \pi_f \otimes \pi_g)$ in the notation of the Langlands' L-function. We have therefore

$$L(s, \pi_f \otimes \pi_g) = \int_{\Gamma \backslash \mathfrak{h}} y^k f(z) g(z) E^*(z, s) \frac{dx\,dy}{y^2}.$$

If we let $f = g$, we get

$$L(s, \pi_f \otimes \pi_f) = \zeta(s) L_2(s)$$

since $\pi \otimes \pi = \text{Alt}(\pi) \oplus \text{Sym}^2(\pi)$. Since the zeta function does not vanish on the line $\text{Re}(s) = 1$, this gives the analytic continuation of $L_2(s)$ until this line.

Though we have not mentioned it explicitly, the above construction is a special case of a more general construction of Langlands of L-functions attached to the tensor product of two automorphic representations. Indeed, Gelbart, Jacquet and Piatetski-Shapiro [11] showed that $L(s, \pi_1 \otimes \pi_2)$ has analytic continuation and functional equation except when π_2 is the contragredient of π_1 in which case it has a simple pole at $s = 1$.

The method however does not give the continuation of $L_2(s)$ for all values of s. This was done by Shimura [30] by making an ingenious observation. He noticed that $L_2(s)$ is essentially

$$\sum_{n=1}^{\infty} \frac{a_{n^2}}{n^s}$$

and so this should be obtained by taking the Rankin-Selberg convolution of f with the classical theta function (which is a modular form of half integral weight). He was able to show that the theory extends to such a situation and obtained an integral representation

$$\sum_{n=1}^{\infty} \frac{a_{n^2}}{n^s} = (fudge\ factor) \int_{\Gamma_0(4)\backslash \mathfrak{h}} y^{k+1/2} f(z) \overline{\theta(z)} E_{\Gamma_0(4)}(z, s) \frac{dx\,dy}{y^2}$$

where $E_{\Gamma_0(4)}(z, s)$ is a half-integral weight Eisenstein series belonging to $\Gamma_0(4)$. Therefore, we obtain the analyticity of $L_2(s)$.

But can we establish Langlands' conjecture for $L_2(s)$, namely that it is an L-series attached to an automorphic representation of GL_3 ? As mentioned earlier, Gelbart and Jacquet proved a converse theorem for GL_3 which requires twisting our original series by Dirichlet characters. The method of Shimura extends to give analytic continuation and functional equations for all these twists and so Gelbart and Jacquet were able to prove Langlands' conjecture for $L_2(s)$. We state their theorem more precisely:

Theorem. If π is an irreducible unitary automorphic representation of $GL_2(\mathbb{A}_K)$, then the map $\pi \mapsto \mathrm{Sym}^2(\pi)$ (Gelbart-Jacquet lift) is a map into irreducible unitary automorphic representations on $GL_3(\mathbb{A}_K)$.

For further discussion, let us introduce the notation $\mathcal{A}(GL_n)$ to denote the space of irreducible unitary automorphic representations of $GL_n(\mathbb{A}_K)$. With the information at hand, we can take Rankin-Selberg convolutions:

$$\pi \otimes \mathrm{Sym}^2(\pi) = \pi \oplus \mathrm{Sym}^3(\pi)$$
$$\mathrm{Sym}^2(\pi) \otimes \mathrm{Sym}^2(\pi) = 1 \oplus \mathrm{Sym}^2(\pi) \oplus \mathrm{Sym}^4(\pi).$$

From the first equation, we deduce that $L_3(s)$ is analytic for $\mathrm{Re}(s) \geq 1$ upon showing that $L(s, \pi)$ does not vanish on this line. This is a minor modification of the classical argument of Hadamard and de la Vallée

Poussin. The second equation shows that $L_4(s)$ is analytic for $\mathrm{Re}\,(s) \geq 1$ by a similar process. If we can consider $\mathrm{Sym}^2(\pi) \otimes \mathrm{Sym}^3(\pi)$ then we would obtain information on $L_5(s)$. This however we cannot do because we do not know that $\mathrm{Sym}^3(\pi) \in \mathcal{A}(GL_4)$. We need a converse theorem for GL_4.

Piatetski-Shapiro showed that the analog of Weil's theorem does not hold for GL_4 and we need the analyticity of additional twists. It turns out that these additional twists are representations of GL_2. Thus, the converse theory for GL_4 requires analytic continuations of $L(s, \phi \otimes \pi)$ as π ranges over $\mathcal{A}(GL_2)$. It is in this context that the result of Paul Garrett [8] in 1985 on triple convolutions was intriguing. (We shall see later other reasons for the intrigue.) Namely, Garrett [8] showed that if f, g and h are modular cusp forms of weight k, with Fourier coefficients a_n, b_n and c_n respectively, then

$$\sum_{n=1}^{\infty} \frac{a_n b_n c_n}{n^s}$$

can be written as an integral transform of Rankin-Selberg type and an analytic continuation can be obtained for it. In the notation of Langlands, this corresponds to $L(s, \pi_f \otimes \pi_g \otimes \pi_h)$, an L-function attached to $GL_2 \times GL_2 \times GL_2$. In the context of converse theory, this result implies that we will have a map

$$\mathcal{A}(GL_2) \times \mathcal{A}(GL_2) \rightarrow \mathcal{A}(GL_4).$$

From this, we would obtain the analytic continuation of $L_5(s), L_6(s), L_7(s)$ and $L_8(s)$ up to $\mathrm{Re}\,(s) = 1$. At the end of this paper, we will see another fundamental reason for the interest in this map.

For the general converse theory on GL_n, one needs to twist by representations of GL_{n-2} though this has not been formally written up. There are informal notes of Piatetski-Shapiro which prove this theorem for function fields.

9. Base change and automorphic induction.

Let K/k be a Galois extension and let $G = \mathrm{Gal}\,(K/k)$. If ρ is an irreducible representation of G, then $L(s, \rho, K/k)$ was conjectured by Artin to be entire if $\rho \neq 1$. We can extend the definition of Artin L-series to an arbitrary representation of G by additivity:

$$L(s, \rho_1 \oplus \rho_2, K/k) = L(s, \rho_1, K/k)L(s, \rho_2, K/k).$$

If now ψ is a representation of a subgroup H of G, then $L(s, \psi, K/K^H)$ is the Artin L-series belonging to the extension K/K^H where K^H denotes the field fixed by H. A simple calculation shows that Artin L-series are invariant under induction:

$$L(s, \operatorname{Ind}_H^G \psi, K/k) = L(s, \psi, K/K^H).$$

This last property implies a corresponding property for L-series attached to automorphic representations. Recall the Langlands' reciprocity conjecture: for each ρ, there is $\pi(\rho) \in \mathcal{A}(GL_n(\mathbb{A}_K))$ so that

$$L(s, \rho, K/k) = L(s, \pi).$$

The natural question is, how does the map $\rho \mapsto \pi(\rho)$ behave under restriction to a subgroup? By the reciprocity conjecture, we should have $\rho|_H \mapsto \pi(\rho|_H) \in \mathcal{A}(GL_n(\mathbb{A}_{K^H}))$. What is $L(s, \rho|_H, K/K^H)$? A standard group theoretic result is

$$\operatorname{Ind}_H^G(\rho|_H \otimes \psi) = \rho \otimes \operatorname{Ind}_H^G \psi.$$

Thus,

$$L(s, \operatorname{Ind}_H^G(\rho|H \otimes \psi), K/k) = L(s, \rho \otimes \operatorname{Ind}_H^G \psi, K/k).$$

But the former L-function is $L(s, \rho|_H \otimes \psi, K/K^H)$ by the invariance of Artin L-series. Therefore,

$$L(s, \rho|_H, K/K^H) = L(s, \rho \otimes \operatorname{Ind}_H^G 1, K/k).$$

But $\operatorname{Ind}_H^G 1 = reg_{G/H}$ is the permutation representation on the cosets of H. This suggests that we make the following definition. Let $\pi \in \mathcal{A}(GL_n(\mathbb{A}_K))$. For each π_v, we associated an $A_v \in GL_n(\mathbb{C})$. Define,

$$L_v(s, B(\pi)) = \det\left(1 - A_v \otimes reg_{G/H}(\sigma_v) N v^{-s}\right)^{-1},$$

where σ_v is the Artin symbol of v. Set

$$L(s, B(\pi)) = \prod_v L_v(s, B(\pi)).$$

$B(\pi)$ should correspond to an element of $\mathcal{A}(GL_n(\mathbb{A}_M))$ where $M = K^H$. The problem of base change is to determine when this map exists.

For $n = 2$, this was done by Langlands [14] when M/k is cyclic. He then used these ideas to deal with the tetrahedral case of Artin's conjecture. For arbitrary n, it is the recent work of Arthur and Clozel [1] Again, the situation is for M/k cyclic.

Now suppose the ψ is a representation of H. Corresponding to ψ there should be a $\pi \in \mathcal{A}(GL_n(\mathbb{A}_M))$ where $n = deg\, \psi$. But the invariance of

Artin L-series under induction implies that there should be an $I(\pi) \in \mathcal{A}(GL_{nr}(\mathbb{A}_k))$ $(r = [G : H])$ so that

$$L(s, I(\pi)) = L(s, \operatorname{Ind}_H^G \psi, K/k).$$

This map $\pi \mapsto I(\pi)$, called the automorphic induction map, is conjectured to exist. Again, Arthur and Clozel showed this exists when M/k is cyclic and arbitrary n. Thus, if M/k is contained in a solvable extension of k, the base change maps and automorphic induction maps exist.

Main Theorem. Let K be an algebraic number field of degree n over $\mathbb{Q}$, and ψ a Hecke character of K. If there exists an automorphic representation $I(\psi)$ of $\mathcal{A}(GL_n(\mathbb{A}_\mathbb{Q}))$ such that

$$L(s, \psi) = L(s, I(\psi))$$

for every Hecke character ψ of K, then both Artin's conjecture and the Langlands reciprocity conjecture are true.

Proof. Let $L/\mathbb{Q}$ be a Galois extension of finite degree and Galois group G. If ρ is an irreducible representation of G, and χ is its character, we can write by the Brauer induction theorem,

$$\chi = \sum_i m_i \operatorname{Ind}_{H_i}^G \chi_i,$$

where m_i are integers, χ_i are abelian characters of certain subgroups H_i of G. By the abelian reciprocity law, $L(s, \chi_i, L/L^{H_i})$ is equal to a Hecke L-series $L(s, \psi_i)$ where ψ_i is a Hecke character of L^{H_i}. As such, we can therefore write

$$L(s, \rho, L/\mathbb{Q}) = \prod_i L(s, \pi_i)^{m_i}$$

where π_i are automorphic representations of $\mathbb{Q}$. By a theorem of Langlands [15], we can further decompose the product so that each constituent L-function corresponds to an irreducible automorphic representation of $\mathbb{Q}$. Thus, without loss, in the above product, we can suppose that each π_i is irreducible and distinct from π_j when $i \neq j$. The identity reveals that

$$\chi(p) = \sum_i m_i a_p^{(i)}$$

where $a_p^{(i)}$ denotes the p-th coefficient in the Dirichlet series of $L(s, \pi_i)$. On one hand, the Chebotarev density theorem shows that

$$\sum_p \frac{|\chi(p)|^2}{p^s}$$

has a simple pole at $s = 1$. On the other hand, the series

$$\sum_p |\sum_i m_i a_p^{(i)}|^2/p^s$$

is on expanding

$$\sum_{i,j} m_i m_j \sum_p a_p^{(i)} \overline{a_p^{(j)}} p^{-s}.$$

By a result of Jacquet, Piatetski-Shapiro and Shalika [11], the inner sum has a simple pole if and only if $i = j$. Thus, the order of the pole is

$$\sum_i m_i^2$$

which must equal 1. This forces m_1 (say) to be ± 1 and the remaining ones to be zero. The possibility $m_1 = -1$ implies that $L(s, \chi)$ has no trivial zeroes which is not the case. Hence, $m_1 = 1$ and $L(s, \chi, L/\mathbb{Q}) = L(s, \pi_1)$ as desired.

Unfortunately, the hypothesis of the theorem is known to be satisfied only if K is a solvable extension of $\mathbb{Q}$ by the theorem of Arthur and Clozel [1]. In the next section, we apply these ideas to the theory of elliptic curves.

10. Application to elliptic curves.

Let E/k be an elliptic curve defined over a number field k. One can think of this as the set of solutions of an equation of the form

$$E: \qquad y^2 = x^3 + ax + b, \qquad a, b \in k.$$

Let $E(k)$ be the set of k-rational points of E. Then, $E(k)$ can be given the structure of an additive abelian group. We have the classical:

Mordell-Weil theorem. $E(k)$ is finitely generated.

Thus,

$$E(k) \simeq \mathbb{Z}^r \oplus E(k)_{tors}.$$

There is a folklore conjecture that $|E(k)_{tors}| \leq C_k$ where C_k is a constant depending only on the field. For $k = \mathbb{Q}$, Mazur proved that $C_{\mathbb{Q}} = 16$. Recently, Kamienny established this for quadratic fields. Given any curve, there is a finite algorithm which enables us to find $E(k)_{tors}$.

More intriguing is to find $r = r_k$, called the rank of E over k. It is unknown at present whether the rank of E over $\mathbb{Q}$ is unbounded though this is known in the function field case. Birch and Swinnerton-Dyer made an analytic conjecture about the rank in the following way: for each prime ideal $\mathfrak{p}$ of k, let $E(O_k/\mathfrak{p})$ have cardinality $N\mathfrak{p} + 1 - a_\mathfrak{p}$. We know, $|a_\mathfrak{p}| \leq (N\mathfrak{p})^{1/2}$. Define

$$L(E/k, s) = \prod_\mathfrak{p} \left(1 - \frac{a_\mathfrak{p}}{N\mathfrak{p}^s} + \frac{1}{N\mathfrak{p}^{2s-1}}\right)^{-1},$$

where the product is defined over primes where E has good reduction. This defines an analytic function for $\mathrm{Re}\,(s) > 3/2$. We have the famous:

Birch and Swinnerton-Dyer conjecture: $L(E/k, s)$ extends to an analytic function for all $s \in \mathbb{C}$ and

$$\mathrm{ord}\,_{s=1} L(E/k, s) = \mathrm{rank}\ E(k).$$

If K/k is an arbitrary extension, then we can consider E over K and it is easy to see that $\mathrm{rank}\ E(K) \geq \mathrm{rank}\ E(k)$. This raises two questions. First, does $L(E/K, s)$ have analytic continuation given that $L(E/k, s)$ does. Second, if it does, then can we prove that

$$\mathrm{ord}\,_{s=1} L(E/K, s) \geq \mathrm{ord}\,_{s=1} L(E/k, s).$$

If we consider the ring of endomorphisms of E, then it is not difficult to show that there are two possibilities for this. Either it is isomorphic to $\mathbb{Z}$ or an order in an imaginary quadratic field. In the latter case, we say E has CM (complex multiplication). The latter case is easier to deal with. For instance, we shall see that the answer to the first question is affirmative in this case.

Deuring [5] showed that if E has CM by an order in F, then $L(E/k, s)$ is a product of two Hecke L-series of k if $k \supset F$ or equal to a Hecke L-series of kF which is a quadratic extension of k.

Thus if E has CM, then $L(E/k, s)$ extends to an entire function of s. But recall Taniyama's conjecture, namely that $L(E/k, s)$ should be an $L(s, \pi)$ for some $\pi \in \mathcal{A}(GL_2(\mathbb{A}_k))$. We can verify this conjecture for E with CM by using automorphic induction.

Suppose that Taniyama's conjecture is true. Let K/k be any Galois extension with group G. Then, the theory of ℓ-adic representations shows that

$$L(E/K, s) = \prod_{\rho\,irred} L(E/k, s, \rho)^{\rho(1)},$$

where $L(E/k, s, \rho)$ is the "twist" by ρ. On Taniyama's conjecture, this corresponds to $L(s, \pi \otimes reg_G)$ where reg_G is the regular representation of G. By what we have seen in the previous section, this is the base change of π, $B(\pi)$. We can prove the following theorem:

Theorem.   (joint with Kumar Murty) ([18])

(1) If E over k has CM then, $L(E/k, s)$ extends to an entire function for all values of s.

(2) Suppose that E/k satisfies Taniyama's conjecture and K is a finite extension of k. If K is a solvable extension of k, then, $L(E/K, s)$ extends to an entire function of s.

(3) In both cases, if K is a solvable extension, then $L(E/K, s)/L(E/k, s)$ is entire. In particular,

$$\operatorname{ord}_{s=1} L(E/K, s) \geq \operatorname{ord}_{s=1} L(E/k, s).$$

If the base change map always exists, one would not have to make the assumption that K is a solvable extension of k for part (3) of the theorem. In the CM case, one could strengthen the result by allowing K to be contained in a solvable extension.

This is highly reminiscent of the Brauer-Aramata theorem that states that if K/k is Galois, then $\zeta_K(s)/\zeta_k(s)$ is entire, where $\zeta_k(s)$ denotes the zeta function of k. There is a classical conjecture of Dedekind that predicts that this should be true even if K/k is not Galois. This is of course implied by Artin's conjecture. All these would follow from the development of the theory of base change.

Our method of proof deviates from the method of Brauer or Aramata. Instead, we use a device exploited by Kumar Murty in his exceedingly simple proof of the Brauer-Aramata theorem (see [6]).

Fix $s_0 \in \mathbb{C}$. Let K/k be Galois with group G. For each character χ of G, let

$$n_\chi = \operatorname{ord}_{s=s_0} L(s, \chi, K/k).$$

Theorem.   (Kumar Murty) $\sum_\chi n_\chi^2 \leq (n_{reg})^2$.

Corollary. $\zeta_K(s)/\zeta_k(s)$ is entire.

Proof.   From the inequality, we get

$$n_1^2 \leq n_{reg}^2$$

from which the result follows.

Proof of the theorem.   Let $H \leq G$. For each character ψ of H define

$$n_\psi = \operatorname{ord}_{s=s_0} L(s, \psi, K/K^H).$$

Define

$$\theta_H = \sum_{\psi \in \hat{H}} n_\psi \psi,$$

where $\hat{H}$ denotes the set of irreducible characters of H. Then,

$$\theta_G|_H = \sum_{\chi \in \hat{G}} n_\chi \chi|_H = \sum_{\chi \in \hat{G}} n_\chi \sum_{\psi \in \hat{H}} (\chi|_H, \psi)\psi.$$

The inner sum is by Frobenius reciprocity

$$= \sum_{\psi \in \hat{H}} (\chi, \operatorname{Ind}_H^G \psi)\psi.$$

Inserting this and interchanging summation we obtain

$$\theta_G|_H = \sum_{\psi \in \hat{H}} \left(\sum_{\chi \in \hat{G}} n_\chi(\chi, \operatorname{Ind}_H^G \psi) \right) \psi.$$

We recognise the inner sum as n_ψ because Artin L-series are invariant under induction. Thus, $\theta_G|_H = \theta_H$. Now,

$$(\theta_G, \theta_G) = \sum_{\chi \in \hat{G}} n_\chi^2 = \frac{1}{|G|} \sum_{g \in G} |\theta_G(g)|^2.$$

But by what was shown above, $\theta_G(g) = \theta_{(g)}(g)$. If H is cyclic,

$$|\theta_H(g)| \leq \sum_{\psi \in \hat{H}} n_\psi = n_{reg}$$

by the abelian reciprocity law. This gives

$$(\theta_G, \theta_G) \leq (n_{reg})^2.$$

This same proof carries over with a few modifications to show that if ρ is an arbitrary representation, then

$$L(s, \rho \otimes reg_G, K/k)/L(s, \rho, K/k)$$

is entire. By a similar method, one can show that if π is an automorphic cuspidal representation and $B(\pi)$ is the base change of it, then the L-function $L(s, B(\pi))/L(s, \pi)$ is entire.

11. Concluding remarks.

We therefore see that Artin Galois representations should correspond to automorphic representations. The Taniyama conjecture predicts the same should happen for elliptic curves. But there are more automorphic representations than Galois representations or elliptic curves. This led Langlands to conjecture a super-reciprocity law in the following form.

If K is a field and G is a group, we can consider the category of all K representations of G. This is a category with tensor products and fibre functor (namely the map that takes each representation to its underlying vector space). Langlands drew attention to the fact that there is a result of Rivano Saavedra [24] that the converse of this is true. That is, if C is a category with tensor products and fibre functors (together with a few other minor compatibility assumptions), then C is the category of representations of a group G. Such a category with tensor products is called a Tannakian category. Is the category of automorphic representations Tannakian ? If so, then there is a giant (or monstrous) group which we can call the automorphic Galois group from which all L-functions arise. Such a situation is not outlandish. In the context of converse theory, it requires us to show that the Rankin-Selberg convolution is the L-series of an automorphic representation. More precisely, there should be a map

$$\mathcal{A}(GL_n) \times \mathcal{A}(GL_m) \to \mathcal{A}(GL_{mn}).$$

By the predicted converse theory of GL_{mn}, we must twist the Rankin-Selberg convolution by automorphic representations of GL_r for all $r \leq mn - 2$, and so we are in the case of triple convolutions. That is why it is tantalizing to understand Garrett's work and extend it (if possible) to higher GL_n. Such a step would be a major one in establishing a super-reciprocity law and which would resolve a huge chunk of the classical unsolved problems of number theory.

REFERENCES

[1] J. Arthur and L. Clozel, Simple algebras, base change and the advanced theory of the trace formula, *Annals of Math. Studies 120,* Princeton University Press (1990).

[2] J. Buhler, Icosahedral Galois representations, *Springer Lecture Notes* **654** Springer-Verlag, Berlin - New York, 1978.

[3] D. Bump and D. Ginzburg, Symmetric square L-functions on $GL(r)$, *Annals of Math.,* **136** (1992) 137-205.

[4] P. Deligne, La conjecture de Weil, I, *Publ. Math. I.H.E.S.,* **43** (1974) 273-307.

[5] M. Deuring, Die Zetafunktion einer algebraischen Kurve vom Geschlechte Eins, *Nach. Akad. Wiss. Göttingen,* Math.-Phys. Kl. I - III, 1953-1956.

[6] R. Foote and K. Murty, Zeros and poles of Artin L-series, *Math. Proc. Camb. Phil. Soc.,* **105**, 5 (1989) 5 - 11.

[7] G. Frey, Links between elliptic curves and solutions of $A - B = C$, *J. Indian Math. Soc.,* **51** (1987) 117-145.

[8] P. Garrett, Decompositions of Eisenstein series: Rankin triple products, *Annals of Math.*, **125** (1987) 209-235.

[9] I. Gelfand and S. Fomin, Geodesic glows on manifolds of constant negative curvature, *Uspekhi Mat. Nauk.*, **7** (1952) no. 1, 118-137.

[10] S. Gelbart and H. Jacquet, A relation between automorphic representations of $GL(2)$ and $GL(3)$, *Annales Sci. École Normale Sup.*, **11** (1978) 471-552.

[11] E. Hecke, Mathematische Werke, Herausgegeben im Auftrage der Akademie der Wissenschaften zu Göttingen, Vandenhoeck and Ruprecht, Göttingen, 1959, pp. 955.

[12] H. Jacquet, I. Piatetskii-Shapiro and J. Shalika, Rankin-Selberg convolutions, *Amer. J. Math.*, **105** (1983) 367-464.

[13] H. Jacquet and R. Langlands, Automorphic forms on $GL(2)$, *Lecture notes in Mathematics*, **114** Springer-Verlag, Berlin-New York 1970.

[14] R. Langlands, Base change for $GL(2)$, *Annals of Math. Studies* **96**, Princeton University Press (1980).

[15] R. Langlands, Automorphic representations, Shimura varieties and motives, in *Automorphic forms, representations and L-functions* (ed. A. Borel and W. Casselman) Proc. Symp. Pure Math., Parts I and II, Vol. 33, pp. 205-246, American Math. Society, Providence, 1979.

[16] H. Maass, Über eine neue Art von nichtanalytischen automorphen Funktionen und die Bestimmung Dirichletscher Reihen durch Funktionalgleichungen, *Math. Ann.* **121** (1949) 141-183.

[17] K. Murty, On the Sato-Tate conjecture, in *Number Theory related to Fermat's last theorem*, (ed. N. Koblitz) (1982) 195-205, Birkhauser-Verlag, Boston.

[18] K. Murty and R. Murty, Base change and the Birch-Swinnerton-Dyer conjecture, *to appear in the Grosswald volume.*

[19] I. Piatetski-Shapiro, Converse theorem for $GL(n)$, Festschrift in honour of I. Piatetski-Shapiro, *Israel Mathematical Conference Proceedings* (Ed. S. Gelbart, R. Howe, P. Sarnak) Weizmann Science Press of Israel, 1990.

[20] S. Ramanujan, On certain arithmetical functions, *Trans. Camb. Phil. Soc.*, **22** (1916) 159-184.

[21] R. Rankin, Contributions to the theory of Ramanujan's function $\tau(n)$ and similar arithmetical functions, II. The order of the Fourier coefficients of integral modular forms, *Proc. Camb. Phil. Soc.*, **35** (1939) 351-372.

[22] K. Ribet, On modular representations of $\mathrm{Gal}(\overline{\mathbb{Q}}/\mathbb{Q})$ arising from modular forms, *Inv. Math.*, **100** (1990) 431-476.

[23] A. Robert, Formes automorphes sur GL_2, *Lecture notes in mathematics* **317** pp. 295-318, Springer-Verlag, Berlin-New York, 1973.

[24] R. Saavedra, Categories Tannakienne, *Lecture notes in mathematics* **265**, Springer-Verlag, Berlin-New York 1972.

[25] A. Selberg, On the estimation of Fourier coefficients of modular forms, Proc. Sympos. Pure Math., Vol. VIII, 1-15, Amer. Math. Soc., Providence, R.I., 1965.

[26] A. Selberg, Bemerkung über eine Dirichletsche Reihe die mit der Theorie der Modulformen nahe verbunden ist, *Arch. Math. Naturvid.* **43** (1940) 47-50.

[27] J.-P. Serre, Abelian ℓ-adic representations and elliptic curves, McGill University lecture notes, (written in collaboration with W. Kuyk and J. Labute), W. A. Benjamin Inc., New York - Amsterdam, 1968.

[28] J.-P. Serre, Sur les representations modulaires de degré 2 de Gal($\overline{\mathbb{Q}}/\mathbb{Q}$), *Duke Math. J.* **54** (1987) 179-230.

[29] F. Shahidi, On certain L-functions, *Amer. J. Math.*, **103** (1981) 297-355.

[30] G. Shimura, On the holomorphy of certain Dirichlet series, *Proc. London Math. Soc.* **(3) 31** (1975) no. 1, 79-98.

[31] G. Shimura, Yukata Taniyama and his time, *Bull. London Math. Soc.*, **21** (1989) 186-196.

[32] Y. Taniyama, Jacobian varieties and number fields, *Proceedings of the international symposium on algebraic number theory*, Tokyo and Nikko, 1955, Science Council of Japan, Tokyo, 1956, pp. 31-45.

[33] J. Tunnell, Artin's conjecture for representations of octahedral type, *Bull. Amer. Math. Soc. N.S.* 5(1981) no. 2, 173-175.

[34] A. Weil, Über die Bestimmung Dirichletscher Reihen durch Funktionalgleichungen, *Math. Annalen* **168** (1967) 149-156.

M. Ram Murty
Department of Mathematics
McGill University
Montreal H3A 2K6, Canada
e-mail murty@shiva.math.mcgill.ca

Transcendental Numbers and Functions of Several Variables

Michel Waldschmidt
Université P. et M. Curie (Paris VI)

SUMMARY. We start by two density problems: one was raised by Sansuc and deals with algebraic number fields, the other is a question of Mazur and deals with abelian varieties. While the first has just been completely solved by D.Roy, only partial answers have been obtained so far for the second. In both cases the main tool is the theorem of the algebraic subgroup.

We show that this theorem enables one to recover Baker's result on linear combinations of logarithms of algebraic numbers in at least four different ways. Another consequence, once more due to R. Roy, concerns matrices whose entries are linear combinations of logarithms of algebraic numbers.

The proof of the algebraic subgroup theorem is briefly explained, by means of an interpolation determinant, following an idea of M. Laurent. Finally we show how the transposition of the interpolation matrix is connected with a duality in transcendence proofs.

1. Two density problems

$\boxed{1}$ *A question of Sansuc*

Let k be a number field of degree $[k : \mathbb{Q}] = n = r_1 + 2r_2$, where, as usual, r_1 (resp. r_2) denotes the number of real (resp. the number of pairs of complex conjugate) embeddings of k into $\mathbb{R}$ (resp. into $\mathbb{C}$). The tensor product $\left(\mathbb{R} \otimes_{\mathbb{Q}} k\right)^{\times}$ is nothing else than $\mathbb{R}^{\times r_1} \times \mathbb{C}^{\times r_2}$. Since the topological group $\mathbb{R}^{\times}$ (resp. $\mathbb{C}^{\times}$) is isomorphic to $\mathbb{R}_+^{\times} \times \mathbb{Z}/2\mathbb{Z}$ (resp $\mathbb{R}_+^{\times} \times \mathbb{R}/\mathbb{Z}$), and since $\mathbb{R}_+^{\times}$ is isomorphic to $\mathbb{R}$, the canonical embedding of $k^{\times}$ into $\left(\mathbb{R} \otimes_{\mathbb{Q}} k\right)^{\times}$ gives rise to an injective homomorphism

$$k^{\times} \longrightarrow \mathbb{R}^{r_1 + r_2} \times (\mathbb{R}/\mathbb{Z})^{r_2} \times (\mathbb{Z}/2\mathbb{Z})^{r_1}.$$

The image of $k^{\times}$ is dense; Colliot-Thélène and Sansuc asked whether there exists a finitely generated subgroup of $k^{\times}$ whose image is dense; then Sansuc [Sa] asked for the smaller rank of such a finitely generated subgroup. The complete answer to this question has just been obtained by D. Roy [R3]:

Theorem 1. *There exists a finitely generated subgroup of $k^{\times}$ of rank $r_1 + r_2 + 1$ which is dense in $\left(\mathbb{R} \otimes_{\mathbb{Q}} k\right)^{\times}$.*

The proof splits in two parts. The first one is a <u>topological argument</u>: let Y_0 be a discrete subgroup of $\mathbb{R}^n$ and Y a finitely generated subgroup of $\mathbb{R}^n$ containing Y_0. If no subgroup of Y of rank $n+1$ containing Y_0 is dense in $\mathbb{R}^n$, then there exists a finitely generated subgroup of Y of rank $\geq \mathrm{rk}Y - n + 1$ which contains Y_0 and is dense in $\mathbb{R}^n$. In the special case we are dealing with, we take for Y_0 the kernel of the exponential map

$$\exp : \mathbb{R} \otimes_{\mathbb{Q}} k \longrightarrow (\mathbb{R} \otimes_{\mathbb{Q}} k)^{\times}.$$

We refer to Roy's paper [R3] for more detailed information on the topological argument. We shall be mainly concerned here with the second part of the proof, which is a <u>transcendence argument</u>: let $\alpha_1, \ldots, \alpha_{\ell}$ in $k^{\times}$ be such that the ℓn numbers $\sigma_i \alpha_j$, $(1 \leq i \leq n, 1 \leq j \leq \ell)$ are multiplicatively independent. Choose $y_1, \ldots, y_{\ell}$ in $\mathbb{R} \otimes_{\mathbb{Q}} k$ such that

$$\exp y_j = (\sigma_1 \alpha_j, \ldots, \sigma_n \alpha_j), \qquad 1 \leq j \leq \ell.$$

If $\ell \geq n^2 - n + 2$, then any subgroup of $Y = \mathbb{Z}y_1 + \cdots + \mathbb{Z}y_{\ell}$ of rank $\geq n^2 - n + 2$ is dense $\mathbb{R} \otimes_{\mathbb{Q}} k$. In particular if $\ell \geq n^2 + 1$, then any subgroup of Y of rank $\geq \mathrm{rk}Y - n + 1$ is dense $\mathbb{R} \otimes_{\mathbb{Q}} k$.

We discuss more thoroughly this statement below.

$\boxed{2}$ *A question of Mazur*

The following question is a special case of a much more general situation considered by B. Mazur [M]. Let A be a simple abelian variety over $\mathbb{Q}$ of dimension $d \geq 1$ with Mordell-Weil group $A(\mathbb{Q})$ of positive rank. Does the connected component of 0 in the topological closure of $A(\mathbb{Q})$ coincide with the connected component $A(\mathbb{R})_0$ of 0 in $A(\mathbb{R})$?

We can provide a positive answer if we assume that the rank of $A(\mathbb{Q})$ is sufficiently large:

Theorem 2. *Assume* $\mathrm{rk}_{\mathbb{Z}} A(\mathbb{Q}) \geq d^2 - d + 1$. *Then* $A(\mathbb{Q}) \cap A(\mathbb{R})_0$ *is dense in* $A(\mathbb{R})_0$.

Remark. Using the above mentioned topological argument of D.Roy, one can also show that if $\mathrm{rk}_{\mathbb{Z}} A(\mathbb{Q}) \geq d^2$, then there is a $u \in A(\mathbb{Q})$ such that $\mathbb{Z}u$ is dense in $A(\mathbb{R})_0$.

In both examples, we are dealing with a finitely generated subgroup of a real vector space. The following easy lemma will reduce the topological problem to an algebraic question: how many independent points of this subgroup can lie in a given hyperplane?

Lemma 3. *Let* $Y = \mathbb{Z}y_1 + \cdots + \mathbb{Z}y_\ell$ *be a finitely generated subgroup of a* $\mathbb{R}$*-vector space* E *of dimension* d. *Then* Y *is dense in* E *if and only if for each hyperplane* V *of* E,

$$\mathrm{rk}_{\mathbb{Z}} Y/Y \cap V \geq 2.$$

Writing V as the kernel of a linear form, we can state this condition as follows: *for each non-zero* $\varphi \in \mathrm{Hom}_{\mathbb{R}}(E, \mathbb{R})$, *one has* $\varphi(Y) \not\subset \mathbb{Z}$.

Let us come back to our two special cases.

boxed{1} *First example* (Sansuc).

We denote by $\bar{\mathbb{Q}}$ the field of algebraic numbers (algebraic closure of $\mathbb{Q}$ into $\mathbb{C}$), and by $\mathbb{L}$ the $\mathbb{Q}$–vector space of logarithms of algebraic numbers:

$$\mathbb{L} = \{\log \alpha; \alpha \in \bar{\mathbb{Q}}^{\times}\} = \{z \in \mathbb{C}; e^z \in \bar{\mathbb{Q}}^{\times}\}.$$

Michel Emsalem first noticed that if V is an hyperplane of $\mathbb{C}^d$ with $V \cap \mathbb{Q}^d \neq \{0\}$, then

$$\dim_{\mathbb{Q}} V \cap \mathbb{L}^d = \infty;$$

indeed, assume $0 \neq (b_1, \ldots, b_d) \in V \cap \mathbb{Q}^d$; then for all $\log \alpha \in \mathbb{L}$,

$$(b_1 \log \alpha, \ldots, b_d \log \alpha) \in V \cap \mathbb{L}^d,$$

which proves the claim. Emsalem [E] also proved the converse: *if* V *is an hyperplane of* $\mathbb{C}^d$ *with* $V \cap \mathbb{Q}^d = \{0\}$, *then*

$$\dim_{\mathbb{Q}} V \cap \mathbb{L}^d < \infty.$$

Using lemma 3, we now show that the first part of the proof of theorem 1 is a consequence of the following:

<u>Transcendence result</u>: if $V \cap \mathbb{Q}^d = \{0\}$, then

$$\dim_{\mathbb{Q}} V \cap \mathbb{L}^d \leq d(d-1).$$

This transcendence result, due to M. Emsalem [E], will be deduced later from the theorem of the algebraic subgroup.

Consequence: let Y' be a subgroup of $\mathbb{L}^d$ of rank $\ell' \geq d^2 - d + 2$; for each hyperplane V with $V \cap \mathbb{Q}^d = \{0\}$ we obtain the bound $\mathrm{rk}_{\mathbb{Z}}(Y' \cap V) \leq d(d-1)$, hence $\mathrm{rk}_{\mathbb{Z}}(Y'/Y' \cap V) \geq 2$. We need to consider also the hyperplanes V with $V \cap \mathbb{Q}^d \neq \{0\}$. Assume Y' is contained in a finitely generated subgroup Y of $\mathbb{L}^d$, and Y is generated by ℓ elements of $\mathbb{L}^d$ such that the ℓd components of the generators are linearly independent over $\mathbb{Q}$. Let S be the subspace of $\mathbb{C}^d$ generated by $V \cap \mathbb{Q}^d$; this is the maximal subspace of $\mathbb{C}^d$ which is rational over $\mathbb{Q}$ and contained in V. We write S as intersection of

δ hyperplanes $\varphi_1 = 0, \ldots, \varphi_\delta = 0$, where $\varphi_1, \ldots, \varphi_\delta$ are linear forms on $\mathbb{C}^d$ with rational coefficients. Hence $\varphi = (\varphi_1, \ldots, \varphi_\delta)$ is a surjective linear map from $\mathbb{C}^d$ onto $\mathbb{C}^\delta$ with kernel S, and $V' = \varphi(V)$ satisfies $V' \cap \mathbb{Q}^\delta = 0$. On the other hand our assumption on Y is more than enough to ensure $Y \cap S = 0$, hence $Y'' = \varphi(Y')$ is again of rank ℓ'. We apply the transcendence result to $Y'' \subset \mathbb{L}^d$ and deduce $\mathrm{rk}(Y'' \cap V') \leq \delta(\delta-1)$. Therefore $\mathrm{rk}(Y' \cap V) \leq d(d-1)$ and $\mathrm{rk}(Y'/Y' \cap V) \geq 2$ for all hyperplanes V of $\mathbb{C}^d$.

[2] *Second example* (Mazur).

Let A be an abelian variety which is defined over $\bar{\mathbb{Q}}$. We denote by T_A the tangent space of A at the origin, and by $\exp : T_A(\mathbb{C}) \to A(\mathbb{C})$ the exponential map of the Lie group $A(\mathbb{C})$. The complex vector space $T_A(\mathbb{C})$ is of dimension $d = \dim A$ and its kernel (periods of the exponential map) is a lattice in $T_A(\mathbb{C})$ (discrete subgroup of rank $2d$).

We consider the $\mathbb{Q}$-vector space of logarithms of algebraic points on A:

$$\Lambda = \exp_A^{-1} A(\bar{\mathbb{Q}}) \subset T_A(\mathbb{C}).$$

We now show that theorem 2 is a consequence of the following statement, which we deduce below from the theorem of the algebraic subgroup:
Transcendence result: if V is a hyperplane of $T_A(\mathbb{C})$ containing $d - 1$ independent elements of $\ker \exp_A$, then $\dim_\mathbb{Q} V \cap \Lambda \leq d^2 - 1$.

Consequence: assume A is defined over $\mathbb{Q}$; then $T_A(\mathbb{R})$ contains d periods $\omega_1, \ldots, \omega_d$, which are linearly independent over $\mathbb{R}$. Consider m points in $A(\bar{\mathbb{Q}})$ which are linearly independent over $\mathbb{Z}$; we can write these points $\exp_A(u_j) \in A(\bar{\mathbb{Q}})$, $(1 \leq j \leq m)$, and the subgroup

$$Y = \mathbb{Z}\omega_1 + \cdots + \mathbb{Z}\omega_d + \mathbb{Z}u_1 + \cdots + \mathbb{Z}u_m \subset T_A(\mathbb{R}) \simeq \mathbb{R}^d$$

is contained in Λ, of rank $\mathrm{rk}_\mathbb{Z} Y = \ell = m + d$. If $m \geq d^2 - d + 1$, then $\ell \geq d^2 + 1$, and we conclude

$$\mathrm{rk}_\mathbb{Z} Y/Y \cap V \geq d^2 + 1 - (d^2 - 1) = 2.$$

2. Theorem of the algebraic subgroup

Let G be an algebraic group defined over $\bar{\mathbb{Q}}$. We denote by Λ the $\mathbb{Q}$-vector space of the logarithms of algebraic points on G:

$$\Lambda = \exp_G^{-1}\big(G(\bar{\mathbb{Q}})\big).$$

Further, let V be a complex subspace of $T_G(\mathbb{C})$. We consider the intersection $V \cap \Lambda$, which is a $\mathbb{Q}$-vector space, and we denote by ℓ its dimension. A necessary condition for ℓ to be finite is that V does not contain any non-zero tangent subspace of an algebraic subgroup H of G, defined over $\bar{\mathbb{Q}}$:

$$V \not\supset T_H(\mathbb{C}) \neq \{0\}.$$

Examples: firstly, let $G = \mathbb{G}_m^d$, where $\mathbb{G}_m$ is the multiplicative group; in this case $\Lambda = \mathbb{L}^d$, and the above condition reads $V \cap \mathbb{Q}^d = \{0\}$. Secondly, let $G = A$ be a simple abelian variety; then the condition is just $V \neq T_A(\mathbb{C})$.

Theorem 4. *If V is a hyperplane in $T_G(\mathbb{C})$ which does not contain any non-zero subspace of the form $T_H(\mathbb{C})$, (H algebraic subgroup of G defined over $\bar{\mathbb{Q}}$), then $\ell < \infty$.*

We now want to give an explicit upper bound for ℓ, assuming that it is finite. A first estimate, which is valid in the general case, is $\ell \leq 2d(d-1)$. This estimate is sufficient for theorem 1; however our *transcendence argument* related to Sansuc's problem above claimed a stronger upper bound. Indeed, when G is a linear algebraic group then one can sharpen this estimate and get $\ell \leq d(d-1)$.

One can produce a bound which is valid in the general case, and includes the refinement in the linear case: we write G as a product $G = G_0 \times G_1 \times G_2$, with $G_0 = \mathbb{G}_a^{d_0}$, $G_1 = \mathbb{G}_m^{d_1}$, $\dim G_2 = d_2$, and $d = d_0 + d_1 + d_2$. This is no loss of generality: we can choose for instance $d_0 = d_1 = 0$, $d_2 = d$. With this notation the following bound for ℓ holds: $\ell \leq (d_1 + 2d_2)(d-1)$.

This estimate is not still sharp enough for theorem 2. In order to have a more precise bound we introduce a further notation: we take into account the number κ of independent periods of $\exp_G$ which sit in V:

$$\kappa = \mathrm{rk}_{\mathbb{Z}}\left(V \cap \ker \exp_G\right).$$

Then $\ell \leq (d_1 + 2d_2 - \kappa)(d-1)$.

For Mazur's problem we choose $d_0 = d_1 = 0$, $d_2 = d$, $\kappa = d-1$, and we conclude $\ell \leq d^2 - 1$.

A further refinement will be needed below: we count how many independent points of $T_G(\mathbb{C})$ which are rational over $\bar{\mathbb{Q}}$ lie inside V (recall that T_G has a natural $\bar{\mathbb{Q}}$–structure, since it is the tangent space of the algebraic group G which is defined over $\bar{\mathbb{Q}}$).

Theorem 4 (continued). *If $V \supset W$ with W rational over $\bar{\mathbb{Q}}$, then*

$$\ell \leq (d_1 + 2d_2 - \kappa) \dim_{\mathbb{C}}(V/W).$$

There is a more general result ([W1] theorem 4.1) but the statement is slightly more complicated.

This theorem of the algebraic subgroup includes essentially all known transcendence results on analytic subgroups of commutative group varieties (apart from results of algebraic independence). The first statements of this form dealing with several parameter subgroups go back to Schneider's works in the first half of this century. In particular his paper [S], where he proves the transcendence of values of the Beta function at rational points, introduces for the first time functions of several complex variables in transcendental number theory. The subject was taken up again by Lang in the 60's [L]; like in Schneider's work, the analytic argument is an interpolation formula for a cartesian product in $\mathbb{C}^n$. The main transcendence result (*Schneider–Lang criterion*) concerns the values on a cartesian product of analytic functions satisfying differential equations; one main point is that the cartesian product does not involve the same system of coordinates than the differential equations, and this is why the result does not reduce to its one dimensional analog; it is useful for applications to so-called *normalized* n-parameter subgroups of algebraic groups [L]. A deeper Schwarz lemma in higher dimension, using Lelong's measure for the area of the analytic hypersurface of zeros of an analytic function, was proved by Bombieri and Lang in [B-L], and they used it to study non-normalized n-parameter subgroups of algebraic groups; but the result they obtain involves very strong repartition results which seem out of reach of the present theory of diophantine approximations. Then Bombieri [B] introduced in the theory the L^2-estimates of Hörmander and replaced the cartesian product by a more natural condition (conjectured by Nagata) that the considered points do not lie in an algebraic hypersurface; but it turns out that these tools do not yield further transcendence results in connection with algebraic groups.

Nowadays, these ingredients from the theory of functions in several variables are no longer useful: the analytic argument is reduced to the very easy Schwarz lemma for functions of a single variable [W2]; on the other hand a very important role is played by a geometrical tool, the zero estimate [P].

In the case $V = W$, the conclusion of theorem 4 is $\ell = 0$. This is a result of Wüstholz [Wü] which rests on Baker's method. The proof of this special case does not involve functions of several variables, since it is sufficient to work with one point and its multiples, which all lie on a complex line in $T_G(\mathbb{C})$.

3. Linear combinations of logarithms of algebraic numbers

In this section we deduce from the algebraic subgroup theorem the following well known result of Baker:

Theorem 5. *Let* $\log \alpha_1, \ldots, \log \alpha_n$ *be elements in* $\mathbb{L}$, *not all of which are zero, and let* $\beta_1, \ldots, \beta_n$ *be algebraic numbers, not all zero. If we have* $\beta_1 \log \alpha_1 + \cdots + \beta_n \log \alpha_n = 0$, *then* $\beta_1, \ldots, \beta_n$ *are* $\mathbb{Q}$-*linearly dependent, and* $\log \alpha_1, \ldots, \log \alpha_n$ *are* $\mathbb{Q}$-*linearly dependent.*

We give four sketches of proof of theorem 5 as a consequence of theorem 4 with $d_2 = \kappa = 0$. We start with a non-trivial relation of minimal length n

between logarithms of algebraic numbers with algebraic coefficients; since n is minimal, it is plainly sufficient to prove either that the β's are $\mathbb{Q}$-linearly dependent, or that the $\log \alpha$'s are $\mathbb{Q}$-linearly dependent.

☐1 *Baker's method*

Let us choose $d_0 = 0$ and $d = d_1 = n$; in this case $\Lambda = \mathbb{L}^n$. We take for V the hyperplane of equation $\beta_1 z_1 + \cdots + \beta_n z_n = 0$ in $\mathbb{C}^n$, and we choose $W = V$. In this case

$$V \cap \Lambda \ni (\log \alpha_1, \ldots, \log \alpha_n) \neq 0,$$

which shows that $\ell > 0$. The theorem of the algebraic subgroup implies $V \cap \mathbb{Q}^n \neq \{0\}$ and therefore $\beta_1, \ldots, \beta_n$ are $\mathbb{Q}$-linearly dependent.

☐2 *Hirata's method*

The choice here is $d_0 = 1$, $d_1 = n$, $d = n + 1$, hence $\Lambda = \bar{\mathbb{Q}} \times \mathbb{L}^n$. Let $V = W$ be the hyperplane of equation $z_0 = \beta_1 z_1 + \cdots + \beta_n z_n$ in $\mathbb{C}^{n+1}$. Since

$$V \cap \Lambda \ni (0, \log \alpha_1, \ldots, \log \alpha_n) \neq 0,$$

we have $\ell > 0$, and theorem 4 implies $V \supset T_H \neq \{0\}$, which means that $\beta_1, \ldots, \beta_n$ are $\mathbb{Q}$-linearly dependent.

☐3 *Dual of Baker's method*

We now choose $d_0 = n - 1$, $d_1 = 1$, $d = n$, so that $\Lambda = \bar{\mathbb{Q}}^{n-1} \times \mathbb{L}$. Further let $W = \{0\}$, and let V be the hyperplane kernel of the linear form $z_1 \log \alpha_1 + \cdots + z_{n-1} \log \alpha_{n-1} + z_n$. For $(h_1, \ldots, h_n) \in \mathbb{Z}^n$ we have

$$V \cap \Lambda \ni \left(h_1, \ldots, h_{n-1}, h_1 \log \alpha_1 + \cdots + h_{n-1} \log \alpha_{n-1}\right)$$

and

$$V \cap \Lambda \ni \left(h_n \frac{\beta_1}{\beta_n}, \ldots, h_n \frac{\beta_{n-1}}{\beta_n}, h_n \log \alpha_n\right),$$

Therefore $\ell \geq n$, hence $V \supset T_H \neq \{0\}$. One therefore deduces that $\log \alpha_1, \ldots, \log \alpha_{n-1}$ are $\bar{\mathbb{Q}}$-linearly dependent, which contradicts our choice of n minimal.

☐4 *Dual of Hirata's method*

The duality that we shall describe at the end of this lecture suggests to choose $d_0 = n$, $d_1 = 1$, $d = n + 1$, hence $\Lambda = \bar{\mathbb{Q}}^n \times \mathbb{L}$, and then $W =$

$\mathbb{C}(\beta_1, \ldots, \beta_n, 0)$, while V is once more an hyperplane: $z_{n+1} = z_1 \log \alpha_1 + \cdots + z_n \log \alpha_n$. For $(h_1, \ldots, h_n) \in \mathbb{Z}^n$ we have

$$V \cap \Lambda \ni (h_1, \ldots, h_n, h_1 \log \alpha_1 + \cdots + h_n \log \alpha_n);$$

therefore $\ell \geq n$, and $V \supset T_H \neq \{0\}$. But here we already know that V contains a non-zero subspace of this form, namely W which is the tangent space of the additive group, image of $\mathbb{G}_a \longrightarrow \mathbb{G}_a^n \times \mathbb{G}_m$ by

$$z \longmapsto (\beta_1 z, \ldots, \beta_m z, 1).$$

Instead of this we work with the quotient V/W; we are back to a situation similar to that in method 3, except that we do not loose the symmetry between $\log \alpha_1, \ldots, \log \alpha_n$, and also the choice of the parameters is substantially different (see [W3]).

It is possible to develop the four methods in order to provide explicit estimates. This is well known for the first method (Baker's theory). The second method has been introduced by N. Hirata–Kohno in her study of lower bounds for linear forms on algebraic commutative groups [H]. The third and fourth method have been used in [W3] and provide sharp estimates; here is an example which has been worked out using method 3:

Theorem 6. *Let* $a_1, \ldots, a_n$ *be rational integers,* $a_i \geq 2$*, and let* $b_1, \ldots, b_n$ *be rational integers with* $a_1^{b_1} \cdots a_n^{b_n} \neq 1$*. Define* $B = \max\{2, |b_1|, \ldots, |b_n|\}$*; then*

$$\left| a_1^{b_1} \cdots a_n^{b_n} - 1 \right| \geq \exp\left\{ -2^{4n+16} n^{3n+5} \log B \log a_1 \cdots \log a_n \right\}.$$

In methods 1 and 2, the points we consider in $V \cap \Lambda$ all lie on a complex line (subspace of dimension 1); this fact has been an essential feature of Baker's method up to recently (see for instance [Wü], and also [H]): it enables one to use interpolation techniques, and to work out the proof without introducing functions of several variables. The situation is quite different for the two dual methods, where the points span a vector space of higher dimension (as soon as $n \geq 3$ for method 3, and also for $n = 2$ in method 4). In this case there is so far no available extrapolation argument, and because of this the method is more primitive. The main fact which enables us to derive sharp bounds is that there is a single factor $\mathbb{G}_m$ (i.e. only one exponential function); cf [W2] and [W3].

4. Matrices whose entries are linear combinations of logarithms

One motivation for the results in this section is Leopoldt's conjecture on the p-adic rank of units of an algebraic number field. When one considers an algebraic number field which is Galois over $\mathbb{Q}$, the action of the Galois group gives rise to a block decomposition of the the p-adic regulator into square matrices whose entries are linear combinations with algebraic coefficients of p-adic logarithms of algebraic numbers. The problem is to give a lower bound for the rank of such matrices.

We consider here only the complex case, but the results can be extended also to non-archimedean fields.

Let M be a $d \times \ell$ matrix whose entries are linear combinations of (complex) logarithms of algebraic numbers. The coefficients of M belong to a $\bar{\mathbb{Q}}$-vector space generated by $\mathbb{Q}$-linearly independent elements $\log \alpha_1, \ldots, \log \alpha_s$ of L ; hence one can write

$$M = \left(\beta_{0ij} + \sum_{s=1}^{S} \beta_{sij} \log \alpha_s \right)_{\substack{1 \le i \le d, \\ 1 \le j \le \ell}} = B_0 + \sum_{s=1}^{S} B_s \log \alpha_s$$

with $\log \alpha_s$ linearly independent over $\mathbb{Q}$ (or over $\bar{\mathbb{Q}}$) and $B_0, \ldots, B_S$ matrices with algebraic entries.

Following D.Roy [R1], one defines the *structural rank* $r_{\text{conj}}(M)$ of M as the rank of the matrix

$$B_0 + \sum_{s=1}^{S} B_s X_s$$

in $\bar{\mathbb{Q}}(X_1, \ldots, X_S)$. From the standard conjecture stating that $\mathbb{Q}$-linearly independent elements of $\mathbb{L}$ are algebraically independent over $\mathbb{Q}$ (hence over $\bar{\mathbb{Q}}$), one immediately deduces:

Conjecture 7.

$$\text{rk}(M) = r_{\text{conj}}(M).$$

It happens often in this subject that one is able to prove half of what one expects; in the present case, D.Roy proved [R1]:

Theorem 8.

$$\text{rk}(M) \ge \frac{1}{2} r_{\text{conj}}(M).$$

Here is a connection with the situation we considered in theorem 4: if $\text{rk}(M) < d$, then the ℓ columns vectors

$$u_j = \left(\beta_{0ij} + \sum_{s=1}^{S} \beta_{sij} \log \alpha_s \right)_{1 \le i \le d}, \qquad (1 \le j \le \ell)$$

belong to a hyperplane of $\mathbb{C}^d$; denote by $\mathcal{L}$ the $\bar{\mathbb{Q}}$-vector space spanned by 1 and $\mathbb{L}$ in $\mathbb{C}$:

$$\mathcal{L} = \{ \beta_0 + \beta_1 \log \alpha_1 + \cdots + \beta_n \log \alpha_n \; ; \; \beta_i \in \bar{\mathbb{Q}}, \; \log \alpha_i \in \mathbb{L} \}.$$

Let V be a complex vector subspace of $\mathbb{C}^d$; if $V \cap \bar{\mathbb{Q}}^d \ni (\beta_1, \ldots, \beta_d) \neq 0$, then

$$(\beta_1 \log \alpha, \ldots, \beta_d \log \alpha) \in V \cap \mathcal{L}^d$$

for all $\log \alpha \in \mathbb{L}$; hence $\dim_{\bar{\mathbb{Q}}} V \cap \mathcal{L}^d = \infty$. As shown by D. Roy [R1], the converse is true: if $\dim_{\bar{\mathbb{Q}}} V \cap \mathcal{L}^d = \infty$, then $V \cap \bar{\mathbb{Q}}^d \ni (\beta_1, \ldots, \beta_d) \neq 0$. More precisely:

Theorem 9. *If $V \cap \bar{\mathbb{Q}}^d = \{0\}$, then*

$$\dim_{\bar{\mathbb{Q}}} V \cap \mathcal{L}^d \leq d(d - 1).$$

This result is in fact a consequence of theorem 4 (this is not obvious). In a joint work with Damien Roy [R-W], we develop the idea of [R1] and study the following generalization of theorem 9: let again G be an algebraic group defined over $\bar{\mathbb{Q}}$; we consider once more the $\mathbb{Q}$-vector space $\Lambda = \exp_G^{-1} G(\bar{\mathbb{Q}})$ in $T_G(\mathbb{C})$. Now let K be a number field; define Λ_K as the K-vector space spanned by Λ in $T_G(\mathbb{C})$:

$$\Lambda_K = \{\beta_1 u_1 + \cdots + \beta_n u_n \,;\, n \geq 0,\, \beta_i \in K,\, \exp_G u_i \in G(\bar{\mathbb{Q}})\}.$$

For $m \geq 1$ and $\beta = (\beta_1, \ldots, \beta_m) \in K^m$, consider the map

$$\varphi_\beta : \quad \begin{array}{ccc} T_{G^m} & \longrightarrow & T_G \\ (z_1, \ldots, z_m) & \longmapsto & z_1 \beta_1 + \cdots + z_m \beta_m \end{array}.$$

If G' is an algebraic subgroup of G^m defined over $\bar{\mathbb{Q}}$ of positive dimension such that $\varphi_\beta(T_{G'}) \subset V$, then for each $u \in T_{G'}(\mathbb{C})$ with $\exp_{G'} u \in G'(\bar{\mathbb{Q}})$, we have $\varphi_\beta(u) \in V \cap \Lambda_K$. Therefore $\dim_K V \cap \Lambda_K = \infty$. Here is the converse:

Theorem 10. *If V does not contain any non-zero subspace of the form $\varphi_\beta(T_{G'})$, then*

$$\dim_K V \cap \Lambda_K \leq (d_1 + 2d_2) \dim_{\mathbb{C}}(V/W).$$

5. Interpolation determinants

This last section is devoted to a sketch of proof of the algebraic subgroup theorem. The original proof [W1] used the classical transcendence method of Gel'fond-Schneider, with a construction of an auxiliary function performed thanks to the Dirichlet box principle (lemma of Thue-Siegel). Here, we follow an idea of Michel Laurent [La] and consider an interpolation determinant:

$$\Delta = \det\left(\varphi_\lambda(\zeta_\mu)\right)_{1 \leq \lambda, \mu \leq N}.$$

The main analytic argument is the following:

Lemma 11. Let $\varphi_1, \ldots, \varphi_N$ be analytic functions in a polydisk $\{z \in \mathbb{C}^n \, ; \, |z| \leq R\}$ in $\mathbb{C}^n$, and let $\zeta_1, \ldots, \zeta_N$ in $\mathbb{C}^n$ satisfy $|\zeta_\mu| \leq r < R$. Then

$$|\Delta| \leq \left(\frac{R}{r}\right)^{-T} N! \prod_{\lambda=1}^{N} |\varphi_\lambda|_R$$

with

$$T \geq \frac{n}{6e} N^{(n+1)/n} \quad \text{for} \quad N \geq 2^n e^{n+1}.$$

This lemma is fundamental for the proof below; its proof just uses the classical Schwarz lemma for functions of one complex variable with a high order multiplicity of zero at the origin.

Here is a sketch of proof of the algebraic subgroup theorem using lemma 11. We take linearly independent points $y_1, \ldots, y_\ell$ in $V \cap \Lambda$; the interpolation points will be indexed by $h = (h_1, \ldots, h_\ell) \in \mathbb{Z}^\ell$ with $0 \leq h_j < H$, $(1 \leq j \leq \ell)$:

$$\xi_h = h_1 y_1 + \cdots + h_\ell y_\ell.$$

We consider the functions $\varphi_\lambda = f_0^{\lambda_0} \cdots f_M^{\lambda_M} / f_0^L$, where $(\lambda_0, \ldots, \lambda_M) \in \mathbb{Z}^{M+1}$, $\lambda_0 + \cdots + \lambda_M = L$, and $(f_0, \ldots, f_M)$ are entire functions which give a representation the exponential map $\exp_G : T_G(\mathbb{C}) \longrightarrow G(\mathbb{C})$ composed with an embedding of $G(\mathbb{C})$ into a projective space $\mathbb{P}_M(\mathbb{C})$. We restrict the values of $\lambda = (\lambda_0, \ldots, \lambda_M)$ so that the φ_λ are linearly independent. Further, the vector subspace W of $T_G(\mathbb{C})$ in theorem 4 provides derivations which we write D_W^τ, $(\tau \in \mathbb{N}^t$ with $t = \dim_{\mathbb{C}} W)$ which have the crucial property:

$$D_W^\tau \varphi_\lambda(\xi_h) \in \bar{\mathbb{Q}}, \quad \text{for all} \quad \tau \in \mathbb{N}^t.$$

We shall use this information only for $\|\tau\| \leq T$, where T is a new parameter.

The main steps of the proof are as follows:

1. One starts by choosing the parameters L, T and H. In practice, one first writes the conditions that these parameters have to satisfy in such a way that the four steps below work, then one optimizes the choice of these parameters, and finally one writes down the proof starting with this choice. The final result depends heavily on these estimates, and there is no other justification than these computations to explain the explicit upper bound we finally obtain.

2. Here comes the geometric part of the proof, with the zero estimate of Philippon; it enables one to choose a subset $\mathcal{H}$ of $\{h = (h_1, \ldots, h_\ell) \in \mathbb{Z}^\ell \, ; \, 0 \leq h_j < H, \, (1 \leq j \leq \ell)\}$, such that the determinant

$$\Delta = \det\left(D_W^\tau \varphi_\lambda(\xi_h)\right)_{\substack{\lambda, \tau \\ h \in \mathcal{H}}}$$

does not vanish.

3. The analytic part of the proof is provided by lemma 11 of M.Laurent (the fact that the φ_λ are meromorphic rather than analytic is not a serious difficulty); one deduces an upper bound for $|\Delta|$, say $|\Delta| \leq \epsilon_1$.

4. The arithmetic argument is nothing else than the classical Liouville estimate; since Δ is a non-zero algebraic number, it cannot be too small: $|\Delta| \geq \epsilon_2$.

5. Now comes the conclusion: $\epsilon_2 \leq \epsilon_1$. There is still some work to do to get the desired statement; in fact the zero estimate yields a non-zero Δ provided that not too many points of $\mathbb{Z}y_1 + \cdots + \mathbb{Z}y_\ell$ lie in the tangent space of a proper algebraic algebraic subgroup of G. But we shall not give more details here (see [W1] for a complete proof).

6. Duality

We end this lecture by a discussion of a duality in transcendence proofs. We consider the following easy identity:

$$(d/dz)^\tau \left(z^\sigma e^{xz}\right)_{z=y} = (d/dz)^\sigma \left(z^\tau e^{yz}\right)_{z=x}$$

where x, y, z are complex numbers while σ and τ are non-negative integers.

There is a generalization of this identity to several variables [W2]: let n, t, s be positive integers, $x, y, z, w_1, \ldots, w_t, u_1, \ldots, u_s \in \mathbb{C}^n$, and $\sigma \in \mathbb{N}^s$, $\tau \in \mathbb{N}^t$; define ζz as the standard scalar product in $\mathbb{C}^n$, namely $\zeta z = \zeta_1 z_1 + \cdots + \zeta_n z_n$, and similarly $D_\zeta = \zeta_1(\partial/\partial z_1) + \cdots + \zeta_n(\partial/\partial z_n)$; then

$$D_{w_1}^{\tau_1} \cdots D_{w_t}^{\tau_t} \left((u_1 z)^{\sigma_1} \cdots (u_s z)^{\sigma_s} e^{xz}\right)_{z=y} =$$

$$= D_{u_1}^{\sigma_1} \cdots D_{u_s}^{\sigma_s} \left((w_1 z)^{\tau_1} \cdots (w_t z)^{\tau_t} e^{yz}\right)_{z=x}$$

This relation occurs in the proofs of theorem 5 as follows: assume we have a non-trivial linear relation $\log \alpha_n = \beta_1 \log \alpha_1 + \cdots + \beta_{n-1} \log \alpha_{n-1}$ and define, for $h = (h_1, \ldots, h_n) \in \mathbb{Z}^n$, $\tau = (\tau_1, \ldots, \tau_{n-1}) \in \mathbb{Z}^{n-1}$ and $\lambda \in \mathbb{Z}$:

$$\gamma_{h;\tau,\lambda} = \prod_{i=1}^{n-1} (h_i + h_n \beta_i)^{\tau_i} \prod_{j=1}^{n} \alpha_j^{\lambda h_j};$$

these numbers $\gamma_{h;\tau,\lambda}$ are interpolation values of analytic functions in two different ways: on one hand

$$\gamma_{h;\tau,\lambda} = \left(\frac{\partial}{\partial z_1}\right)^{\tau_1} \cdots \left(\frac{\partial}{\partial z_{n-1}}\right)^{\tau_{n-1}} \psi_h(\lambda \log \alpha_1, \ldots, \lambda \log \alpha_{n-1}),$$

with

$$\psi_h(z_1, \ldots, z_{n-1}) = e^{(h_1 + h_n \beta_1)z_1} \cdots e^{(h_{n-1} + h_n \beta_{n-1})z_{n-1}};$$

on the other

$$\gamma_{h;\tau,\lambda} = \varphi_{\tau,\lambda}(h_1 + h_n\beta_1, \ldots, h_{n-1} + h_n\beta_{n-1})$$

where

$$\varphi_{\tau,\lambda}(z_1, \ldots, z_{n-1}) = z_1^{\tau_1} \cdots z_{n-1}^{\tau_{n-1}} \left(\alpha_1^{z_1} \cdots \alpha_{n-1}^{z_{n-1}}\right)^{\lambda}.$$

If one specializes the preceding proof to the special case of theorem 5, depending on whether one express $\gamma_{h;\tau,\lambda}$ in terms of $\varphi_{\tau,\lambda}$ or in terms of ψ_h, one recognizes Baker's method or its dual.. For Hirata's method and its dual, one writes, for $\tau \in \mathbb{N}^n$, $h \in \mathbb{N}^{n+1}$ and $\lambda \in \mathbb{Z}$

$$\left(\frac{\partial}{\partial z_1}\right)^{\tau_1} \cdots \left(\frac{\partial}{\partial z_n}\right)^{\tau_n} \tilde{\psi}_h(\lambda \log \alpha_1, \ldots, \lambda \log \alpha_n) =$$

$$\left(\sum_{i=1}^{n} \beta_i \left(\frac{\partial}{\partial z_i}\right)\right)^{h_0} \tilde{\varphi}_{\tau,\lambda}(h_1, \ldots, h_n),$$

where

$$\tilde{\psi}_h(z_1, \ldots, z_n) = (\beta_1 z_1 + \cdots + \beta_n z_n)^{h_0} e^{h_1 z_1} \cdots e^{h_n z_n}$$

and

$$\tilde{\varphi}_{\tau,\lambda}(z_1, \ldots, z_n) = z_1^{\tau_1} \cdots z_n^{\tau_n} \left(\alpha_1^{z_1} \cdots \alpha_n^{z_n}\right)^{\lambda}$$

References

[B] Bombieri, E. Algebraic values of meromorphic maps. Invent. Math. **10** (1970), 267–287 and **11** (1970), 163–166.

[B-L] Bombieri, E, and Lang, S. Analytic subgroups of group varieties. Invent. Math. **11** (1970), 1–14.

[E] Emsalem, M. Sur les idéaux dont l'image par l'application d' Artin dans une $\mathbb{Z}_p$-extension est triviale. J. reine angew. Math. **382** (1987), 181–198.

[H] N. Hirata. Formes linéaires de logarithmes de points algébriques sur les groupes algébriques. Invent. Math. **104** (1991), 401–433.

[L] Lang, S. *Introduction to transcendental numbers*. Addison Wesley, 1966.

[La] M. Laurent. Hauteur de matrices d'interpolation. in *Approximations Diophantiennes et Nombres Transcendants*, éd. P. Philippon, Proc. Conf. Luminy 1990, W. de Gruyter, à paraître.

[M] B. Mazur. The topology of rational points. Experimental Mathematics **1** (1992), 35–45.

[P] P. Philippon. Lemmes de zéros dans les groupes algébriques commutatifs. Bull. Soc. Math. France **114** (1986), 355–383 et **115** (1987), 397–398.

[R1] D. Roy. Matrices whose coefficients are linear forms in logarithms. J. Number Theory, to appear.

[R2] D. Roy. Transcendance et questions de répartition dans les groupes algébriques. in *Approximations Diophantiennes et Nombres Transcendants*, éd. P. Philippon, Proc. Conf. Luminy 1990, W. de Gruyter, à paraître.

[R3] D. Roy. Simultaneous approximation in number fields

[R-W] D. Roy et M. Waldschmidt. Autour du théorème du sous-groupe algébrique

[Sa] J-J. Sansuc. Descente et principe de Hasse pour certaines variétés rationnelles. Sém. Théorie des Nombres Delange-Pisot-Poitou, Paris 1980–81, Birkhäuser Verlag, Progress in Math. **22** (1982), 253–271.

[S] Th. Schneider. Zur Theorie des Abelschen Funktionen und Integrale. J. reine angew. Math. **183** (1941), 110–128.

[W1] M. Waldschmidt. On the transcendence methods of Gel'fond and Schneider in several variables. in *New Advances in Transcendence Theory*, ed. A. Baker, Cambridge Univ. Press. 1988, Chap. 24, 375–398.

[W2] M. Waldschmidt. Fonctions auxiliaires et fonctionnelles analytiques. J. Analyse Math., **56** 1991, 231–279.

[W3] M. Waldschmidt. Nouvelles méthodes pour minorer des combinaisons linéaires de logarithmes de nombres algébriques. Sém. Th. Nombres Bordeaux **3** (1991), 129–185; (II), in *Problèmes Diophantiens 1989-1990*, Publ. Univ. P. et M. Curie (Paris VI) **93**, N°8, 36 p.; Minorations de combinaisons linéaires de logarithmes de nombres algébriques. à paraître.

[Wü] G. Wüstholz. Some remarks on a conjecture of Waldschmidt. in *Approximations Diophantiennes et Nombres Transcendants*, éd. D. Bertrand and M. Waldschmidt, Proc. Conf. Luminy 1982, Birkhäuser Verlag, Progress in Math. **31** (1983), 329–336.

Michel Waldschmidt
Université P. et M. Curie (Paris VI)
Problèmes Diophantiens, URA 763 du C.N.R.S.
Institut Henri Poincaré
11, rue P. et M. Curie
F-75231 PARIS Cedex 05
e-mail : miw@frunip62.bitnet

Square roots of central values of Hecke L-series

Fernando Rodriguez Villegas and Don Zagier

Princeton University

and

Max-Planck-Institut für Mathematik, Bonn

§1. Introduction

In [2] numerical examples were produced suggesting that the "algebraic" part of central values of certain Hecke L-series are perfect squares. More precisely, let ψ_1 be the grossencharacter of $\mathbb{Q}(\sqrt{-7})$ defined by

$$\psi_1(\mathfrak{a}) = \left(\frac{m}{7}\right)\alpha \quad \text{if} \quad \mathfrak{a} = (\alpha), \quad \alpha = \frac{m + n\sqrt{-7}}{2} \in \mathbb{Z}\left[\frac{1 + \sqrt{-7}}{2}\right]$$

and consider the central value $L(\psi_1^{2k-1}, k)$ of the L-series associated to an odd power of ψ_1. This value vanishes for k even by virtue of the functional equation, but for k odd one has

$$(1) \qquad L(\psi_1^{2k-1}, k) = 2 \frac{(2\pi/\sqrt{7})^k \, \Omega^{2k-1}}{(k-1)!} A(k), \qquad \Omega = \frac{\Gamma(\frac{1}{7})\Gamma(\frac{2}{7})\Gamma(\frac{4}{7})}{4\pi^2},$$

with $A(1) = \frac{1}{4}$, $A(3) = A(5) = 1$, $A(7) = 9$, $A(9) = 49$, ..., $A(33) = 44762286327255^2$, suggesting the conjecture

$$(2) \qquad L(\psi_1^{2k-1}, k) \stackrel{?}{=} 2 \frac{(2\pi/\sqrt{7})^k \, \Omega^{2k-1}}{(k-1)!} B(k)^2,$$

with $B(1) = \frac{1}{2}$ and $B(k) \in \mathbb{Z}$ for all $k \geq 3$.

In this paper we will prove this conjecture and analogous results for other grossencharacters. The method will be a modification of the method of [8], where it was shown that the central values of weight one Hecke L-series are essentially the squares of certain sums of values of weight 1/2 theta series at CM points. The new ingredient for higher weight is that we have to use (non-holomorphic) *derivatives* of modular forms. For instance, the value of $B(k)$ in (2) will turn out to be essentially the value of a certain derivative of

$$(3) \qquad \theta_{1/2}(z) = \sum_{\substack{n \geq 1 \\ n \text{ odd}}} e^{\pi i n^2 z/4} \qquad (z \in \mathcal{H} = \text{ upper half-plane})$$

at $z = (1 + i\sqrt{7})/2$. Because the derivatives of modular forms can be computed recursively, this leads to *a simple recursive formula* for the special values of Hecke L-series and their square roots. In particular, for the example above the result is:

Theorem. *Define sequences of polynomials $a_{2n}(x)$, $b_n(x)$ by the recursions*

$$a_{n+1}(x) = \sqrt{(1+x)(1-27x)}\left(x\frac{d}{dx} - \frac{2n+1}{3}\right)a_n(x)$$

(4)
$$- \frac{n^2}{9}(1-5x)a_{n-1}(x)$$

$$21\,b_{n+1}(x) = \left((32nx - 56n + 42) - (x-7)(64x-7)\frac{d}{dx}\right)b_n(x)$$

(5)
$$- 2n(2n-1)(11x+7)b_{n-1}(x)$$

with initial conditions $a_0(x) = 1$, $a_1(x) = -\frac{1}{3}\sqrt{(1-x)(1+27x)}$, $b_0(x) = 1/2$, $b_1(x) = 1$. Then the values $A(2n+1) = a_{2n}(-1)/4$ and $B(2n+1) = b_n(0)$ satisfy equations (1) and (2).

For numerical values of the first few $a_{2n}(x)$, $b_n(x)$, and $B(k)$, see §7. Surprisingly, there seems to be no simple direct proof of the identity $a_{2n}(-1) = 4\,b_n(0)^2$!

We mention briefly several applications of the theorem. First, it gives us a specific choice of the square root of the numbers $A(k)$ occurring in (1). Now it is well-known that the existence of a p-adic L-function imply that the squares $A(k)$ satisfy congruences modulo certain prime powers, and it has been conjectured [4] that there should be analogous p-adic interpolation properties for appropriately chosen square roots. Testing this on the square roots produced by the theorem, we do indeed find that these satisfy certain congruences of the desired type, e.g.

(6) $B(k) \equiv -k \pmod{4}$, $B(k+10) \equiv 7B(k) \pmod{11}$ $(k \geq 3)$.

This is the topic of a forthcoming thesis by A. Sofer. Notice, by the way, that either of the congruences (6) implies the non-vanishing of $L(\psi_1^{2k-1}, k)$, which is not a priori obvious.

The second "application" is that one can compute the numbers $A(k)$ and $B(k)$ much more easily than was previously possible. The method of computation in [4] was to compute $L(\psi_1^{2k-1}, k)$ as

$$2\sum_{n=1}^{\infty} a_n^{(k)}\left(\sum_{j=0}^{k}\frac{1}{j!}\left(\frac{2\pi n}{7}\right)^j\right)e^{-2\pi n/7},$$

where $L(\psi_1^{2k-1}, s) = \sum_{n=1}^{\infty} a_n^{(k)} n^{-s}$, but this becomes unmanageable for large k. The recurrences are easier to work with and can also be used to compute the numbers in question modulo high powers of a prime p, without computing the numbers themselves, which grow very rapidly. This is useful both for testing the above-mentioned conjectures of Koblitz and in connection with a beautiful recent result of Rubin [6], proved by him modulo the Birch–Swinnerton-Dyer conjecture, which gives a transcendental construction of points on certain elliptic curves via p-adic interpolation of values of Hecke L-series.

The third application is that the central values of the L-series under consideration are always non-negative, which in turn by a remark of Greenberg implies a result on their average value (cf. Corollary in §5 and the following comments).

In a different direction, it was also found in [2] that the values of the *twisted* L-series $L(\psi_1^{2k-1}, \left(\frac{-}{p}\right), s) = \sum_{n=1}^{\infty} \left(\frac{n}{p}\right) a_n^{(k)} n^{-s}$ ($p \equiv 1 \pmod 4$ prime) at $s = k$ were again essentially perfect squares:

$$(7) \qquad L(\psi_1^{2k-1}, \left(\tfrac{-}{p}\right), k) \stackrel{?}{=} \frac{3 + \left(\frac{p}{7}\right)}{2} \frac{(\sqrt{7}/2\pi)^{k-1} (\Omega/\sqrt{p})^{2k-1}}{(k-1)!} B(k, p)^2;$$

The well-known theorem of Waldspurger, of course, tells us that the central values of the twists of the L-series of a given modular (Hecke eigen) form f are essentially square multiples of one another, the square-roots being proportional to the Fourier coefficients of the half-integral weight form attached to f by the Shimura lifting, but it does not tell us the values themselves. In §8 we propose a formula for the numbers $B(k, p)$ which when combined with Waldspurger's theorem may eventually give an explicit formula for the coefficients of the Shimura lifting of a modular form attached to a grossencharacter.

The main result of this paper (like that of [8]) involves a "factorization formula" which expresses the value (or derivative) of a weight one theta series at a CM point as a product of values (or derivatives) of weight $1/2$ theta series at other CM points. This formula will be proved in §4, while §5 gives the application to grossencharacters and §§6–7 describe recurrence relations like the ones in the sample theorem above.

The first author was partially supported by NSF grant 2156272. He would also like to thank the Max-Planck-Institut für Mathematik, where part of this work was done.

§2. Derivatives of modular forms

The differential operator

$$D = \frac{1}{2\pi i} \frac{d}{dz} = \frac{1}{q} \frac{d}{dq} \qquad (q = e^{2\pi i z})$$

maps holomorphic functions to holomorphic functions and functions with
a Fourier expansion of the form $\sum a(n)\, q^n$ to functions with a Fourier
expansion of the form $\sum na(n)\, q^n$ and with Fourier coefficients in the same
field, but it destroys the property of being a modular form. As is well-
known, this can be corrected by introducing the modified differentiation
operator

$$\partial_k = D - \frac{k}{4\pi y}$$

which satisfies $\partial_k(f|_k\gamma) = (\partial_k f)|_{k+2}\gamma$ for all $\gamma \in SL_2(\mathbb{R})$, where as usual
$f|_k\begin{pmatrix} a & b \\ c & d \end{pmatrix}(z) = (cz + d)^{-k} f(\frac{az + b}{cz + d})$. In particular, if $\Gamma \subset SL_2(\mathbb{R})$ is
some modular group and $M_k^*(\Gamma)$ denotes the space of differentiable modular
forms of weight k on Γ, possibly with some character or multiplier system
v (i.e., of f satisfying $f|_k\gamma = v(\gamma)f$ for all $\gamma \in \Gamma$), then $\partial_k f$ belongs to
$M_{k+2}^*(\Gamma)$ and more generally $\partial_k^h f$ to $M_{k+2h}^*(\Gamma)$, where

$$\partial_k^h = \partial_{k+2h-2} \circ \partial_{k+2h-4} \circ \ldots \circ \partial_{k+2} \circ \partial_k .$$

In this situation we will often drop the subscript and write simply $\partial^h f$ for
$\partial_k^h f$, since f determines its weight k uniquely.

An easy induction shows that

$$(8) \qquad \partial_k^h = \sum_{j=0}^{h} \binom{h}{j} \frac{\Gamma(h+k)}{\Gamma(j+k)} \left(\frac{-1}{4\pi y}\right)^{h-j} D^j .$$

In particular,

$$(9) \qquad \partial_k^h \left(\sum_{n=0}^{\infty} a(n)\, e^{2\pi i n z} \right) = \frac{(-1)^h h!}{(4\pi y)^h} \sum_{n=0}^{\infty} a(n)\, L_h^{k-1}(4\pi n y)\, e^{2\pi i n z} ,$$

where

$$(10) \qquad L_h^\alpha(z) = \sum_{j=0}^{h} \binom{h+\alpha}{h-j} \frac{(-z)^j}{j!} \qquad (h \in \mathbb{Z}_{\geq 0}, \alpha \in \mathbb{C})$$

denotes the h-th *generalized Laguerre polynomial*. In the special case $k = 1/2$ we have the identity $L_h^{-1/2}(z) = (-1/4)^h H_{2h}(\sqrt{z})/h!$, where

$$(11) \qquad H_p(z) = \sum_{0 \leq j \leq p/2} \frac{p!}{j!(p-2j)!} (-1)^j (2z)^{p-2j} \qquad (p \in \mathbb{Z}_{\geq 0})$$

is the *pth Hermite polynomial*. In particular, the non-holomorphic derivatives of the weight $1/2$ theta series $\theta_{1/2}$ defined in (3) are given by

$$\partial^h \theta_{1/2}(z) = \frac{1}{(16\pi y)^h} \sum_{\substack{n>0 \\ n \text{ odd}}} H_{2h}(n\sqrt{\pi y/2})\, e^{\pi i n^2 z/4} .$$

A similar calculation applies to the weight $3/2$ theta series

$$(12) \qquad \theta_{3/2}(z) = \sum_{n=1}^{\infty} \left(\frac{-4}{n}\right) n\, e^{\pi i n^2 z/4} \qquad (z \in \mathcal{H})$$

and shows that the Fourier expansions of the functions $\theta_{p+1/2}$ $(p \in \mathbb{Z}_{\geq 0})$ defined by

$$(13) \qquad \theta_{p+1/2}(z) = \begin{cases} 8^h\, \partial^h_{1/2}\theta_{1/2}(z) & \text{if } p = 2h,\ h \geq 0 \\ 8^h\, \partial^h_{3/2}\theta_{3/2}(z) & \text{if } p = 2h+1,\ h \geq 0 \end{cases}$$

can be given by the uniform formula

$$(14) \qquad \theta_{p+1/2}(z) = \frac{1}{(2\pi y)^{p/2}} \sum_{\substack{n \geq 1 \\ n \text{ odd}}} \left(\frac{-4}{n}\right)^p H_p(n\sqrt{\pi y/2})\, e^{\pi i n^2 z/4} .$$

We remark that the holomorphic theta series defined by (3) and (12) can be expressed in terms of the Dedekind eta-function $\eta(z) = e^{\pi i z/12} \prod(1-e^{2\pi i n z})$ as

$$(15) \qquad \theta_{1/2}(z) = \frac{\eta(2z)^2}{\eta(z)}, \qquad \theta_{3/2}(z) = \eta(z)^3 .$$

Finally, from (8) and the binomial theorem we immediately get the identity

$$(16) \qquad \partial^h_k \left(\frac{1}{(mz+n)^k}\right) = \frac{\Gamma(h+k)}{\Gamma(k)} \left(\frac{-1}{4\pi y}\, \frac{m\bar{z}+n}{mz+n}\right)^h \frac{1}{(mz+n)^k}$$

which will be used in the next section to express the values of Hecke L-series at critical points as values of non-holomorphic derivatives of holomorphic Eisenstein series at CM points.

§3. L-series and Eisenstein series

From now on we fix the following notation: K is an imaginary quadratic field of odd discriminant $-d$, $\mathcal{O}_K$ its ring of integers, $\mathfrak{d} = (\sqrt{-d})$ its different, and Cl_K its class group. We suppose $d \neq 3$ so that $\mathcal{O}_K^* = \{\pm 1\}$; later we will also suppose that d is prime to avoid complications due to genus characters. We denote by $\varepsilon(n) = \left(\frac{-d}{n}\right) = \left(\frac{n}{d}\right)$ the Dirichlet character associated to K. We can extend it via the isomorphism $\mathbb{Z}/d \xrightarrow{\sim} \mathcal{O}_K/\mathfrak{d}$

to a quadratic character of K of conductor $\mathfrak{d}$. (Explicitly, we can write any $\mu \in \mathcal{O}_K$ as $\mu = \frac{1}{2}(m + n\sqrt{-d})$ with m and n of the same parity, and then $\varepsilon(\mu) = \varepsilon(2m)$.) Finally, we define $\delta = 0$ or 1 by $(d+1)/4 \equiv \delta \pmod 2$ or $\varepsilon(2) = (-1)^\delta$.

We fix a positive integer k and consider Hecke characters ψ of K satisfying

$$(17) \qquad \psi((\alpha)) = \varepsilon(\alpha)\,\alpha^{2k-1}$$

for $\alpha \in \mathcal{O}_K$ prime to $\mathfrak{d}$. Clearly the number of such ψ equals $h(-d)$, the class number of K, and each one has conductor $\mathfrak{d}$. Associated to ψ is the Hecke L-series

$$L(\psi, s) = \sum_{\mathfrak{a}} \frac{\psi(\mathfrak{a})}{N(\mathfrak{a})^s}.$$

If we define for any ideal $\mathfrak{a}$ prime to $\mathfrak{d}$ a partial Hecke series by

$$Z(2k-1, \mathfrak{a}, s) = \frac{1}{2} \sum_{\lambda \in \mathfrak{a}}{}' \frac{\varepsilon(\lambda)\,\bar{\lambda}^{2k-1}}{|\lambda|^{2s}} = \sum_{\substack{\lambda \in \mathfrak{a} \\ \varepsilon(\lambda) = +1}} \frac{\bar{\lambda}^{2k-1}}{|\lambda|^{2s}},$$

where the prime indicates that 0 is excluded, then $\psi(\mathfrak{a})N(\mathfrak{a})^{s-2k+1}Z(2k-1, \mathfrak{a}, s)$ depends only on the class $[\mathfrak{a}]$ of $\mathfrak{a}$ in Cl_K and a standard one-line calculation gives the decomposition

$$(18) \qquad L(\psi, s) = \sum_{[\mathfrak{a}] \in \mathrm{Cl}_K} \frac{\psi(\mathfrak{a})}{N(\mathfrak{a})^{2k-1-s}}\, Z(2k-1, \mathfrak{a}, s)\,.$$

The series $L(\psi, s)$ and $Z(2k-1, \mathfrak{a}, s)$ converge only for $\Re(s) > k + \frac{1}{2}$, but can be analytically continued to the whole s-plane and satisfy a functional equation under $s \mapsto 2k - s$, with root number $w_k = (-1)^{k+1+\delta}$. Their critical values correspond to $s = k + r$ for integers $0 \le r \le k-1$ and their reflections $s = k - r$. In particular, their center value corresponds to $r = 0$, and $L(\psi, k) = 0$ if $k \equiv \delta \pmod 2$.

Any primitive ideal $\mathfrak{a}$ (i.e., one not divisible by rational integers > 1) can be written as $\mathbb{Z}a + \mathbb{Z}\frac{b+\sqrt{-d}}{2}$ where $a = N(\mathfrak{a})$ and b is an integer, determined mod $2a$, satisfying $b^2 \equiv -d \bmod 4a$. The number $(b + \sqrt{-d})/2a$ in $\mathcal{H}$ is then well-defined modulo $\mathbb{Z}$ and its class in $\mathcal{H}/SL_2(\mathbb{Z})$ depends only the ideal class of $\mathfrak{a}$. However, we will be wanting to evaluate modular forms of level d and for this we have to require additional congruences modulo d. We will choose $\mathfrak{a}$ prime to $\mathfrak{d}$ and then choose b divisible by d (which we can do since $2a$ is prime to d). We set

$$(19) \qquad z_{\mathfrak{a}}^{(d)} = \frac{b + \sqrt{-d}}{2ad} \in \mathcal{H}\,,$$

where

$$\mathfrak{a} = \left[a, \frac{b + \sqrt{-d}}{2}\right], \quad (a, d) = 1, \quad b \equiv 0 \pmod{d}.$$

Then $z_{\mathfrak{a}}^{(d)}$ is well-defined modulo $\mathbb{Z}$ and its image in $\mathcal{H}/\Gamma_0(d)$ depends only on the class $[\mathfrak{a}]$.

Each $\lambda \in \mathfrak{a}$ can be written as $a(mdz_{\mathfrak{a}}^{(d)} + n)$ for some integers $m, n \in \mathbb{Z}$, and by virtue of our choice of b we have $\varepsilon(\lambda) = \varepsilon(n)$ (note that $\varepsilon(a) = 1$ automatically). Hence

$$Z(2k - 1, \mathfrak{a}, s) = \frac{a^{2k-1-2s}}{2} {\sum_{m,n}}' \frac{\varepsilon(n)\,(md\bar{z}_{\mathfrak{a}}^{(d)} + n)^{2k-1}}{|mdz_{\mathfrak{a}}^{(d)} + n|^{2s}} \qquad \left(\Re(s) > k + \frac{1}{2}\right).$$

For each integer $r \geq 0$ we define an Eisenstein series of weight $2r + 1$ and character ε on $\Gamma_0(d)$ by

$$(20) \qquad G_{2r+1,\varepsilon}(z) = \frac{1}{2} {\sum_{m,n}}' \frac{\varepsilon(n)}{(mdz + n)^{2r+1}} \qquad (z \in \mathcal{H})$$

(if $r = 0$ this does not converge absolutely and has to be summed by the usual Hecke trick), with Fourier expansion given by

$$(21) \quad G_{2r+1,\varepsilon}(z) = L(2r + 1, \varepsilon) + \frac{(-1)^r (2\pi)^{2r+1}}{(2r)!\, d^{2r+1/2}} \sum_{n \geq 1} \left(\sum_{m|n} \varepsilon(m)\, m^{2r}\right) q^n.$$

($q = e^{2\pi i z}$). As an immediate consequence of (16) we find

$$\partial^{k-r-1} G_{2r+1,\varepsilon}(z) = \frac{1}{2} \frac{(k + r - 1)!}{(2r)!} \left(\frac{-1}{4\pi y}\right)^{k-r-1} {\sum_{m,n}}' \frac{\varepsilon(n)\,(md\bar{z} + n)^{2k-1}}{|mdz + n|^{2k+2r}}$$

and hence finally

Proposition. Let $\mathfrak{a}$ and $z_{\mathfrak{a}}^{(d)}$ be as in (19) and $0 \leq r \leq k - 1$. Then

$$(22)$$
$$Z(2k - 1, \mathfrak{a}, k + r) = \frac{(2r)!}{(k + r - 1)!} \frac{(-2\pi/\sqrt{d})^{k-r-1}}{N(\mathfrak{a})^{k+r}} \partial^{k-r-1} G_{2r+1,\varepsilon}\left(z_{\mathfrak{a}}^{(d)}\right).$$

This formula is true also for $r = 0$, the case of primary interest to us, because the summation via Hecke's trick used to define $G_{1,\varepsilon}$ commutes with the differentiation operator ∂.

§4. A factorization identity for theta series

Recall that H_p and L_h^α denote the Hermite and Laguerre polynomials. For $\mu, \nu \in \mathbb{Q}, p \in \mathbb{Z}_{\geq 0}$ and $z = x + iy \in \mathcal{H}$ we define

$$\theta_{(p)}\begin{bmatrix}\mu\\\nu\end{bmatrix}(z) = i^{-p}\,(2\pi y)^{-p/2}\sum_{n\in\mathbb{Z}+\mu}H_p(n\sqrt{2\pi y})\,e^{\pi i n^2 z + 2\pi i\nu n}\ .$$

For $z = x + iy \in \mathcal{H}$ we set $Q_z(m,n) = |mz - n|^2/2y$, the general positive definite binary quadratic form of discriminant -1 with real coefficients.

Theorem (Factorization Formula). *For $a \in \mathbb{N}$, $z \in \mathcal{H}$, $\mu,\nu \in \mathbb{Q}$, and $p, \alpha \in \mathbb{Z}_{\geq 0}$,*

(23)
$$\sum_{m,n\in\mathbb{Z}} e^{2\pi i(m\nu + n\mu)}\left(\frac{mz-n}{ay}\right)^{\alpha} L_p^{\alpha}\left(\frac{2\pi}{a}Q_z(m,n)\right) e^{\pi(imn - Q_z(m,n))/a}$$
$$= \frac{(-\pi)^p}{p!}\,2^{1/2}a^{\alpha+1/2}y^{p+\alpha+1/2}\,\theta_{(p)}\begin{bmatrix}a\mu\\\nu\end{bmatrix}(a^{-1}z)\cdot\theta_{(p+\alpha)}\begin{bmatrix}\mu\\-a\nu\end{bmatrix}(-a\bar{z})\ .$$

Remark. For the simplest case $a = 1$, $\alpha = 0$ the right-hand side of the (23) becomes $(-1)^p\sqrt{2y}\,\left|\theta_{(p)}\begin{bmatrix}\mu\\\nu\end{bmatrix}(z)\right|^2$, so the sum on the left is nonnegative, which is not clear a priori.

Proof. For u_1, $u_2 \in \mathbb{C}$ we have the identity

$$\left(e^{\pi u_2^2/2ay}\sum_{n\in\mathbb{Z}+\mu}e^{-\pi i a n^2\bar{z} + 2\pi in(u_2 - a\nu)}\right)\left(e^{\pi a u_1^2/2y}\sum_{\ell\in\mathbb{Z}+a\mu}e^{\pi i\ell^2 z/a + 2\pi i\ell(u_1+\nu)}\right)$$

$$= e^{\pi(a^2 u_1^2 + u_2^2)/2ay}\sum_{m\in\mathbb{Z}}e^{\pi im^2 z/a + 2\pi im(u_1+\nu)}\sum_{n\in\mathbb{Z}+\mu}e^{-2\pi an^2 y + 2\pi in(mz + au_1 + u_2)}$$

$$= \frac{e^{-\pi u_1 u_2/y}}{\sqrt{2ay}}\sum_{m,n\in\mathbb{Z}}e^{2\pi i(m\nu + n\mu) + \pi[imn - Q_z(m,n)]/a - \pi[(m\bar{z}-n)u_1 + (mz-n)u_2/a]/y}$$

the first equality being obtained by the substitution $\ell = an + m$ and the second by applying the formula

$$\sum_{n\in\mathbb{Z}+\mu}e^{-\pi n^2 A + 2\pi in B} = \frac{1}{\sqrt{A}}\sum_{n\in\mathbb{Z}}e^{-\pi(n-B)^2/A + 2\pi in\mu}\qquad (A > 0,\quad B \in \mathbb{C}),$$

which is a standard consequence of the Poisson summation formula, to the inner sum. Identity (23) follows by comparing the Taylor coefficients of $u_1^p\,u_2^{p+\alpha}$ on both sides. ∎

Remarks. 1. This formula in the case $a = 1$ is essentially contained in Kronecker's work ([5], Chapter III). It is also a special case of a general transformation formula for products of Fourier series which is used in the field of radar signal design (cf. Chapter 8 of [7], especially the Corollary to Theorem 8.18).

2. The proof can be expressed in an essentially equivalent but somewhat different way by using the transformation formula of the genus 2 theta series

$$\Theta^{(2)}(u, Z) = \sum_{n \in \mathbb{Z}^2} e^{\pi i \, ^t n Z n} e^{2\pi i \, ^t n u} \qquad (z \in \mathcal{H}_2, \quad u \in \mathbb{C}^2)$$

under the action of the matrix

$$M = \begin{pmatrix} 1 & 0 & 0 & 0 \\ 0 & 0 & 0 & -1 \\ 0 & 0 & 1 & -a \\ a & 1 & 0 & 0 \end{pmatrix} \in Sp_4(\mathbb{Z})$$

to relate the values of $\Theta^{(2)}$ at the two points

$$(u, Z) = \left(\begin{pmatrix} u_1 + \mu z + \nu \\ u_2 - a(\mu \bar{z} + \nu) \end{pmatrix}, \begin{pmatrix} a^{-1} z & 0 \\ 0 & -a\bar{z} \end{pmatrix} \right)$$

(where it obviously splits into a product of genus one series) and

$$(u, Z) = \left(\frac{i}{2ay} \begin{pmatrix} u_1 a\bar{z} + u_2 z \\ -a u_1 - u_2 \end{pmatrix} + \begin{pmatrix} \nu \\ \mu \end{pmatrix}, \quad \frac{i}{2ay} \begin{pmatrix} |z|^2 & -z \\ -z & 1 \end{pmatrix} \right).$$

This is connected with an interesting interpretation of the factorization formula in terms of the geometry of the Siegel modular variety of genus 2 which will be discussed in a later publication.

We now apply the theorem to the case when z is an appropriately chosen CM point in the upper half-plane, in which case the left-hand side of (23) becomes the value at a CM point of a non-holomorphic derivative of a holomorphic theta series of weight $\alpha + 1$. We let $K = \mathbb{Q}(\sqrt{-d})$ and ε have the same meanings as in §3 and define for each ideal $\mathfrak{a}$ and odd integer $h \geq 1$ a theta series

$$\Theta_{\mathfrak{a}}^{(h)}(z) = \frac{1}{2} \sum_{\lambda \in \mathfrak{a}} \lambda^{h-1} q^{N(\lambda)/N(\mathfrak{a})} \qquad (z \in \mathcal{H}, \quad q = e^{2\pi i z})$$

which is a modular form in $M_h(\Gamma_0(d), \varepsilon)$ and a cusp form if $h > 1$. Clearly $\Theta_{\lambda \mathfrak{a}}^{(h)}(z) = \lambda^{h-1} \Theta_{\mathfrak{a}}^{(h)}(z)$ (the $\Theta_{\mathfrak{a}}^{(h)}$ are the modular forms corresponding to the partial zeta functions for unramified Hecke L-series of even weight $h - 1$). In particular, $\Theta_{\mathfrak{a}}^{(1)}(z)$ depends only on the class $[\mathfrak{a}]$ of $\mathfrak{a}$ in Cl_K.

On the half-integral weight side we will be evaluating modular forms of level a power of 2 at our CM points, so we have to impose further congruence conditions on the bases of our ideals modulo powers of 2, just as we did modulo d in §3. We set
(24)

$$z_{\mathfrak{a}}^{(2)} = \frac{b + \sqrt{-d}}{2a} \in \mathcal{H} \qquad (\mathfrak{a} = [a, \frac{b + \sqrt{-d}}{2}], \ 2 \nmid a, \ b \equiv 1 \pmod{16}).$$

which is well defined modulo $8\mathbb{Z}$. (Any other fixed odd choice of b modulo 16 would be just as good.)

Theorem. *Let* $\mathfrak{a}$, $\mathfrak{a}_1$ *be coprime ideals of* K *prime to* $2\mathfrak{d}$ *and to* 2, *respectively. Then for* k, $h \geq 1$ *satisfying* $k \equiv 1 + \delta \pmod{2}$, $h \equiv 1 \pmod{2}$ *we have*

$$\partial^{k-1}\Theta^{(h)}_{\mathfrak{a}\mathfrak{a}_1}\left(z^{(d)}_{\mathfrak{a}}\right) = (-1)^\delta \left(\frac{-4}{N(\mathfrak{a})}\right)^\delta \frac{d^{k+h/2-5/4}}{2^{2k+h-4}} \frac{N(\mathfrak{a})^{h-1}}{N(\mathfrak{a}_1)^{k-1/2}}$$

$$(25) \qquad\qquad \cdot \, \theta_{k-1/2}\left(z^{(2)}_{\mathfrak{a}^2\mathfrak{a}_1}\right) \cdot \overline{\theta_{k+h-3/2}\left(z^{(2)}_{\mathfrak{a}_1}\right)},$$

where $\theta_{*+1/2}$ *are the theta series defined by* (3), (12) *and* (13) *and* $z^{(d)}_{\mathfrak{a}}$ *and* $z^{(2)}_{\mathfrak{a}}$ *have the meanings given in* (19) *and* (24), *respectively.*

Proof. We will choose $p = k - 1$, $\alpha = h - 1$, $\mu = 1/2$, $\nu = \delta/2$ in (23), observing that

$$\theta_{(p)}\begin{bmatrix} r/2 \\ s/2 \end{bmatrix}(z) = 2\,(-1)^{(p-s)/2}\,\theta_{p+1/2}(z) \qquad \text{for} \quad r \equiv 1, \; s \equiv p \pmod{2}$$

by (14) and the definition of $\theta_{(p)}\begin{bmatrix} \mu \\ \nu \end{bmatrix}$. We further choose $b \equiv 1 \pmod{16}$ such that $(b + \sqrt{-d})/2$ belongs to $\mathfrak{a}^2\mathfrak{a}_1\mathfrak{d}$ (in particular, b is divisible by d and $c := (b^2 + d)/4aa_1$ is divisible by a and congruent to δ modulo 2) and set $z = (b + \sqrt{-d})/2aa_1$ in (23). Then

$$\mathfrak{a}\mathfrak{a}_1 = aa_1\,[1, z], \quad z^{(d)}_{\mathfrak{a}} = a_1 z/d, \quad z^{(2)}_{\mathfrak{a}^2\mathfrak{a}_1} = a^{-1}z, \quad z^{(2)}_{\mathfrak{a}_1} = az$$

and

$$\pi \frac{imn - Q_z(m, n)}{a} \equiv 2\pi i\left[(m^2 c - bmn + aa_1 n^2)z^{(d)}_{\mathfrak{a}} + \frac{m\delta + n}{2}\right] \pmod{2\pi i\mathbb{Z}},$$

and substituting all this into (23) gives the assertion of the theorem. ∎

Remark. Notice that the theorem relates values of modular forms of different levels. This forces ratios of such values to be in smaller fields than one would suspect a priori. Also from the transformation properties of $\partial^p\Theta^{(h)}_{\mathfrak{a}}(z)$ under homotheties of $\mathfrak{a}$ and the action of $\Gamma_0(d)$ on z one immediately gets the following:

Corollary. *Let* $\mathfrak{a}$ *and* $\mathfrak{a}_1$ *be as in the theorem and* ψ *a Hecke character of* K *satisfying* (17). *Then*

$$\left(\frac{-4}{N(\mathfrak{a})}\right)^\delta \overline{\psi(\mathfrak{a})}^{-1} \theta_{k-1/2}\left(z^{(2)}_{\mathfrak{a}^2\mathfrak{a}_1}\right)$$

depends only on ψ, $\mathfrak{a}_1$ *and the ideal class of* $\mathfrak{a}$.

This transformation property is quite non-obvious and was proved in [8] by a long calculation (pp. 559–562) of quadratic symbols. The point is that the theta series $\theta_{k-1/2}$ has level 4 and has nothing to do with d at all, so that the Kronecker symbol $(d/\cdot)$ implicit in the factor ψ has to come from the transformation behavior of the CM point $z^{(2)}_{\mathfrak{a}^2\mathfrak{a}_1}$ under change of ideals and from the automorphy factor of $\theta_{k-1/2}$.

§5. Final formula for the central value of $L(\psi, k)$

The proposition of §3 and the theorem just proved in general involve different modular forms: non-holomorphic derivatives of Eisenstein series (of odd weight and character ε) in one case and non-holomorphic derivatives of theta series (again of odd weight and character ε) in the other. There is one case where these overlap. Namely, for the Eisenstein series of weight one we have

$$(26) \qquad \frac{\sqrt{d}}{2\pi} G_{1,\varepsilon}(z) = \frac{h(-d)}{2} + \sum_{n=1}^{\infty} \left(\sum_{m|n} \varepsilon(m) \right) q^n$$

by (21), and since the coefficient of n is the number of integral ideals of K of norm n this is simply $\sum_{[\mathfrak{a}]} \Theta^{(1)}_{\mathfrak{a}}(z)$. Hence we can combine equations (18), (22) (with $r = 0$) and (25) (with $h = 1$) to get

$$L(\psi, k) = \frac{(2\pi/\sqrt{d})^k}{(k-1)!} \sum_{[\mathfrak{a}],[\mathfrak{a}_1]\in \mathrm{Cl}_K} \overline{\psi(\mathfrak{a})}^{-1} \partial^{k-1} \Theta^{(1)}_{\mathfrak{a}_1}(z^{(d)}_{\mathfrak{a}})$$

$$= \frac{(-1)^{\delta} \pi^k d^{k/2-3/4}}{2^{k-1}(k-1)!} \sum_{[\mathfrak{a}],[\mathfrak{a}_1]\in \mathrm{Cl}_K} \frac{\overline{\psi(\mathfrak{a})}^{-1}}{N(\mathfrak{a}_1)^{k-1/2}} \theta_{k-1/2}\big(z^{(2)}_{\mathfrak{a}^2\mathfrak{a}_1}\big) \overline{\theta_{k-1/2}\big(z^{(2)}_{\mathfrak{a}_1}\big)} .$$

If we assume that d is prime, so that the class number $h(-d)$ is odd, then we can replace first $\mathfrak{a}_1$ by $\mathfrak{a}_1^2$ and then $\mathfrak{a}$ by $\mathfrak{a}\mathfrak{a}_1^{-1}$ to obtain:

MAIN THEOREM. *Let $d > 3$ be a prime $\equiv 3 \mod 4$, k a positive integer satisfying $k \equiv \delta + 1 \pmod 2$ and ψ a Hecke character of $K = \mathbb{Q}(\sqrt{-d})$ satisfying (17). Then*

$$(27) \qquad L(\psi, k) = \frac{\pi^k d^{k/2-3/4}}{2^{k-1}(k-1)!} \left| \sum_{[\mathfrak{a}]\in \mathrm{Cl}_K} \left(\frac{-4}{N(\mathfrak{a})}\right)^{\delta} \overline{\psi(\mathfrak{a})}^{-1} \theta_{k-1/2}\big(z^{(2)}_{\mathfrak{a}^2}\big) \right|^2 .$$

Notice that the terms of the sum are well-defined by virtue of the corollary in §4.

Corollary. *Under the assumptions of the theorem, $L(\psi, k)$ is non-negative.*

Remark. According to Greenberg ([1], p. 258), the corollary has the application that the values of $L(\psi_1^{2k-1}, k)$ for a fixed Hecke character ψ_1 of weight one have a well-defined average value, equal to $L(1, \varepsilon)$, as $k \to \infty$.

He points out that this implies the rather weak estimate $L(\psi_1^{2k-1}, k) = o(k)$ and asks whether this can be improved. It might be of interest to see whether this can be done using the above theorem.

§6. Recurrences

The results of the preceding sections imply that both the central values of odd weight Hecke L-series and their square roots can be expressed in terms of non-holomorphic derivatives of classical theta series evaluated at CM points. In particular, for the odd powers of the weight one character ψ_1 introduced at the beginning of the paper, for which $d = 7$ with class number $h(-d) = 1$ and $\delta = 0$ (the only such case!), they give

$$L(\psi_1^{2k-1}, k) = \frac{(2\pi/\sqrt{7})^k}{(k-1)!} \, \partial^{k-1} \Theta \left(\frac{7 + \sqrt{-7}}{14} \right)$$

(28)
$$= \frac{7^{k/2-3/4} \pi^k}{2^{k-3}(k-1)!} \left| \theta_{k-1/2} \left(\frac{1 + \sqrt{-7}}{2} \right) \right|^2,$$

for $k \geq 1$ odd, where $\theta_{1/2}$ is the function (3) and

$$(29) \quad \Theta(z) = \frac{1}{2} \sum_{m,n \in \mathbb{Z}} q^{m^2+mn+2n^2} = \frac{1}{2} + \sum_{n=1}^{\infty} \left(\frac{-7}{n} \right) \frac{q^n}{1 - q^n} \qquad (q = e^{2\pi i z}).$$

In this section we show how to obtain the values $\{\partial^n f(z_0)\}_{n \geq 0}$ of the non-holomorphic derivatives of a modular form f at a CM point z_0 as (essentially) the constant terms of a sequence of polynomials satisfying a recurrence relation. We will illustrate with the case of the full modular group, treating other groups, and the functions occurring in (28), in the next section.

As well as the differential operator D and $\partial_k = D - k/4\pi y$ of §2, we will use the operator

$$\vartheta_k = D - \frac{k}{12} E_2 = \frac{1}{2\pi i} \frac{d}{dz} - \frac{k}{12} E_2(z),$$

where $E_2(z) = 1 - 24 \sum_{n=1}^{\infty} \sigma_1(n) q^n$ is the Eisenstein series of weight 2 on $SL_2(\mathbb{Z})$. As is well-known, this Eisenstein series is not quite modular, but transforms instead by

$$E_2 \left(\frac{az+b}{cz+d} \right) = (cz+d)^2 E_2(z) + \frac{6}{\pi i} c(cz+d) \qquad \left(\begin{pmatrix} a & b \\ c & d \end{pmatrix} \in SL_2(\mathbb{Z}) \right).$$

Equivalently, the function $E_2^*(z) = E_2(z) - 3/\pi y$ $(y = \Im(z))$, though not holomorphic, transforms under the action of $SL_2(\mathbb{Z})$ like a holomorphic

modular form of weight two. It follows that $\vartheta_k f = \partial_k f - k E_2^* f/12$ transforms like a modular form of weight $k + 2$ if $f \in M_k^*(\Gamma)$ for any subgroup
Γ of $SL_2(\mathbb{Z})$. On the other hand, $\vartheta_k f$ is clearly holomorphic if f is, so ϑ_k
maps $M_k(\Gamma)$ to $M_{k+2}(\Gamma)$. The operator D, which does not preserve the
ring $M_*(SL_2(\mathbb{Z})) = \mathbb{C}[E_4, E_6]$, does preserve the larger ring $\mathbb{C}[E_2, E_4, E_6]$.
We have
(30)
$$D(E_2) = \frac{E_2^2 - E_4}{12}, \quad D(E_4) = \frac{E_2 E_4 - E_6}{3}, \quad DE_6(z) = \frac{E_2 E_6 - E_4^2}{2}$$

and hence—since D clearly acts as a derivation—

$$D = \frac{E_2^2 - E_4}{12} \frac{\partial}{\partial E_2} + \frac{E_2 E_4 - E_6}{3} \frac{\partial}{\partial E_4} + \frac{E_2 E_6 - E_4^2}{2} \frac{\partial}{\partial E_6}$$

as a differential operator $\mathbb{C}[E_2, E_4, E_6] \to \mathbb{C}[E_2, E_4, E_6]$. If f is a holomorphic modular form on $SL_2(\mathbb{Z})$, thought of as a weighted homogeneous
polynomial of degree k in E_4 and E_6 (where E_h has weight h), then from
$\partial f/\partial E_2 = 0$ and the Euler equation $4E_4 \partial f/\partial E_4 + 6E_6 \partial f/\partial E_6 = kf$ we
get

(31) $\quad \vartheta f = -\dfrac{E_6}{3} \dfrac{\partial f}{\partial E_4} - \dfrac{E_4^2}{2} \dfrac{\partial f}{\partial E_6} \qquad (f \in M_*(SL_2(\mathbb{Z})) = \mathbb{C}[E_4, E_6])$.

Each of the three differentiation operators D, ∂_k and ϑ_k has advan-
tages over the others: the first preserves holomorphicity and acts in a
simple way on Fourier expansions, but destroys modularity; the second
preserves modularity and acts in a simple way on Fourier expansions, but
destroys holomorphicity; and the third preserves both the properties of
holomorphicity and modularity but has a complicated action on Fourier
expansions. The nicest way to understand the action of these operators
and their iterates is to put them together into three generating series. The
first is the *Kuznetsov-Cohen series*

$$f_D(z, X) = \sum_{n=0}^{\infty} \frac{D^n f(z)}{k(k+1)\cdots(k+n-1)} \frac{X^n}{n!} \qquad (z \in \mathcal{H}, \ X \in \mathbb{C}),$$

where $f \in M_k(\Gamma)$, and the other two are

(32) $\quad f_\partial(z, X) = e^{-X/4\pi y} f_D(z, X) = \displaystyle\sum_{n=0}^{\infty} \frac{\partial^n f(z)}{k(k+1)\cdots(k+n-1)} \frac{X^n}{n!}$

(the second equality is a restatement of equation (8)) and

(33) $\quad f_\vartheta(z, X) = e^{-E_2(z)X/12} f_D(z, X) = e^{-E_2^*(z)X/12} f_\partial(z, X)$.

The Kuznetsov-Cohen series transforms under $\left(\begin{smallmatrix} a & b \\ c & d \end{smallmatrix}\right) \in \Gamma$ by

$$f_D\left(\frac{az+b}{cz+d}, \frac{X}{(cz+d)^2}\right) = (cz+d)^k \, e^{cX/2\pi i(cz+d)} \, f_D(z, X),$$

and the transformation properties of y^{-1} and $E_2(z)$ under Γ imply that f_∂ and f_ϑ satisfy a similar equation but without the exponential factor, which simply says that the nth Taylor coefficient in each of these series transforms like a holomorphic modular form of weight $k+2n$. These Taylor coefficients of f_∂ are the $\partial^n f$, by (32). For f_ϑ they are given by:

Proposition. *Let* $f \in M_k(\Gamma)$ *for some* $\Gamma \subset SL_2(\mathbb{Z})$. *Then*

$$(34) \qquad f_\vartheta(z, X) = \sum_{n=0}^{\infty} \frac{F_n(z)}{k(k+1)\cdots(k+n-1)} \frac{X^n}{n!},$$

where the modular forms $F_n \in M_{k+2n}(\Gamma)$ *are defined recursively by*

$$(35) \qquad F_{n+1} = \vartheta F_n - \frac{n(n+k-1)}{144} E_4 \, F_{n-1} \quad (n \geq 1)$$

with initial conditions $F_0 = f$, $F_1 = \vartheta f$.

Proof. Using (33) we find $F_n = \sum_{\ell=0}^{n} \frac{n!}{\ell!} \binom{n+k-1}{n-\ell} \left(-\frac{E_2}{12}\right)^{n-\ell} D^\ell f$, and the result follows using (30) and (31). ∎

We now illustrate how to get the recursions for the numbers $\{\partial^n f(z_0)\}$ in the simplest case $\Gamma = SL_2(\mathbb{Z})$, $f = E_4$, $z_0 = i$. By the transformation property of E_2^* under Γ and the fact that i is fixed under $z \mapsto -1/z$, we deduce that $E_2^*(i) = 0$ and hence by (33) that $f_\partial(i, X) = f_\vartheta(i, X)$ for any $f \in M_k(SL_2(\mathbb{Z}))$, so $\partial^n E_4(i) = F_n(i)$ where the polynomials $F_0 = E_4$, $F_1 = -\frac{1}{3} E_6$, $F_2 = \frac{5}{36} E_4^2$, $F_3 = -\frac{5}{72} E_4 E_6$, $F_4 = \frac{5}{288} E_4^2 + \frac{5}{216} E_6^2$, ... can be computed by (35) and (31). By homogeneity we can write F_n as $E_4^{n/2+1} f_n(E_6/E_4^{3/2})$ where $f_n \in \mathbb{Z}[\frac{1}{6}][t]$ is a polynomial in one variable (even if n is even and odd if n is odd, and of degree $\leq (n+2)/3$) given recursively by

$$(36) \quad f_0 = 1, \quad f_1 = -\frac{t}{3}, \quad f_{n+1} = \frac{t^2-1}{2} f_n' - \frac{n+2}{6} t f_n - \frac{n(n+3)}{144} f_{n-1}.$$

Since $E_6(i)$ vanishes, we obtain finally

Example. *For* $n \geq 0$ *we have* $\partial^n E_4(i) = f_n(0) \omega^{n/2+1}$ ($= 0$ *if* n *is odd*), *where* $\omega = E_4(i)$ ($= 3\Gamma(\frac{1}{4})^8/(2\pi)^6$) *and* $\{f_n(t)\}$ *are the polynomials defined by (36).*

§7. Examples

To calculate the non-holomorphic derivatives for modular forms on other groups Γ than $SL_2(\mathbb{Z})$ and for other CM points z_0, we choose a function $\phi(z)$ satisfying

 i) $\phi(z)$ is holomorphic;

 ii) $\phi^*(z) = \phi(z) - 1/4\pi y$ transforms like a holomorphic modular form of weight 2 on Γ;

 iii) $\phi(z_0) = 0$.

Condition ii) is equivalent to

 ii') $\phi\left(\dfrac{az+b}{cz+d}\right) = (cz+d)^2\,\phi(z) + \dfrac{c(cz+d)}{2\pi i}$ for all $\left(\begin{smallmatrix} a & b \\ c & d \end{smallmatrix}\right) \in \Gamma$,

and a short calculation then shows that

 iv) the function $\Phi = D\phi - \phi^2$ belongs to $M_4(\Gamma)$, and

 v) if $f \in M_k(\Gamma)$, then $\vartheta_\phi f := Df - k\phi f$ belongs to $M_{k+2}(\Gamma)$.

(In §6, we had $\phi = \frac{1}{12}E_2$, $\Phi = \frac{-1}{144}E_4$, $\vartheta_\phi = \vartheta$.) The analogue of the proposition in §6 is then that the series $e^{-\phi(z)X} f_D(z, X)$ has an expansion as in (35) with $F_0 = f$, $F_1 = \vartheta_\phi f$, and $F_{n+1} = \vartheta_\phi F_n + n(n+k-1)\,\Phi\,F_{n-1}$ for $n \geq 1$.

If we choose an explicit set of generators for the ring $M_*(\Gamma)$ and then express the differential operator ϑ_ϕ in terms of these generators, as was done in (31) for the case of $SL_2(\mathbb{Z})$, we obtain an explicit recursion for the polynomials F_n. The fact that ϕ vanishes at z_0 then shows that $\partial^n f(z_0) = F_n(z_0)$, so we get the numbers $\partial^n f(z_0)$ as special values of a sequence of polynomials satisfying a recursion.

We give the details of this for the functions $\Theta(z)$ and $\theta_{1/2}(z)$ occurring in equation (28), which are modular forms (with multiplier system) on $\Gamma_0(7)$ and $\Gamma_0(2)$, respectively. We give the details $\theta_{1/2}$, since this is the one needed to prove (3) and also because the structure of the corresponding ring of modular forms is simpler.

We abbreviate $\theta = \theta_{1/2}$. By (3), θ^8 is a modular form without character on $\Gamma_0(2)$. (In fact it is the Eisenstein series $\sum_{n\geq 1} n^3 q^n/(1-q^{2n})$.) The ring of modular forms on $\Gamma_0(2)$ is generated by the functions

$$A = A(z) = -E_2(z) + 2E_2(2z) = 1 + 24 \sum_{\substack{n \geq 1 \\ n \text{ odd}}} \frac{nq^n}{1-q^n}$$

and θ^8, of weight 2 and 4, respectively. For instance, we have

$$E_4(z) = A^2 + 192\theta^8, \qquad E_4(2z) = A^2 - 48\,\theta^8.$$

We are interested in the point $z_0 = \dfrac{1+\sqrt{-7}}{2}$. By standard complex multiplication theory,

(37) $\qquad E_2^*(z_0) = \dfrac{3}{7}\Omega^2, \qquad A(z_0) = \dfrac{3}{2}\Omega^2, \qquad \theta(z_0)^8 = -\dfrac{1}{287}\Omega^4,$

with Ω as in equation (1). We therefore choose

$$\phi(z) = \frac{1}{12}\left(E_2(z) - \frac{2}{7}A(z)\right) = \frac{3}{28}E_2(z) - \frac{1}{21}E_2(2z)$$

so that $\phi^*(z) = \phi(z) + 1/4\pi y$ transforms like a form of weight 2 on $\Gamma_0(2)$ and vanishes at z_0. The ϑ_ϕ-derivatives of θ and A are found to be

$$\vartheta_\phi \theta = \frac{2}{21}\theta A, \qquad \vartheta A = 32\,\theta^8 - \frac{5}{42}A^2,$$

so (since ϑ is a derivation) the action of ϑ on either $M_*(\Gamma_0(2)) = \mathbb{C}[\theta^8, A]$ or its 8th degree extension $\mathbb{C}[\theta, A]$ is given by

(38) $\qquad\qquad\qquad \vartheta_\phi = \dfrac{2}{21}\theta A\,\dfrac{\partial}{\partial\theta} + (32\,\theta^8 - \dfrac{5}{42}A^2)\,\dfrac{\partial}{\partial A}\,.$

Also, $\Phi = D\phi - \phi^2$ equals $-(\dfrac{5}{84})^2 A^2 - \dfrac{44}{21}\theta^8$. The above discussion then shows that $\partial^n\theta(z_0) = F_n(z_0)$, where F_n is a weighted homogeneous polynomial in θ and A of weight $2n + \frac{1}{2}$ given inductively by

$$F_0 = \theta, \qquad F_1 = \frac{2}{21}A\theta, \qquad F_{n+1} = \vartheta_\phi F_n + n(n - \frac{1}{2})\Phi F_{n-1} \quad (n \geq 1)$$

with ϑ_ϕ as in (38). By homogeneity we have $F_n = \theta^{4n+1}f_n(A/\theta^4)$ where f_n is a polynomial of the same parity as n, and in terms of these the recurrence becomes

$$f_{n+1}(t) = (32 - \frac{1}{2}t^2)f_n'(t) + \frac{8n+2}{21}t\,f_n(t) - n(n - \frac{1}{2})(\frac{5^2}{84^2}t^2 + \frac{44}{21})f_{n-1}(t),$$

the first few values being

$$f_0 = 1, \quad f_1 = \frac{2}{21}t, \quad f_2(t) = -\frac{19}{4704}t^2 + 2, \quad f_3(t) = -\frac{43}{98784}t^3 + \frac{6}{7}t,$$

and (28) and (37) tell us that the numbers

$$B(2n + 1) = \frac{(-\sqrt{-7})^n}{2^{3n+1}}f_n(24\sqrt{-7}) = \frac{1}{2},\ 1,\ -1,\ -3,\ 7,\ -315,\ -609,\ \ldots$$

satisfy equation (2). A slightly more convenient choice of normalization turns out to be $F_n = 2\,(A/21)^n\,\theta\,b_n(441\,\theta^8/A^2 + 7/64)$ rather than $F_n = \theta^{4n+1}f_n(A/\theta^4)$, so that

$$b_0 = \frac{1}{2}, \quad b_1 = 1, \quad b_2 = x - 1, \quad b_3 = 9x - 3, \quad b_4 = -2x^2 + 133x + 7, \quad \ldots\ .$$

With this choice the $B(2n + 1)$ are simply the constant terms $b_n(0)$ and the $b_n(x)$ satisfy (5).

For the other assertion of the theorem in §1, giving the numbers $A(k)$ of (1) in terms of the recursion (4), we must use the formula $A(k) = \partial^{k-1}\Theta(z_1)/2\Omega^{2k-1}$, where $z_1 = (7 + \sqrt{-7})/14$ (this is just a restatement of (1) and (28)). The calculations here are more complicated because the ring of modular forms with character $M_*\left(\Gamma_0(7), \left(\frac{-7}{\cdot}\right)^*\right)$ is not free, but is generated by the function Θ defined in (29) together with the two weight three forms

$$\Theta^{(3)}(z) = \frac{1}{8} \sum_{\substack{r,\,s \in \mathbb{Z} \\ r \equiv s \;(\mathrm{mod}\ 2)}} (r^2 - 7s^2)\, q^{(r^2+7s^2)/4}\,,$$

$$E(z) = 1 - \frac{7}{8} \sum_{n=1}^{\infty} \left(\sum_{d|n} (d^2 + 7\frac{n^2}{d^2})\left(\frac{-7}{d}\right)\right) q^n$$

subject to the relation $E^2 = (\Theta^3 + \Theta^{(3)})(\Theta^3 - 27\Theta^{(3)})$. When we write the non-holomorphic derivatives $\partial^n\Theta(z_1)$ as the values of a sequence of polynomials, then these are polynomials in three algebraically dependent variables. We can use the homogeneity to write them as simple factors times polynomials in $x = \Theta^{(3)}/\Theta^3$ and $\sqrt{(1+x)(1-27x)} = E/\Theta^3$, and after some computation we obtain (4). We leave the details to the reader.

Remarks. 1. According to the recursion (4), the polynomials a_{2n} have coefficients in $\mathbb{Z}[\frac{1}{3}]$, and this is actually true: the first values are 1, $(2 - 34x)/9$, and $(8-218x+314x^2+432x^3)/27$. It is not clear on an elementary level why their values at $x = 1$ are integers (let alone squares). Similarly, the recursion (5) involves dividing by 21 at each stage, but in fact the polynomials $b_n(x)$ apparently belong to $\mathbb{Z}[x]$.

2. It is now clear that we cannot expect any simple relations between the polynomials $a_{2n}(x - 1)$ and $b_n(x)$, even though the constant terms of one are the squares of the constant terms of the other: the variables "x" have completely different meanings in the two polynomials, being a modular function on $X_0(7)$ in the one case and a modular function on $X_0(2)$ in the other. The identity of the constant terms has to do with the way these two modular curves intersect in the Siegel 3-fold $\mathcal{H}_2/Sp_4(\mathbb{Z})$.

Finally, we mention that in the 5 cases $d = 11, 19, 43, 67$ and 163 with $h(-d) = 1$ and $\delta = 1$ the calculation is even easier than the case treated here with $\delta = 0$, since now the function whose non-holomorphic derivatives must be computed at CM points is $\theta_{3/2}$, which is a modular form (with multiplier system) for the full modular group (cf. equation (15)). Here the recursions obtained are essentially the classical recursions for the Taylor coefficients of the Weierstrass σ-function, as given for instance in [3], Chap. VII, p. 237.

§8. Twists

We finish by mentioning a result on the central values of quadratic twists of Hecke L-series. For simplicity we consider only the case of equation (7), i.e. twists of the L-series for the grossencharacters of conductor $(\sqrt{-7})$ of $\mathbb{Q}(\sqrt{-7})$ by Legendre symbols attached to primes $p \equiv 1 \pmod 4$. Then one of us (F.V.) has found a formula giving integers $B(k,p)$ such that (7) holds. In the simplest case $k = 1$ this formula is

$$B(1,p) = \frac{\sum_n \chi(n^2 + 7)\,\theta_{1/2}\big(\dfrac{n + \sqrt{-7}}{2p}\big) + \sqrt{p}\,\theta_{1/2}\big(\dfrac{p + p\sqrt{-7}}{2}\big)}{S(\chi)\,G(\chi)\,\theta_{1/2}\big(\dfrac{1 + \sqrt{-7}}{2}\big)},$$

with $\theta_{1/2}(z)$ as in (3); here the sum is over $n \pmod{16p}$ satisfying $n \equiv 1 \pmod{16}$, $\chi(n)$ is one of the two quartic characters modulo p, $G(\chi) = \sum_{n \pmod p} \chi(n)e^{2\pi i n/p}$ the associated Gaussian sum, and $S(\chi) = 2,\, 1 - i,\, 2i$ or $1 + i$ according as $\chi(7) = 1,\, i,\, -1$ or $-i$ (so that $\chi(7)S(\chi)^2 = 3 + \big(\dfrac{7}{p}\big)$). For $p < 200$ this gives the values

p	5	13	17	29	37	41	53	61	73	89	97	101
$B(1,p)$	1	−1	−1	−1	1	1	0	1	3	3	−1	−1

p	109	113	137	149	157	173	181	193	197
$B(1,p)$	−1	2	1	0	−3	1	1	0	0

This will be discussed in a later publication. We have not succeeded in correlating these numbers with the Fourier coefficients of a modular form of weight $3/2$.

REFERENCES

[1] R. Greenberg, On the Birch and Swinnerton-Dyer conjecture, Invent. math. **72** (1983), 241–265.

[2] B.H. Gross and D. Zagier, On the critical values of Hecke L-series, in *Fonctions abéliennes et nombres transcendants* (eds. D. Bertrand and M. Waldschmidt), Mém. Soc. Math. de France **108** (1980), 49–54.

[3] G.H. Halphen, *Traité des fonctions elliptiques et de leurs applications*, Gauthier-Villars, Paris, 1896.

[4] N. Koblitz, p-adic congruences and modular forms of half integral weight, Math. Annalen **274** (1986), 199–220.

[5] L. Kronecker, Zur Theorie der Elliptischen Funktionen, Sitzungsber. d. Königl. Akad. d. Wiss. (1883), L. Kronecker's *Mathematische Werke*, Leipzig, 1895, Vol. IV, 345–496.

[6] K. Rubin, p-adic variants of the Birch and Swinnerton-Dyer conjecture for elliptic curves with complex multiplication, preprint, Ohio State Mathematical Research Institute (1992).

[7] W. Schempp, *Harmonic analysis on the Heisenberg nilpotent Lie group, with appli-cations to signal theory*, Longman, Harlow 1986.

[8] F. Rodriguez Villegas, On the square root of special values of certain L-series, Inv. math. **106** (1991), 549–573.

Fernando Rodriguez Villegas
Department of Mathematics
Fine Hall
Princeton University
Princeton, NJ 08544
USA
e-mail villegas@math.Princeton.edu

and

Don Zagier
Max-Planck-Institut für Mathematik
Gottfried-Claren-Strasse 26
D-5300 Bonn 3
Germany
e-mail zagier@mpim-bonn.mpg.de

Chapter II

Analytic Number Theory

Invited Addresses

and

Contributed Papers

A Formula for the Fourier Coefficients of Maass Cusp Forms

Henryk Iwaniec

Rutgers University

1 Introduction

In his seminal paper [5] H. Maass extended the ghetto of classical cusp forms by automorphic forms which are not holomorphic but, instead, are eigenfunctions of the Laplace operator Δ. For a group with parabolic elements those Maass forms which decay exponentially to zero at every cusp are called cusp forms. The Maass cusp forms remain wrapped in a shroud of mystery; it is not even known today whether they exist for a generic group. However, for any congruence group the spectral resolution of Δ abounds with Maass cusp forms. This is shown by a counting argument using Selberg's trace formula, thus it is not constructive. Surprisingly enough, not a single Maass cusp form has ever been constructed for the modular group.

A convenient access to a Maass cusp form $f(z)$ goes via its Fourier expansion

$$(1) \qquad f(z) = \sum_{m \neq 0} f_m W(mz) \quad,$$

where $W(z)$ is the Whittaker function. If $f(z)$ is an eigenfunction of Δ with eigenvalue $\lambda = \frac{1}{4} - \nu^2$ then $W(z)$ is given by

$$(2) \qquad W(z) = 2|y|^{\frac{1}{2}} K_\nu \left(2\pi|y|\right) e(x) \quad,$$

it has the property

$$(3) \qquad W(z + x) = e(x)W(z) \quad \text{for all} \quad x \in \mathbb{R}$$

and it satisfies the asymptotic formula

$$(4) \qquad W(z) \sim e(z) \quad \text{as} \quad y \to \infty \quad.$$

The complex numbers f_m are called the Fourier coefficients of $f(z)$, they received most attention in the whole theory. Here the basic question is how large the coefficients can be. The Ramanujan-Petersson conjecture asserts (in our notation) that

$$(5) \qquad f_m \ll m^{-\frac{1}{2}} \tau(m) \ ,$$

while the sharpest bound established so far in [1] is

$$(6) \qquad f_m \ll m^{-\frac{9}{28}} \tau(m) \ .$$

Another interesting problem about the Fourier coefficient f_m comprises finding an explicit, exact formula. In the case of the classical cusp forms of weight k, say, the problem reduces to that for the Poincaré series since the latter span the whole space of cusp forms (it is a finite-dimensional space). Let $P_n(z)$ be the n-th Poincaré series of weight k for the modular group and let

$$(7) \qquad P_n(z) = \sum_{1}^{\infty} P_{mn} e(mz)$$

be its Fourier expansion. H. Petersson [6], R. Rankin [8] and A. Selberg [12] independently have expressed the m-th coefficient P_{mn} as a sum of Kloosterman sums

$$(8) \qquad S(m, n; c) = \sum_{ad \equiv 1 (\mathrm{mod}\, c)} e\left(\frac{am + dn}{c}\right) \ .$$

More precisely they showed the formula (in our notation)

$$(9) \qquad P_{mn} = \left(\frac{m}{n}\right)^{\frac{k-1}{2}} \left\{ \delta_{mn} + 2\pi i^k \sum_{c=1}^{\infty} c^{-1} S(m, n; c) J_{k-1}\left(\frac{4\pi\sqrt{mn}}{c}\right) \right\} \ ,$$

where δ_{mn} is the diagonal symbol of Kronecker and $J_{k-1}(x)$ is the Bessel function.

The Petersson-Rankin-Selberg formula (9) has helped to establish a great variety of results about the Fourier coefficients of classical cusp forms. It suffices to mention a few most recent applications by R. Rankin [9], Z. Rudnick [10] and N. Pitt [7].

By striking analogy with the original Poincaré series A. Selberg [13] has introduced non-holomorphic Poincaré series for which he developed a Fourier expansion similar to (9). The Selberg-Poincaré series live in the Hilbert space of square-integrable automorphic functions (they are not Maass forms since they barely fail to be eigenfunctions of Δ). As in the classical case the projections of the Selberg-Poincaré series on a Maass cusp form yield the Fourier coefficients of the cusp form. This fact does not suffice to express a single coefficient as a sum of Kloosterman sums because the Hilbert space has infinite dimension. Nevertheless, using the spectral theorem it was possible for N.V. Kuznetsov [4] to get a relevant expression by allowing a suitable spectral averaging. Kuznetsov's formula is derived by processing Selberg's data with certain integral transforms (see [11]) to create a class of handy test functions. A considerable flexibility was achieved this way, making the formula ready for combinations with various arithmetical means.

In this paper we give a formula for an individual Fourier coefficient f_m of a Maass cusp form $f(z)$ for the modular group. The method would work for the Maass forms as well as for the holomorphic forms with respect to the Hecke congruence group $\Gamma_0(q)$ but we abandoned such generality for the sake of simplicity. Governed by practical values of the formula we have chosen to show only a good approximation rather than an exact identity, for it is easier to round off the intermediate expressions than the resulting terms of the identity. Our approach is direct and it can be viewed as a simpler alternative to the circle method applied in [3].

2　The delta-symbol

To select the n-th term of the series (1) we need a Fourier expansion for the symbol

$$(10) \qquad \delta(n) = \begin{cases} 1 & \text{if } n = 0 \\ 0 & \text{if } n \neq 0 \end{cases} .$$

Let $\omega(t)$ be a smooth and compactly supported function on $\mathbb{R}$ such that

$$(11) \qquad \omega(t) = \omega(-t)$$

$$(12) \qquad \omega(0) = 0$$

and

$$(13) \qquad \sum_{1}^{\infty} \omega(k) = 1 .$$

Put

$$(14) \qquad \delta_t(n) = \omega(t) - \omega\left(\frac{n}{t}\right) .$$

We then have by (12) and (13),

$$(15) \qquad \delta(n) = \sum_{k|n} \delta_k(n) .$$

We detect the divisibility $k|n$ by means of additive characters and get

$$\delta(n) = \sum_k k^{-1} \sum_{a(\mathrm{mod}\, k)} e\left(\frac{an}{k}\right) \delta_k(n) \ .$$

Hence, using (3)

$$
\begin{aligned}
f_n W(nz) &= \sum_m \delta(n-m) f_m W(mz) \\
&= \sum_k k^{-1} \sum_{a(\mathrm{mod}\, k)} e\left(\frac{an}{k}\right) \sum_m f_m W\left(m\left(z - \frac{a}{k}\right)\right) \delta_k(n-m) \ .
\end{aligned}
$$

Here we have

$$\delta_k(n-m) = \omega(k) - \int_{\mathbb{R}} \hat\omega(u) e\left(u\frac{n-m}{k}\right) du \ ,$$

where

$$\hat\omega(u) = \int_{\mathbb{R}} \omega(t) e(ut)\, dt$$

is the Fourier transform of $\omega(t)$. Thus the innermost sum is equal to

$$\sum_m = \omega(k) f\left(z - \frac{a}{k}\right) - \int_{\mathbb{R}} \hat\omega(u) e\left(\frac{un}{k}\right) f\left(z - \frac{a+u}{k}\right) du \ .$$

Put

$$r(z,u) = f(z-u) - f(z) + u f_x(z) \ ,$$

where f_x denotes the partial derivative in the x variable. We get

$$
\begin{aligned}
\sum_m = \delta_k(n) f\left(z - \frac{a}{k}\right) &- \frac{1}{2\pi i k}\omega'\left(\frac{n}{k}\right) f_x\left(z - \frac{a}{k}\right) \\
&+ \int_{\mathbb{R}} \hat\omega(u) e\left(\frac{nu}{k}\right) r\left(z - \frac{a}{k}, \frac{u}{k}\right) du \ .
\end{aligned}
$$

Accordingly, we write $f_n W(nz) = P_n(z) + Q_n(z) + R_n(z)$, say, where

$$
\begin{aligned}
P_n(z) &= \sum_k k^{-1} \delta_k(n) \sum_{a(\mathrm{mod}\, k)} e\left(\frac{an}{k}\right) f\left(z - \frac{a}{k}\right) \ , \\
Q_n(z) &= \sum_k (2\pi i k^2)^{-1} \omega'\left(\frac{n}{k}\right) \sum_{a(\mathrm{mod}\, k)} e\left(\frac{an}{k}\right) f_x\left(z - \frac{a}{k}\right) \ , \\
R_n(z) &= \int_{\mathbb{R}} \hat\omega(u) \sum_k k^{-1} \sum_{a(\mathrm{mod}\, k)} e\left(n\frac{a+u}{k}\right) r\left(z - \frac{a}{k}, \frac{u}{k}\right) du \ .
\end{aligned}
$$

3 Evaluation of $P_n(z)$

Reducing the fractions $\frac{a}{k}$ we write

$$P_n(z) = \sum_c c^{-1}\Delta_c(n) \sum_{a(\mathrm{mod}\,c)}^{*} e\left(\frac{an}{c}\right) f\left(z - \frac{a}{c}\right) ,$$

where

(16) $$\Delta_c(n) = \sum_r r^{-1}\delta_{cr}(n) = \sum_r r^{-1}\left(\omega(cr) - \omega\left(\frac{n}{cr}\right)\right) .$$

Then we apply the modular relation

$$f\left(z - \frac{a}{c}\right) = f\left(\frac{-1}{c^2 z} + \frac{d}{c}\right) ,$$

where $ad \equiv 1 \pmod c$, and the Fourier expansion (1) giving sums of Kloosterman sums

(17) $$P_n(z) = \sum_m f_m \sum_c c^{-1} S(m, n; c) \Delta_c(n) W\left(\frac{-m}{c^2 z}\right) .$$

4 Estimates for $Q_n(z)$

One could evaluate $Q_n(z)$ as $P_n(z)$ using the modular relation but we prefer to give good estimates instead. Suppose the test function $w(t)$ satisfies the following conditions:

(18) $$w(t) \text{ is supported in } T < |t| < 2T$$

and

(19) $$w^{(j)}(t) \ll T^{-j-1} \text{ if } 0 \leq j \leq 4 .$$

Then we have trivially

(20) $$Q_n(z) \ll T^{-2} .$$

We can do better on average, namely that

(21) $$\sum_{N<|n|<2N} |Q_n(z)|^2 \ll N^{-1}T^{-2} + T^{-4} .$$

Proof. Reducing the fractions $\frac{a}{k}$ we get

$$Q_n(z) = \sum_c (2\pi i c^2)^{-1} \gamma_c(n) \sideset{}{^*}\sum_{a(\bmod c)} e\left(\frac{an}{c}\right) f_x\left(z - \frac{a}{c}\right) ,$$

where

$$\gamma_c(n) = \sum_r r^{-2} \omega'\left(\frac{n}{cr}\right) .$$

Next we execute the summation in r by the Euler-Maclaurin formula

$$\sum_r F(r) = \int F(t)\, dt + \int \psi(t)\, dF(t) ,$$

where $\psi(t) = t - [t] - \frac{1}{2}$, which is valid for any smooth, compactly supported function $F(t)$. Since the mean-value of $F(t) = t^{-2} \omega'\left(\frac{n}{ct}\right)$ vanishes this yields

$$\gamma_c(n) = c^2 \int \psi\left(\frac{t}{c}\right) d\left(t^{-2}\omega'\left(\frac{n}{t}\right)\right)$$

by a change of variables. Hence we obtain

$$Q_n(z) = \frac{1}{2\pi i} \int \sum_c \sideset{}{^*}\sum_{a(\bmod c)} e\left(\frac{an}{c}\right) f_x\left(z - \frac{a}{c}\right) \psi\left(\frac{t}{c}\right) d\left(t^{-2}\omega'\left(\frac{n}{t}\right)\right) .$$

Note that $\gamma_c(n)$ vanishes if $cT > |n|$. Thus we can restrict the summation to $c \leq C = 2NT^{-1}$. We get

$$\sum_n |Q_n(z)|^2 \ll N^{-4} \sum_n \left| \sum_c \sideset{}{^*}\sum_{a(\bmod c)} \lambda\left(\frac{a}{c}\right) e\left(\frac{an}{c}\right) \right|^2$$

with some bounded complex numbers $\lambda\left(\frac{a}{c}\right)$. By the large sieve inequality this gives

$$\ll N^{-4}\left(N + C^2\right) C^2 \ll N^{-1} T^{-2} + T^{-4}$$

completing the proof of (21).

5   An estimate for $R_n(z)$

Since $\hat{\omega}(u) \ll T^{-1}(1 + |u|T)^{-4}$ by (18) and (19) it follows trivially that

(22) $$R_n(z) \ll T^{-3} .$$

Remarks. Taking more terms in the Taylor expansion for $f\left(z - \frac{a+u}{k}\right)$ one can establish an asymptotic expansion for R_n. One can also establish an exact formula by integrating over z in the upper half-plane against a suitable kernel function, however, the attempts made so far did not lead us to anything useful and elegant simultaneously. Concerning the estimate (22) there is no point to improve it further because it matches (21) when $N = T^2$ which is the best choice at any rate.

6 Main Results

Gathering together the above evaluations we get our main result

Theorem 1. *Let $f(z)$ be a Maass cusp form for the modular group and $\omega(t)$ be any test function satisfying the conditions (11)-(13) and (18)-(19). We have*

$$(23) \qquad f_n W(nz) = \sum_m f_m \sum_c c^{-1} S(m, n; c) \Delta_c(n) W\left(\frac{-m}{c^2 z}\right) + \rho_n(z) \ ,$$

where $\Delta_c(n)$ is given by (16) and the error term $\rho_n(z)$ satisfies the bound

$$(24) \qquad \sum_{N < |n| < 2N} |\rho_n(z)|^2 \ll N^{-1} T^{-2} + N T^{-6} \ ,$$

which is uniform in z.

The free variable z in (23) makes the formula flexible. Taking $z = iyn^{-1}$ gives

Corollary 1. *For any $n \neq 0$ and $y > 0$ we have*

$$(25) \qquad f_n W(iy) = \sum_m f_m \sum_c c^{-1} S(m, n; c) \Delta_c(n) W\left(\frac{imn}{yc^2}\right) + \rho_n \ ,$$

where the error term ρ_n satisfies the bound (24).

Corollary 2. *For any $n \neq 0$ we have*

$$(26) \qquad f_n = \sum_m f_m \sum_c c^{-1} S(m, n; c) \Delta_c(n) K\left(\frac{4\pi\sqrt{|mn|}}{c}\right) + \rho_n$$

where

$$(27) \qquad\qquad K(x) = x K_{2\nu}(x) \nu^{-1} \sin \pi\nu$$

and the error term ρ_n satisfies the bound (24).

Proof. Multiply both sides of (25) by $W(iy)$ and then integrate. On the left-hand side we get

$$\int_0^\infty W(iy)^2 \, dy = \int_0^\infty 4y K_\nu(2\pi y)^2 \, dy = \nu(2\pi \sin \pi\nu)^{-1}$$

and on the right-hand side we apply the duplication formula

$$\int_0^\infty K_\nu(y) K_\nu\left(\frac{a}{y}\right) \, dy = \pi K_{2\nu}\left(2\sqrt{a}\right) \ ,$$

(see 6.514.8 on p. 669 of [2]) giving (26).

Remarks. In some situations it may be convenient to transform $\Delta_c(n)$ by the Euler-Maclaurin formula to separate c from n, it gives

$$(28) \qquad \Delta_c(n) = c \int_0^\infty \psi\left(\frac{t}{c}\right) dt^{-1}\delta_t(n) \ \text{ if } \ n \neq 0 \ .$$

This also shows that

$$(29) \qquad \Delta_c(n) \ll cn^{-1} + cT^{-2} \ .$$

Note that $\Delta_c(n)$ vanishes for $c > \max\left(2T, |n|T^{-1}\right)$. Choosing $T^2 \sim n$ we are left with $c \ll \sqrt{n}$. On the other hand we have $W(iy) \ll \exp\left(-2\pi|y|\right)$ by (4), so the terms in (25) with $c^2 \ll |mn|(\log|mn|)^{-1}$ can be ignored at a small cost. Therefore, the only significant contribution in (25) comes from low frequencies $m \ll \log n$ and moduli in the interval

$$n^{\frac{1}{2}}(\log n)^{-\frac{1}{2}} \ll c \ll n^{\frac{1}{2}} \ .$$

The formulas (25) and (26) are as powerful as the Petersson-Rankin-Selberg formula (9) for the coefficients of holomorphic Poincaré series and they may serve as a substitute for Kuznetsov's formula when no spectral averaging is desired.

Applying Weil's bound for the Kloosterman sum

$$|S(m,n;c)| \leq (m,n,c)^{\frac{1}{2}}c^{\frac{1}{2}}\tau(c)$$

the formula (26) gives

$$f_n \ll n^{-\frac{1}{4}}\log n \ .$$

References

1. D. Bump, W. Duke, J. Hoffstein and H. Iwaniec, An estimate for the Hecke eigenvalues of Maass forms, *Duke Math. J.* **66** (1) (1992), Research Notices, 75-81.

2. I.S. Gradshteyn and I.M. Ryzhik, Table of Integrals, Series and Products, Academic Press, 1965.

3. H. Iwaniec, The circle method and the Fourier coefficients of modular forms, in *Number Theory and Related Topics*, Tata Institute and Oxford University Press, 1989, pp. 47-55.

4. N.V. Kuznetsov, Petersson hypothesis for parabolic forms of weight zero and Linnik hypothesis. Sums of Kloosterman sums, *Math. Sbornik* **111** (1980), 334-383.

5. H. Maass, Über eine neue Art von nichtanalytischen automorphen Funktionen, *Math. Ann.* **121** (1949), 141-183.

6. H. Petersson, Über die Entwicklungskoeffizienten der automorphen Formen, *Acta Math.* **58** (1932), 169-215.

7. N.J.E. Pitt, Shifted convolutions of ζ^3 with automorphic L-series, Talk given at the Workshop on Discrete Groups, Number Theory and Ergodic Theory, MSRI, Berkeley, 1991.

8. R.A. Rankin, Modular Forms of Negative Dimensions, Dissertation, Clare College Cambridge, 1940.

9. R. Rankin, The vanishing of Poincaré series, *Proc. Edinburgh Math. Soc.* **23** (1980), 151-161.

10. Z. Rudnick, Poincaré Series, Dissertation, Yale University, 1990.

11. D.B. Sears and E.C. Titchmarsh, Some eigenfunction formulae, *Quart. J. Math. Oxford (2)*, **1** (1950), 165-75.

12. A. Selberg, Über die Fourierkoeffizienten elliptischer Modulformen negativer Dimension, *C.R. Neuvième Congrès Math*, Scandinaves, Helsingfors (1938), 320-322. Mercatorin Kirjapaino, Helsinki 1939.

13. A. Selberg, On the estimation of Fourier coefficients of modular forms, *Proc. Symposia in Pure Math. VIII*, AMS, Providence 1965, 1-15.

Henryk Iwaniec
Department of Mathematics
Rutgers University
New Brunswick
NJ 08903, USA
e-mail bgonzal@math.rutgers.edu

The Additive Divisor Problem and Exponential Sums

M. Jutila

University of Turku

1   Introduction

In its simplest form, the Weyl-van der Corput method for ordinary exponential sums $\sum e(f(n))$, where $e(\alpha) = e^{2\pi i \alpha}$, consists of Weyl's "smooth
and square" argument followed by a transformation by Poisson's summa-
tion formula. The exponential sums to be transformed are of the form

$$\sum_n e(f(n+k) - f(n)), \tag{1.1}$$

where the shift k is essentially smaller than n. Our object in this paper is
to carry out an analogous procedure for exponential sums

$$\sum_{N_1}^{N_2} d(n)g(n)e(f(n)) \tag{1.2}$$

involving the divisor function $d(n)$ and some functions f and g satisfying
certain regularity conditions. Because it is easier to work with smoothed
sums, we equip the outer sum in Weyl's method - taken over squares of
short sums - with a suitable weight function $\eta(n)$, which is a smoothed
version of the characteristic function of the interval of summation. Then,
sums of the type

$$\sum_n \eta(n)d(n)d(n+k)\overline{g(n)}g(n+k)e(f(n+k) - f(n)) \tag{1.3}$$

will play the role of the sums (1.1). Compared with (1.2), the sum (1.3) is
analytically easier because the exponential factor is a more slowly oscillating
function of n. On the other hand, this sum is *arithmetically* more complicated because of the appearance of the arithmetic function $d(n)d(n+k)$
familiar from the additive divisor problem.

   Exponential sums involving the Weyl shift appear in a natural way
in mean value problems. For if the mean square of an exponential sum

$\sum e(f(n,t))$ with respect to the parameter t over an interval is calculated as

$$\sum_{m,n} \int e(f(m,t) - f(n,t))\, dt,$$

then typically only those terms with m and n sufficiently near each other will be of significance. Putting now $m = n+k$, sums of the type (1.1) appear under the integral sign. Note that a sum $\sum e(f(n))$ without a parameter can be understood as $\sum e((t/T)f(n))$ for $t = T$, and its mean square over an interval $[T - T_0, T + T_0]$ leads to sums essentially of the form (1.1). Thus the underlying idea of Weyl's trick consists in estimating a function at a point by the mean square of this function over a neighbourhood of this point. Of course, the same reasoning applies to exponential sums involving any coefficients, in particular $d(n)$. We developed recently (see [7–10]) a method to deal with mean value problems for such (and related) exponential sums. Our argument was in principle of classical and relatively elementary nature, being based on devices such as Voronoï's summation formula (a generalized version), Farey intervals, and two innovations of E. Bombieri and H. Iwaniec (the multidimensional large sieve and a lemma on pairs of rationals). In contrast, the method to be adopted in the following is less elementary for it makes essential use of the spectral theory of the hyperbolic Laplacian. On the other hand, it leads to deeper results, for instead of mere *estimates* it gives *explicit formulae*, curiously similar to those occurring in prime number theory. In this analogy, eigenvalues of the hyperbolic Laplacian correspond in a mysterious way to the zeros of Riemann's zeta-function. Our approach was motivated and inspired by the recent important work of Y. Motohashi [14] on the fourth power moment of Riemann's zeta-function over a short segment on the critical line (for applications of Motohashi's formula, see [4]). The fourth moment of $|\zeta(1/2 + it)|$ is closely related to the mean square of Dirichlet polynomials $\sum d(n)n^{-1/2-it}$, so it is natural to expect an analogue of Motohashi's formula to hold for the latter, and then for more general exponential sums as well. This will be our point of view, besides attempting to find an approach as simple, general, and straightforward as possible.

Actually there is a spectral theoretic result for the transformation of the sum (1.3), namely a very interesting identity due to N. V. Kuznetsov [13] for the sum

$$\sum_{n=1}^{\infty} d(n)d(n+k)W(n) \tag{1.4}$$

where k is a positive integer and W is a smooth function supported in the positive real numbers; see also Motohashi [15] for an alternative and more detailed proof. (Strictly speaking, Kuznetsov formulates his result in terms of $W(n/k)$ instead of $W(n)$, but this is of course just a technicality).

The proof of this identity proceeds via Kloosterman sums and Kuznetsov's trace formula. We take a somewhat different line of argument avoiding these tools. Our goal is to derive an approximate transformation formula for exponential sums of the type

$$S = \sum \eta(n)d(n)d(n+k)g(n)e(f(n)) \tag{1.5}$$

in terms of the discrete spectrum of the hyperbolic Laplacian (for notational simplicity, we keep on using the same symbols f and g even in this new context different from (1.2)). By the Mellin inversion formula, this sum can be written as a complex integral involving the Dirichlet series

$$\zeta_k(s) = \sum_{n=1}^{\infty} d(n)d(n+k)n^{-s}. \tag{1.6}$$

However, the automorphic function theory produces more naturally a modified function $\zeta_k^*(s)$ which is more convenient to work with. A function closely related to our $\zeta_k^*(s)$ was analyzed by L. A. Tahtadjan and A. I. Vinogradov [17]. The relevant properties of $\zeta_k^*(s)$ will be considered in §2. It can be written in two forms (Lemmas 1 and 2), one of these representations being of number theoretic and the other of spectral theoretic nature. Combining these two aspects is, in fact, the key of the whole argument.

Proceeding to summation formulae, we first derive, in §3, an identity (Lemma 3) for the sum (1.4). It shows that $\zeta_k^*(s)$ really may take the role of $\zeta_k(s)$. This identity is applied to the sum S in §4, and the main term of the resulting formula (Lemma 4) is a complex integral which is evaluated by the theorem of residues in terms of the spectrum of the hyperbolic Laplacian. The essential contribution is given by the discrete spectrum, that of the continuous spectrum being estimated as an error term. The resulting formula is stated as a theorem in the end of §4. Then, in §5, this formula is used to carry out the transformation step in the Weyl - van der Corput method. The case where the original sum (1.2) is a Dirichlet polynomial is worked out explicitly as an illustration. The expression for S we end up with in this special case is analogous to that of Y. Motohashi [14] for a weighted average of $|\zeta(1/2+it)|^4$ over a short interval. Thus it is seen that our approach and that of Kuznetsov - Motohashi lead to similar results. However, we appeal only to rather general principles, without making any explicit use of the arithmetic meaning of the function $d(n)$ (it just appears in Fourier coefficients of an Eisenstein series), so there are chances for generalizations. Our previous method could be carried over as such to exponential sums involving Fourier coefficients $a(n)$ of a cusp form, and the same seems to be the case with the present spectral theoretic method as well. Indeed, spectral methods have been applied by A. Good [3] to the sum function of $a(n)a(n+k)$ (an analogue of the additive divisor problem concerning the sum function of $d(n)d(n+k)$), and in principle exponential sums with $a(n)a(n+k)$ or $d(n)d(n+k)$ as coefficients can be related to such sum functions simply by summation by parts. We could follow this line of argument, but for technical reasons we prefer working with Mellin transforms of smooth functions. However, the final result can be interpreted in a natural way from the point of view of the additive divisor problem.

Acknowledgment. I am most grateful to Professors Aleksandar Ivić and Yoichi Motohashi for kindly putting unpublished material at my disposal and for their valuable comments concerning the present work.

2   The function $\zeta_k^*(s)$

2.1   Eisenstein series and the function $\zeta_k^*(s)$

The latter function will be defined in terms of the (non-holomorphic) *Eisenstein series*

$$
\begin{aligned}
E(z,s) &= \sum_{\gamma \in \Gamma_\infty \backslash \Gamma} y^s(\gamma(z)) \\
&= y^s + \frac{1}{2} \sum_{(c,d)=1,\ c \neq 0} \frac{y^s}{|cz+d|^{2s}},
\end{aligned}
$$

where Re $s = \sigma > 1$, $z = x + yi$ lies in the upper half-plane $y > 0$, Γ is the modular group, Γ_∞ is the stabilizer of the cusp $i\infty$ in Γ, and $y(\gamma(z))$ denotes the imaginary part of $\gamma(z)$. Let $\xi(s) = \pi^{-s/2}\Gamma(s/2)\zeta(s)$, and define $E^*(z,s) = \xi(2s)E(z,s)$. It is well-known that $E^*(z,s)$ can be analytically continued to a meromorphic function of s satisfying the functional equation $E^*(z,1-s) = E^*(z,s)$. Indeed, these properties readily follow from the Fourier series (see [11], p. 46)

$$
E^*(z,s) = \xi(2s)y^s + \xi(2-2s)y^{1-s} + 2\sqrt{y} \sum_{\substack{n=-\infty \\ n \neq 0}}^{\infty} \tau_s(n)K_{s-1/2}(2\pi|n|y)e(nx),
$$

(2.1)

where $\tau_s(n) = |n|^{s-1/2}\sigma_{1-2s}(n)$ with $\sigma_\alpha(n) = \sum_{d|n} d^\alpha$, and the K - Bessel function is defined by

$$
K_\nu(x) = \int_0^\infty e^{-x\cosh t}\cosh(\nu t)\,dt. \tag{2.2}
$$

The Fourier series of the function $E^*(z) = E^*(z,1/2)$ involves the divisor function $d(n)$, for (2.1) gives

$$E^*(z) = \sqrt{y}(\log y - c) + 2\sqrt{y} \sum_{\substack{n=-\infty \\ n \neq 0}}^{\infty} d(|n|)K_0(2\pi|n|y)e(nx) \qquad (2.3)$$

with $c = 2\log 2 + \log \pi - \gamma$, where γ is Euler's constant. Define now, for $\sigma > 1$,

$$\zeta_k^*(s) = \frac{(4\pi)^{s-1}s}{2\Gamma(s)} \int_{\Pi} |E^*(z)|^2 e(kz)y^s\, d\mu(z), \qquad (2.4)$$

where $d\mu(z) = dx\,dy/y^2$ is the invariant hyperbolic measure and Π is the strip $|x| \leq 1/2$, $y > 0$. Following [17], we are going to derive two expressions for $\zeta_k^*(s)$ by the Rankin - Selberg method.

2.2 An arithmetic representation for $\zeta_k^*(s)$

The first expression follows from (2.3), which gives evidently

$$\zeta_k^*(s) = \sum_{n=1}^{\infty} d(n)d(n+k)\psi(s; n+k, n, k)$$

$$+ \frac{1}{2} \sum_{n=1}^{k-1} d(n)d(k-n)\psi(s; n, k-n, k) + d(k)\psi_0(s, k), \qquad (2.5)$$

where

$$\psi(s; m, n, k) = \frac{2^s s}{\pi\Gamma(s)} \int_0^{\infty} K_0(my)K_0(ny)e^{-ky}y^{s-1}dy \qquad (2.6)$$

and

$$\psi_0(s, k) = \frac{2^s s}{2\pi k^s \Gamma(s)} \int_0^{\infty} K_0(y)(\log y - c - \log(2\pi k))e^{-y}y^{s-1}\, dy. \qquad (2.7)$$

The functions ψ and ψ_0 here are analogous to φ and φ_0 in [17], up to certain factors. The next lemmas show that $\zeta_k^*(s)$ is indeed related to $\zeta_k(s)$.

Lemma 1. *For* Re $s = \sigma > 0$ *we have*

$$\psi(s; m, n, k) = \frac{m+n+k}{2\sqrt{mn}} \left(\frac{m+n+k}{2}\right)^{-s} + \rho(s; m, n, k), \qquad (2.8)$$

where

$$\rho(s; m, n, k) = \frac{m+n+k}{\pi\sqrt{mn}} \int_0^{\infty} \int_0^{\pi/2} \left(\frac{m+n+k}{2} + mr^2\right)^{-s} Q_r\, dr\, d\theta \qquad (2.9)$$

with $Q = Q(r, \theta; m, n, k)$ *defined by*

$$Q = (1 + r^2 \cos^2 \theta)^{-1/2} (1 + r^2 (m/n) \sin^2 \theta)^{-1/2} \left(1 + \frac{2mr^2}{m+n+k} \right).$$

Also,

$$\psi_0(s, k) = \frac{\Gamma(s+1)}{2\sqrt{\pi} k^s \Gamma(s + \frac{1}{2})} \left(2\frac{\Gamma'}{\Gamma}(s) - \frac{\Gamma'}{\Gamma}(s + \frac{1}{2}) - c - \log(4\pi k) \right). \quad (2.10)$$

Proof. By (2.6) and (2.2) we may write $\psi(s; m, n, k)$ as

$$\frac{2^s s}{\pi \Gamma(s)} \int_0^\infty \left(\int_0^\infty \int_0^\infty e^{-(m \cosh u + n \cosh v + k)y} \, du \, dv \right) y^{s-1} \, dy$$

$$= \pi^{-1} 2^s s \int_0^\infty \int_0^\infty (m \cosh u + n \cosh v + k)^{-s} \, du \, dv.$$

Further, after the substitutions $\cosh u = 1 + 2x^2$, $\cosh v = 1 + 2(m/n)y^2$, this becomes

$$\psi(s; m, n, k) = \frac{2^{s+2} s \sqrt{m/n}}{\pi (m+n+k)^s} \times$$

$$\times \int_0^\infty \int_0^\infty \frac{1}{\sqrt{(1+x^2)(1+(m/n)y^2)}} \left(1 + \left(\frac{2m(x^2 + y^2)}{m+n+k} \right) \right)^{-s} \, dx \, dy.$$

In polar coordinates, the double integral here equals

$$\int_0^\infty \int_0^{\pi/2} Q(r, \theta; m, n, k) \left(1 + \frac{2mr^2}{m+n+k} \right)^{-s-1} r \, dr \, d\theta.$$

Partial integration with respect to r then yields (2.8).

The formula (2.10) follows from the Mellin transform (see [2], p. 331)

$$\int_0^\infty K_0(y) e^{-y} y^{s-1} \, dy = \frac{\sqrt{\pi} \Gamma^2(s)}{2^s \Gamma(s + \frac{1}{2})}$$

together with its differentiated variant.

2.3 A spectral representation for $\zeta_k^*(s)$

The second expression for $\zeta_k^*(s)$, in addition to (2.5), is obtained by decomposing the strip Π into translates of a fundamental domain F for the group Γ by elements of Γ modulo Γ_∞. Since $E^*(z)$ is automorphic under the action of Γ and the measure $d\mu(z)$ is Γ - invariant, we find that

$$\zeta_k^*(s) = \frac{(4\pi)^{s-1} s}{2\Gamma(s)} \int_F |E^*(z)|^2 P_k(z, s) d\mu(z), \quad (2.11)$$

where

$$P_k(z, s) = \sum_{\gamma \in \Gamma_\infty \backslash \Gamma} y^s (\gamma(z)) e(k\gamma(z))$$

is the (non-holomorphic) *Poincaré series*.

In [17], the function corresponding to our $\zeta_k^*(s)$, say $\zeta_k^{**}(s)$, is defined a bit differently, for the factor standing in front of the integral in (2.11) is $\pi^s \Gamma(s)/\Gamma^4(s/2)$. Hence

$$\zeta_k^*(s)/\zeta_k^{**}(s) = \frac{4^s s \Gamma^4(s/2)}{8\pi \Gamma^2(s)}. \tag{2.12}$$

This is $1 + O((|t| + 1)^{-1})$ by Stirling's formula

$$\log \Gamma(s) = \frac{1}{2}\log 2\pi + (\sigma - \frac{1}{2})\log t - \frac{\pi}{2}t + i((\sigma - \frac{1}{2})\frac{\pi}{2} + t\log t - t) + O(t^{-1}), \tag{2.13}$$

valid for $t \to \infty$ if σ is bounded.

The integral in (2.11) is of the form of a scalar product

$$(f, g) = \int_F f(z)\overline{g(z)}d\mu(z)$$

of two automorphic functions. However, it is not a scalar product in the sense of L^2-theory, because $|E^*(z)|^2$ is not square integrable. But this function can be made square integrable by subtracting a suitable automorphic function. Such an "automorphic regularization", which is explained in detail in [17], yields a complicated but perfectly explicit main term, say $Z_k(s)$, for $\zeta_k^*(s)$. The function $Z_k(s)$, which can be read from Theorem 1 in [17] as the leading term on the right of (0.18) multiplied by the function (2.12), involves the gamma-function together with its first and second derivatives and has a pole of third order at $s = 1$.

The difference $\zeta_k^*(s) - Z_k(s)$ can be written as a scalar product of two L^2-functions, so that Parseval's formula is available. For automorphic functions f and g, this reads as follows:

$$(f, g) = \frac{1}{4\pi}\int_{-\infty}^{\infty} (f, E(\cdot, 1/2 + it))\overline{(g, E(\cdot, 1/2 + it))}\, dt$$

$$+ \sum_{j=1}^{\infty}(f, f_j)\overline{(g, f_j)} + \frac{3}{\pi}(f, 1)\overline{(g, 1)},$$

where the f_j are the *Maass wave forms* and 1 denotes the constant function 1. The cusp forms f_j are eigenfunctions of the hyperbolic Laplacian

$$L = -y^2\left(\frac{\partial^2}{\partial x^2} + \frac{\partial^2}{\partial y^2}\right),$$

thus

$$Lf_j = \lambda_j f_j,$$

where λ_j is the corresponding eigenvalue. It is well-known that the λ_j are real (because L is Hermitian) and that $\lambda_j > 1/4$, so $\kappa_j = \sqrt{\lambda_j - \frac{1}{4}}$ is a positive real number. The Fourier series of f_j is of the form (see [12], eq. (3.38))

$$f_j(z) = \sqrt{y} \sum_{\substack{n=-\infty \\ n \neq 0}}^{\infty} \rho_j(n) K_{i\kappa_j}(2\pi|n|y)e(nx),$$

where $\rho_j(n) = \overline{\rho_j(-n)}$. Also, by definition, the Maass wave forms have the following properties: they constitute an orthonormal set with respect to the scalar product defined above, they are eigenfunctions of all Hecke operators T_n, and they are either even or odd functions with respect to x. Since f_j is an eigenfunction T_n, there are eigenvalues $t_j(n)$ such that

$$T_n f_j = t_j(n)f_j.$$

These eigenvalues are real multiplicative functions of n which are related to the Fourier coefficients $\rho_j(n)$ by the formula $\rho_j(n) = \rho_j(1)t_j(n)$. Moreover, as $f_j(x + yi)$ is either an even or odd function of x for given y, we have

$$f_j(-x + yi) = \varepsilon_j f_j(x + yi)$$

with $\varepsilon_j = \pm 1$. The *Maass L-function*

$$H_j(s) = \sum_{n=1}^{\infty} t_j(n)n^{-s} \qquad (\mathrm{Re}\ s > 1)$$

admits an analytic continuation to an entire function and satisfies the functional equation

$$H_j(s) = 2^{2s-1}\pi^{2(s-1)}\Gamma(1 + i\kappa_j - s)\Gamma(1 - i\kappa_j - s) \times$$
$$\times (\varepsilon_j \cosh(\pi\kappa_j) - \cos(\pi s))H_j(1 - s)$$

(see [1] for the case $\varepsilon_j = 1$, the case $\varepsilon_j = -1$ being analogous). Note that this implies $H_j(1/2) = 0$ if $\varepsilon_j = -1$. Hence $H_j(1/2)$ can be non-zero only if $f_j(x + yi)$ is an even function of x. In this case, the coefficients $\rho_j(n)$ are real. In particular, $\rho_j = \rho_j(1)$ is real.

The next lemma, due to L. A. Tahtadjan and A. I. Vinogradov [17], gives a decomposition of $\zeta_k^*(s)$ into three parts: the main term $Z_k(s)$ and the contributions from the discrete and continuous spectrum of L. This lemma is essentially Theorem 1 of [17], where the contribution of the continuous spectrum should be, however, revised. This term is a certain integral involving a function $\eta(w)$ which is not explicitly given. As a matter of fact, this function vanishes identically as can be seen by going through the calculations in §5 in [17] and by correcting the last term in (5.11), where a factor π should be lifted from the denominator to the numerator. (In the course of the calculations, observe a misprint in (1.27), where $(1 - iu)^{s_2}$

should read $(1+iu)^{s_2}$). The actual contribution of the continuous spectrum comes from the term (5.9) in [17]. Moreover, in Theorem 1 in [17], the term $R_k(s)$ standing for $\zeta_k(s) - \zeta_k^{**}(s)$ can be omitted if the result is formulated for $\zeta_k^{**}(s)$. As a final modification, the expression for $\zeta_k^{**}(s)$ emerging from the calculations in [17] should be multiplied by the quotient (2.12).

Lemma 2. *In the half-plane $\sigma > 1/2$, we have*

$$\zeta_k^*(s) = Z_k(s) +$$

$$\frac{k^{1/2-s}s}{8\pi\Gamma^2(s)} \sum_{j=1}^{\infty} |\rho_j|^2 t_j(k) H_j^2(\tfrac{1}{2}) |\Gamma(z_j/2)|^4 \Gamma(s-z_j)\Gamma(s-\bar{z}_j) +$$

$$\frac{k^{1/2-s}s}{8\Gamma^2(s)} \int_{-\infty}^{\infty} \frac{k^{-iu}\sigma_{2iu}(k)|\xi(\tfrac{1}{2}+iu)|^4 \Gamma(s-\tfrac{1}{2}+iu)\Gamma(s-\tfrac{1}{2}-iu)}{|\Gamma(\tfrac{1}{2}+iu)|^2|\zeta(1+2iu)|^2}\, du,$$

$$(2.14)$$

where $z_j = 1/2 + i\kappa_j$, $\xi(s) = \pi^{-s/2}\Gamma(s/2)\zeta(s)$, and $Z_k(s)$ is a certain meromorphic function with a triple pole at $s = 1$ which can be explicitly given in terms of the gamma-function.

Remark 1. The sum over j in (2.14) looks formally slightly different from that in [17], where $(\mathrm{Re}\,\rho_j(1))\rho_j(k)$ stands in place of $|\rho_j|^2 t_j(k)$. However, as we pointed out above, these numbers coincide if $H_j(1/2) \neq 0$, and only such terms are of significance in (2.14).

Remark 2. The principal part of the Laurent expansion of $\zeta_k^*(s)$ at the pole $s = 1$ is different from that of $\zeta_k(s)$. Therefore the residue of $\zeta_k^*(s)x^s s^{-1}$ at this point does not give the true main term in the additive divisor problem. However, for our present purposes, the behaviour of $\zeta_k^*(s)$ for s lying far away from the real axis will be more relevant, and for such values the coincidence of $\zeta_k(s)$ and $\zeta_k^*(s)$ is better.

Remark 3. The equation (2.14) gives an analytic continuation of $\zeta_k^*(s)$ to a meromorphic function. This is clear for the first two terms on the right; in particular, the poles of the series are $1/2 \pm i\kappa_j - \nu$ for $\nu = 0, 1, \ldots$ The analytic continuation of the integral can be done by using the Sohotskij-Plemelj formulae (as in [17], p. 110); then the poles of the last term in the half-plane $\sigma > 0$ occur at the zeros of $\zeta(2s)$. However, we are not going to need an analytic continuation of this term beyond the half-plane $\sigma > 1/2$.

3   A general summation formula

This formula will serve as a substitute for Perron's formula, which is not as such at our disposal because we are not working with the function $\zeta_k(s)$. As usual, we write $\int_{(a)}$ to denote integration over the vertical line $\mathrm{Re}\, s = a$. The function $Q = Q(r, \theta; m, n, k)$ is as defined in Lemma 1.

Lemma 3. *Let $W(x)$ be a continuously differentiable function with a compact support in $(0, \infty)$, and let $M_k(s)$, for a positive integer k, be the Mellin transform of the function $\sqrt{\frac{x-k}{x}} W(x-k)$. Then, for any $a > 1$, we have*

$$\sum_{n=1}^{\infty} d(n)d(n+k)W(n) = \frac{1}{2\pi i} \int_{(a)} \zeta_k^*(s) M_k(s)\, ds \tag{3.1}$$

$$-\frac{2}{\pi} \sum_{n=1}^{\infty} d(n)d(n+k)W_1(n,k) - \frac{k}{\pi} \sum_{n=1}^{k-1} d(n)d(k-n)W_2(n,k) +$$

$$\frac{d(k)i}{4\pi^{3/2}} \int_{(a)} \frac{\Gamma(s+1)}{k^s \Gamma(s+\frac{1}{2})} \left(2\frac{\Gamma'}{\Gamma}(s) - \frac{\Gamma'}{\Gamma}\left(s+\frac{1}{2}\right) - c - \log(4\pi k) \right) M_k(s)\, ds,$$

where

$$W_1(n,k) =$$

$$\int_0^{\infty} \int_0^{\pi/2} \sqrt{1 + \frac{kr^2}{n(1+r^2)}} W(n + (n+k)r^2) Q_r(r,\theta; n+k, n, k)\, dr\, d\theta$$

and

$$W_2(n,k) = \frac{1}{\sqrt{k-n}} \int_0^{\infty} \int_0^{\pi/2} \frac{rW(nr^2)}{\sqrt{k+nr^2}} Q_r(r,\theta; n, k-n, k)\, dr\, d\theta.$$

Proof. By definition,

$$M_k(s) = \int_0^{\infty} \sqrt{\frac{x-k}{x}} W(x-k) x^{s-1}\, dx,$$

and by the Mellin inversion formula

$$\sqrt{\frac{x-k}{x}} W(x-k) = \frac{1}{2\pi i} \int_{(a)} M_k(s) x^{-s}\, ds \tag{3.2}$$

for all real a and all positive x. We substitute $x = n+k$ into the latter equation, multiply both sides by $d(n)d(n+k)\sqrt{\frac{n+k}{n}}$, and sum over all positive integers n. Then, supposing (for convergence) that $a > 1$, we obtain

$$\sum_{n=1}^{\infty} d(n)d(n+k)W(n)$$

$$= \frac{1}{2\pi i} \int_{(a)} \left(\sum_{n=1}^{\infty} d(n)d(n+k)\sqrt{\frac{n+k}{n}}(n+k)^{-s} \right) M_k(s)\, ds. \tag{3.3}$$

By (2.5) and Lemma 1,

$$\sum_{n=1}^{\infty} d(n)d(n+k)\sqrt{\frac{n+k}{n}}(n+k)^{-s} = \zeta_k^*(s) - \sum_{n=1}^{\infty} d(n)d(n+k)\rho(s;n+k,n,k)$$

$$-\frac{1}{2}\sum_{n=1}^{k-1} d(n)d(k-n)\psi(s;n,k-n,k) - d(k)\psi_0(s,k).$$

Now the formula (3.1) follows if this is substituted into (3.3), the functions $\rho(s;m,n,k)$ and $\psi_0(s,k)$ are written according to (2.9) and (2.10), the order of the integrations with respect to (r,θ) and s is inverted, and finally the innermost integrals over s are evaluated by the inversion formula (3.2). In addition to those terms given in (3.1), there appear some others involving the integral (3.2) for $x = k$, which vanishes however like the respective terms.

Remark 1. If $W(x)$ is an oscillating function, then the other terms besides the leading integral on the right of (3.1) are usually small, as will be seen in the next section. On the other hand, if $W(x)$ is a nonoscillating function, say the characteristic function of an interval with appropriate smoothing at the ends, then even the terms involving $W_1(n,k)$ will give a significant contribution comparable with the left hand side of (3.1).

Remark 2. If $\zeta_k^*(s)$ is replaced by $\zeta_k(s)$ in (3.1), and $M_k(s)$ is replaced by the Mellin transform of $W(x)$, then only the leading integral remains on the right. Thus the additional terms account for the reverse replacement, allowing us to work with a function having a nice spectral representation. The estimation of these terms is in practice usually a less tedious problem a the direct analysis of the effect of this replacement. Also, one may then virtually dispense with the function $\zeta_k(s)$.

4 Transformation of the sum S

4.1 Assumptions on the sum S

This sum was defined in (1.5). Suppose that the support of the weight function $\eta(x)$ is the interval $[N,N']$, where $N'-N \asymp N$ and N is sufficiently large. Here the relation $\asymp$ means the simultaneous validity of $\ll$ and $\gg$. Further, suppose that

$$k \ll N^{1/2}. \tag{4.1}$$

At this stage, we specify the weight function $\eta(x)$. Let us construct it as in [6], §2.1, by averaging the characteristic function of an interval repeatedly with respect to its end points. Then the resulting function $\eta(x)$ vanishes outside the interval (N,N') and equals 1 in the interval $[N+U,N'-U]$ for a certain number U. We suppose that $U \asymp N$. Depending on the number of the averaging steps, the derivatives of $\eta(x)$ up to a certain order vanish at N and N'. Also, $\eta^{(j)}(x) \ll N^{-j}$ for all derivatives up to a sufficiently high order.

Next we impose certain regularity conditions on the functions f and g occurring in the sum S. Let us suppose that $f(z)$ and $g(z)$ are holomorphic in the neighbourhood

$$D = \{z \in \mathbf{C} \mid |z - x| < bN \text{ for some } x \in [N, N']\} \tag{4.2}$$

of the interval $[N, N']$. Here b is a positive constant. Suppose, moreover, that $f(x)$ is real and

$$|f'(x)| \asymp FN^{-1}, \tag{4.3}$$

$$|(xf'(x))'| \asymp FN^{-1} \tag{4.4}$$

for real $x \in D$. The parameter F is a positive number such that

$$N^\varepsilon \ll F \ll N^{1/2-\varepsilon}. \tag{4.5}$$

The assumption (4.4) will not be needed until we invoke the saddle point method. Finally, let us suppose that

$$f'(z) \ll FN^{-1} \text{ for } z \in D \tag{4.6}$$

and

$$g(z) \ll G \text{ for } z \in D. \tag{4.7}$$

The condition (4.3) implies that $f'(x)$ is does not change its sign for real x. Suppose, to be specific, that $f'(x)$ is *negative*.

Remark. The condition (4.4) will be needed later in connection with exponential integrals. Its purpose is to secure that for any given $t \asymp F$ we have either

$$|(f(x) + t \log x)''| \asymp FN^{-2}, \tag{4.8}$$

or

$$|(f(x) + t \log x)'| \asymp FN^{-1} \tag{4.9}$$

in the whole interval $[N, N']$, at least if we suppose that $N' \leq (1+\delta)N$ for a suitable positive constant δ (as we may on splitting the original sum into shorter pieces if necessary). Indeed, if

$$f'(x) + t/x = \theta FN^{-1}$$

at a point $x \in [N, N']$ with $|\theta|$ sufficiently small, then

$$|(f(x) + t \log x)''| = |x^{-1}((xf'(x))' - \theta FN^{-1})| \gg FN^{-2}$$

at the same point, and this is seen to be valid in the whole integral by observing that the third derivative of our function is $\asymp FN^{-3}$.

4.2   A preliminary transformation

As a preparation for the proof of the spectral transformation formula, we first express the sum S approximately in terms of the function $\zeta_k^*(s)$. In the following, we let ε and A generally stand for fixed positive numbers, with ε arbitrarily small and A arbitrarily large, which need not be the same at each occurrence.

Lemma 4. *We have*

$$\sum \eta(n)d(n)d(n+k)g(n)e(f(n))$$

$$= \frac{1}{2\pi i} \int_{a+iT_1}^{a+iT_2} \zeta_k^*(s)M_k(s)\,ds + O(GN^{1+\varepsilon}F^{-1}), \qquad (4.10)$$

where $T_i \asymp F$, a *is an arbitrarily fixed real number exceeding 1, and* $M_k(s)$ *is the Mellin transform of the function* $\sqrt{\frac{x-k}{x}}W(x-k)$ *for* $W(x) = \eta(x)g(x)e(f(x))$.

Proof. We write the sum on the left of (4.10) according to the identity (3.1). Then the assertion will follow if we truncate the first integral on the right and estimate satisfactorily the truncation error together with the other three terms in (3.1).

To begin with, we estimate the Mellin transform

$$M_k(s) = \int_0^\infty \eta(x)g(x)x^{1/2}(x+k)^{\sigma-3/2}e(f(x)+(t/2\pi)\log(x+k))\,dx \quad (4.11)$$

on vertical lines $\mathrm{Re}\,s = \sigma$ with σ bounded. This is a smoothed exponential integral of the type considered in [6], Ch. II. Its possible saddle point is the root of the equation

$$f'(x) + \frac{t}{2\pi(x+k)} = 0 \qquad (4.12)$$

in the interval $[N, N']$. It may exist only if $t \asymp F$, and otherwise, if t lies outside a certain interval $[T_1, T_2]$ with $T_i \asymp F$, the modulus of the left hand side of (4.12) is $\asymp (F + |t|)/N$. Then, by Theorem 2.3 of [6], we have

$$M_k(s) \ll GN^\sigma(F + |t|)^{-A} \qquad (4.13)$$

for any given $A > 0$ provided that the function $\eta(x)$ is sufficiently smooth. This shows that the above mentioned truncation error is indeed negligible. On the other hand, if $T_1 \leq t \leq T_2$, then either one of the conditions (4.8) or (4.9) is fulfilled (and clearly the same holds if $\log x$ in these conditions is replaced by $\log(x + k)$. In the former case, the estimate (4.13) (with $t \asymp F$) is valid again, and in the latter case we have

$$M_k(s) \ll GN^\sigma F^{-1/2} \qquad (4.14)$$

by the familiar "second derivative test".

Returning to the equation (3.1), we next estimate the last term on the right. Here the integral can be truncated as the one in the leading term, and the integration can be moved to the interval $[iT_1, iT_2]$ because $M_k(s)$ is small on the horizontal lines $t = T_i$ by (4.13). Now, by (4.14), Stirling's formula, and (4.5), the new integral is $\ll GF \log F \ll GN^{1+\varepsilon}F^{-1}$.

It remains to estimate the terms involving $W_1(n, k)$ or $W_2(n, k)$. The integral over r in the latter terms can be written as the integral of $\eta(x)\tilde{g}(x)e(f(x))$ for a certain function $\tilde{g}(x)$, and for the same reasons as above, this integral is negligibly small. The former terms can be dealt with in the same manner if $n \leq N/2$, and they vanish if $n \geq N'$. Otherwise, if $N/2 < n < N'$, the relevant range for r in the definition of $W_1(n, k)$ is $r \ll 1$, and then $Q_r \ll r$. The integral over $0 \leq r \leq F^{-1/2}$ is thus $\ll GF^{-1}$ by a trivial estimation, and the same is true for the integral over $R \leq r \leq 2R$ for any $F^{-1/2} \leq R \ll 1$ by the "first derivative test". Thus the contribution of the terms in question is $\ll GN^{1+\varepsilon}F^{-1}$ as desired, and the proof of the lemma is complete.

Remark. The error term $O(GN^{1+\varepsilon}F^{-1})$ in (4.10) is crude if F is not large. However, by a suitable iterative argument, one may reduce this error term to $O(GN^{1/2+\varepsilon})$ on allowing a finite number of additional main terms similar to the leading integral in (4.10). Namely, resuming the preceding discussion of the sum involving $W_1(n, k)$ in (3.1), observe that the integral over r can be truncated to the interval $[0, F^{-1/2+\varepsilon}]$ if a suitable smooth weight function is introduced; this can be seen by using again Theorem 2.3 of [6]. Now, for given r and θ, this sum is of the same type as that standing on the left of (4.10) except that the function playing the role of g is smaller (by a factor $F^{-1/2+\varepsilon}$) than that in the original sum. In this way, the iteration gets started, and after a finite number of steps, each producing another main term, we end up with a sum which can be estimated trivially.

4.3   The transformation formula

To derive the final formula for the sum S, we substitute the expression (2.14) for $\zeta_k^*(s)$ into (4.10). To begin with, we dispose of the contribution of $Z_k(s)$ as a negligible error term. The oscillatory nature of this function is determined by the factor k^{-s}. Therefore it is easily seen by repeated integration by parts that the integral of $Z_k(s)M_k(s)$ is small.

Turning to the spectral contributions in (2.14), note that, for s as in (4.10), the terms in the series decay exponentially when κ_j exceeds F, and likewise the integral converges rapidly for $|u| > F$. So the series and the integral can be truncated at the level $F \log^2 N$, say, with a negligible error. To justify this precisely, one may appeal to the mean value estimates (4.16) - (4.17) below. In (4.10), the spectral integral gives rise to a double integral with respect to s and u, where we invert the order of the integrations.

The integral over s in (4.10) is next moved left to a line $\sigma = -\nu_0$, where ν_0 is a sufficiently large positive integer, taking into account the residues at some poles. By (4.1) and (4.5), the shifted integral will be small, because the integrand decays comparably with $(N/(kt))^\sigma$ as σ decreases avoiding half-integer values (for $M_k(s)$ decays like N^σ by (4.14), whereas the spectral expressions increase like $(kt)^{-\sigma}$). Thus the choice of ν_0 is inversely proportional to $\log(N/(kF))/\log N$. Also, the integrals over the horizontal segments joining the old and new path of integration are small by (4.13). If κ_j or $\pm u$ lies near to T_1 or T_2, small indentations in the horizontal parts of the contour are needed.

It remains to consider the residues at the poles. For $u \in (T_1, T_2)$, there are poles at $s = s(u, \nu) = 1/2 + iu - \nu$ for $\nu = 0, 1, \ldots, \nu_0$, and also there are poles at $s = 1/2 - iu - \nu$ for $u \in (-T_2, -T_1)$. For given ν, the contribution of these is $(4\pi\nu!)^{-1}(-k)^\nu$ times the integral

$$\int_{T_1}^{T_2} \frac{s(u,\nu)\sigma_{-2iu}(k)\Gamma(2iu - \nu)|\Gamma(\tfrac{1}{4} + \tfrac{iu}{2})|^4|\zeta(\tfrac{1}{2} + iu)|^4 M_k(s(u,\nu))}{\Gamma^2(s(u,\nu))|\Gamma(\tfrac{1}{2} + iu)|^2|\zeta(1 + 2iu)|^2} \, du.$$

By using Stirling's formula, (4.14), and well-known results on the zeta-function (namely a lower bound on the line $\sigma = 1$ and an estimate for the fourth moment on the critical line), it is easily seen that the sum of these over ν is $\ll GN^{1/2+\varepsilon}$, which can be viewed as an error term.

The second term in (2.14), depending on the discrete spectrum, has poles at $s = z_j - \nu$. The residues at these points contribute

$$(8\pi)^{-1}\sum_{\nu=0}^{\nu_0}(-1)^\nu(\nu!)^{-1}\sum_{T_1 < \kappa_j < T_2}(z_j - \nu)k^{-i\kappa_j+\nu} \times$$

$$\times|\rho_j|^2 t_j(k)H_j^2(\tfrac{1}{2})|\Gamma(z_j/2)|^4\Gamma(2i\kappa_j - \nu)M_k(z_j - \nu)/\Gamma^2(z_j - \nu), \quad (4.15)$$

where

$$(z_j - \nu)|\Gamma(z_j/2)|^4\Gamma(2i\kappa_j - \nu)/\Gamma^2(z_j - \nu)$$

$$= 4e(1/8)\pi^{3/2}\kappa_j^{-1/2}4^{i\kappa_j}(i\kappa_j/2)^\nu(\cosh(\pi\kappa_j))^{-1}(1 + O(\kappa_j^{-1})).$$

by Stirling's formula (2.13). We end up with an approximate formula for our sum S on substituting the main term into (4.15).

The approximation error is $O(GN_1^{1+\varepsilon}F^{-1})$, as can be seen by (4.14) and the following estimations: putting $\alpha_j = |\rho_j|^2/\cosh(\pi\kappa_j)$, we have (see [12], Theorem 6)

$$\sum_{\kappa_j \le x} \alpha_j t_j^2(k) \ll x^2 + xk^\varepsilon + k^{1/2+\varepsilon} \tag{4.16}$$

and (see [16], Theorem 5)

$$\sum_{\kappa_j \le x} \alpha_j H_j^4(1/2) \ll x^2 \log^{20} x. \tag{4.17}$$

The preceding considerations may be summarized as follows.

Theorem. *On the assumptions made in sec. 4.1, we have*

$$\sum \eta(n)d(n)d(n+k)g(n)e(f(n)) = \frac{1}{2}\sqrt{\pi}e(1/8)\sum_{\nu=0}^{\nu_0}\frac{(-ik/2)^\nu}{\nu!} \times$$

$$\times \sum_{T_1 < \kappa_j < T_2} \alpha_j t_j(k)H_j^2(1/2)\kappa_j^{-1/2}(4/k)^{i\kappa_j}\kappa_j^\nu M_k(z_j - \nu) + O(GN^{1+\varepsilon}F^{-1}),$$

$$\tag{4.18}$$

where $T_i \asymp F$, ν_0 is sufficiently large, and M_k is as in Lemma 4.

5   The Weyl - van der Corput method

5.1   The general case

We now return to the sum (1.2),

$$S_0 = \sum_{N_1}^{N_2} d(n)g(n)e(f(n)), \tag{5.1}$$

where N_1, N_2 are large positive numbers such that $N_2 \le 2N_1$. We suppose that the functions $f(z)$ and $g(z)$ are holomorphic in a neighbourhood D of the interval $[N_1, N_2]$, defined as in (4.2). Further, suppose that for real $x \in D$, $f(x)$ and $g(x)$ are real, and

$$f''(x) \asymp TN_1^{-2}, \tag{5.2}$$
$$(xf''(x))' \asymp TN_1^{-2}, \tag{5.3}$$

where T is a parameter such that

$$T^{2/3+\varepsilon} \ll N_1 \ll T. \tag{5.4}$$

If S_0 is a Dirichlet polynomial, then T will have its usual meaning in this context. Finally, suppose that

$$f'(z) \ll TN_1^{-1} \text{ for } z \in D, \tag{5.5}$$

and

$$g(z) \ll G \text{ for } z \in D. \tag{5.6}$$

Let K be a positive integer such that

$$N_1^\varepsilon \ll K \ll N_1^{3/2-\varepsilon} T^{-1}. \tag{5.7}$$

Weyl's "process A" consists in writing

$$S_0 = K^{-1} \sum_{n=N_1}^{N_2} \sum_{k=1}^{K} d(n+k)g(n+k)e(f(n+k)) + O(GKN_1^\varepsilon),$$

followed by an application of Cauchy's inequality to the outer sum. Then

$$S_0^2 \ll K^{-2}N_1 \sum_{n} \eta(n) |\sum_{k=1}^{K} d(n+k)g(n+k)e(f(n+k))|^2 + G^2 K^2 N_1^\varepsilon,$$

where $\eta(x)$ is a smooth weight function whose support is an interval $[N_1', N_2']$ containing $[N_1, N_2]$ and being contained in $[N_1/2, 2N_2]$. Suppose that $\eta(x) = 1$ for $N_1 \le x \le N_2$, and that $\eta^{(j)}(x) \ll N_1^{-j}$ for all derivatives up to a sufficiently high order. Squaring out, we obtain

$$S_0^2 \ll K^{-2}N_1 \sum_{k_1,k_2=1}^{K} \sum_{n} \eta(n)d(n+k_1)d(n+k_2) \times$$

$$\times g(n+k_1)g(n+k_2)e(f(n+k_1) - f(n+k_2)) + G^2 K^2 N_1^\varepsilon.$$

The contribution of the diagonal terms is $\ll G^2 K^{-1} N_1^{2+\varepsilon}$. As to non-diagonal terms, let us consider those with $k_1 > k_2$; the others give complex conjugate values. We may replace $\eta(n)$ by $\eta(n+k_2)$ with an error $\ll G^2 K N_1^{1+\varepsilon}$. Then, putting $k = k_1 - k_2$ and writing n for $n + k_2$, we may rewrite the preceding inequality as

$$S_0^2 \ll \text{Re } S_1 + G^2 K^{-1} N_1^{2+\varepsilon}, \tag{5.8}$$

where

$$S_1 = K^{-2}N_1 \sum_{k=1}^{K-1}(K-k) \sum_{n} \eta(n)d(n)d(n+k)g(n)g(n+k)e(f(n+k)-f(n)). \tag{5.9}$$

For convenience, we restrict here k to an interval $[K_0, 2K_0)$, where $T^\varepsilon \le K_0 < K$; the terms with $k < T^\varepsilon$ can be treated trivially like the diagonal terms.

The inner sum in (5.9) can be written as

$$\sum \eta(n)d(n)d(n+k)g_k(n)e(f_k(n)) \qquad (5.10)$$

with $g_k(z) = g(z)g(z+k)$ and $f_k(z) = f(z+k) - f(z)$. We are going to apply the formula (4.18) to this sum. Therefore we have to check that the conditions in sec. 4.1 are satisfied for certain parameters F_k and G_k (depending on k) in place of F and G. Clearly we may choose $G_k = G^2$. The functions f_k and g_k should be holomorphic in a certain domain, say $\tilde{D}$, which is of the type (4.2) and contains the interval $[N_1', N_2']$. Such a domain with $\tilde{D} \subset D$ can be constructed if the interval $[N_1', N_2']$ is suitably chosen.

Turning to the verification of the conditions, we note first that k satisfies (4.1) (with $N \asymp N_1$) by (5.7) and (5.4). Next, by (5.2), we have

$$f_k'(x) = f'(x+k) - f'(x) \asymp kTN_1^{-2}$$

for real $x \in \tilde{D}$. Then, if we choose

$$F_k = K_0 N_1^{-1} T \qquad (5.11)$$

for $K_0 \le k < 2K_0$, the condition (4.3) holds, and (4.6) is also easily verified. Moreover, F_k satisfies (4.5) by (5.4), (5.7), and our choice of K_0. Finally, to verify the condition (4.4), note that $f_k(x) \approx kf'(x)$, whence the validity of (4.4) for f_k essentially amounts to (5.3).

We assumed earlier $f'(x)$ to be negative, and this will the case for f_k, too, if $f''(x)$ is negative, as we may suppose.

Now, having checked the necessary conditions, we may transform the sum in (5.10) by the formula (4.18). The contribution of its error term to S_1 is $\ll G^2K^{-1}N_1^{3+\varepsilon}T^{-1} \ll G^2K^{-1}N_1^{2+\varepsilon}$, and a term like this occurred already in (5.8). So the inner sum in (5.9) can be replaced, with sufficient accuracy, by the corresponding spectral sum in (4.18).

In this sum, only the Mellin transform

$$M_k(s) = \int_0^\infty \eta(x)g(x)g(x+k)x^{1/2}(x+k)^{\sigma-3/2} \times$$
$$\times e(f(x+k) - f(x) + (t/2\pi)\log(x+k))\, dx \qquad (5.12)$$

depends on the functions f and g. This can be calculated approximately by the saddle-point method. Note that if a saddle point exists, then, by the remark made in sec. 4.1, the second derivative of the function in the exponential is of the expected order.

In general, the saddle-point method gives for the exponential integral

$$\int_a^b \varphi(x)e(\psi(x))\,dx \tag{5.13}$$

the main term

$$\varphi(x_0)\psi''(x_0)^{-1/2}e(\psi(x_0) + 1/8) \tag{5.14}$$

if ψ'' is positive and there exists a saddle point, that is to say a zero of ψ' in the interval (a, b). The error term can be diminished on introducing a weight function $\eta(x)$ to the integrand in (5.13). A saddle-point theorem for exponential integrals with a weight function is given in [6] (Theorem 2.2). It involves certain error terms, one of which depends on x_0. If the integrand is a sufficiently rapidly oscillating function, in other words if f is not too stationary, then only the last mentioned error term is of significance. This term can be estimated in different ways. The version given in the above mentioned theorem is not quite suitable for our present application. Instead, in the notation of [6], the estimate $\ll (\mu(x_0)/U)^J G(x_0)\mu(x_0)F(x_0)^{-3/2}$ for the error term in question is more appropriate. This follows immediately if one applies the argument used in case 1) of the proof in the cases 3) and 4) as well. In this variant of the result, the weight factor $\eta(x_0)$ then appears in (5.14) if $\eta(x)$ is inserted into the integrand in (5.13).

In practical terms, if the functions f and g satisfying the conditions in sec. 4.1 stand in place of ψ and φ, then the main term will be of the order $GN_1F^{-1/2}$, whilst the error term is $\ll GN_1F^{-3/2}$, thus it is smaller by a factor F^{-1}. In the case of $M_k(z_j - \nu)$, the main term is of the order $G^2N^{1-\nu}(K_0T)^{-1/2}$, and the error term is smaller by a factor F_k^{-1}. Such a relative error was admissible in the transformation formula (4.17). Therefore we may approximate $M_k(z_j - \nu)$ by the "saddle-point term" of the type (5.14).

The problem of the estimation of our sum S_0 has now been translated into the language of spectral theory. The resulting expression is a triple sum over k, κ_j, and ν, but the last mentioned summation is less significant; to fix ideas, one may take $\nu = 0$, for the other terms are similar but in general smaller. So, essentially, an oscillating double sum over k and κ_j is to be estimated. Thus the situation is similar to the ordinary Weyl-van der Corput method except that the spectral sum is now less precisely understood; in the classical case, the spectral sum produced by Poisson's summation formula runs simply over integers. Because the oscillatory nature of the spectral sum is unknown, one may just try to gain some saving in the sum over k, in the mean over κ_j. Such estimates of the large sieve type are indeed known since the pioneering work of H. Iwaniec [5]. These show that a typical saving factor in the sum of $t_j(k)$ over $k \in [K_0, 2K_0]$ is about $K_0^{-1/2}$. However, for the present purposes, Iwaniec's large sieve estimates should be generalized, which can be done by a modification of his arguments.

Assuming the expected cancellation in the sum over k, the preceding discussion of the sum S_1 and (5.8) give the estimate

$$S_0^2 \ll G^2 N_1^\varepsilon (K^{1/2} N_1^{1/2} T + K^{-1} N_1^2).$$

Choosing K optimally, we obtain

$$S_0 \ll G N_1^{1/2+\varepsilon} T^{1/3}.$$

This is actually what we proved already in [8] by our previous method. It may seem disappointing that the spectral methods do not give any new estimate. However, as we pointed out in the introduction, the main interest of the spectral theoretic approach lies in structural aspects.

5.2   An application to Dirichlet polynomials

For illustration, we work out the preceding considerations for the Dirichlet polynomial

$$S_0 = \sum_{N_1}^{N_2} d(n) n^{-1/2+iT}$$

corresponding to the specifications $f(x) = (T/2\pi) \log x$ and $g(x) = x^{-1/2}$. It is easily verified that these functions satisfy the necessary conditions; in particular, $f''(x) < 0$, as we supposed.

The Mellin transform $M_k(s)$, defined in (5.12), takes now the form

$$M_k(s) = \int_0^\infty \eta(x)(x+k)^{\sigma-2} e((T/2\pi) \log(\frac{x+k}{x}) + (t/2\pi) \log(x+k)) \, dx$$

with the saddle point $x_0 = kT/t$. The saddle-point term can be calculated according to (5.14). As to the required accuracy of the calculations, observe that relative errors of the proportion $F_k^{-1} = N_1(K_0 T)^{-1}$ are admissible, as we realized in the discussion of the saddle-point method above.

Turning to the calculation of (5.14) for $\varphi(x) = (x+k)^{\sigma-2}$, $\psi(x) = (T/2\pi) \log(\frac{x+k}{x}) + (t/2\pi) \log(x+k)$, and $x_0 = kT/t$, note that

$$\psi''(x_0) = t^3 (2\pi k^2 (T+t)T)^{-1}$$

and

$$e(\psi(x_0)) = (\frac{keT}{t})^{it} e(\frac{t^2}{4\pi T})(1 + O(t^3 T^{-2})).$$

Hence, for $s = z_j - \nu$, the expression (5.14) is approximately

$$\sqrt{2\pi} k^{-1/2} T^{-1/2} \left(\frac{keT}{\kappa_j}\right)^{i\kappa_j} e(\frac{1}{8}) e\left(\frac{\kappa_j^2}{4\pi T}\right) \left(\frac{\kappa_j}{kT}\right)^\nu. \qquad (5.15)$$

We may now write down an approximate transformation formula for that part, say $S_1(K_0)$, of the sum S_1 in (5.9), where the range for k is $[K_0, 2K_0)$. By (4.18) and (5.15), we obtain

$$\mathrm{Re}\, S_1(K_0) = \frac{\pi N_1}{K\sqrt{2T}} \sum_j \alpha_j \left(\sum_{K_0 \le k < 2K_0} (1 - \frac{k}{K}) \eta(kT/\kappa_j) t_j(k) k^{-1/2} \right) \times$$

$$\times H_j^2(1/2) \kappa_j^{-1/2} \sin\left(\kappa_j \log\left(\frac{\kappa_j}{4eT}\right)\right) + O(K^{-1} N_1^{1+\varepsilon}). \qquad (5.16)$$

This formula should be compared with the following result of Y. Motohashi ([14], corollary to the main theorem): for $T^{1/2} < \Delta < T(\log T)^{-1}$

$$(\Delta\sqrt{\pi})^{-1} \int_{-\infty}^{\infty} |\zeta(1/2 + i(T+t))|^4 e^{-(t/\Delta)^2}\, dt = \frac{\pi}{\sqrt{2T}} \sum_{j=1}^{\infty} \alpha_j H_j^3(1/2) \times$$

$$\times \kappa_j^{-1/2} \sin\left(\kappa_j \log\left(\frac{\kappa_j}{4eT}\right)\right) \exp\left(-\left(\frac{\Delta\kappa_j}{2T}\right)^2\right) + O(\log^B T), \qquad (5.17)$$

where B is a constant. The sum over k in (5.16) corresponds to the third factor $H_j(1/2)$ in (5.17). Further, the right hand side of (5.8) represents the mean square of S_0 over an interval of length about $\Delta = N_1/K$, so both sides of (5.16) should be divided by this number in order to make (5.16) and (5.17) more precisely comparable, and then the analogy is quite perfect.

As an application of (5.16) or (5.17), one may reprove Iwaniec's [5] estimate for the fourth moment of $\zeta(1/2 + it)$ over a set of short intervals. Moreover, Motohashi's main theorem (a more precise version of (5.17)) can be used to establish best known upper and Ω-estimates for the error term in the asymptotic formula for the fourth moment of $\zeta(1/2 + it)$ (see [4]); such applications lie beyond the scope of (5.16).

It is now an interesting problem to try to generalise (5.16) and (5.17) to other Dirichlet polynomials and series, for instance to Dirichlet, Hecke, or Maass L-functions.

References

1. C. Epstein, J. L. Hafner and P. Sarnak (1985). Zeros of L-functions attached to Maass forms. *Math. Z.*, **190**, 113–128.

2. A. Erdelyi, W. Magnus, F. Oberhettinger and F. G. Tricomi (1953). *Higher Trancendental Functions*, vol. II. McGraw-Hill, New York-Toronto-London.

3. A. Good (1983). On Various Means Involving the Fourier Coefficients of Cusp Forms. *Math. Z.*, **183**, 95 –129.

4. A. Ivić and Y. Motohashi. On the fourth moment of the Riemann zeta-function (to appear).

5. H. Iwaniec (1979/80). Fourier coefficients of cusp forms and the Riemann zeta-function. *Séminaire de Théorie des Nombres Bordeaux*, exposé no **18**.

6. M. Jutila (1987). *Lectures on a Method in the Theory of Exponential Sums*. Tata Institute of Fundamental Research, Bombay.

7. M. Jutila (1990). The fourth power moment of the Riemann zeta-function over a short interval. *Proc. Coll. János Bolyai*, **51**,*Number Theory, Budapest 1987*, 221–244. Elsevier, Amsterdam - New York.

8. M. Jutila (1989). Mean value estimates for exponential sums. *Number Theory, Ulm 1987*, 120–136. Lecture Notes in Mathematics 1380, Springer-Verlag, Berlin-Heidelberg-New York.

9. M. Jutila (1990). Mean value estimates for exponential sums. II. *Arch. Math.*, **55**, 267–274.

10. M. Jutila (1991). Mean value estimates for exponential sums with applications to L-functions. *Acta Arith.*, **57**, 93–114.

11. T. Kubota (1973). *Elementary theory of Eisenstein series*. John Wiley & Sons, New York.

12. N. V. Kuznetsov (1981). Petersson's conjecture for cusp forms of weight zero and Linnik's conjecture. Sums of Kloosterman sums (in Russian). *Mat. Sb.*, **111**, 334-383.

13. N. V. Kuznetsov (1983). Convolution of the Fourier coefficients of the Eisenstein-Maass series (in Russian). *Zap. Nauch. Sem. LOMI, AN SSSR*, **129**, 43–84.

14. Y. Motohashi. An Explicit Formula for the Fourth Power Mean of the Riemann Zeta-function. *Acta Math.* (to appear).

15. Y. Motohashi. On a binary additive divisor problem (to appear).

16. Y. Motohashi. Spectral mean values of Maass waveform L-functions (to appear).

17. L. A. Tahtadjan and A. I. Vinogradov (1984). The zeta-function of the additive divisor problem and the spectral decomposition of the automorphic Laplacian (in Russian). *Zap. Nauch. Sem. LOMI, AN SSSR*, **134**, 84-116.

Matti Jutila
Department of Mathematics
University of Turku
SF–20500 Turku
Finland
e-mail jutila@konfu.utu.fi

Distribution of Small Powers
of a Primitive Root

Hugh L. Montgomery

University of Michigan

1 Statement of Results

Let g be an arbitrary primitive root of an odd prime p, and set $\mathcal{N} = \{g, g^2, \ldots, g^N\}$ where N is some positive integer. Our object is to determine the extent to which the residue classes comprising $\mathcal{N}$ are well-distributed. For this purpose we borrow a definition from the theory of uniform distribution: If $U = \{u_n\}_{n=1}^N$ is a finite sequence of members of the circle group $\mathbf{T} = \mathbf{R}/\mathbf{Z}$, then we define the *discrepancy* of U to be

$$D(U) = \sup_{0 \le \alpha \le 1} \left| \frac{1}{N} \operatorname{card} \{n : 1 \le n \le N, \, u_n \in [0, \alpha) \pmod 1\} - \alpha \right|.$$

Trivially $0 \le D(U) \le 1$, and we regard U as being approximately uniformly distributed if $D(U)$ is small.

Theorem 1. *Let g be an arbitrary primitive root of an odd prime p, and take $U = \{g^n/p\}_{n=1}^N$. Then*

$$D(U) \ll N^{-1} p^{1/2} (\log p)^2.$$

From the above we see that the residue classes $\mathcal{N}$ are approximately uniformly distributed if $N/(p^{1/2}(\log p)^2)$ is large. Presumably more is true. We conjecture that for any $\epsilon > 0$

$$D(U) \ll N^{-1/2} p^{\epsilon}. \tag{1.1}$$

In order to analyze the running time of a particular sorting algorithm, Martin Tompa has asked for what N is it true that a positive proportion of the intervals $(m, m + H]$ contain a member of $\mathcal{N}$, where $H \approx p/N$. To address this question we establish

This research was supported in part by NSF grant DMS-9107605.

Theorem 2. *Let g be an arbitrary primitive root of an odd prime p, and set $\mathcal{N} = \{g^\nu : 1 \leq \nu \leq N\}$. If m and H are integers and $1 \leq H \leq P$, let $f(m, H) = \mathrm{card}\,\{n \in (m, m + H] : n \equiv n' \pmod{p}, n' \in \mathcal{N}\}$. Then for any $\epsilon > 0$*

$$\sum_{m=1}^{p} (f(m, H) - NH/p)^2 \sim NH \tag{1.2}$$

uniformly for $H \approx p/N$, $p^{5/7+\epsilon} \leq N \leq p$.

If we were to assume the Generalized Lindelöf Hypothesis, which asserts that

$$L(1/2 + it, \chi) \ll (q(|t| + 2))^\epsilon,$$

then the exponent $5/7$ could be replaced by $2/3$, though presumably this still falls considerably short of the truth: One may expect that the above remains true with the exponent $5/7$ deleted.

With $f(m, H)$ as above it is clear that

$$\sum_{m=1}^{p} f(m, H) = NH. \tag{1.3}$$

Thus (1.2) constitutes an estimate of the second moment of f about its mean. By elementary reasoning we deduce a result of the desired shape:

Corollary. *Let $\mathcal{N}$ be as above. Then a positive proportion of the intervals $(m, m + H]$ contain a member of $\mathcal{N}$, provided that $H \approx p/N$ and that $p^{5/7+\epsilon} \leq N \leq p$.*

It would be interesting to know how much the range of N might be enlarged if one were to average over the choice of the primitive root, rather than consider the worst one, as we have done here.

The author is happy to thank Martin Tompa and R. C. Vaughan for stimulating comments on the topic of this paper. The research reported on in this paper was supported in part by National Science Foundation Grand DMS–9107605.

2   Auxiliary Sums

Questions regarding the distribution of $\mathcal{N}$ are easily described in terms of the generating function

$$A(k) = \sum_{n \in \mathcal{N}} e(kn/p). \tag{2.1}$$

(Here we have used Vinogradov's notation, $e(\theta) = e^{2\pi i\theta}$.) Unfortunately, the multiplicative nature of the set $\mathcal{N}$ makes it difficult to treat the $A(k)$ directly. On the other hand, the alternative generating function

$$K(\chi) = \sum_{n \in \mathcal{N}} \chi(n), \tag{2.2}$$

though less conveniently related to the distribution of $\mathcal{N}$, is easily estimated. Let $\chi_r(n) = e(\frac{r \operatorname{ind} n}{p-1})$. Then

$$
\begin{aligned}
K(\chi_r) &= \sum_{n \in \mathcal{N}} e\left(\frac{r \operatorname{ind} n}{p-1}\right) \\
&= \sum_{\nu=1}^{N} e\left(\frac{r\nu}{p-1}\right) \\
&= \frac{e(r(N+1)/(p-1)) - e(r/(p-1))}{e(r/(p-1)) - 1}.
\end{aligned}
$$

Here the numerator has modulus at most 2, while the modulus of the denominator is $2|\sin\frac{\pi r}{p-1}|$. Hence

$$|K(\chi_r)| \le \left|\sin\frac{\pi r}{p-1}\right|^{-1} \le \left(2\left\|\frac{r}{p-1}\right\|\right)^{-1}$$

where $\|\theta\|$ is the natural metric on $\mathbf{T}$, $\|\theta\| = \min_{n \in \mathbf{Z}}|\theta - n|$. We also have the trivial estimate $|K(\chi_r)| \le N$. On combining these, we deduce that

$$|K(\chi_r)| \ll \min\left(N, \left\|\frac{r}{p-1}\right\|^{-1}\right). \tag{2.3}$$

This is quite sharp.

The functions $\chi(k)$ form a basis for the vector space of functions defined on the reduced residue classes modulo p. Hence the function $e(k/p)$, restricted to the reduced residue classes, can be expressed as a linear combination of Dirichlet characters. Indeed, if $(k, p) = 1$ then

$$e(k/p) = \frac{1}{p-1}\sum_{\chi} \tau(\overline{\chi})\chi(k).$$

This change of basis formula is easily verified by invoking the definition of the Gauss sum, $\tau(\chi) = \sum_{a=1}^{p}\chi(a)e(a/p)$, and using the basic orthogonality of the Dirichlet characters (see Davenport [1], p. 31). We replace k above by kn and sum both sides over all $n \in \mathcal{N}$, to find that

$$A(k) = \frac{1}{p-1}\sum_{\chi}\tau(\overline{\chi})K(\chi)\chi(k). \tag{2.4}$$

Now $\tau(\chi_0) = -1$, and $|\tau(\chi)| = \sqrt{p}$ for all non-principal characters χ (see pp. 66–67 of Davenport [1]). Hence by (2.3) we deduce that

$$A(k) \ll N/p + p^{-1/2} \sum_{r=1}^{p-2} \left\| \frac{r}{p-1} \right\|^{-1} \ll p^{1/2} \log p \qquad (2.5)$$

whenever $(k, p) = 1$. This presumably is not sharp. While there are $\approx p/N$ characters χ for which the summand in (2.4) has modulus $\approx N p^{1/2}$, it is not expected that all these terms should pull in the same direction. Indeed, probabilistic reasoning leads us to conjecture that if $(k, p) = 1$ then

$$A(k) \ll N^{1/2} p^\epsilon. \qquad (2.6)$$

3   Proof of Theorem 1

For a given sequence $U = \{u_n\}_{n=1}^N$ of members of the circle group $\mathbf{T}$, put

$$A(k) = \sum_{n=1}^N e(k u_n).$$

An estimate for the discrepancy $D(U)$ in terms of the sums $A(k)$ is provided by the Erdős-Turán inequality: For any positive integer K,

$$D(U) \ll 1/K + \frac{1}{N} \sum_{k=1}^K |A(k)|/k.$$

This was first proved by Erdős and Turán [2]. Additional proofs have been given since, yielding better values for the implicit constant. For a proof with good constants, see §8 of Vaaler [3]. Taking $K = p - 1$, and applying (2.5), we deduce that

$$D(U) \ll N^{-1} p^{1/2} (\log p) \sum_{k=1}^{p-1} 1/k \ll N^{-1} p^{1/2} (\log p)^2,$$

as desired. If we use the conjectural estimate (2.6) instead of (2.5), then we obtain (1.1), which presumably would be best-possible.

4   Transformations of Character Sums

A reader who is content to prove Theorem 2 in the restricted range $p^{3/4+\epsilon} \leq N \leq p$ may skip this section and the next, apart from noting the Pólya-Vinogradov inequality (4.2) below.

Lemma 1. *Suppose that $\{c(n)\}_{n=1}^{q}$ is given, and put*

$$\widehat{c}(k) = \frac{1}{q} \sum_{n=1}^{q} c(n)e(nk/q).$$

If χ is a primitive character modulo q, then

$$\sum_{n=1}^{q} c(n)\chi(n) = \frac{q}{\tau(\overline{\chi})} \sum_{k=1}^{q} \widehat{c}(k)\overline{\chi}(k). \tag{4.1}$$

Proof. Since χ is primitive, it follows (see p. 65 of Davenport [1]) that

$$\chi(n) = \frac{1}{\tau(\overline{\chi})} \sum_{k=1}^{q} \overline{\chi}(k)e(nk/q)$$

for all integers n. We insert this on the left hand side of (4.1), invert the order of summation, and appeal to the definition of $\widehat{c}(k)$, to obtain the stated result.

Lemma 1 is the primary ingredient in the proof of the Pólya-Vinogradov inequality, which asserts that if χ is a non-principal character modulo q then

$$\sum_{a}^{b} \chi(n) \ll q^{1/2} \log q. \tag{4.2}$$

The classical papers regarding this inequality are those of Pólya [4], Vinogradov [5], Landau [6], and Schur [7]. For a modern exposition, see §23 of Davenport [1].

Our next lemma has many formulations and appears in various contexts. In the form below, it was introduced by Jutila [8] in order to establish zero-density estimates for Dirichlet L-functions.

Lemma 2. *For each Dirichlet character modulo q let $w(\chi)$ be a complex number. For each integer n let a_n be a complex number with $\sum |a_n| < \infty$. Put*

$$J(\mathbf{w}, \mathbf{a}) = \sum_{\chi} \sum_{\psi} w(\chi)\overline{w(\psi)} \left| \sum_{n} a_n \chi\overline{\psi}(n) \right|^2.$$

If $|a_n| \leq b_n$ for all n, and if $\sum b_n < \infty$, then

$$J(\mathbf{w}, \mathbf{a}) \leq J(\mathbf{w}, \mathbf{b}). \tag{4.3}$$

Proof. On expanding, we see that

$$J(\mathbf{w}, \mathbf{a}) = \sum_{\chi} \sum_{\psi} \sum_{m} \sum_{n} w(\chi)\overline{w(\psi)} a_m \overline{a_n} \chi(m)\overline{\psi(m)}\psi(n)\overline{\chi}(n)$$

$$= \sum_{m} \sum_{n} a_m \overline{a_n} \left| \sum_{\chi} w(\chi)\chi(m)\overline{\chi}(n) \right|^2.$$

This is

$$\leq \sum_{m} \sum_{n} b_m b_n \left| \sum_{\chi} w(\chi)\chi(m)\overline{\chi}(n) \right|^2,$$

which by the same reasoning is $J(\mathbf{w}, \mathbf{b})$.

5   Moment Estimates of Character Sums

Let $\chi_s(n) = e(s(\operatorname{ind} n)/(p-1))$ where $\operatorname{ind} n$ is calculated with respect to some fixed primitive root g of the odd prime p. We require estimates for moments of character sums, of the form

$$M_k(S, H) = \sum_{s=1}^{S} \left| \sum_{h=-H}^{H} (1 - |h|/H)\chi_s(h) \right|^k. \tag{5.1}$$

The most trivial estimate of this expression is obtained by observing that the inner sum is obviously $\ll H$. Hence for any fixed $k \geq 0$ it follows that

$$M_k(S, H) \ll SH^k. \tag{5.2}$$

If $H > \sqrt{p}$ then we can do better by applying the Pólya-Vinogradov inequality (4.2) to the inner sum. If $S < p - 1$ then all the characters χ_s are non-principal, and hence

$$M_k(S, H) \ll Sp^{k/2}(\log p)^k$$

for any fixed $k \geq 0$. If $S \geq p - 1$ then there are $\approx S/p$ values of s in the interval $1 \leq s \leq S$ for which $(p-1)|s$, i.e. for which χ_s is the principal character. Hence if k is fixed, $k \geq 0$, then

$$M_k(S, H) \ll Sp^{-1}H^k + Sp^{k/2}(\log p)^k \tag{5.3}$$

uniformly for all positive S and H.

We now use Lemma 2 to show that if k is a positive integer then $M_{2k}(S, H)$ can be bounded in terms of $M_2(2S, H^k)$.

Lemma 3. *Let k be a fixed positive integer, and suppose that $\epsilon > 0$. Then*

$$M_{2k}(S, H) \ll H^{2k+\epsilon} + M_2(2S, H^k)H^\epsilon. \tag{5.4}$$

Proof. Since

$$\sum_{h=-H}^{H} (1 - |h|/H)\chi(h) = 2\Re\left(\sum_{h=1}^{H}(1 - h/H)\chi(h)\right),$$

we see that

$$M_{2k}(S, H) \ll \sum_{s=-S}^{S}\left|\sum_{h=1}^{H}(1 - h/H)\chi_s(h)\right|^{2k}$$

$$\ll \frac{1}{S}\sum_{s=-2S}^{2S}(2S - |s|)\left|\sum_{h=1}^{H}(1 - h/H)\chi_s(h)\right|^{2k}$$

$$= \frac{1}{S}\sum_{s=-2S}^{2S}(2S - |s|)\left|\sum_{h=1}^{H^k}a_h\chi_s(h)\right|^{2}$$

$$= \frac{1}{S}\sum_{r=1}^{2S}\sum_{t=1}^{2S}\left|\sum_{h=1}^{H^k}a_h\chi_r\overline{\chi_t}(h)\right|^{2}.$$

In the notation of Lemma 2 this is $S^{-1}J(\mathbf{w}, \mathbf{a})$ where

$$w(\chi_r) = \text{card}\left\{s : 1 \le s \le 2S, \ s \equiv r \pmod{p - 1}\right\}.$$

As $|a_h| \le d_k(h)(1 - h/H^k)$ for $1 \le h \le H^k$, we have $a_h \ll b_h$ where

$$b_h = H^{\epsilon/2}\max(0, 1 - |h|/H^k).$$

Hence by Lemma 2 our expression is $\ll S^{-1}J(\mathbf{w}, \mathbf{b})$, which by the same reasoning is

$$= \frac{1}{S}\sum_{s=-2S}^{2S}(2S - |s|)\left|\sum_{h=-H^k}^{H^k}b_h\chi_s(h)\right|^{2}.$$

Here the term $s = 0$ contributes an amount $\ll H^{2k+\epsilon}$, while the terms $s \ne 0$ are majorized by $M_2(2S, H^k)H^\epsilon$. This gives the stated result.

Lemma 4. *Suppose that H is an integer, $1 \le H \le p/2$, and put $L = [p/(2H)]$. Then*

$$M_2(S, H) \ll H^2 + \frac{H^2}{p}M_2(2S, L). \tag{5.5}$$

Proof. Clearly

$$M_2(S, H) \leq \frac{1}{S} \sum_{s=-2S}^{2S} (2S - |s|) \left| \sum_{h=-H}^{H} (1 - |h|/H)\chi_s(h) \right|^2.$$

In the notation of Lemma 2, this is $S^{-1}J(\mathbf{w}, \mathbf{a})$ where $\mathbf{w}$ is defined as in the preceding proof and $a_h = \max(0, 1 - |h|/H)$. Put

$$b_h = \left(\frac{\sin \pi L h/p}{L \sin \pi h/p} \right)^2.$$

Then $a_h \ll b_h$, so by Lemma 2 our expression is

$$\ll S^{-1}J(\mathbf{w}, \mathbf{b}) \ll \sum_{s=0}^{2S} \left| \sum_{h} b_h \chi_s(h) \right|^2. \tag{5.6}$$

Suppose first that $S < (p-1)/2$. Since $\sum_{h=1}^{p} b_h = p/L \approx H$, it follows that the term $s = 0$ above contributes an amount $\ll H^2$. For $0 < s \leq 2S$, the character χ_s is primitive, and hence we may apply Lemma 1. Since $\hat{c}(k) = L^{-1}\max(0, 1 - |k|/L)$ for $-p/2 < k < p/2$, and since $|\tau(\chi_s)| = \sqrt{p}$, it follows that the sum over $s > 0$ is $= pL^{-2}M_2(2S, L) \approx H^2 p^{-1}M_2(2S, L)$.

Now suppose that $S > (p-1)/2$. In this case the expression (5.6) is

$$\approx \frac{S}{p-1} \sum_{\chi} \left| \sum_{h} b_h \chi(h) \right|^2 = S \sum_{h} |b_h|^2 \approx SH.$$

Here we have used a variant of Parseval's identity for Dirichlet characters (see Theorem 6.2 of Montgomery [9]). Similarly,

$$M_2(2S, L) \approx \frac{S}{p-1} \sum_{\chi} \left| \sum_{k=-L}^{L} (1 - |k|/L)\chi(k) \right|^2$$

$$= S \sum_{k=-L}^{L} (1 - |k|/L)^2 \approx SL.$$

Thus the second term on the right hand side of (5.5) is $\approx SH$, so (5.5) holds in this case also.

Lemma 5. *Suppose that $H < \sqrt{p/2}$ and that $\epsilon > 0$ is given. Then*

$$M_1(S, H) \ll S^{3/4}H p^\epsilon + S^{7/8}p^{1/4+\epsilon} + S p^{1/8+\epsilon} + SH p^{-1/8+\epsilon}.$$

Proof. By Hölder's inequality we see that

$$M_1(S, H) \le S^{3/4} M_4(S, H)^{1/4}.$$

By Lemma 3, we know that

$$M_4(S, H) \ll H^{4+\epsilon} + M_2(2S, H^2) H^\epsilon.$$

From Lemma 4 we find that

$$M_2(2S, H^2) \ll H^4 + H^4 p^{-1} M_2(4S, L)$$

where $L \approx p/H^2$. By Cauchy's inequality,

$$M_2(4S, L) \ll S^{1/2} M_4(4S, L)^{1/2}.$$

By Lemma 3 we discover that

$$M_4(4S, L) \ll L^{4+\epsilon} + M_2(8S, L^2) L^\epsilon.$$

Finally,

$$M_2(8S, L^2) \ll S p^{-1} L^4 + S p (\log p)^2$$

by (5.3). On combining these estimates, we obtain the stated result.

6 Proof of Theorem 2

Clearly

$$f(m, H) = \frac{1}{p} \sum_{k=1}^{p} A(k) e(-km/p) \sum_{h=1}^{H} e(-kh/p).$$

Here the term $k = p$ contributes the mean value of f, so that

$$f(m, H) - NH/p = \frac{1}{p} \sum_{k=1}^{p-1} A(k) B(-k) e(-km/p)$$

where $B(k) = \sum_{h=1}^{H} e(hk/p)$. Hence by Parseval's identity for the finite Fourier transform,

$$\sum_{m=1}^{p} (f(m, H) - NH/p)^2 = \frac{1}{p} \sum_{k=1}^{p-1} |A(k) B(k)|^2. \tag{6.1}$$

Here $|B(k)|$ is large (i.e. $\approx H$) for small values of k (those for which $|k| \le p/H$), and then is smaller for larger values of k. Hence the right hand side above may be regarded as being a weighted partial sum of $|A(k)|^2$. We note that $A(k)$ is a character sum, for by (2.4),

$$A(k) = \sum_\chi a(\chi)\chi(k)$$

where $a(\chi) = \tau(\overline{\chi})K(\chi)/(p-1)$. On inserting this in (6.1), expanding the modulus-squared, and inverting the order of summation, we find that the right hand side of (6.1) is

$$= \frac{1}{p} \sum_\chi \sum_\psi a(\chi)\overline{a(\psi)} \sum_{k=1}^{p-1} |B(k)|^2 \chi \overline{\psi}(k). \tag{6.2}$$

The terms $\chi = \psi$ contribute an amount

$$\frac{1}{p} \left(\sum_\chi |a(\chi)|^2 \right) \left(\sum_{k=1}^{p-1} |B(k)|^2 \right). \tag{6.3}$$

On appealing to the definitions of $a(\chi)$ and of $K(\chi)$, we find that the sum over χ is

$$\frac{N^2}{(p-1)^2} + \frac{p}{(p-1)^2} \sum_{\chi \neq \chi_0} |K(\chi)|^2 = -\frac{N^2}{p-1} + \frac{p}{(p-1)^2} \sum_\chi |K(\chi)|^2.$$

By Parseval's identity for Dirichlet characters this is

$$-\frac{N^2}{p-1} + \frac{pN}{p-1} = \frac{N(p-N)}{p-1}.$$

By appealing to the definition of $B(k)$, and also to Parseval's identity for the additive characters modulo p, we find that the sum over k is

$$-H^2 + \sum_{k=1}^p |B(k)|^2 = -H^2 + pH = H(p-H).$$

Hence the contribution to (6.2) of $\chi = \psi$ is $NH(p-H)(p-N)/(p(p-1))$.

Before completing the proof we present a simple argument that establishes the desired estimate in the restricted range $p^{3/4+\epsilon} \leq N \leq p$. By partial summation we see that if $\chi \neq \chi_0$ and w has bounded variation then

$$\sum w(k)\chi(k) \ll p^{1/2}(\log p)(\sup|w| + \operatorname{var} w).$$

Taking $w(k) = |B(k)|^2$, we deduce that if $\chi \neq \psi$ then the sum over k in (6.2) is $\ll p^{1/2}(\log p)H^2$. Since

$$\sum |a(\chi)| \ll p^{-1/2} \sum |K(\chi)| \ll p^{1/2} \log p$$

by (2.3), it follows that the expression (6.2) is

$$= \frac{(p-H)(p-N)}{p(p-1)} NH + O(p^{1/2} H^2 (\log p)^3). \tag{6.4}$$

This gives the stated result if $H \approx p/N$ and $p^{3/4} (\log p)^2 \le N \le p$.

We now return to the proof of Theorem 2. For $\chi \ne \psi$ the character $\chi\bar{\psi}$ is primitive, so we may appeal to Lemma 1. If $c(k) = |B(k)|^2$ then by the orthogonality property of the additive characters we see that $\hat{c}(h) = \max(0, H - |h|)$ where $-p/2 < h < p/2$. Here we have assumed that $H \le p/2$. This is permissible, since $H \approx p/N < p^{2/7}$. Hence by Lemma 1, if $\chi \ne \psi$ then the sum over k in (6.2) is

$$\frac{p}{\tau(\overline{\chi}\psi)} \sum_{h=-H}^{H} (H - |h|) \overline{\chi}\psi(h).$$

Thus the expression in (6.1) is

$$\frac{(p-H)(p-N)}{p(p-1)} NH + \sum_{\chi \ne \psi} \frac{a(\chi)\overline{a(\psi)}}{\tau(\overline{\chi}\psi)} \sum_{h=-H}^{H} (H - |h|) \overline{\chi}\psi(h). \tag{6.5}$$

The first term is our anticipated main term. From the definition of $a(\chi)$ we see that the second term is

$$\ll p^{-3/2} \sum_{\chi \ne \psi} |K(\chi)K(\psi)| \left| \sum_{h=-H}^{H} (H - |h|) \overline{\chi}\psi(h) \right|. \tag{6.6}$$

In using the triangle inequality here, we have discarded any possible cancellation in the double sum over χ and ψ. If the Generalized Lindelöf Hypothesis is true then the sum over h in (6.5) is $\ll H^{3/2} p^\epsilon$, and so the expression (6.6) is $\ll H^{3/2} p^{1/2+\epsilon}$. This suffices if $p^{2/3+\epsilon} \le N \le p$. On the other hand, the expression (6.6) is virtually certain to be $\gg H^{3/2} p^{1/2}$, so our approach will never succeed for $N \le p^{2/3}$. The sum over h in (6.6) is trivially $\ll H^2$; this gives (6.4) again. We now show that—on average—there is some cancellation in this sum over h. Suppose that $\chi = \chi_r$, $\psi = \chi_{r+s}$. By (2.3) we see that

$$\sum_{r=1}^{p-1} |K(\chi_r)K(\chi_{r+s})| \ll c_s$$

where $c_s = pN$ if $1 \le s \le p/N$, $c_s = p^2 s^{-1} \log(2Ns/p)$ for $p/N \le s \le (p-1)/2$, $c_s = c_{p-1-s}$ for $(p-1)/2 < s \le p-1$. Hence the expression (6.6) is

$$\ll p^{-3/2} H \sum_{s=1}^{(p-1)/2} c_s \left| \sum_{h=-H}^{H} (1 - |h|/H)\chi_s(h) \right|.$$

By partial summation this is

$$\ll p^{-1/2} N H M_1(p/N, H) + p^{1/2} H \sum_{S=p/N}^{p-1} M_1(S, H) S^{-2} \log(2NS/p) \quad (6.7)$$

in the notation of (5.1). By the bound of Lemma 5, the above is

$$\ll H^2 N^{1/4} p^{1/4+\epsilon} + H N^{1/8} p^{5/8+\epsilon} + H p^{5/8+\epsilon} + H^2 p^{3/8+\epsilon}.$$

This is $o(p)$ when $H \approx p/N$, $p^{5/7+2\epsilon} \leq N \leq p$, so the proof is complete.

7   Proof of the Corollary

Let $\mathcal{G} = \{m : 1 \leq m \leq p, f(m, H) > 0\}$ be the set of "good" m. By Cauchy's inequality

$$\left(\sum_{m=1}^{p} f(m, H) \right)^2 \leq \left(\sum_{m \in \mathcal{G}} 1 \right) \left(\sum_{m=1}^{p} f(m, H)^2 \right).$$

By (1.3) the left hand side is $N^2 H^2$. By Theorem 2, the second sum on the right is $\ll N^2 H^2 p^{-1} + N H$ if $H \approx p/N$, $p^{5/7+\epsilon} \leq N \leq p$. Hence under these conditions

$$\operatorname{card} \mathcal{G} \gg \frac{N^2 H^2}{N^2 H^2 p^{-1} + N H} \gg p.$$

That is, $f(m, H) > 0$ for a positive proportion of the residue classes m (mod p).

References

1. Davenport, H. (1980). *Multiplicative Number Theory* (2nd edn). Springer-Verlag, New York.

2. Erdős, P. and Turán, P. (1948). On a problem in the theory of uniform distribution I, II. *Indag. Math.* **10**, 370–378, 406–413.

3. Vaaler, J. D. (1985). Some extremal functions in Fourier analysis. *Bull. Amer. Math. Soc.*, **12**, 183–216.

4. Pólya, G. (1918). Über die Verteilung der quadratischen Reste und Nichtreste. *Nachr. Akad. Wiss. Göttingen*, 21–29.

5. Vinogradov, I. M. (1918). Sur la distribution des résidues et non résidues de puissances. *Perm. Univ. Fiz.-mat. ob-vo Zh.*, **1**, 18–28, 94–98.

6. Landau, E. (1918). Abschätzungen von Charaktersummen, Einheiten und Klassenzahlen. *Nachr. Akad. Wiss. Göttingen*, 79–97.

7. Schur, I. (1918). Einige Bemerkungen zu der vorstehenden Arbeit des Herrn G. Pólya: Über die Verteilung der quadratischen Reste und Nichtreste. *Nachr. Akad. Wiss. Göttingen*, 30–36.

8. M. Jutila (1977). Zero-density estimates for L-functions. *Acta. Arith.*, **32**, 52–62.

9. Montgomery, Hugh L. (1971). *Topics in Multiplicative Number Theory*. Springer Lecture Notes 227. Springer-Verlag, Berlin.

Hugh L. Montgomery
Department of Mathematics
University of Michigan
Ann Arbor, MI 48109–1003
USA
e-mail hugh.montgomery@math.lsa.umich.edu

Ordered Factorizations for Integers and Arithmetical Semigroups

A. Knopfmacher, J. Knopfmacher and R. Warlimont

University of the Witwatersrand, South Africa
and
Universität Regensburg, Germany

1   Introduction

The classical problems of "factorisatio numerorum" concern (i) the total number of ordered factorizations of a natural number $n > 1$ into products of natural numbers larger than 1, and (ii) the corresponding total number of unordered factorizations of the same type. Results of this nature can be found, for example, in [1], [7], [12] and the references therein. Recently, in [8], the present authors considered the analogous problems within the context of additive arithmetical semigroups as treated in the monographs [10], [11]. Concrete cases of these semigroups include (i) the multiplicative semigroup G_q of all monic polynomials in one indeterminate over a finite field $\mathbb{F}_q$, (ii) semigroups of ideals in principal orders within algebraic function fields over $\mathbb{F}_q$, (iii) semigroups formed under direct sum by the isomorphism classes of certain kinds of finite modules or algebras over such principal orders.

The purpose of this paper is to study in both the integer and semigroup cases the numbers of ordered factorizations into elements satisfying certain restrictions. As far as we are aware the only previous investigation of this type is by Erdös [2], for the integer case, who proved the following result (see [3] for the correct statement of this Theorem).

Theorem A   *Let $A : 1 < a_1 < a_2 < \dots$ be a sequence of integers a_i such that for some ρ,*

$$\sum_{i=1}^{\infty} 1/a_i^{\rho} = 1 \quad and \quad \sum_{i=1}^{\infty} \frac{\log a_i}{a_i^{\rho}} \quad converges,$$

Presented at CNTA '91 by A. Knopfmacher.

and such that not all the a_i's are powers of a_1. Denote by $F(n) = F_A(n)$ the number of ordered factorizations of n into the a_i's and let $F(1) = 1$. Then

$$\sum_{m=1}^{n} F(m) = cn^{\rho}(1 + o(1)) . \tag{1.1}$$

The value of the constant c in (1) is *not* however determined. Here we shall show:

Theorem 1. Let $1 < a_1 < a_2 < \ldots$ be an infinite sequence of positive integers and assume that the associated Dirichlet series $B(s) = \Sigma a_n^{-s}$ has abscissa of convergence β, $0 < \beta < \infty$. In addition assume that $\lim_{\sigma \to \beta+} B(\sigma) > 1$ and $\frac{\log a_k}{\log a_\ell}$ is irrational for some indices k, ℓ. Then there exists a unique ρ, $\beta < \rho < \infty$ such that $B(\rho) = 1$ and with the notation of Theorem A,

$$\sum_{m=1}^{n} F(m) \sim -\frac{1}{\rho B'(\rho)} n^{\rho} \qquad as \quad n \to \infty,$$

where $B'(\rho) < 0$.

(Here and subsequently the notation $f \sim g$ means $f = g(1 + o(1))$.) To state the analogous result for semigroups we require some definitions:

An *(additive) arithmetical semigroup* will be understood to be free commutative semigroup G with identity element 1, generated by a (countable) set P of "prime" elements, which admits an integer-valued "degree" mapping ∂ such that

(i) $\partial(1) = 0, \partial(p) > 0$ for all $p \in P$,

(ii) $\partial(ab) = \partial(a) + \partial(b)$ for all $a, b \in G$,

(iii) the total number $G^{\#}(n)$ of elements of degree n in G is finite for all $n \in \mathbb{N}$.

All the concrete cases referred to earlier provide examples of an arithmetical semigroup G which satisfies

AXIOM $A^{\#}$ There exist constants $A > 0, q > 1$, and ν with $0 \leq \nu < 1$, such that

$$G^{\#}(n) = Aq^n + 0(q^{\nu n}) \quad as \quad n \to \infty.$$

In dealing with an arithmetical semigroup G, it is frequently helpful to consider the "zeta" (or "generating") function

$$Z(y) = Z_G(y) = \sum_{n=0}^{\infty} G^{\#}(n) y^n.$$

Under Axiom $A^{\#}$, $Z(y)$ is holomorphic for $y| < q^{-1}$, and extends to a meromorphic function in the disc $|y| < q^{-\nu}$, having exactly one pole (of order 1) at $y = q^{-1}$. In fact, for $|y| < q^{-\nu}$,

$$Z(y) = \prod_{p \in P} (1 - y^{\partial(p)})^{-1} = \frac{A}{1 - qy} + h_1(y)$$

where $h_1(y)$ is holomorphic for $|y| < q^{-\nu}$.

Similarly if $A = \{a_i\}$ is an infinite subset of the elements $a_i \neq 1$, of such a semigroup, denote by $A^{\#}(n)$ the total number of elements of degree n belonging to A, $A^{\#}(0) = 1$, and let

$$Z_A(y) = \sum_{n=0}^{\infty} A^{\#}(n) y^n$$

be the corresponding "zeta" function.

Now consider the function $F(b) = F_A(b)$ of $b \neq 1$ in G which is directly analogous to the factorization-counting function $F(n)$ earlier, and let $F(1) = 1$. Define the associated summatory function $\bar{F}$ by

$$\bar{F}(n) = \sum_{\partial(b)=n} F(b).$$

Theorem 2. *Let $G \neq \{1\}$ be an arithmetical semigroup and $A = \{a_i\}$ be an infinite subset of the elements of $G - \{1\}$ which has a zeta function $Z_A(y)$ which is holomorphic in some disc $|y| < r$. In addition assume that $\lim_{y \to r^-} Z_A(y) > 2$ and that there are $1 \leq k_1 < k_2 < \cdots < k_h$ such that $A^{\#}(k_j) > 0 \quad (1 \leq j \leq h)$ and $(k_1, k_2, \ldots, k_h) = 1$. Then*

$$\bar{F}(n) = \frac{y_0^{-n}}{y_0 Z_A'(y_0)} + 0(t_0^{-n}) \quad as \quad n \to \infty,$$

where $y_0 \in (0, r)$ is the unique real solution of $Z_A(y) - 2 = 0$, and $y_0 < t_0 < r$.

The question of obtaining corresponding results for unordered factorizations is unsettled. In this case results would depend strongly on the quantities $A^{\#}(n)$.

The proofs of the theorems 1 and 2 appear in sections 2 and 3 respectively. A case of particular interest is that of ordered "prime factorizations" for integers and semigroups. In addition to the mean value we determine for this and a few other special cases the normal order of the function $\bar{F}(n)$.

2   Ordered Factorizations of Integers

Our approach to Theorem 1 is based on the following consequence of the Ikehara-Wiener Tauberian Theorem (see e.g. [8, Chapter 6, Theorem 1.4, and Lemma 2.2]).

Theorem B   *Let* $A(s) = \sum_{n=1}^{\infty} a_n n^{-s}$ *be a Dirichlet-series with coefficients* $a_n \geq 0$, *and abscissa of convergence* $\alpha, 0 < \alpha < \infty$. *Assume that* $A(s)$ *can be continued meromorphically onto a region* Ω *which contains the closed half plane* $\operatorname{Re} s \geq \alpha$ *in such a way that there is a unique pole which is located at* α *and has principal part* $\frac{c}{s-\alpha}$. *Then*

$$\sum_{n \leq x} a_n \sim \frac{c}{\alpha} x^{\alpha} .$$

Now for a given infinite sequence of integers

$$A : 1 < a_1 < a_2 < \cdots$$

the associated formal Dirichlet series is

$$\sum_{i=1}^{\infty} a_i^{-s} = \sum_{n=2}^{\infty} b_n n^{-s} = B(s), \quad \text{say,}$$

where $b_n = \begin{cases} 1, & n = a_i, \quad \text{some } i, \\ 0, & \text{otherwise.} \end{cases}$

Formally the generating function for the number of such ordered factorizations $F(n)$ in terms of the sequence $A = \{a_i\}$ is

$$\sum_{n=1}^{\infty} F(n) n^{-s} = 1 + \sum_{k=1}^{\infty} B^k(s) = \frac{1}{1 - B(s)} .$$

Theorem 1 is now a consequence of the more general result proved below.

Theorem 3. *Let* $B(s) = \sum_{n=2}^{\infty} b_n n^{-s}$ *be a Dirichlet series with coefficients* $b_n \geq 0$ *and abscissa of convergence* $\beta, 0 < \beta < \infty$. *Then for any given* r *satisfying* $0 < r < \lim_{\sigma \to \beta+} B(\sigma)$ *there is a unique* $\alpha, \beta < \alpha < \infty$, *such that* $B(\alpha) = r$, *and* $B'(\alpha) < 0$. *Furthermore, for* $\operatorname{Re} s > \alpha$, *the function*

$$A(s) = \frac{1}{r - B(s)}$$

can be represented as a Dirichlet series $A(s) = \sum_{n=1}^{\infty} c_n n^{-s}$. *Assume there are subscripts* k, ℓ *with* $2 \leq k < \ell$ *such that* $b_k > 0$, $b_\ell > 0$ *and* $\log k / \log \ell$ *is irrational. Then one has*

$$\sum_{n \le x} c_n \sim -\frac{1}{\alpha B'(\alpha)} x^\alpha.$$

Proof. $B(s)$ decreases strictly from ∞ to 0 as s increases from β to ∞. Therefore there is a unique α, $\beta < \alpha < \infty$, with $B(\alpha) = r$ and $B'(\alpha) < 0$. For Re $s > \alpha$ one has

$$|B(s)| \le B(\mathrm{Re}\ s) < B(\alpha) = r.$$

Therefore by absolute convergence we get

$$\begin{aligned}
\frac{1}{r - B(s)} &= \frac{1}{r} \frac{1}{\left(1 - \frac{1}{r} B(s)\right)} \\
&= \frac{1}{r} \sum_{k=0}^{\infty} r^{-k} B(s)^k \\
&= \frac{1}{r} \sum_{k=0}^{\infty} r^{-k} \sum_{n=1}^{\infty} b_{k,n} n^{-s} \\
&= \sum_{n=1}^{\infty} c_n n^{-s}
\end{aligned}$$

with

$$c_n = \frac{1}{r} \sum_{k=0}^{\infty} b_{k,n} r^{-k}.$$

Now let there exist k, ℓ with the property formulated in the statement of the theorem. Let $t \in \mathbb{R}$, $t \ne 0$, be given. Then

$$\begin{aligned}
\mathrm{Re}\, B(\alpha + it) &= \sum_{n=2}^{\infty} b_n n^{-\alpha} \cos(t \log n) \\
&\le \sum_{\substack{n=2 \\ n \ne k, \ell}}^{\infty} b_n n^{-\alpha} + b_k k^{-\alpha} \cos(t \log k) + b_\ell \ell^{-\alpha} \cos(t \log \ell).
\end{aligned}$$

Since $\log k / \log \ell$ is irrational either $\cos(t \log k) < 1$ or $\cos(t \log \ell) < 1$. Therefore $\mathrm{Re}\, B(\alpha + it) < B(\alpha) = r$ which implies $B(\alpha + it) \ne r$.

Thus there is a region $W \supset \{\mathrm{Re}\ s \ge \alpha\}$ such that $B(s) \ne r$ for all $s \in W, s \ne \alpha$. Consequently $A(s) = \frac{1}{r - B(s)}$ is holomorphic on $W - \{\alpha\}$ with a pole at α of first order and residue

$$\lim_{s \to \alpha} \frac{s - \alpha}{r - B(s)} = -\lim_{s \to \alpha} \frac{s - \alpha}{B(s) - B(\alpha)} = -\frac{1}{B'(\alpha)}.$$

Now the result follows from the version of the Ikehara-Wiener theorem stated above.

Remark. If k, ℓ are such that $\prod_{p|k} p \neq \prod_{p|\ell} p$ then our condition is fulfilled. In particular it holds if k and ℓ are different primes.

Of particular interest are the following special cases of Theorem 1:

(i) Let $a_n = n$, $n \geq 2$; then we re-obtain the result of Kalmár [7]:
$$\sum_{n=1}^{\infty} F(n)n^{-s} = \frac{1}{2-\zeta(s)}, \text{ where } \zeta(s) \text{ is the Riemann zeta function,}$$
and
$$\sum_{m=1}^{n} F(m) \sim -\frac{1}{\rho_1 \zeta'(\rho_1)} n^{\rho_1},$$

where $1 < \rho_1 < 2$ is the unique root of $\zeta(s) = 2$.

(ii) Let $\{a_n\}$ consist of all the *square free* positive integers > 1. Let $B_2(s) = \sum_{n=1}^{\infty} a_n^{-s} = -1 + \frac{\zeta(s)}{\zeta(2s)}$. Then the average number of ordered factorizations of $n > 1$ into *square-free* factors is
$$\frac{1}{n}\sum_{m=1}^{n} F(m) \sim -\frac{1}{\rho_2 B_2'(\rho_2)} n^{\rho_2-1}$$

where $B_2(\rho_2) = 1$ and $1 < \rho_2 < \rho_1$, since $B_2(s) < \zeta(s) - 1$, for $s > 1$.

(iii) The generating function for ordered factorizations of $n > 1$ into *primes* is
$$\sum_{n=1}^{\infty} \hat{F}(n)n^{-s} = \frac{1}{1 - \sum_{p \text{ prime}} p^{-s}} = \frac{1}{1 - B_3(s)}, \quad \text{say.}$$

It follows from Theorem 1 that the average number of ordered "prime" factorizations is
$$\frac{1}{n}\sum_{m=1}^{n} \hat{F}(m) \sim -\frac{1}{\rho_3 B_3'(\rho_3)} n^{\rho_3-1},$$

where $1 < \rho_3 < \rho_2$ and $B_3(\rho_3) = 1$.

Unlike the previous two cases one can write down a simple exact formula for $\hat{F}(n)$ in terms of the prime factorization of n. If $n = p_1^{\alpha_1} p_2^{\alpha_2} \ldots p_k^{\alpha_k}$ then
$$\hat{F}(n) = \frac{(\alpha_1 + \alpha_2 + \ldots + \alpha_k)!}{\alpha_1! \alpha_2! \ldots \alpha_k!}.$$

This formula is not useful however for the purposes of obtaining the mean value of $\hat{F}(n)$. By using a computer we have obtained the numerical estimates 1.73, 1.58 and 1.40 for ρ_1, ρ_2 and ρ_3, respectively.

In addition to the mean value it is of interest to determine the *normal order* of the respective functions $F(n)$ for these special cases.

Here, like Sklar [13], we define a normal order of an arithmetical function $h(n) : \mathbb{N} \to \mathbb{R}$ to be $H(n)$ if $h(n) \sim H(n)$ as $n \to \infty$ over "almost all" integers (in the sense of Hardy and Ramanujan [4]); this definition is a slight variation of that in [4], Chap.2 .

Theorem 4. *A normal order of* $\log F(n)$ *is* $(\log \log n)(\log \log \log n)$ *in each of the cases (i), (ii) and (iii) above.*

Proof. Denote by $\Omega(n)$ and $\omega(n)$ the arithmetical functions which count the total number (respectively, the distinct number) of irreducible factors of n.

For cases (i) and (ii) we use the fact that for a square free positive integer m with exactly r distinct prime factors $F(m) = B_r$ where B_r denotes the number of ordered partitions of an r-element set. Asymptotically it is known (see e.g. Sklar [13]) that

$$ B_r = \frac{r!}{2(\log 2)^{r+1}} \big(1 + o(1)\big) . $$

Hence

$$ B_{\omega(n)} \le F(n) \le B_{\Omega(n)} $$

which implies by Stirling's formula that

$$ \omega(n) \log \omega(n)\big(1 + o(1)\big) \le \log F(n) \le \Omega(n) \log \Omega(n)\big(1 + o(1)\big) . $$

Since both $\Omega(n)$ and $\omega(n)$ have normal order $\log \log n$ ([4]), the result follows for cases (i) and (ii). Case (iii) follows similarly using the inequality

$$ \omega(n)! \le F(n) \le \Omega(n)! . $$

Remark. The normal order for case (i) is also derived in the paper of Sklar [13].

We conclude this section with a critique of Erdös' theorem.

In his proof Erdös uses his assumption "not all the a_i's are powers of a_i" in the following fashion:

Let a_i be the first element of A which is not a power of a_1.

Now consider the strictly increasing sequence (n_k) of all naturals of the form $a_1^\alpha a_i^\beta$. Then he asserts: "it is well known that $\lim_{k \to \infty} \frac{n_{k+1}}{n_k} = 1$". But take the example where $a_1 = 4$ and $a_2 = 8$.

Then (n_k) is the sequence of powers of 4 and twice powers of 4 and $\lim_{k \to \infty} \frac{n_{k+1}}{n_k} = 2$.

Correction of Erdös' theorem: replace the above assumption with our assumption "$\log a_k / \log a_\ell$ is irrational for some indices k, ℓ. Then the sequence (n_k) of all naturals of the form $a_k^\alpha a_\ell^\beta$ fulfills indeed $\lim_{k \to \infty} \frac{n_{k+1}}{n_k} = 1$ and Erdös' proof goes through after minor modifications.

Let us stay for a last moment with Erdös theorem. We want to evaluate his constant c.

Let $\rho < \sigma \leq \rho + 1$. Then

$$
\begin{aligned}
\frac{1}{\rho - B(\sigma)} &= A(\sigma) = \sigma \int_1^\infty \left(\sum_{n \leq x} f(n) \right) x^{-\sigma-1} \mathrm{d}x \\
&= \sigma \int_1^\infty c x^\rho (1 + o(1)) x^{-\sigma-1} \mathrm{d}x \\
&= \frac{c\sigma}{\sigma - \rho} + o\left(\frac{1}{\sigma - \rho} \right) \quad \text{for } \sigma \to \rho^+ .
\end{aligned}
$$

From this we see that $B'_+(\rho)$ exists and

$$
c = -\frac{1}{\rho B'_+(\rho)} .
$$

This too is in agreement with our Theorem 1.

3   Ordered Factorizations in Arithmetical Semigroups

The proof of Theorem 2 follows immediately from the following Lemmas and the generating function

$$
\sum_{n=0}^\infty \bar{F}(n) y^n = 1 + \sum_{k=1}^\infty \left(Z_A(y) - 1 \right)^k = \left(2 - Z_A(y) \right)^{-1} .
$$

Lemma 5. Let $G \neq \{1\}$ be an arithmetical semigroup and $A = (a_i)$ an infinite sequence of elements of $G - \{1\}$ which has a zeta function $Z_A(y)$ which is holomorphic in some disc $|y| < r$ and takes values > 2 for some real values of $y < r$. In addition assume that there are subscripts $1 \leq k_1 < k_2 < \cdots < k_h$ such that $A^\#(k_j) > 0$ $(1 \leq j \leq h)$ and $(k_1, k_2, \ldots, k_h) = 1$. Then the equation $Z_A(y) - 2 = 0$ has a unique and simple root $y_0 \in (0, r)$ but no other complex root y with $|y| \leq y_0$.

Lemma 6. Let $h(y)$ be holomorphic for $|y| < r$, and let $w_0 = h(y_0)$ for some complex y_0 with $0 < |y_0| < r$. Suppose that $h'(y_0) \neq 0$ and $h(y) \neq w_0$ for $|y| \leq |y_0|, y \neq y_0$. Then, if

$$
\left(w_0 - h(y) \right)^{-1} = \sum_{n=0}^\infty c(n) y^n \quad \text{for } |y| < |y_0|,
$$

there exists t_0 with $|y_0| < t_0 < r$, such that

$$
c(n) = \frac{y_0^{-n}}{y_0 h'(y_0)} + 0(t_0^{-n}) \quad \text{as } n \to \infty.
$$

The proofs of Lemmas 5 and 6 appear in [8] except that the necessary assumptions on the indices k_j in Lemma 5 were incorrectly omitted in the corresponding Lemma 1.1 of [8]. We require these assumptions in order to show that $Z_A(y) \neq 2$ for $|y| = y_0$, $y \neq y_0$: Suppose that $0 < t < 1$. Then

$$\operatorname{Re} Z_A(y_0 e^{2\pi i t}) = \sum_{n=1}^{\infty} A^{\#}(n) y_0^n \cos(2\pi t n)$$

$$\leq \sum_{\substack{n=1 \\ n \neq k_j \, (1 \leq j \leq h)}}^{\infty} A^{\#}(n) y_0^n$$

$$+ \sum_{j=1}^{h} A^{\#}(k_j) y_0^{k_j} \cos(2\pi t k_j) \,.$$

Assume

$$\cos(2\pi t k_j) = 1 \quad (1 \leq j \leq h) \,.$$

Then there would exist integers l_j $(1 \leq j \leq h)$ such that

$$2\pi k_j t = 2\pi l_j \quad (1 \leq j \leq h) \,.$$

However, since $(k_1, k_2, \ldots, k_h) = 1$, there are integers q_j $(1 \leq j \leq h)$ such that $q_1 k_1 + \cdots + q_h k_h = 1$. But then

$$t = \sum_{j=1}^{h} q_j k_j t = \sum_{j=1}^{h} q_j l_j \in \mathbb{Z} \,,$$

a contradiction. We deduce that

$$\operatorname{Re} B(\alpha e^{2\pi i t}) < B(\alpha) = \beta$$

as required.

The above argument replaces the erroneous one used to prove the equivalent result in Lemma 1.1 of [8].

It is now trivial to deduce estimates from Theorem 2 for the special cases analogous to cases (i), (ii), (iii) treated in the previous section. If, however, we consider some concrete cases of semigroups satsifying Axiom $A^{\#}$, as indicated in the introduction, and consider cases for which q is an arbitrarily large prime power then we can make some interesting deductions about the average $\bar{F}(n)/G^{\#}(n)$ as $q \to \infty$. In all the cases below we assume that the constant ν in Axiom $A^{\#}$, $0 \leq \nu < 1$ is independent of q.

(i) If $\bar{F}(n)$ denotes the average in the case of all ordered factorizations, then as shown in [8], $y_0 \sim \frac{1}{(A+1)q}$, $Z_G'(y_0) \sim \frac{q}{a}(A+1)^2$ and $\frac{\bar{F}(n)}{G^{\#}(n)} \sim (A+1)^{n-1}$ as $q \to \infty$. For the earlier special semigroup G_q of Section 1 (for which $A = 1$) we have the exact result $\bar{F}(n) = 2^{n-1} q^n$.

(ii) If $\bar{F}(n)$ denotes the average number of ordered "square-free" factorizations then

$$\sum_{n=0}^{\infty} \bar{F}(n)y^n = \frac{1}{2 - \hat{Z}(y)}$$

where $\hat{Z}(y) = \frac{Z(y)}{Z(y^2)}$ is the generating function for *square-free* elements of G. Now under Axiom $A^{\#}$ for $0 < y < q^{-1}$

$$Z_G(y) = 1 + \frac{Aqy}{1 - qy} + O(q^{-1}) \quad \text{as} \quad q \to \infty.$$

Thus $Z_G(y^2) = 1 + O(q^{-1})$ as $q \to \infty$ and it follows that also, $\hat{Z}_G(y) = 1 + \frac{Aqy}{1-qy} + O(q^{-1}), q \to \infty$. So if y_0 satisfies $\hat{Z}_G(y_0) - 2 = 0$ then

$$y_0 = \frac{1}{(A+1)q} + O(q^{-2}).$$

From this we obtain

$$Z'_G(y_0) = \frac{q}{A}(A+1)^2 + O(1),$$

$$Z'_G(y_0^2) = qA + O(1).$$

Thus

$$\hat{Z}'_G(y_0) = \frac{Z'_G(y_0)}{Z_G(y_0^2)} - \frac{Z_G(y_0)2y_0 Z'_G(y_0^2)}{Z_G^2(y_0^2)}$$

$$= \left(1 + O(q^{-1})\right)\left(\frac{q}{A}(A+1)^2 + O(1)\right) -$$

$$\left(1 + O(q^{-1})\right)\left(\frac{4}{(A+1)q} + O(q^{-2})\right)qA\left(1 + O(q^{-1})\right)$$

$$= \frac{q}{A}(A+1)^2 + O(1).$$

Hence as in case (i) we obtain

$$\frac{\bar{F}(n)}{G^{\#}(n)} \sim (A+1)^{n-1} \quad \text{as} \quad q \to \infty.$$

For the semigroup G_q we can again derive an exact result. Here

$$\hat{Z}_G(y) = \frac{1 - qy^2}{1 - qy} = 1 + qy + \left(1 - \frac{1}{q}\right)\sum_{n=2}^{\infty} q^n y^n.$$

The generating function for ordered "square-free" factorizations becomes

$$\frac{1}{2 - \frac{1-qy^2}{1-qy}} = \frac{1-qy}{1-2qy+qy^2}$$

$$= \frac{q^{-1} - 1 - \sqrt{1-q^{-1}}}{2\sqrt{1-q^{-1}}(y-1-\sqrt{1-q^{-1}})}$$

$$-\frac{q^{-1} - 1 + \sqrt{1-q^{-1}}}{2\sqrt{1-q^{-1}}(y-1+\sqrt{1-q^{-1}})}$$

Thus

$$\bar{F}(n) = \frac{1}{2\sqrt{1-q^{-1}}}\left\{\left(q^{-1} - 1 + \sqrt{1-q^{-1}}\right)\left(1 - \sqrt{1-q^{-1}}\right)^{-n-1}\right.$$

$$\left. -\left(q^{-1} - 1 - \sqrt{1-q^{-1}}\right)\left(1 + \sqrt{1-q^{-1}}\right)^{-n-1}\right\}$$

$$\sim q^n 2^{n-1}$$

as $q \to \infty$.

(iii) In the case of ordered "prime" factorizations we have as generating function

$$\frac{1}{1 - \sum_{n=1}^{\infty} \pi(n)y^n}$$

where $\pi(n)$ denotes the number of irreducible elements of degree n in the semigroup G. Under the assumption of Axiom $A^{\#}$ it is possible to derive an abstract prime number theorem consisting of two distinct cases. In its full generality, following partial contributions by earlier authors, this has only recently been established by Indlekofer, Manstavicius, & Warlimont [5,6] and Zhang [14]. Also for polynomials over finite fields the number of ordered "prime" factorizations has already been investigated by Knopfmacher and Ridley [9].

Case 1: $Z_G(-\frac{1}{q}) \neq 0$. Here

$$\pi(n) = \frac{q^n}{n} + 0(s^n) \quad \text{where} \quad 0 \le s < q.$$

Thus if we let $q \to \infty$, then $\frac{\pi(n)}{q^n} \to \frac{1}{n}$ and we obtain in the limit the generating function $\frac{1}{1-\sum_{n=1}^{\infty}\frac{q^n y^n}{n}} = \frac{1}{1+\log(1-qy)}$. Using the estimates for the coefficients of this function in [9] or the approach used for case 2 below we obtain

$$\frac{\bar{F}(n)}{G^{\#}(n)} = \frac{1}{A(e-1)}\left(\frac{e}{e-1}\right)^n + 0(1), \quad as \quad n \to \infty.$$

Even for the semigroup G_q there seems to be no simple exact result for finite q. However in this case using the exact result

$$\pi(n) = \frac{1}{n} \sum_{d|n} \mu(d) q^{n/d} \quad \text{for small} \quad n,$$

and the bounds

$$\frac{q^n}{n} - \frac{2q^{n/2}}{n} < \pi(n) < \frac{q^n}{n} \quad \text{for larger} \quad n,$$

a computer can be used to obtain the following estimates for the root of $1 - \sum_{n=1}^{\infty} \frac{\pi(m)}{q^m} y^m = 0$ for small q. As expected this root $\to 1 - e^{-1} = 0.6321 \cdots$ as $q \to \infty$.

q	root of $1 - \sum_{m=1}^{\infty} \dfrac{\pi(m)}{q^m} y^m$
2	0.6880
3	0.6640
4	0.6544
5	0.6492
7	0.6438

Case 2: $Z_G\left(-\frac{1}{q}\right) = 0$. Here

$$\pi(n) = \frac{q^n}{n}\left(1 - (-1)^n + 0(s^n)\right) \quad \text{where} \quad \sqrt{q} \le s < q.$$

In the limiting case the generating function now becomes

$$\frac{1}{1 - 2\sum_{n=1}^{\infty} \frac{q^{2n}y^{2n}}{2n}} = \frac{1}{1 + 2\log(1 - q^2 y^2)} = \frac{1}{1 + 2\log(1 - w^2)} = G(w),$$

say, where we set $w = qy$.

This is a holomorphic function of w for $|w| < 1$ except for simple poles at $w = \sqrt{1 - e^{-\frac{1}{2}}}$ and $w = -\sqrt{1 - e^{-\frac{1}{2}}}$, with residues $-\dfrac{e^{-\frac{1}{2}}}{4\sqrt{1-e^{-\frac{1}{2}}}}$ and $\dfrac{e^{-\frac{1}{2}}}{4\sqrt{1-e^{-\frac{1}{2}}}}$ respectively.

In addition this function is bounded on the circle $|w| = 1$ because

$$G(w) = 1 - \frac{1}{\left(2\log(1 - w^2)\right)^{-1} + 1} \to 0 \quad as \quad w \to \pm 1.$$

Let $G(w) = \sum_{n=0}^{\infty} c_n w^n$ for $|w| < 1$ and put

$$J(n) = \frac{1}{2\pi i} \int_{|w|=1} \frac{G(w)}{w^{n+1}} dw.$$

Since $G(w)$ is an even function, $J(n) = 0$ for n odd. Now consider n even. On the one hand $J(n) = 0(1)$ and on the other Cauchy's residue theorem yields

$$J(n) = c_n - \frac{e^{-\frac{1}{2}}}{4\sqrt{1 - e^{-\frac{1}{2}}}} \left(-\sqrt{1 - e^{-\frac{1}{2}}}\right)^{-n-1}$$
$$+ \frac{e^{-\frac{1}{2}}}{4\sqrt{1 - e^{-\frac{1}{2}}}} \left(-\sqrt{1 - e^{-\frac{1}{2}}}\right)^{n-1}$$

Since $G^{\#}(n) \sim Aq^n$ as $n \to \infty$ we deduce that

$$\frac{\bar{F}(n)}{G^{\#}(n)} = \begin{cases} \dfrac{1}{2A\sqrt{e}} \left(\dfrac{1}{\sqrt{1 - e^{-\frac{1}{2}}}}\right)^{n+2} + 0(1), & n \quad \text{even}, \\ \\ 0 & \text{odd}. \end{cases}$$

In analogy to the integer case we consider the question of normal order. Recall that an arithmetical function $h : G \to \mathbb{R}$ is said to be of *normal order* $H(n)$ if for almost all elements b of degree n in G (i.e. for all except $o(G^{\#}(n))$ elements)

$$h(b) = H(n) \left(1 + o(1)\right) \quad \text{as} \quad n \to \infty.$$

Theorem 7. *Let G be a semigroup satisfying Axiom $A^{\#}$. A normal order of $\log F(n)$ is $(\log n)(\log \log n)$ in each of the cases (i), (ii), (iii) above.*

Proof. For each of the three cases and for any $b \in G$ we obtain as in the proof of Theorem 4 that $\omega(b) \log \omega(b)(1 + o(1) \leq \log F(b) \leq \Omega(b) \log \Omega(b)(1 + 0(1))$ where $\Omega(b)$ and $\omega(b)$ denote the total number (respectively, the distinct number) of irreducible factor of b. For semigroups satisfying Axiom $A^{\#}$ we have the result (Theorem 9.7 of J. Knopfmacher [11, p.90]):
For almost all elements b of degree n in G and for any $\delta > 0$,

$$\omega(b) = \log n \left(1 + 0(\log n)^{-\frac{1}{2}+\delta}\right), \Omega(b) = \log n \left(1 + 0(\log n)^{-\frac{1}{2}+\delta}\right).$$

The conclusion now follows easily from the inequality.

Remark. A normal order for case (iii) has previously been established in [9].

Acknowledgement. We thank one of the referees for pointing out an error in the proof of Theorem 3.

4 References

1. E. R. Canfield, P. Erdös & C Pomerance (1983). On a problem of Oppenheim concerning "factorisatio numerorum". *J Number Th.*, **17**, 1-28.

2. P. Erdős (1941). On some asymptotic formulas in the theory of the "Factorisatio Numerorum". *Annals of Math.* **42**, 989-993.

3. P. Erdős (1943). Corrections to two of my papers, *Annals of Math.* 44, 647-651.

4. G.H. Hardy and S. Ramanujan (1920). The normal number of prime factors of a number *n*. *Quart. J. Math.* **48**, 76-92.

5. K.-H. Indlekofer 1989. The abstract prime number theorem for function fields. Preprint.

6. K.-H. Indlekofer, E. Manstavicius, R. Warlimont (1991). On a certain class of infinite products with an application to arithmetical semigroups, *Arch. Math.* **56**, 446-453.

7. L. Kalmár (1931). Über die mittlere Anzahl der Produktdarstellungen der Zahlen. *Acta Litt. Sci. Szeged*, **5**, 95-107.

8. A. Knopfmacher, J. Knopfmacher and R. Warlimont (1992). "Factorisatio Numerorum" in Arithmetical Semigroups, *Acta Arith.* forthcoming.

9. A. Knopfmacher and J.N. Ridley (1990). Reciprocal sums over partitions and compositions. (Preprint).

10. J. Knopfmacher. *Abstract Analytic Number Theory* (North-Holland/ Elsevier, 1975; Dover, 1990).

11. J. Knopfmacher. *Analytic Arithmetic of Algebraic Function Fields* (Dekker, 1979).

12. A. Oppenheim (1927). On an arithmetic function. *J. London Math. Soc.*, **1**(1926), 205-211 and **2**, 123-130.

13. A. Sklar (1952). On the factorization of squarefree integers, *Proc. Amer. Math. Soc.* **3**, 701-705.

14. W. Zhang (1990). The abstract prime number theorem for algebraic function fields, in "Analytic Number Theory", Proc. Conf. in honor of Paul T. Bateman, Urbana, Illinois, 1989, *Prog. Math.* **85**, 529-558.

A. Knopfmacher
Department of Computational and Applied Mathematics
University of the Witwatersrand
P.O. Wits 2050
South Africa
e-mail 076MAIL@witsvma.wits.ac.za

J. Knopfmacher
Department of Mathematics
University of the Witwatersrand
P.O. Wits 2050
South Africa

and

R. Warlimont
Fachbereich Mathematik
Universität Regensburg
Postfach 397
8400 Regensburg
Germany

A Simple Proof for Non-Monogenesis of the Rings of Integers in some Cyclic Fields

Toru Nakahara

Saga University

To the memory of Professor Koichi Yamamoto

1   Introduction

Let K be an algebraic number field of finite degree over the rationals Q and Z_K the ring of integers in K. When Z_K coincides with $Z[\alpha]$ for a number α in K as a free Z−module, we say that Z_K is monogenic, Z_K has a power basis or K has an integral power basis, otherwise Z_K is of non-monogenesis[5, 15].

It is called a problem of Hasse to determine that Z_K is monogenic or of non-monogenesis.

Our aim is to give a simple proof for non-monogenesis of Z_K in cyclic quartic fields K with prime conductor. Namely K has no integral power basis in the case of conductor p congruent to 1 mod 8 in §3, and in some cases of conductor p congruent to 5 mod 8, $p \neq 5$ in §4.

In other fields, we can see the related phenomena in [1, 3–5, 7, 10, 13, 16, 18].

Our first method is controlled by the relative trace and the relative norm from K to the quadratic subfield, where K is generated by the Gauss sum attached to a biquadratic character.

In §5 a second succinct method is shown by using the decomposition law of ideals in the case of cyclic cubic, quartic and sextic fields.

In [15] we showed that any cyclic quartic field with prime conductor

This work was partially supported by Grant-in Aid for Scientific Research 63540060, Ministry of Education, Science and Culture of Japan.

$\neq 5$ does not have a power basis. However for the proof we had to compute the group index of a submodule in Z_K.

2   Gauss sum

Let p be a prime $\equiv 1 \bmod 4$. Let G be the Galois group of a cyclotomic extension k_p/Q. Here k_n denotes the n-th cyclotomic field $Q(\zeta_n)$, and ζ_n a primitive n-th root of unity. We identify G and the reduced residue class group modulo p.

Let $p = \pi\bar{\pi}$ be the prime decomposition of p in k_4, where we normalize $\pi = a + 2bi \equiv 1 \bmod 2(1-i)$, $i = \sqrt{-1}$. Here $\bar{\alpha}$ means the complex conjugate of α. Let χ be a pure quartic character of conductor p determined by the biquadratic residue symbol $\chi(x) = \left(\dfrac{x}{\pi}\right)_4$. Let

$$\tau(\chi) = \sum_{x \in G} \chi(x)\zeta^x$$

be the Gauss sum attached to χ for $\zeta = \zeta_p$. Then the following relations of Gauss sums are well known[15]:

$$\tau(\chi)\tau(\bar{\chi}) = \chi(-1)p,$$

$$\tau(\chi)^2 = \chi(-1)\pi\sqrt{p},$$

$$\tau(\chi^2) = \sqrt{p}.$$

Herein $\bar{\chi}$ is the conjugate character of χ .

The subgroup $< \chi >$ in the character group of G is a cyclic group of order 4. Let H be the kernel of χ and K the subfield of k_p corresponding to H. Then by the Gauss period of $(p-1)/4$ terms

$$\eta = \sum_{x \in H} \zeta^x$$

we have $K = Q(\eta)$ [2, 17]. We fix a representative σ of $< \sigma H >= G/H$ such that $\chi(\sigma) = i$, and we denote $\sigma^j(\xi)$ by $\xi^{(j)}$ for $\xi \in K$, $j \bmod 4$. Then we get

$$\eta = \left(-1 + \tau(\chi) + \tau(\chi^2) + \tau(\bar{\chi})\right)/4,$$

$$\tau(\chi^k)^{(j)} = (-i)^{kj}\tau(\chi^k).$$

3 The case of conductor $p \equiv 1 \mod 8$

For a prime $p \equiv 1 \mod 8$ we have a unique sum of two squares $p = a^2 + 4b^2$, $\pi = a + 2bi \equiv 1 \mod 4$. Then $\chi(-1) = 1$, i.e. χ is an even character. By [8] the ring Z_R of integers in K coincides with the Z–module $Z[1, \eta, \eta', \eta'']$ with the Gauss period η.

We shall give a new proof of the next theorem in § 5

Theorem 1. *Let K be a cyclic quartic field with prime conductor $\equiv 1 \mod 8$ over Q. Then there exists explicitly an integer in K outside of $Z[\alpha]$ for any integer α in K. Then Z_K is of non-monogenesis.*

Remark 1. In [12] this theorem is shown in the case of prime conductors are congruent to 1 or 25 mod 32.

Its proof depends on

Lemma 2. *With notation as above, if $Z_K = Z[\alpha]$, then $Z_K = Z[\alpha']$, for some α' which is congruent to η, $\eta + \eta'$ or $\eta + \eta''$ modulo 2.*

We can find a number $\beta \in Z_K \backslash Z[\alpha']$ for any integer α' in Lemma 2 in the same way as in the first proof of Theorem 4.

Remark 2. For the maximal real subfield of any cyclotomic field k_n, a simple proof of monogenesis is given in [19], Proposition 2.16 contrary to [14].

4 The case of conductor $p \equiv 5 \mod 8$

In this section we consider the case of $p = a^2 + 4$, $\pi = a + 2i$, $a \equiv 3 \mod 12$. Then $\chi(-1) = -1$, i.e. χ is an odd character. It is enough to calculate the following five cases:

Lemma 3. *With notation as above, if $Z_K = Z[\alpha]$, then $Z_K = Z[\alpha']$, for some α' which is congruent to η, $\eta \pm \eta'$ or $\eta \pm \eta''$ modulo 3.*

Proof. We write $\alpha \sim \beta$ for α, β in Z_K if $\alpha - \beta$ in Z, $\sigma^j(\beta) = \alpha$ or $\alpha - \beta = 3\gamma$ for some γ in Z_K. Adding to the cases

$$\eta^{(j)} \sim \eta, \ \eta + \eta''' \sim \eta'' + \eta' \sim \eta' + \eta, \ \eta + \eta' + \eta'' = -1 - \eta''' \sim \eta''' \sim \eta$$

in Lemma 2, we have

$$\eta + \eta' - \eta'' = -1 - 2\eta'' - \eta''' \sim -2\eta'' - \eta''' \sim \eta'' - \eta''' \sim \eta - \eta',$$
$$\eta - \eta' + \eta'' = -1 - 2\eta' - \eta''' \sim -2\eta' - \eta''' \sim \eta' - \eta''' \sim \eta - \eta'',$$
$$\eta - \eta' - \eta'' = -1 + 2\eta' + \eta''' \sim 2\eta' + \eta''' \sim -\eta' + \eta.$$

Thus we obtain the lemma.

Now let β_1 be the number $(\alpha - \alpha^3)/3$ in the outside of $\mathbf{Z}[\alpha]$. In the case of $\alpha \equiv \eta \mod 9$, $T\beta_1 = (T\eta - T\eta^3)/3 = (\theta - \theta^3 + 3\theta N\eta)/3 \equiv (\theta - 1)(1 - p)/12 + \theta N\eta \mod 3$, where T and N mean the relative trace and the norm from K to $\mathbf{Q}(\theta)$ and $\theta = T\eta = (-1 + \sqrt{p})/2$, respectively. Then $\theta^2 + \theta + (1 - p)/4 = 0$. Thus $T\beta_1$ is an integer. Next it holds that $N\beta_1 \equiv N\eta N(1 - \eta^2)/9 \mod 3$ and

$$N\eta = \left(-1 + \sqrt{p} + \tau(\chi) + \tau(\bar{\chi})\right)\left(-1 + \sqrt{p} - \tau(\chi) - \tau(\bar{\chi})\right)/16$$
$$= (1 + 3p - 2(a + 1)\sqrt{p})/16.$$

Then $N\eta \equiv 2 - \theta \mod 3$ and $N(1 - \eta^2) \equiv 3\theta(1 + \gamma) \mod 9$ for some integer γ, hence $N\beta_1$ is an integer. In the cases of $\alpha = \eta \pm \eta^{(j)}$ by virtue of $\beta = (\alpha - \alpha^3)/3 \sim (\eta \pm \eta^{(j)} - (\eta^3 \pm \eta^{(j)^3}))/3 \sim \beta_1 \pm \beta_1^{(j)}$, β is an integer. Therefore we have

Theorem 4. *For a prime $p = (3 + 12t)^2 + 4$, let K be the cyclic quartic field with conductor p over $\mathbf{Q}$. Then there exists explicitly an integer in K outside of $\mathbf{Z}[\alpha]$ for any integer α in K. Thus $\mathbf{Z}_K$ is of non-monogenesis.*

5   Second proof of the theorems

We can show another succinct proof of Theorem 1 and Theorem 2. However we are obliged to use the decomposition law of ideals in a cyclic quartic field K.

Lemma 5. *Let K be a galois extension of degree n over $\mathbf{Q}$. Assume that L is an unramified prime ideal in K over a prime number ℓ whose norm is not larger than n. Namely $NL = \ell^f \le n$. Then the ring $\mathbf{Z}_K$ of integers is of non-monogenesis.*

Proof. Let α be a primitive element of K in $\mathbf{Z}_K$. If α and $\alpha + 1$ are not prime to L, then $0 \equiv (\alpha + 1) - \alpha \mod L$, which is a contradiction. So we may choose a number α with $(\alpha, L) = 1$. For any prime $L \mid \ell$, we have
$$\alpha^{NL} \equiv \alpha \mod \ell.$$
Therefore $(1/\ell)\alpha^{NL-1} - (1/\ell)$ is an integer outside of $\mathbf{Z}[\alpha]$. This completes the proof.

Let $p = a^2 + 4b^2 > 5$ be a prime conductor of K. When $4 \mid b$ holds, then $\chi(2) = i^{-b} = 1$, namely 2 is completely decomposed in K. Thus for any prime ideal $L \mid 2$ we have $N\ell = 2$. On the other hand, when $2 \parallel b$ holds, then $\chi(2) = -1$, namely 2 is not ramified and does not remain prime in K. Then for any prime ideal $L \mid 2$ we have $NL = 4$. Similarly in the case that a prime $p = (3 + 24t)^2 + 4$ is the conductor of a quartic field K, $\chi(3) = 1$ holds, because we have $\left(\dfrac{\pi}{-3}\right)_4\left(\dfrac{-3}{\pi}\right)_4 = 1$ for $\pi = (3 + 24t) + 2i$ from the

Gauss' biquadratic reciprocity law [8]. Namely 3 is completely decomposed in K. Then, for any prime ideal $L \mid 3$, it follows that $NL = 3$. Therefore by Lemma 5 we obtain the second proof of the theorems.

Finally let K be a cyclic sextic field with prime conductor $p \equiv 1$ mod 3. Suppose that $p = c^2 + 27d^2$. Then from Eisenstein's cubic reciprocity law [11] the prime 2 is completely decomposed in the unique cubic subfield k in K. Consequently it holds that $N_k\ell = 2$ for any prime ideal $\ell \mid 2$ in k, and $N_K L = 2^e$ for any prime $L \mid 2$ in K, where e is equal to 2 and 1 for the prime 2 inert in k and otherwise respectively. Here N_F denotes the ideal norm with respect to F/Q for a number field F.

Therefore by virtue of Lemma 5 we have

Theorem 6. *For a prime $p = c^2 + 27d^2$, let K be the cyclic sextic field with conductor p over Q and k be the cubic subfield of K. Then both of the rings Z_K and Z_k are of non-monogenesis.*

Remark 3. Combining Theorem 1 and Theorem 6, we may observe the same phenomena as in Theorem 3 for a cyclic field K with conductor p of degree 12 over Q. In fact let $p = c^2 + 27d^2$ be a prime congruent to 1 mod 24 and F be any proper subfield of K, whose degree is larger than 2. Then it follows that the ring Z_F is of non-monogenesis.

Remark 4. There exists the opposite fact to Theorem 1 in the former paper[15], Proposition 2. Namely for the cyclic quartic field K with a composite conductor n, the ring Z_K has a power basis, where

$$n = lm = 64z^6 + 192z^5 + \left\{ \begin{array}{l} 336z^4 + 352z^3 + 268z^2 + 124z + 39 \\ 272z^4 + 224z^3 + 108z^2 + 28z - 13 \end{array} \right\},$$

with $l = m^2 + 4, m = (2z + 1)^2 \pm 2$ and n is square-free.

We propose the following:

Problem. Do there exist infinitely many cyclic quartic fields whose rings of integers are monogenic?

References

1. Bremner, A. (1988). On power bases in cyclotomic number fields, *J. Number Theory*, **28**, 288–298.

2. Dickson, L. E. (1935). Cyclotomy, higher congruences and Waring's problem, *Am. J. Math.*, **57**, 391–424. (= *The collected Mathematical Papers* Vol. III, 227–260, New York, Chelsea Pub. Co. 1975).

3. Dummit, D. S. and Kisilevsky, H. (1977). Indices in cyclic cubic fields, *Number Theory and Algebra*; Collect. Pap. Dedic. H. B. Mann, A. E. Ross and O. Taussky-Todd, New York San Francisco London, Academic Press , 29–42.

4. Gaál, I. Pethö, A. and Pohst, M. (1991). On the resolution of index form equations in biquadratic number fields, II, *J. Number Theory*, **38**, 35–51.

5. Gras, M. -N. (1986). Non monogénéité de l' anneau des entiers des extensions cycliques de *Q* de degré premier $\ell \geq 5$, *J. Number Theory*, **23**, 347–353.

6. Györy, K. (1981). On discriminants and indices of integers of an algebraic number field, *J Reine Angew. Math.*, **324**, 114–125.

7. Györy, K. (1983-1984). Sur les générateurs des ordres monogènes des corps de nombres algébriques, *Séminaire de Théorie des Nombres de Bordeaux*, Année ,no. **32**, 1–12.

8. Hasse, H. (1950). Arithmetische Bestimmung von Grundeinheit und Klassenzahl in zyklischen und biquadrartischen Zahlkörpern, *Abh. Deutsch. Akad. Wiss. Berlin, Math-Nat.*, Kl. 2, 3–95 (= *Math. Abhandlungen* Bd. 3, 289–379 Berlin-New York, Walter de Gruyter 1979).

9. Hasse, H. (1970). *Zahlbericht*, Würzburg-Wien, Physica-Verlag .

10. Huad, J. G. (1979). Cyclic cubic fields that contain an integer of given index, *Lecture Note in Mathematics*, Springer-Verlag, **751**,195–199.

11. Ireland, K. and Rosen, M. (1990). *A Classical Introduction to Modern Number Theory*(2nd edn). GTM 84, New York Heidelberg Berlin, Springer-Verlag.

12. Koga, Y. (1990). A simple proof of nonexistence of any integral power basis of a cyclic quartic field with some prime conductor, *Master's thesis*(in Japanese), Saga University.

13. Kovács, B. (1981). Canonical number systems in algebraic number fields, *Acta Math. Acad. Sci. Hungar.*, **37**, 405–407.

14. Liang, J. J. (1976). On the integral basis of the Maximal real subfield of a cyclotomic field, *J. Reine Angew. Math.*, **286/87**, 223–226.

15. Nakahara, T. (1982). On cyclic biquadratic fields related to a problem of Hasse, *Mh. Math.*, **94** , 125–132.

16. Nakahara, T. (1987). On the minimum index of a cyclic quartic field, *Arch. Math.*, **48**, 322–325.

17. Storer, T. (1967). *Cyclotomy and Difference Sets*, Chicago, Markhum Pub. Co.

18. Tanoé, F. (1990). Monogénéité des corps biquadratiques, *Thèse, Diplôme de docteur de l'université de Franche-Comté*.

19. Washington, L. C. (1982). *Introduction to Cyclotomic fields*, GTM **83**, New York Heidelberg Berlin, Springer-Verlag.

Toru Nakahara
Department of Mathematics
Faculty of Science and Engineering
Saga University
Saga 840
Japan
e-mail nakahara@math.ms.saga-u.ac.jp

Representing integers as sums of squares and Hypothesis H

Jukka Pihko

University of Helsinki

Lagrange's Theorem states that every positive integer can be written as the sum of four squares of integers. However it is known that we don't need to use all possible squares of integers, and that even some infinite sets can be missed out. (For example, we can miss out $\{(5^m)^2 \mid m \geq 2\}$—see [1], p. 2).

Here we ask the analogous question for two squares: "Are all squares necessary in representing all integers representable by the sum of two squares?" We believe so:

Conjecture 1. *For every positive integer h there exists an integer z whose only representation as the sum of two squares of integers is $z = h^2 + k^2$, for some integer k.*

Fermat stated, and Euler proved ([2], p. 480):

Lemma 1. *The only positive integers z which have a unique representation as the sum of two squares of integers, are those for which all prime factors $\equiv 3 \pmod 4$ appear with even exponents and the largest odd divisor of z, free of prime factors $\equiv 3 \pmod 4$, is either 1 or a prime.*

Now write any given h as $2^t mn$ with n odd and m the largest odd divisor of h free of prime factors $\equiv 1 \pmod 4$. Suppose that $q = a^2 + (a - n)^2$ is prime for some integer a. Then

$$z := (2^t m)^2 2q = (2^t m)^2 \{n^2 + (2a - n)^2\} = h^2 + k^2$$

where $k = 2^t m(2a - n)$. By Lemma 1 this is the unique representation of z. So we have proved

Theorem. *Conjecture 1 is true provided that, for each integer n consisting only of prime factors $\equiv 1 \pmod 4$, the polynomial $f_n(x) = 2x^2 - 2nx + n^2$ is prime for some integer x.*

Of course $f_n(x)$ will take on infinitely many prime values from Schinzel's Hypothesis H ([3], p. 194), so we deduce

Corollary. *If Schinzel's Hypothesis H is true then our Conjecture 1 is true, that is that every square of an integer is necessary to write every representable integer as the sum of two squares.*

We have verified Conjecture 1, on a computer, for each $h \leq 223620$, using the Theorem. In fact we found values of a, $1 \leq aJ \leq n$ such that $f_n(a)$ is prime, for each odd n, $3J \leq n \leq 223605$; and also values of $b > n$ such that $f_n(b)$ is prime, for each odd n, $1 \leq n \leq 158101$.

Acknowledgments: I thank Kalevi Suominen for useful discussions and Martti Nikunen for helping me with computers.

References

1. Härtter, E., Zöllner, J. (1977) Darstellungen natürlicher Zahlen als Summe und als Differenz von Quadraten. *Norske Vid. Selsk. Skr. (Trondheim)* no. **1** 8 pp.

2. Lehmer, D.H. (1948) On the partition of numbers into squares. *Amer. Math. Monthly*, **55**, 476–481.

3. Schinzel, A., Sierpiński, W. (1958) Sur certaines hypothèses concernant les nombres premiers. *Acta Arithmetica*, **4**, 185–208. Corrigendum (1960) *ibidem*, **5**, p. 259.

Jukka Pihko
Department of Mathematics
University of Helsinki
Hallituskatu 15
SF-00100, Helsinki
Finland
e-mail pihko@opmvax.csc.fi

Chapter III

Arithmetical Algebraic Geometry

Invited Addresses

and

Contributed Papers

The Tate Conjecture for Almost Ordinary Abelian Varieties over Finite Fields

Hendrik W. Lenstra, Jr. and Yuri G. Zarhin

University of California at Berkeley
and
Russian Academy of Sciences, Pushchino

1 Introduction

Let K be a finite field of characteristic p consisting of q elements. Let $K(a)$ denote the algebraic closure of K and $G(K) := \mathrm{Gal}(K(a)/K)$ the Galois group of K. Let Y be a smooth projective variety over K and set $Y(a) := Y \times K(a)$. Let ℓ be a rational prime different from p. The Galois group $G(K)$ acts on the (twisted) ℓ-adic étale cohomology groups $H^{2m}(Y(a), \mathbb{Q}_\ell)(m)$. In [T1] Tate conjectured that the subspace fixed under the Galois action is spanned by the cohomology classes of co-dimension m algebraic cycles on Y.

This conjecture has been proved in certain cases, e.g., Fermat hypersurfaces satisfying certain numerical conditions [Sh, Y], elliptic $K3$ surfaces [AS], $K3$ surfaces of finite height [N, NO] and powers of ordinary $K3$ surfaces [Z3].

Now, let $Y = X$ be a g-dimensional Abelian variety. In this case, Tate [T2] has proved his conjecture for $m = 1$. Notice that the well known interpretation of ℓ-adic étale cohomology groups of X as skew-symmetric multilinear forms on the Tate module $V_\ell(X)$ allows us to identify the Galois invariant subspace $H^{2m}(X(a), \mathbb{Q}_\ell)(m)^{G(K)}$ with the space of all skew-symmetric $2m$-linear forms E on $V_\ell(X)$ such that

$$E(\mathrm{Fr}\, x_1, \cdots, \mathrm{Fr}\, x_{2m}) = q^m\, E(x_1, \cdots, x_{2m})$$

for all $x_1, \cdots, x_{2m} \in V_\ell(X)$. Here Fr is the Frobenius endomorphism of X acting on the Tate module. In view of the result of Tate mentioned above, in order to prove the Tate conjecture for arbitrary m, it suffices to check that each E can be presented as a linear combination of exterior products of skew-symmetric bilinear forms $\varphi : V_\ell(X) \times V_\ell(X) \to \mathbb{Q}_\ell$ such that

$$\varphi(\mathrm{Fr}\, x, \mathrm{Fr}\, y) = q\, \varphi(x, y) \quad \text{for all } x, y \in V_\ell(X).$$

For example, this method makes it possible to verify easily the Tate conjecture for a supersingular Abelian variety X (in fact, in this case, if K is sufficiently large, then all even-dimensional cohomology classes are algebraic). Tankeev [T] verified the Tate conjecture for an absolutely simple Abelian variety X of prime dimension g. The papers [Z1, Z2] contain the proof of the Tate conjecture for the "almost supersingular" case (that is, for Abelian varieties with the set of slopes $\{0, 1/2, 1\}$ and such that the slopes 0 and 1 have length 1).

The aim of this paper is to prove the Tate conjecture for the "almost ordinary" case. Namely, we consider a simple Abelian variety X having the same Newton polygon as one of the product of a supersingular elliptic curve and an ordinary $(g-1)$-dimensional Abelian variety [O]. This means that the set of slopes of X is $\{0, 1/2, 1\}$ and the slope $1/2$ has length 2. A general result of [LO] guarantees that such Abelian varieties exist in all dimensions $g > 1$ and all prime characteristics p.

Our proof is based on the study of multiplicative relations between the eigenvalues of Fr.

Acknowledgments.   We are deeply grateful to Noriko Yui for her interest in this paper and invaluable help during the preparation of the manuscript. The research of the first author was supported by the National Science Foundation under Grant No. DMS 9002939. Part of the research we report on was done while the second author was a Visiting Professor at Ohio State University and was finished while he was a Visiting Scholar at Harvard University; he would like to thank both universities for their hospitality.

2   Tate modules

Let X be a g-dimensional Abelian variety defined over a commutative field K. Let us fix a separable algebraic closure $K(s)$ of K and let $G(K) :=$ $\mathrm{Gal}(K(s)/K)$ denote the Galois group of K. For a positive integer m, we denote by X_m the group of elements $x \in X(K(s))$ such that $mx = 0$. It is well known that if $\mathrm{char}(K)$ does not divide m, then X_m is a free $\mathbb{Z}/m\mathbb{Z}$-module of rank $2g$ [M]. Let us fix a prime number $\ell \neq \mathrm{char}(K)$. Then one may define the $\mathbb{Z}_\ell$-Tate module $T_\ell(X)$ as the projective limit of the groups X_m where m runs through the set of all powers ℓ^i and the transition map is just multiplication by ℓ. It is well known that $T_\ell(X)$ is a free $\mathbb{Z}_\ell$-module of rank $2g$ [M]. Clearly, all X_m are finite Galois submodules of $X(K(s))$, and the Galois actions for $m = \ell^i$ glue together to yield a continuous homomorphism [S,R] $\rho_\ell = \rho_{\ell,X} : G(K) \to \mathrm{Aut}\, T_\ell(X)$.

The image $G_\ell = G_{\ell,X} = \mathrm{Im}\, \rho_{\ell,X} \subset \mathrm{Aut}\, T_\ell(X)$ is a compact ℓ-adic Lie subgroup in $\mathrm{Aut}\, T_\ell(X)$. Let us put $V_\ell(X) := T_\ell(X) \otimes \mathbb{Q}_\ell$ (the tensor product is taken over $\mathbb{Z}_\ell$). Clearly, $V_\ell(X)$ is a $\mathbb{Q}_\ell$-vector space of dimension

$2g$ [M] and one may naturally identify $T_\ell(X)$ with a certain $\mathbb{Z}_\ell$-lattice of rank $2\dim(X)$ in $V_\ell(X)$. In particular, $\operatorname{Aut} T_\ell(X)$ becomes an open compact subgroup in $\operatorname{Aut} V_\ell(X)$. This allows us to regard ρ_ℓ as an ℓ-adic representation [S,R],

$$\rho_\ell = \rho_{\ell,X} : G(K) \to \operatorname{Aut} T_\ell(X) \subset \operatorname{Aut} V_\ell(X).$$

We have

$$G_\ell \subset \operatorname{Aut} T_\ell(X) \subset \operatorname{Aut} V_\ell(X).$$

Let $\operatorname{End}(X)$ denote the ring of all K-endomorphisms of X. Clearly, all X_m are $\operatorname{End}(X)$-invariant subgroups of $X(K(s))$, and their invariance gives rise to a canonical ring homomorphism

$$\operatorname{End}(X) \otimes \mathbb{Z}/m\mathbb{Z} \to \operatorname{End}(X_m)$$

which is an embedding if $\operatorname{char}(K)$ does not divide m. The image of the homomorphism lies in $\operatorname{End}_{G(K)}(X_m)$. For m running through powers of ℓ, these homomorphisms glue together to the embedding [R, T2]

$$(2.1) \qquad\qquad \operatorname{End}(X) \otimes \mathbb{Z}_\ell \to \operatorname{End} T_\ell(X).$$

The image of this embedding lies in $\operatorname{End}_{G(K)} T_\ell(X)$. Extending (2.1) by $\mathbb{Q}_\ell$-linearity, we obtain a canonical embedding [R, T2]

$$(2.2) \qquad\qquad \operatorname{End}(X) \otimes \mathbb{Q}_\ell \to \operatorname{End} V_\ell(X).$$

The image of this embedding lies in $\operatorname{End}_{G(K)} V_\ell(X)$.

We write $\chi_\ell : G(K) \to \mathbb{Z}_\ell^*$ for the cyclotomic character defining the Galois action on all the ℓ-power roots of unity. Let $\mathcal{L}$ be an invertible ample sheaf on X [M]. One may associate to $\mathcal{L}$ a skew-symmetric non-degenerate bilinear form

$$\mathcal{H}_{\mathcal{L}} : V_\ell(X) \times V_\ell(X) \to \mathbb{Q}_\ell,$$

called the Riemann form [M] . This form is uniquely defined up to multiplication by a constant in $\mathbb{Z}_\ell^*$, and enjoys the following property:

$$\mathcal{H}_{\mathcal{L}}(\rho_\ell(\sigma)\, x,\, \rho_\ell(\sigma)\, y) = \chi_\ell(\sigma) \mathcal{H}_{\mathcal{L}}(x, y)$$

for all $\sigma \in G(K)$ and $x, y \in V_\ell(X)$. It follows that G_ℓ lies in the group $\operatorname{Gp}(V_\ell(X))$ of symplectic similitudes of $V_\ell(X)$ with respect to $\mathcal{H}_{\mathcal{L}}$, and there is a continuous homomorphism $\chi'_\ell : G_\ell \to \mathbb{Z}_\ell^*$ such that

$$\mathcal{H}_{\mathcal{L}}(gx, gy) = \chi'_\ell(g)\mathcal{H}_{\mathcal{L}}(x, y) \quad \text{for all } g \in G_\ell \text{ and } x, y \in V_\ell(X).$$

If $\mathcal{L}'$ is another (ample) invertible sheaf on X, then there exists $u \in \operatorname{End}(X) \otimes \mathbb{Q}_\ell$ such that

$$\mathcal{H}_{\mathcal{L}'}(x, y) = \mathcal{H}_{\mathcal{L}}(ux, y) \quad \text{for all } x, y \in V_\ell(X).$$

This implies that χ' does not depend on the choice of $\mathcal{L}$.

On the other hand, each skew-symmetric bilinear form of the type

$$V_\ell(X) \times V_\ell(X) \to \mathbb{Q}_\ell, \quad x, y \mapsto \mathcal{H}_\mathcal{L}(ux, y) \quad \text{where} \quad u \in \text{End}(X) \otimes \mathbb{Q}_\ell$$

can be presented as a $\mathbb{Q}_\ell$-linear combination of the forms of type $\mathcal{H}_{\mathcal{L}'}$ (see [M], Section 20).

We write $\text{End}^0 X$ for the tensor product $\text{End}(X) \otimes \mathbb{Q}$. It is well known that $\text{End}^0 X$ is a finite-dimensional semi-simple $\mathbb{Q}$-algebra. We have $\text{End}(X) \otimes \mathbb{Q}_\ell = \text{End}^0 X \otimes_\mathbb{Q} \mathbb{Q}_\ell$.

Notice that X is a simple Abelian variety if and only if $\text{End}^0 X$ is a division algebra.

3   Abelian varieties over finite fields

Let us assume that K is a finite field of characteristic p with q elements. Clearly, q is a a power of p. The Galois group $G(K)$ is pro-cyclic and its canonical topological generator is the Frobenius automorphism:

$$\sigma_K : K(a) \to K(a), \quad x \mapsto x^q.$$

One may easily check that $\chi_\ell(\sigma_K) = q$. If K' is a finite algebraic extension of K, then K' is also finite and $\sigma_{K'} = \sigma_K^{[K':K]}$. Let X be an Abelian variety defined over K. We denote by

$$\text{Fr} = \text{Fr}_K \in \text{End}(X)$$

the Frobenius endomorphism Fr of X [T2, M, R]. It is known that the action of Fr on $T_\ell(X)$ coincides with the action of σ_K, i.e., we may write

$$\text{Fr} = \text{Fr}_K := \rho_\ell(\sigma_K) \in G_\ell \subset \text{Aut}\, T_\ell(X) \subset \text{Aut}\, V_\ell(X).$$

Clearly,

$$\chi'_\ell(\text{Fr}) = \chi_\ell(\sigma_K) = q.$$

According to a well known result of A. Weil [M], the linear operator

$$\text{Fr} : V_\ell(X) \to V_\ell(X)$$

is semi-simple and its characteristic polynomial

$$\mathbb{P}(t) = \mathbb{P}_X(t) := \det(t\, \text{id} - \text{Fr}\,|\, V_\ell(X)) \in \mathbb{Z}_\ell[t]$$

lies in $\mathbb{Z}[t]$ and does not depend on the choice of ℓ. (Here

$$\text{id} : V_\ell(X) \to V_\ell(X)$$

is the identity map.) In addition, all eigenvalues of Fr have archimedean absolute values equal to $q^{1/2}$. More precisely, let L denote the splitting field of the polynomial $\mathbb{P}(t) \in Z[t] \subset \mathbb{Q}[t]$. Then L is a finite Galois extension of $\mathbb{Q}$. Let $R = R_X$ denote the collection of roots of $\mathbb{P}(t)$ in L, i.e., the set of eigenvalues of Fr. By the very definition of L, we have $R \subset L$. Then the last statement of a theorem of Weil asserts the following: *For each embedding of L into the field $\mathbb{C}$ of complex numbers, we have $|\alpha| = q^{1/2}$ for all $\alpha \in R$.* It follows easily that $R \subset L^*$ and the map $\alpha \to q/\alpha$ is a permutation of R [M] (because q/α is just the complex conjugate of α) and furthermore, the multiplicities of roots α and q/α coincide. If $\alpha = q/\alpha \in R$, then its multiplicity is even, since the action of Fr multiplies by q a non-degenerate skew-symmetric bilinear (Riemann) form. This means that it is possible to rearrange the collection R of all roots (with multiplicities) of $\mathbb{P}(t)$ in such a way that it will be of the form

$$\alpha_1, \alpha_2, \cdots, \alpha_g; q/\alpha_1, q/\alpha_2, \cdots, q/\alpha_g.$$

3.0. Remark. By definition, L is a number field obtained by adjoining to $\mathbb{Q}$ all elements of R. Let us fix an embedding $L \subset \mathbb{C}$. Then the complex conjugation $\iota : \mathbb{C} \to \mathbb{C}$ maps each root α into q/α. It follows that $\iota(L) \subset L$, i.e., $\iota(L) = L$ and one may view ι as an element of $\mathrm{Gal}(L/\mathbb{Q})$. Clearly, this element does not depend on the choice of embedding of L into $\mathbb{C}$.

3.0.1. Lemma. ι lies in the center of $\mathrm{Gal}(L/\mathbb{Q})$.

Proof. It suffices to check that $\iota\sigma(\alpha) = \sigma\iota(\alpha)$ for all roots α and automorphisms σ of L.

We have $\sigma\iota(\alpha) = \sigma(q/\alpha) = q/\sigma(\alpha)$. Since $\sigma(\alpha) \in R$, we have $\iota\sigma(\alpha) = q/\sigma(\alpha)$, i.e, $\iota\sigma(\alpha) = \sigma\iota(\alpha)$.

3.1. Let us consider the multiplicative subgroup Γ of L^* generated by all the eigenvalues of Fr, i.e., by all elements of R. It is a finitely generated commutative group. Clearly, Γ is generated by q and $\alpha_1, \alpha_2, \cdots, \alpha_g$. It follows that the rank of Γ is always less or equal to $g + 1$; the equality holds if and only if all the roots of $\mathbb{P}(t)$ are simple and (in the notation above) the set $\{\alpha_1, \alpha_2, \cdots, \alpha_g; q\}$ consists of multiplicatively independent elements. Incidentally, if the set $\{\alpha_1, \alpha_2, \cdots, \alpha_g; q\}$ consists of multiplicatively independent elements, then the description of R given above implies that Γ is a free commutative group of rank $g+1$. This means, in particular, that Γ does not contain non-trivial roots of unity, that is, roots of unity different from 1.

Example [T]. If X is an absolutely simple Abelian variety of prime dimension $g > 3$, all endomorphisms of X are defined over K and the endomorphism algebra $\mathrm{End}^0 X$ of X is a commutative field, then Γ is a commutative group of rank $g + 1$ or g. (The result remains true for $g = 1, 2, 3$.)

3.2. Remark. In our notation, the assertion "all roots of $\mathbb{P}(t)$ are simple" means that $\alpha_i \neq \alpha_j$ while $i \neq j$ and $\alpha_i \neq q/\alpha_j$ for all pairs (i, j), including $i = j$.

3.3. Remark. If X is simple, then $\mathbb{P}(t)$ is either irreducible over $\mathbb{Q}$ or a power of a $\mathbb{Q}$-irreducible polynomial [T2, R].

3.4. Theorem of Tate [T2, R].
(a) $\mathrm{End}^0 X$ is commutative if and only if all roots of $\mathbb{P}(t)$ are simple, and in this case, $\mathrm{End}^0 X$ is generated by Fr as $\mathbb{Q}$-algebra.

(b) $\mathrm{End}^0 X \otimes_{\mathbb{Q}} \mathbb{Q}_\ell = \mathrm{End}_{G_\ell} V_\ell(X) = \mathrm{End}_{\mathrm{Fr}} V_\ell(X)$.

(c) Each skew-symmetric bilinear form $\varphi : V_\ell(X) \times V_\ell(X) \to \mathbb{Q}_\ell$ such that

$$\varphi(\mathrm{Fr}\, x, \mathrm{Fr}\, y) = q\, \varphi(x, y) \quad \text{for all } x, y \in V_\ell(X)$$

is a $\mathbb{Q}_\ell$-linear combination of forms of type $\mathcal{H}_{\mathcal{L}'}$.

3.5. Remark. Let us assume that $\mathbb{P}(t)$ is irreducible over $\mathbb{Q}$, i.e., the endomorphism algebra $\mathrm{End}^0 X$ of X is a commutative field. Then the Galois group $\mathrm{Gal}(L/\mathbb{Q})$ acts transitively on R.

For each root $\alpha \in R$ the subfield $\mathbb{Q}(\alpha)$ of L is canonically isomorphic to $\mathrm{End}^0 X$; under this isomorphism α is mapped to Fr. Let us consider the commutative group U consisting of all functions $f : R \to \mathbb{Z}$ enjoying

$$f(\alpha) = -f(q/\alpha) \quad \text{for all} \quad \alpha \in R.$$

Clearly, U is a free commutative group of rank g. (There are no square roots of q in R, because all elements of R are of (odd) multiplicity one.)

There is a natural $\mathrm{Gal}(L/\mathbb{Q})$-action on U induced by the action on R. Namely, $(\sigma f)(\alpha) = f(\sigma^{-1}\alpha)$ for all $\sigma \in \mathrm{Gal}(L/\mathbb{Q}), \alpha \in R$. Notice that under this action, complex conjugation ι acts on U as multiplication by -1. Now assume that there exists an infinite cyclic $\mathrm{Gal}(L/\mathbb{Q})$-invariant subgroup Δ of U. Since $\mathrm{Aut}(\Delta) = \{1, -1\}$, the Galois action on Δ is defined by quadratic character

$$\kappa_\Delta : \mathrm{Gal}(L/\mathbb{Q}) \to \mathrm{Aut}(\Delta) = \{1, -1\}.$$

This character gives rise to imaginary quadratic subfield B of L: notice that the complex conjugation acts as multiplication by -1 on $\Delta \subset U$. The Galois group $\mathrm{Gal}(L/B)$ coincides with the kernel of the character. The transitivity of the $\mathrm{Gal}(L/\mathbb{Q})$-action on R implies that

$$f(\alpha) = \pm f(\beta) \quad \text{for all} \quad \alpha, \beta \in R.$$

In other words, for each $f \in \Delta$ there exists a non-negative integer m such that $f(\alpha) = \pm m$ for all roots α. Indeed, let σ be an automorphism of L such that $\sigma\beta = \alpha$. Then $(\sigma f)(\alpha) = f(\beta)$. On the other hand, since $f \in \Delta$, $\sigma f = \pm f$. This implies that $f(\alpha) = \pm f(\beta)$. Notice, that B lies in $E := \mathbb{Q}(\alpha)$ for each root α. Indeed, one has only to check that if $\sigma\alpha = \alpha$ for some $\sigma \in \text{Gal}(L/\mathbb{Q})$, then σ acts identically on Δ. In order to prove this, notice that $(\sigma f)(\alpha) = f(\alpha)$ for all $f \in \Delta$ (even for all $f \in U$). Since $\sigma f = \pm f$ and $f(\alpha)$ does not vanish for non-zero f, we obtain that $\sigma f = f$.

One may easily check that R consists of exactly two $\text{Gal}(L/B)$-orbits; a non-zero f is constant on each of these orbits and takes different values on different orbits. More precisely, $f(\alpha) = -f(\beta)$ if roots α and β belong to different orbits. As above, $f \in \Delta$.

As a corollary, we obtain that under our assumption (the existence of Galois invariant infinite cyclic subgroup of U), $\text{End}^0 X$ contains an imaginary quadratic subfield (namely, B).

Let us put $\alpha_i' = \alpha_i^2/q$ for all i and denote by Γ' the multiplicative group generated by all α_i'. One may easily check that the rank of Γ is equal to $1 + $ the rank of Γ'.

Consider the homomorphism

$$u : U \to \Gamma', u(f) = \prod_{\alpha \in R} \alpha^{f(\alpha)} = \prod_{i=1}^{g} \alpha_i'^{f(\alpha_i)}.$$

Using the assumption $f(\alpha) = -f(q/\alpha)$, one may easily check that u is a surjective homomorphism from U to Γ'. It is also clear that u is $\text{Gal}(L/\mathbb{Q})$-equivariant.

3.5.1. Remark. Under the assumptions and notations of the previous Remark let β be a root and

$$\text{Norm}_{E/B} : E = \mathbb{Q}(\beta) \to B$$

is the norm map. Clearly, $g = [E : B]$. Assume that $f \in \Delta$ and $f(\beta) = m$. Then one may easily check that

$$u(f) = (\text{Norm}_{E/B}(\beta^2/q))^m.$$

3.6. Theorem. *Assume that $\mathbb{P}(t)$ is irreducible over $\mathbb{Q}$, i.e., the endomorphism algebra $\text{End}^0 X$ of X is a commutative field. Let us assume also that the rank of Γ is equal to g.*

Then

(a) The endomorphism algebra of X, $\text{End}^0 X$, contains an imaginary quadratic subfield;

(b) There exists a g-tuple $(n_1, n_2, \cdots, n_g) \in \mathbb{Z}^g$ with all $|n_i| = 1$ such that $\prod_{i=1}^{g} \alpha_i'^{n_i}$ is a root of unity;

(c) Put $\Delta = \{f \in U \mid u(f) \text{ is a root of unity}\}$. Then Δ is infinite cyclic $\mathrm{Gal}(L/\mathbb{Q})$-invariant subgroup of U. If f is a generator of Δ, then $f(\alpha) = \pm 1$ for all $\alpha \in R$ and $\prod_{i=1}^{g} \alpha_i'^{f(\alpha_i)} = u(f)$ is a root of unity.

Proof of Theorem 3.6. Clearly, $\Gamma' = u(U)$, and Δ is a commutative free group of finite rank with rank equal to

$$g - \mathrm{rank}(\Gamma') = g - (\mathrm{rank}(\Gamma) - 1) = 1.$$

So Δ is a $\mathrm{Gal}(L/\mathbb{Q})$-invariant infinite cyclic group. Let f be its generator. Let us put $n_i = f(\alpha_i)$. We have $f(q/\alpha_i) = -n_i$. From the definition of a generator, if there are any integers $m_1, m_2, \cdots, m_g$ such that

$$\prod \alpha_i'^{m_i} = \prod (\alpha_i^2/q)^{m_i}$$

is a root of unity, then there exists an integer d such that $m_i = dn_i$ for every i. Indeed, one has only to notice that the function

$$h : R \to \mathbb{Z}, \quad h(\alpha_i) = m_i, \; h(q/\alpha_i) = -m_i$$

belongs to Δ. Clearly, the greatest common divisor of $n_1, n_2, \cdots, n_g$ is equal to 1.

Now it is time to apply results of Section 3.5. Recall that all $f(\alpha)$ have the same absolute value. Since

$$n_1 = f(\alpha_1), \; n_2 = f(\alpha_2), \cdots, n_g = f(\alpha_g)$$

have the greatest common divisor equal to 1, this absolute value is equal to 1, i.e., all $n_i = \pm 1$.

This yields the proof of the second and the third assertion. The first assertion also follows from the results of Section 3.5, because Δ is a cyclic Galois–invariant subgroup of U.

3.6.1. Remark. We keep the notations and assumptions of Theorem 3.6. Let $\beta \in R$. Let us identify canonically the endomorphism algebra of X with $E = \mathbb{Q}(\beta)$ and let B be the imaginary quadratic subfield of E. Let f be a generator of Δ normalized by the condition $f(\beta) = 1$. Now, it follows easily from Remark 3.5.1 and the definition of Δ, that $\mathrm{Norm}_{E/B}(\beta^2/q)$ is a root of unity, i.e., the norm of β to the imaginary quadratic field B is a root of unity times a power of the square root of q.

4   Newton polygons

We keep all notation and assumptions of the previous Section. Let $\mathcal{O}_L$ be the ring of all algebraic integers in L. Clearly, $R \subset \mathcal{O}_L$. Therefore, we have

$\alpha \in \mathcal{O}_L$ and $q/\alpha \in \mathcal{O}_L$ for all $\alpha \in R$. Since q is a power of p, we obtain that if $\mathfrak{Q}$ is a maximal ideal in $\mathcal{O}_L$ not lying over p, then

$$\mathrm{ord}_{\mathfrak{Q}}(\alpha) = 0 \quad \text{for all } \alpha \in R$$

where $\mathrm{ord}_{\mathfrak{Q}} : L^* \to \mathbb{Z}$ is a discrete valuation attached to $\mathfrak{Q}$. It follows that $\mathrm{ord}_{\mathfrak{Q}}(\gamma) = 0$ for all $\gamma \in \Gamma$. This is because Γ is, by definition, generated by elements of R.

We denote by $\mathcal{S}$ the set of maximal ideals $\mathfrak{p}$ in $\mathcal{O}_L$, that lie over p. For $\mathfrak{p} \in \mathcal{S}$, we denote by $\mathrm{ord}_{\mathfrak{p}} : L^* \to \mathbb{Q}$ the discrete valuation attached to $\mathfrak{p}$ and normalized by the condition that $\mathrm{ord}_{\mathfrak{p}}(q) = 1$.

Now we are ready to recall the definition of the Newton polygon of X. The *Newton polygon* of X consists of a finite set $\mathrm{Slp}_X \subset \mathbb{Q}$ called the set of slopes and a positive integral-valued function $\mathrm{length}_X : \mathrm{Slp}_X \to \mathbb{Z}_+$, which assigns to each slope its length. For each $\mathfrak{p} \in \mathcal{S}$, Slp_X is given by

$$\mathrm{Slp}_X = \mathrm{ord}_{\mathfrak{p}}(R),$$

and for each $c \in \mathrm{Slp}_X$, $\mathrm{length}_X(c)$ is given by

$$\mathrm{length}_X(c) = \#\{\text{roots } \alpha \in R \text{ (with multiplicities) such that } \mathrm{ord}_{\mathfrak{p}}(\alpha) = c\}.$$

Since L is a Galois extension of $\mathbb{Q}$, the Galois group $\mathrm{Gal}(L/\mathbb{Q})$ acts transitively on $\mathcal{S}$. This implies that the definition of Slp_X and length_X does not depend on the choice of $\mathfrak{p}$, because R is the collection of roots of $\mathbb{P}(t)$, and the polynomial $\mathbb{P}(t)$ has rational coefficients (i.e., R is $\mathrm{Gal}(L/\mathbb{Q})$-invariant).

Since the multiplicities of roots α and q/α coincide and

$$\mathrm{ord}_{\mathfrak{p}}(q/\alpha) = \mathrm{ord}_{\mathfrak{p}}(q) - \mathrm{ord}_{\mathfrak{p}}(\alpha) = 1 - \mathrm{ord}_{\mathfrak{p}}(\alpha),$$

the following assertion holds true: If $c \in \mathrm{Slp}_X$, then $1 - c \in \mathrm{Slp}_X$ and $\mathrm{length}_X(c) = \mathrm{length}_X(1 - c)$. If $1/2$ is a slope then its length is even.

4.1. Remark. Assume that the polynomial $\mathbb{P}(t)$ is a power of a $\mathbb{Q}$-irreducible polynomial, i.e., there exists a polynomial $P(t)$ and a positive integer i such that $\mathbb{P}(t) = P(t)^i$. Then, it is easily seen that $\mathrm{length}_X(c)$ is divisible by i for all $c \in \mathrm{Slp}_X$.

4.2. Definition. An Abelian variety X is called *ordinary* if $\mathrm{Slp}_X = \{0, 1\}$ and $\mathrm{length}_X(0) = \mathrm{length}_X(1) = g$.

4.3. Definition. An Abelian variety X is called *supersingular* if $\mathrm{Slp}_X = \{1/2\}$ and $\mathrm{length}_X(1/2) = 2g$.

If X is supersingular, then one may easily check (see [Z1]) that the rank of Γ is 1.

4.4. Definition. ([Z1, Z2].) An Abelian variety X is of $K3$ *type* if $\mathrm{Slp}_X = \{0, 1\}$ or $\{0, 1/2, 1\}$ and $\mathrm{length}_X(0) = \mathrm{length}_X(1) = 1$.

If X is a simple Abelian variety of $K3$ type then the rank of Γ is equal to $g + 1$ [Z1, Z2].

4.5. Example. Let us assume that the endomorphism algebra of X is a commutative field. Let us define positive integer f by the formula $q = p^f$. Let e be the ramification index of L at p, i.e., the principal ideal (p) is equal to $(\prod_{\mathfrak{p} \in S} \mathfrak{p})^e$. Let $\beta \in R$. Let us split the principal ideal (β) into the product of maximal ideals in the ring of integers in L. Clearly,

$$(\beta) = \prod_{\mathfrak{p} \in S} \mathfrak{p}^{a(\mathfrak{p})}$$

for a certain non-negative integral valued function a on S. Now, using the transitivity of the Galois action on R and S, one may easily get the following description of the Newton polygon of X.

$$\mathrm{Slp}_X = \{\frac{a(\mathfrak{p})}{ef} \mid \mathfrak{p} \in S\},$$

$$\mathrm{length}_X\left(\frac{b}{ef}\right) = \frac{\#\{\mathfrak{p} \in S \text{ with } a(\mathfrak{p}) = b\}}{\#(S)} 2g$$

for all slopes $\frac{b}{ef}$.

5   Almost ordinary Abelian varieties

Throughout this section we assume that X is a simple g-dimensional Abelian variety, $\mathrm{Slp}_X = \{0, 1/2, 1\}$ and $\mathrm{length}_X(1/2) = 2$. These varieties were introduced and studied by F. Oort [O]. In particular, he proved that the endomorphism algebra $\mathrm{End}^0 X$ of X is a commutative field, that is, the polynomial $\mathbb{P}(t)$ is irreducible over the field of rational numbers. This implies that all its roots have multiplicity 1.

5.1. Remark. It follows from the very definition of X that for each maximal ideal $\mathfrak{p} \in S$ there exist exactly two roots $\alpha \in R$ such that $\mathrm{ord}_{\mathfrak{p}}(\alpha)$ are not integers. Clearly, if α is one of these roots, then q/α is other.

5.2. Corollary. In the notation above, for each maximal ideal $\mathfrak{p} \in S$ there exists exactly one root $\alpha \in \{\alpha_1, \alpha_2, \cdots, \alpha_g\}$ such that $\mathrm{ord}_{\mathfrak{p}}(\alpha)$ is not an integer.

5.3. Corollary. If for some $\beta \in \{\alpha_1, \alpha_2, \cdots, \alpha_g\}$ and $\mathfrak{p} \in S$, $\mathrm{ord}_{\mathfrak{p}}(\beta) = 1/2$, then $\mathrm{ord}_{\mathfrak{p}}(\alpha)$ is an integer (namely, 0 or 1) for all $\alpha \in \{\alpha_1, \alpha_2, \cdots, \alpha_g\} - \{\beta\}$.

5.4. Proposition. *For each* $\alpha \in \{\alpha_1, \alpha_2, \cdots, \alpha_g\}$, *there exists a maximal ideal* $\mathfrak{p} \in S$ *such that* $\mathrm{ord}_\mathfrak{p}(\alpha) = 1/2$. *In particular,* $\mathrm{ord}_\mathfrak{p}(\alpha)$ *is not an integer.*

Proof. Indeed, let us assume that there exists $\beta \in \{\alpha_1, \alpha_2, \cdots, \alpha_g\}$ such that $\mathrm{ord}_\mathfrak{p}(\beta) = 0$ or 1 for all $\mathfrak{p} \in S$. Since $\mathbb{P}(t)$ is irreducible over $\mathbb{Q}$, the group $\mathrm{Gal}(L/\mathbb{Q})$ acts transitively. Since $\mathrm{ord}_{\sigma\mathfrak{p}}(\beta) = \mathrm{ord}_\mathfrak{p}(\sigma^{-1}\beta))$ is an integer for all $\sigma \in \mathrm{Gal}(L/\mathbb{Q})$, we obtain that $\mathrm{ord}_\mathfrak{p}(\alpha)$ is an integer for all $\alpha \in R$. This contradicts Corollary 5.2.

5.5. Main Theorem. *Assume that there exist integers* M *and* $m_1, m_2, \cdots, m_g$ *such that*

$$\prod \alpha_i^{m_i} = q^M.$$

Then all $m_1, m_2, \cdots, m_g$ *are even.*

Proof. Assume that for some j the integer m_j is odd. According to the Proposition 5.4, there exists $\mathfrak{p} \in S$ such that $\mathrm{ord}_\mathfrak{p}(\alpha_j) = 1/2$. It follows from the Corollary 5.3 that for all α_i with $i \neq j$ we have that $\mathrm{ord}_\mathfrak{p}(\alpha_i)$ is an integer. Applying the homomorphism $\mathrm{ord}_\mathfrak{p}$ to $\prod \alpha_i^{m_i} = q^M$, we see that

$$M = \mathrm{ord}_\mathfrak{p}(q^M) = \mathrm{ord}_\mathfrak{p}(\prod \alpha_i^{m_i}) = \text{some integer} + (m_j/2)$$

is not an integer, because m_j is odd. This leads us to a contradiction. Therefore, m_j must be even for every j. This completes the proof.

5.5.1. Remark. The same arguments combined with the theorem of A. Weil prove the following generalization of the Theorem 5.5.

Assume that there exist a root of unity $\varepsilon \in L^*$ *and integers* M *and* $m_1, m_2, \cdots, m_g$ *such that*

$$\prod \alpha_i^{m_i} = \varepsilon q^M.$$

Then all $m_1, m_2, \cdots, m_g$ *are even and* $M = (\sum m_i)/2$.

5.5.2. Remark. Assume that there exist a *root of unity* $\varepsilon \in L^*$, integers $M, m_1, m_2, \cdots, m_g$ and a non-negative integer m such that
(a) $|m_i| = m$ for all i, and
(b) $\prod \alpha_i^{m_i} = \varepsilon q^M$.

In this situation, if g is even, then $m = m_i = M = 0$ and $\varepsilon = 1$.

Proof. It follows from a theorem of A. Weil that

$$M = (\sum_{i=1}^{g} m_i)/2 = m(g/2) + \sum_{i=1}^{g} (m_i - m)/2$$

is divisible by m. So, if m is not equal to zero, then we may assume, without loss of generality, that $m = 1$. This implies that

$$\prod \alpha_i^{m_i} = \varepsilon' q^M .$$

where all $m_i = \pm 1$ and ε' is a root of unity. Recall the Remark 5.2 that for each $\mathfrak{p} \in \mathcal{S}$, there exists exactly one root $\alpha = \alpha_j \in \{\alpha_1, \alpha_2, \cdots, \alpha_g\}$ such that $\mathrm{ord}_{\mathfrak{p}}(\alpha)$ is not an integer. Applying the homomorphism $\mathrm{ord}_{\mathfrak{p}}$ to $\prod \alpha_i^{m_i} = \varepsilon' q^M$, we obtain that

$$
\begin{aligned}
M &= \mathrm{ord}_{\mathfrak{p}}(\varepsilon' q^M) = \mathrm{ord}_{\mathfrak{p}}(\prod \alpha_i^{m_i}) = \text{some integer} + m_j \mathrm{ord}_{\mathfrak{p}}(\alpha_j) \\
&= \text{some integer} \pm \mathrm{ord}_{\mathfrak{p}}(\alpha_j)
\end{aligned}
$$

and this is not an integer. This is a contradiction.

5.5.3. Remark. Assume that there exist a root of unity $\varepsilon \in L^*$, integers $m_1, m_2, \cdots, m_g$ and a non-negative integer m such that
(a) $|m_i| = m$ for all i, and
(b) $\prod \alpha_i'^{m_i} = \varepsilon$ where $\alpha_i' = \alpha_i^2/q$ for all i.

Under this situation, if g is even, then $m = m_i = 0$ and $\varepsilon = 1$.

Proof. We obtain $\prod \alpha_i^{2m_i} = \varepsilon q^M$ with $M = \sum m_i$. One has only to apply the Remark 5.5.2.

5.6. Corollary to Theorem 5.5. *Assume that there exist integers M and $m_1, n_1, m_2, n_2, \cdots, m_g, n_g$ such that*

$$\prod \alpha_i^{m_i} (q/\alpha_i)^{n_i} = q^M .$$

Then all $m_1 - n_1, m_2 - n_2, \cdots, m_g - n_g$ are even integers. In particular, if all n_i and m_i are equal either to 0 or to 1, then $n_i = m_i$ for all i.

5.7. Theorem. *The set $\{\alpha_1, \alpha_2, \cdots, \alpha_g\}$ consists of multiplicatively independent elements. In particular, the rank of Γ is greater or equal than g, i.e., it is equal either to g or to $g + 1$.*

Proof of Theorem 5.7. Let $m_1, m_2, \cdots, m_g$ be integers such that $\prod \alpha_i^{m_i} = 1$. Assume that not all $m_1, m_2, \cdots, m_g$ are equal to zero. Let d denote the greatest common divisor of $m_1, m_2, \cdots, m_g$. Then $\varepsilon = \prod \alpha_i^{m_i/d}$

is a root of unity. Applying the Remark 5.5.1, we obtain that all m_i/d are even integers. But this contradicts the definition of d.

5.8 Theorem. *Assume that g is even. Then the set $\{\alpha_1, \alpha_2, \cdots, \alpha_g, q\}$ consists of multiplicatively independent elements, i.e., Γ is a free commutative group of rank $g + 1$.*

Proof of Theorem 5.8. We will use notation of Remark 3.5 and Theorem 3.6. It follows from Theorem 3.6 combined with Remark 5.5.3 that $\Delta = \{0\}$. In particular, u is an embedding. So $u(\mathbb{Z}^g) = \Gamma'$ is the group generated by all α_j'. Since u is an embedding, we obtain that Γ' is a commutative free group of rank g. On the other hand, the rank of Γ is equal to $1 +$ the rank of Γ' (3.5). Hence the rank of Γ is equal to $1 + g$.

6   Skew-symmetric multilinear forms on the Tate module

6.1. Theorem. *Let X be a a simple g-dimensional Abelian variety over K. Let us assume that $\mathrm{Slp}_X = \{0, 1/2, 1\}$ and $\mathrm{length}_X(1/2) = 2$. Then for each positive integer m and skew-symmetric $2m$-linear form*

$$E : V_\ell(X) \times \cdots \times V_\ell(X) \to \mathbb{Q}_\ell,$$

the following condition holds true: If

$$E(\mathrm{Fr}\, x_1, \cdots, \mathrm{Fr}\, x_{2m}) = q^m E(x_1, \cdots, x_{2m})$$

for all $x_1, \cdots, x_{2m} \in V_\ell(X)$, then E can be presented as a linear combination of exterior products of Riemann forms

$$\mathcal{H}_{\mathcal{L}'} : V_\ell(X) \times V_\ell(X) \to \mathbb{Q}_\ell.$$

6.2. Remark. In view of Tate's theorem 3.4 (c), it suffices to check that E can be presented as a linear combination of exterior products of skew-symmetric bilinear forms $\varphi : V_\ell(X) \times V_\ell(X) \to \mathbb{Q}_\ell$ such that

$$\varphi(\mathrm{Fr}\, x, \mathrm{Fr}\, y) = q\varphi(x, y) \quad \text{for all} \quad x, y \in V_\ell(X).$$

6.3. Remark. Using the identification between the space of all skew-symmetric $2m$-linear forms on $V_\ell(X)$ and the $2m$-th ℓ-adic étale cohomology group of X (over $K(s)$), we obtain that each Galois-invariant $2m$-dimensional (suitably twisted) ℓ-adic étale cohomology class on X is a linear combination of the exterior products of divisor classes (recall that under this identification, $\mathcal{H}_{\mathcal{L}'}$ becomes the cohomology class of any divisor whose linear equivalence class corresponds to $\mathcal{L}'$). This proves the algebraicity of Galois-invariant cohomology classes for X satisfying the assumption of the Theorem 6.1. This means that the Tate conjecture [T1] holds true for such X. We refer the reader to [Z2] for the details.

6.4.   Elementary Lemma.   Let r be a positive integer, V a $2r$-dimensional vector space over a commutative field k of characteristic zero, $\phi : V \times V \to k$ a skew-symmetric non-degenerate bilinear form, $F : V \to V$ an invertible semi-simple linear operator in V such that

$$\phi(Fx, Fy) = q\phi(x, y) \quad \text{for all} \quad x, y \in V.$$

Here q is a non-zero element of k which is not a root of unity. Suppose that all the eigenvalues of F have multiplicity one. Let us present the spectrum of F in the form

$$\alpha_1, \alpha_2, \cdots, \alpha_r \; ; \; q/\alpha_1, q/\alpha_2, \cdots, q/\alpha_r.$$

Further assume that the set $\{\alpha_1, \alpha_2, \cdots, \alpha_r; q\}$ enjoys the following properties: Suppose that there exist integers M and $m_1, n_1, m_2, n_2, \cdots, m_g, n_g$ such that

$$\prod \alpha_i^{m_i} (q/\alpha_i)^{n_i} = q^M.$$

Under this situation, if all n_i and m_i are equal either to 0 or to 1, then $n_i = m_i$ for all i.

Then for each positive integer M, the following conditions hold true: Assume that a skew-symmetric $2M$-linear form

$$E : V \times \cdots \times V \to k$$

enjoys the following property:

$$E(Fx_1, \cdots, Fx_{2M}) = q^M E(x_1, \cdots, x_{2M}) \quad \text{for all} \quad x_1, \cdots, x_{2M} \in V.$$

Then E can be presented as a linear combination of exterior products of skew-symmetric bilinear forms $\varphi : V \times V \to k$ such that

$$\varphi(Fx, Fy) = q\varphi(x, y) \quad \text{for all} \quad x, y \in V.$$

6.5. Proof of Theorem 6.1.   Using the Corollary 5.6, one has only to apply the Elementary Lemma 6.4. to $r = g$, $V = V_\ell(X)$, $k = \mathbb{Q}_\ell$ and $F = \mathrm{Fr}$.

6.6. Remark.   One may deduce from the Theorem 5.8 (see [Z2]) the validity of the Tate conjecture for all powers X^n of X if $g = \dim(X)$ is even. If g is odd, then one may easily deduce from the Theorem 3.6 and the Theorem 5.7 that the Tate conjecture holds true for X^n if one restricts oneself by co-dimension $m < g$.

6.7. Remark. One may easily generalize results of this paper to the case of simple Abelian varieties X enjoying the following properties:

(a) The endomorphism algebra $\text{End}^0 X$ of X is a commutative field,

(b) $1/2 \in \text{Slp}_X$ and $\text{length}_X(1/2) = 2$, and

(c) each slope c different from $1/2$ can be presented as a rational fraction with odd denominator.

In particular, the Tate conjecture holds true for such an Abelian variety X.

7   Sketch of a proof of Elementary Lemma

We may assume that k is algebraically closed. Let $\{\, e_1, \cdots, e_r \,;\, e_{-1}, \cdots, e_{-r}\}$ be a basis for V consisting of eigenvectors with respect to F.

$$F e_i = \alpha_i\, e_i, \quad F\, e_{-i} = (q/\alpha_i) e_{-i}$$

for all $i \in \{1, 2, \cdots, r\}$. Let $e_1^*, \cdots, e_r^*; e_{-1}^*, \cdots, e_{-r}^*$ be the basis for the dual space V^*. Now it is clear that if a skew–symmetric bilinear form is multiplied by q under the action of F, then it is a linear combination of exterior products $e_i^* \wedge e_{-i}^*$. For each subset $D \subset \{1, 2, \cdots, r; -1, -2, \cdots, -r\}$ consisting of $2M$ elements, one may define (up to sign) a skew–symmetric multilinear form E^D on V which is the exterior product of all e_i^*, $(i \in D)$. Clearly, all E^D constitute an eigenbasis with respect to F of the space of all skew–symmetric $2M$-linear forms on V. One may easily check that E^D is multiplied by q^M under the action of F if and only if there exists $D_{\{+\}} \subset \{1, 2, \cdots, r\}$ such that $D = D_{\{+\}} \cup -D_{\{+\}} = D_{\{+\}} \cup \{-i \,|\, i \in D_{\{+\}}\}$. Clearly, if D enjoys this property then E^D is equal (up to sign) to the exterior product of bilinear forms $E^{\{(i,-i)\}} = e_i^* \wedge e_{-i}^*$. The rest is plain.

References

[AS] M. Artin and H. P. F. Swinnerton-Dyer, *The Tate Shafarevich conjecture for pencils of elliptic curves on K3 surfaces*, Invent. Math. **20** (1973), pp. 279–296.

[LO] H. W. Lenstra Jr. and F. Oort, *Simple Abelian varieties having a prescribed formal isogeny type*, J. Pure Appl. Algebra, 4(1974), pp. 47–53.

[M] D. Mumford, *Abelian varieties*, second edition, Oxford University Press, 1974.

[N] N. O. Nygaard, *The Tate conjecture for ordinary K3 surfaces over finite fields*, Invent. Math. **74** (1983), pp. 213–237.

[NO] N. O. Nygaard and A. Ogus, *The Tate conjecture for K3 surfaces of finite height*, Ann. Math. **122** (1985), pp. 461–507.

[O] F. Oort, *CM-liftings of Abelian varieties*, J. Algebraic Geometry **1** (1992), pp. 131–146.

[R] C. P. Ramanujan, *The Theorem of Tate*, Appendix 1 to [M].

[S] J.-P. Serre, *Abelian ℓ-adic representations and elliptic curves*, second edition, Addison-Wesley, 1989.

[Sh] T. Shioda, *The Hodge conjecture and the Tate conjecture for Fermat varieties*, Proc. Japan Academy **55** (1979), pp. 111–114.

[T] S. G. Tankeev, *On cycles on Abelian varieties of prime dimension over finite or number fields*, Math. USSR Izvestya **22** (1984), pp. 329–337.

[T1] J. Tate, *Algebraic cycles and poles of zeta functions*, Arithmetical Algebraic Geometry, Harper and Row, New York, 1965, pp. 93–110.

[T2] J. Tate, *Endomorphisms of Abelian varieties over finite fields*, Invent. Math. **2** (1966), pp. 134–144.

[T3] J. Tate, *Classes d'isogénie des variétés abéliennes sur un corps fini* (d'aprés T. Honda), Séminaire Bourbaki **352** (1968), Springer Lecture Notes in Mathematics **179** (1971), pp. 95–110.

[Y] N. Yui, *Special values of zeta functions of Fermat varieties over finite fields*, Number Theory New York Seminar 1989-1990, Springer Verlag, New York (1991), pp. 251–275.

[Z1] Yu. Zarhin, *Abelian varieties of K3 type and ℓ-adic representations*, Algebraic Geometry and Analytic Geometry, ICM-90 Satellite Conference Proceedings, Springer-Verlag, Tokyo 1991, pp. 231–255.

[Z2] Yu. Zarhin, *Abelian varieties of K3 type*, preprint, 1992.

[Z3] Yu. Zarhin, *The Tate conjecture for powers of ordinary K3 surfaces over finite fields.* in preparation.

Hendrik W. Lenstra, Jr.
Department of Mathematics
University of California
Berkeley, CA 94720
USA
e-mail hwl@math.berkeley.edu

and

Yuri G. Zarhin
Institute for Mathematical Problems in Biology
Russian Academy of Sciences
Pushchino, Moscow Region 142292
Russia
e-mail IMPB@venus.ibioc.serpukhov.su

The Notion of a Shimura Variety

V. Kumar Murty

University of Toronto

To the memory of Prof. R. Sitaramachandrarao

§1. Motivation.

The purpose of this note is to introduce the number theorist to Shimura varieties. These are a class of algebraic varieties associated to certain reductive algebraic groups, and studied extensively by G. Shimura over a period of many years. They embody deep arithmetic and geometric information. In recent years, Deligne has added some new insights and provided an intrinsic formulation. In this note, we shall introduce the terminology and give some examples. In particular, we barely scratch the surface of the theory, and we shall not discuss the deeper aspects such as canonical models, cohomology and L-functions. The reader interested in a more thorough discussion as well as references to the literature is advised to consult the article of Milne [Mi].

The theory of Shimura varieties has its origins in explicit class field theory. Let L/K be an abelian extension of number fields and let $G = \mathrm{Gal}(L/K)$. For any ideal $\mathfrak{f}$ of the ring of integers $\mathcal{O}_K$ of K, we may consider the free abelian group $I(\mathfrak{f})$ generated by prime ideals which do not divide $\mathfrak{f}$. Inside $I(\mathfrak{f})$, there is a subgroup $P(\mathfrak{f})$ of principal ideals which are prime to $\mathfrak{f}$, and which have a generator $\equiv 1 (\mathrm{mod}\ ^{\times}\mathfrak{f})$. The main theorem of class field theory is the reciprocity law of Artin (and Takagi) which asserts that there is an ideal $\mathfrak{f}$ and a canonical surjection

$$I(\mathfrak{f})/P(\mathfrak{f}) \longrightarrow G$$

given by

$$\mathcal{P} \mapsto \sigma_{\mathcal{P}}$$

where $\sigma_{\mathcal{P}}$ denotes the Frobenius automorphism associated to $\mathcal{P}$. In particular, if $K = \mathbb{Q}$ then $\mathfrak{f} = f\mathbb{Z}$ for some $0 < f \in \mathbb{Z}$ and

$$I(\mathfrak{f})/P(\mathfrak{f}) \simeq (\mathbb{Z}/f\mathbb{Z})^{\times}.$$

Moreover, the isomorphism tells us that p splits completely if and only if $p \equiv 1 (\mathrm{mod}\ f)$. Since the primes that split completely characterize the

---

Research partially supported by a grant from NSERC

number field, it follows that $L \subseteq \mathbb{Q}(\zeta_f)$. This is the classical result of Kronecker and Weber.

Kronecker was also interested in classifying the abelian extensions when K is an imaginary quadratic extension of $\mathbb{Q}$. Let E be an elliptic curve with complex multiplication by $\mathcal{O}_K$. Every such curve has the property that

$$E(\mathbb{C}) = \mathbb{C}/\mathfrak{a}$$

where $\mathfrak{a}$ is a (fractional) ideal of $\mathcal{O}_K$. Moreover, two such curves, corresponding to ideals $\mathfrak{a}$ and $\mathfrak{b}$ say, are isomorphic if and only if $\mathfrak{a}$ and $\mathfrak{b}$ lie in the same ideal class. The Hilbert class field of K (that is, the maximal abelian unramified extension) is generated over K by the j-invariant $j(E)$ of E. Let $1 < N \in \mathbb{Z}$ and denote by $E[N]$ the set of points (in $E(\bar{\mathbb{Q}})$) of order dividing N. Then, the extension $K(j(E), E[N])$ obtained by adjoining to K the j-invariant of E and the coordinates of all the points in $E[N]$ is an abelian extension, and any abelian extension of K is contained in one of these. This was proved by Weber and Hasse. Thus, $E[N]$ plays the same role as the roots of unity played in the case $K = \mathbb{Q}$ and the curve E plays the role played by the multiplicative group $\mathbb{G}_m$.

The $\mathfrak{f}$-ideal class group $I(\mathfrak{f})/P(\mathfrak{f})$ can be interpreted adelically. Let

$$K_\infty^{\times+} = (K \otimes \mathbb{R})^{\times+}$$

be the connected component of the identity of $(K \otimes \mathbb{R})^\times$. We see that

$$I(\mathfrak{f})/P(\mathfrak{f}) \simeq K_\mathbb{A}^\times / K^\times K_\infty^{\times+} U$$

where

$$U = \prod_{v < \infty} U_v.$$

Here v ranges over the non-archimedean places of K and

$$U_v = \begin{cases} \mathcal{O}_v^\times & \text{if } \mathcal{P}_v \nmid \mathfrak{f} \\ 1 + \mathcal{P}_v^i \mathcal{O}_v & \text{if } \mathcal{P}_v^i \| \mathfrak{f} \end{cases}$$

where we have set $\mathcal{P}_v$ for the prime corresponding to v.

In particular, when K is an imaginary quadratic field, and $\mathfrak{f} = 1$ the ideal class group is

$$K_\mathbb{A}^\times / \mathbb{C}^\times K^\times \prod_v \mathcal{O}_v^\times.$$

By what we have said above, this coset space also classifies isomorphism classes of elliptic curves with complex multiplication by $\mathcal{O}_K$.

This theme can be pursued further along these lines. We may seek to explicitly generate the abelian extensions of other number fields using points of finite order on group theoretic or geometric objects. If K is a CM

field (i.e. a totally imaginary quadratic extension of a totally real number field) then Shimura and Taniyama [Sh-T] develop a theory which produces some abelian extensions of K using points of finite order on abelian varieties with complex multiplication. If K is an abelian extension of $\mathbb{Q}$ then some extensions can be produced by adjoining points of finite order from certain abelian varieties (without complex multiplication).

However, it is not this theme that we will pursue here. Rather, let us begin by considering the space of all elliptic curves. Every elliptic curve over $\mathbb{C}$ is isomorphic to a complex torus $\mathbb{C}/L$ where

$$L = \mathbb{Z} \oplus \mathbb{Z}\tau$$

with a $\tau \in \mathbb{C}$ having positive imaginary part. The modular group $SL_2(\mathbb{Z})$ acts on such lattices: if

$$\gamma = \begin{pmatrix} a & b \\ c & d \end{pmatrix}$$

then

$$\gamma L = \mathbb{Z} \oplus \mathbb{Z}\tau', \quad \tau' = (a\tau + b)/(c\tau + d).$$

Two such lattices L_1, L_2 give rise to isomorphic curves if and only if $\gamma L_1 = L_2$ for some $\gamma \in SL_2(\mathbb{Z})$. Thus, the set of all elliptic curves over $\mathbb{C}$ may be identified with the quotient space

$$SL_2(\mathbb{Z})\backslash\mathfrak{h}$$

where $\mathfrak{h}$ denotes the upper half plane (i.e. the set of complex numbers with positive imaginary part). Now identifying $\mathfrak{h}$ with the symmetric space of $SL_2(\mathbb{R})$ we have

$$SL_2(\mathbb{Z})\backslash\mathfrak{h} \simeq SL_2(\mathbb{Z})\backslash SL_2(\mathbb{R})/SO_2(\mathbb{R}).$$

To write this adelically, we use the strong approximation theorem

$$SL_2(\mathbb{A}) = SL_2(\mathbb{R})SL_2(\mathbb{Q})\prod SL_2(\mathbb{Z}_p)$$

and deduce that

$$SL_2(\mathbb{Z})\backslash SL_2(\mathbb{R})/SO_2(\mathbb{R}) \simeq SL_2(\mathbb{Q})\backslash SL_2(\mathbb{A})/SO_2(\mathbb{R})\prod SL_2(\mathbb{Z}_p).$$

We can also consider the collection of elliptic curves with level structure. (For example, we may fix an integer $N \geq 1$, and consider the family of pairs (E, C_N) consisting of elliptic curves together with a cyclic subgroup of order N.) This corresponds to replacing $SL_2(\mathbb{Z})$ with a congruence subgroup Γ, or equivalently, to replacing $\prod SL_2(\mathbb{Z}_p)$ with a smaller open compact subgroup of $SL_2(\mathbb{A}_f)$. (Here $\mathbb{A}_f$ denotes the finite adeles).

The quotients $\Gamma \backslash \mathfrak{h}$ have analytic structure and may be viewed as open Riemann surfaces. They can be compactified by the adjunction of a finite number of points and the resulting compact Riemann surface may be thought of as a nonsingular algebraic curve. Fundamental results of Eichler and Shimura [Sh1] show that these curves can be defined over number fields and their zeta function can be written as a product of L-functions associated to modular forms.

The theory of Shimura varieties embodies, in part, a grand higher dimensional generalization of these facts. The space $\mathfrak{h}$ is replaced by a Hermitian symmetric domain and Γ by an arithmetic group of congruence type. We introduce and discuss these notions in the next section.

I would like to thank the referee for some helpful comments on an earlier version of this note.

§2. Arithmetic quotients of bounded symmetric domains.

A *domain* is an open connected subset of $\mathbb{C}^m$ for some m. Let D be a bounded domain. We say that D is symmetric if for each point $x \in D$, there is an automorphism σ_x of D of order 2 and having x as isolated fixed point. The simplest example of a bounded symmetric domain is the unit disc

$$B_m = \{x \in \mathbb{C}^m : |x| < 1\}.$$

A complex manifold which is isomorphic to a bounded symmetric domain is called a *Hermitian symmetric domain*. A simple example is given by the upper half plane

$$\mathfrak{h} = \{z \in \mathbb{C} : \operatorname{Im} z > 0\}.$$

It is an open Riemann surface (complex manifold of dimension one) and

$$\mathfrak{h} \xrightarrow{\sim} B_1$$

is given by

$$z \longmapsto \frac{z - i}{z + i} \ .$$

Let D be a Hermitian symmetric domain. Then the set of automorphisms $Aut(D)$ of the complex manifold D is a real semisimple Lie group G with trivial center and having only a finite number of components. Moreover, the connected component of the identity G^+ of G acts transitively on D and the stabilizer of any point is a compact subgroup K of G^+. Thus, we may identify $D \simeq G^+/K$. (See [S, Ch II, §1] for example).

Let G be a real semisimple Lie group with finite center. Then G^+ (as well as G) has maximal compact subgroups, and any two are conjugate. Let K be a maximal compact subgroup of G^+. Then

$$D = G^+/K$$

is a Hermitian symmetric domain only in a very few cases. A complete list of the simple groups G such that D is a Hermitian symmetric domain was given by E. Cartan and the details of the proof were given partly by him and completely by Harish-Chandra.(See Helgason[H] for references and further details. The material below is taken from [H, ch. X, §2, §6.3, §6.4]). In order to give this list, it is necessary to introduce some notation.

We introduce the matrices

$$I_{p,q} = \begin{pmatrix} -I_p & 0 \\ 0 & I_q \end{pmatrix}, \quad J_m = \begin{pmatrix} 0 & I_m \\ -I_m & 0 \end{pmatrix}$$

where for any $0 < n \in \mathbb{Z}$, we denote by I_n the $n \times n$ identity matrix. Consider the following groups:

(a)

$$U(p,q) = \{g \in GL(p+q,\mathbb{C}) : {}^t g I_{p,q} \bar{g} = I_{p,q}\}.$$

It is the automorphisms of the Hermitian form

$$-z_1 \bar{z}_1 - \cdots - z_p \bar{z}_p + z_{p+1} \overline{z_{p+1}} + \cdots + z_{p+q} \overline{z_{p+q}}.$$

We also set $U(n) = U(n,0) = U(0,n)$.

(b)

$$SU(p,q) = U(p,q) \cap SL(p+q,\mathbb{C})$$

$$S(U(p) \times U(q)) = \left\{ \begin{pmatrix} g_1 & \\ & g_2 \end{pmatrix} : g_1 \in U(p), g_2 \in U(q), (\det g_1)(\det g_2) = 1 \right\}.$$

We also set $SU(n) = SU(n,0) = SU(0,n)$.

(c)

$$SO(p,q) = \{g \in SL(p+q,\mathbb{R}) : {}^t g I_{p,q} g = I_{p,q}\}.$$

It is the automorphisms of the quadratic form

$$-x_1^2 - \cdots - x_p^2 + x_{p+1}^2 + \cdots + x_{p+q}^2.$$

We also set $SO(n) = SO(n,0) = SO(0,n)$.

(d)
$$SO(n, \mathbb{C}) = \{g \in SL_n(\mathbb{C}) : {}^t g g = I_n\}.$$

It is the automorphisms of the form

$$z_1^2 + \cdots + z_n^2.$$

(e)
$$Sp(n, \mathbb{R}) = \{g \in GL(2n, \mathbb{R}) : {}^t g J_n g = J_n\}.$$

It is the automorphisms of the exterior form

$$x_1 \wedge x_{n+1} + x_2 \wedge x_{n+2} + \cdots + x_n \wedge x_{2n}.$$

(f)
$$Sp(n, \mathbb{C}) = \{g \in GL(2n, \mathbb{C}) : {}^t g J_n g = J_n\}.$$
$$Sp(n) = Sp(n, \mathbb{C}) \cap U(2n).$$

(g)
$$SO^*(2n) = \{g \in SO(2n, \mathbb{C}) : {}^t g g = I_{2n}, \; {}^t g J_n \bar{g} = J_n\}.$$

The groups $SU(p, q), SO^*(2n), Sp(n, \mathbb{R}), SU(n), SO(n)$ and $Sp(n)$ are all connected. The group $SO(p, q)$ has two connected components if $p \neq 0$ and $q \neq 0$. It is connected otherwise.

If G is simple and $D = G/K$ is Hermitian symmetric, then apart from two exceptional cases (of dimension 32 and 54), it is one of the following:

D	dimension (over $\mathbb{C}$)
$SU(p, q)/S(U(p) \times U(q))$	pq
$SO^*(2n)/U(n)$	$\frac{1}{2}n(n-1)$
$SO(p, 2)^+/(SO(p) \times SO(2))$	p
$Sp(n, \mathbb{R})/U(n)$	$\frac{1}{2}n(n+1)$

Notice that $Sp(n, \mathbb{R})$ is a subgroup of $GL(2n, \mathbb{R})$ and that $SO^*(2n)$ is a subgroup of $SO(2n, \mathbb{C})$. On the other hand, $U(n)$, which is a subgroup of $GL(n, \mathbb{C})$, is embedded into $Sp(n, \mathbb{R})$ and $SO^*(2n)$ as follows: write a general element of $U(n)$ as $A + iB$ with A and B real matrices. Then

$$A + iB \mapsto \begin{pmatrix} A & B \\ -B & A \end{pmatrix}.$$

In low dimensions, there are some coincidences amongst the above. For example, if D is one dimensional, then it is one of

$$SU(1, 1)/S(U(1) \times U(1)), \; SO^*(4)/U(2),$$

$$SO(1,2)^+/(SO(1) \times SO(2)), \text{ or } Sp(1,\mathbb{R})/U(1).$$

All of these are isomorphic to the upper half plane $\mathfrak{h}$. If D is two dimensional and irreducible, then it is isomorphic to

$$SU(2,1)/S(U(2) \times U(1)) \simeq B_2.$$

If it is not irreducible, then it is isomorphic to

$$SO(2,2)^+/(SO(2) \times SO(2)) \simeq \mathfrak{h} \times \mathfrak{h}.$$

If D is three dimensional and irreducible, then either it is isomorphic to

$$SU(3,1)/S(U(3) \times U(1)) \simeq SO^*(6)/U(3) \simeq B_3$$

or to

$$SO(3,2)^+/(SO(3) \times SO(2)) \simeq Sp(2,\mathbb{R})/U(2)$$

which is the Siegel upper half plane consisting (in this case) of 4×4 complex matrices $Z = X + iY$ with ${}^tZ = Z$ and Y positive definite. If it is not irreducible, then it is isomorphic to

$$\mathfrak{h} \times \mathfrak{h} \times \mathfrak{h}, \text{ or } \mathfrak{h} \times B_2.$$

Let $D = G^+/K$ be a Hermitian symmetric domain. Let $\mathfrak{g}$ (resp. $\mathfrak{k}$) denote the Lie algebra of G^+ (resp. K). As G is semisimple, the Killing form provides an orthogonal decomposition

$$\mathfrak{g} = \mathfrak{k} \oplus \mathfrak{p}.$$

The tangent space at the point o of D corresponding to the identity coset may be identified with $\mathfrak{p}$. The complex structure on D gives a decomposition

$$\mathfrak{p} \otimes \mathbb{C} = \mathfrak{p}^+ \oplus \mathfrak{p}^-.$$

Let $P^\pm$ denote the subgroup of $G_\mathbb{C}$ corresponding to $\mathfrak{p}^\pm$. They are unipotent groups and are normalized by $K_\mathbb{C}$. It is known (see for example, Satake [S, ch. II, §4] or Helgason [H, ch. VIII, §7]) that

$$P^+ \cap K_\mathbb{C}P^- = \{1\}$$
$$G^+ \subset P^+K_\mathbb{C}P^-$$
$$G^+ \cap K_\mathbb{C}P^- = K.$$

Hence,

$$G^+/K \simeq G^+K_\mathbb{C}P^-/K_\mathbb{C}P^- \hookrightarrow P^+K_\mathbb{C}P^-/K_\mathbb{C}P^- \simeq P^+ \simeq \mathfrak{p}^+.$$

This map

$$D = G^+/K \hookrightarrow \mathfrak{p}^+$$

is called the Harish-Chandra embedding and it realizes D as a bounded symmetric domain.

Example. If $G = Sp(1, \mathbb{R})$ then D is the upper half plane $\mathfrak{h}$ and $\mathfrak{p}^+ \simeq \mathbb{C}$. The Harish-Chandra embedding identifies $\mathfrak{h}$ with the unit disc B_1. More generally, if $G = Sp(n, \mathbb{R})$, then the Harish-Chandra embedding identifies D with

$$\{ \ z \in M_n(\mathbb{C}), \ ^t z = z, \ I_n - z\bar{z} \ \text{is positive definite} \ \}.$$

If $G = SU(n, 1)$ then the Harish-Chandra embedding identifies D with the unit disc B_n in $\mathbb{C}^n$. (See [S, Appendix] for explicit realizations of the D in the table above as bounded symmetric domains.)

If we compose the Harish-Chandra embedding with the map

$$P^+ K_{\mathbb{C}} P^- / K_{\mathbb{C}} P^- \ \hookrightarrow \ G_{\mathbb{C}}/K_{\mathbb{C}} P^-$$

we obtain the Borel embedding of D. The space $\check{D} = G_{\mathbb{C}}/K_{\mathbb{C}} P^-$ is a *compact* Hermitian symmetric domain and is called the compact dual of D. It can also be realized as U/K where U is a compact real form of G. In the above cases, it is as follows:

D	$\check{D}$
$SU(p, q)/S(U(p) \times U(q))$	$SU(p + q)/S(U(p) \times U(q))$
$SO^*(2n)/U(n)$	$SO(2n)/U(n)$
$SO(p, 2)^+/(SO(p) \times SO(2))$	$SO(p + 2)/(SO(p) \times SO(2))$
$Sp(n, \mathbb{R})/U(n)$	$Sp(n)/U(n)$

For each of the above, it is possible to give a concrete description of $\check{D}$. For example, if $G = SU(n, 1)$, the compact dual is complex projective space $\mathbb{P}^n$. (See [PS] for realizations of D as so-called Siegel domains.)

If G is an algebraic subgroup of GL_n defined over $\mathbb{Q}$, a subgroup $\Gamma \subseteq G(\mathbb{Q})$ is *arithmetic* if it is commensurable with $G(\mathbb{Z}) = G(\mathbb{Q}) \cap GL_n(\mathbb{Z})$.

Now let G be an algebraic group defined over $\mathbb{Q}$ and suppose that $D = G(\mathbb{R})^+/K$ is a Hermitian symmetric domain. If Γ is an arithmetic subgroup of $G(\mathbb{Q})$, then it is a discrete subgroup of $G(\mathbb{R})$ and so acts properly discontinuously on D. (This means that every compact subset of D meets only a finite number of its Γ translates. In particular, the stabilizer of any point is finite.) Hence, if Γ is also torsion free, then it acts freely on D and so

$$S = \Gamma \backslash D$$

is again a complex manifold. (For any arithmetic Γ, there is a normal subgroup Γ' of Γ of finite index such that Γ' is torsion free.) A theorem of Baily and Borel implies that it has algebraic structure also. Namely, they show that S has the structure of quasi-projective variety (i.e. a Zariski open subset of a projective variety). Later, Borel showed that this structure is unique in the following sense: let V be any complex algebraic variety and $V^{an} \longrightarrow S$ a holomorphic map of the underlying analytic spaces. Then this map is a morphism of algebraic varieties.

The variety S is called a *locally symmetric variety*. Satake and Baily-Borel found a compactification $\bar{S}$ of S. (References for this result and others quoted above may be found in Milne [Mi]). It is obtained by adding components at the boundary corresponding to classes of parabolic subgroups of G. It is a minimal compactification in the sense that if V is a smooth algebraic variety containing S as an open subset and such that $V - S$ has only normal crossings as singularities, then there is a map $V \longrightarrow \bar{S}$ which is the identity on S. In general, $\bar{S}$ is a singular variety. A smooth compactification $\tilde{S}$ is obtained by the method of torus embeddings [AMRT].

For a real Lie group M let us say that a discrete subgroup Γ is *arithmetic* if there is an algebraic group G defined over $\mathbb{Q}$, and a morphism

$$\phi \; : \; G(\mathbb{R}) \longrightarrow M$$

such that the kernel is compact and the image is open and such that the image of any arithmetic subgroup of G is commensurable with Γ.

Remark. A deep theorem of Margulis [M] asserts that if M is simple and $\mathrm{rank}_{\mathbb{R}} M \geq 2$, then any lattice in M (i.e. any discrete subgroup Γ such that $\mathrm{vol}(\Gamma \backslash M) < \infty$) is necessarily arithmetic. If $\mathrm{rank}_{\mathbb{R}} M = 1$, there exist non-arithmetic lattices. Recent work of Gromov and Piatetski-Shapiro [G-PS] produces such lattices in $SO(n,1)$ for any n. Mostow [Mo] has produced some non-arithmetic lattices in $SU(2,1)$ and $SU(3,1)$. The article of Prasad [P] provides an excellent overview of the status of these and related problems.

A subgroup Γ of $G(\mathbb{R})$ is a *congruence subgroup* if it contains

$$\mathrm{Ker}(G(\mathbb{Z}) \longrightarrow G(\mathbb{Z}/N\mathbb{Z}))$$

for some $1 \leq N \in \mathbb{Z}$. There exist some arithmetic groups which are not congruence subgroups. (For example, the fundamental group of the Fermat curve is of this kind.) The theory of Shimura varieties is developed, however, only for arithmetic groups which are congruence subgroups.

Shimura [Sh2] showed that the locally symmetric varieties $\Gamma \backslash D$, with Γ a congruence subgroup and D as in the table above, can be defined over a number field k_Γ. This field k_Γ is an abelian extension of a number field E which depends only on G and D but not on Γ. In many cases, the locally symmetric varieties are parameter spaces for families of abelian varieties. For example,

$$Sp(n, \mathbb{Z}) \backslash Sp(n, \mathbb{R}) / U(n)$$

is the parameter space of n−dimensional principally polarized abelian varieties. (For $n = 1$, this is the example considered in §1.) However, it is not the case that all locally symmetric varieties arise in this way. An abelian variety may be viewed as a polarized Hodge structure of weight 1. Deligne's point of view is to consider the locally symmetric varieties which are parameter spaces for a variation of uniformly polarized Hodge structures. In general, these Hodge structures may not have any geometric interpretation. In some cases, however, Deligne expects them to be Hodge structures of motives.

§3. Deligne's axioms.

Let $\mathbb{S} = \operatorname{Res}_{\mathbb{C}/\mathbb{R}}(\mathbb{G}_m)$. Thus, $\mathbb{S}$ is an $\mathbb{R}$–algebraic group with $\mathbb{S}(\mathbb{R}) = \mathbb{C}^{\times}$ and $\mathbb{S}(\mathbb{C}) = \mathbb{C}^{\times} \times \mathbb{C}^{\times}$. A (mixed) Hodge structure on a $\mathbb{R}$–vector space $V_{\mathbb{R}}$ is a decomposition

$$V_{\mathbb{R}} \otimes \mathbb{C} = \oplus V^{p,q}$$

with $\overline{V^{p,q}} = V^{q,p}$. Giving a Hodge structure on $V_{\mathbb{R}}$ is equivalent to giving a representation

$$h : \mathbb{S}(\mathbb{R}) = \mathbb{C}^{\times} \longrightarrow GL(V_{\mathbb{R}}).$$

Indeed, given such a map as above, we define $V^{p,q}$ to be the subspace of $V_{\mathbb{R}} \otimes \mathbb{C}$ on which $\mathbb{C}^{\times}$ acts by the character $z \mapsto z^{-p}\bar{z}^{-q}$. Associated to a Hodge structure, we have the Hodge filtration. Define

$$F^a = \oplus_{p \geq a} V^{p,q}.$$

Then

$$F^0 \supseteq F^1 \supseteq \cdots$$

is a decreasing filtration.

A *weight* of the Hodge structure is an integer n such that there exists $V^{p,q} \neq 0$ and $p + q = n$. In general, there will be more than one weight in the Hodge structure of $V_{\mathbb{R}}$. Consider the inclusion of $\mathbb{R}$–algebraic groups $\mathbb{G}_m \subset \mathbb{S}$ and set $w_h = h|_{\mathbb{G}_m}$. Then the weights that occur are just the characters that occur in w_h. If the Hodge structure has a unique weight, then, we may recover the Hodge structure from the Hodge filtration by

$$F^p \cap \overline{F^q} = V^{p,q}.$$

A polarisation ψ of an $\mathbb{R}$–Hodge structure $V_{\mathbb{R}}$ of weight n is a morphism

$$\psi : V_{\mathbb{R}} \times V_{\mathbb{R}} \longrightarrow \mathbb{R}(-n) = (2\pi i)^{-n}\mathbb{R}$$

such that the form $(x, y) \mapsto (2\pi i)^n \psi(x, h(i)y)$ is symmetric and positive definite. Here, $\mathbb{R}(-n)$ is the real Hodge structure $(2\pi i)^{-n}\mathbb{R} \subseteq \mathbb{C}$ of type (n, n).

Let us fix a finite number of $\mathbb{R}$ vector spaces $\{V_i\}$ and some tensors $\{s_j\}$ in $\otimes_i(V_i \otimes V_i^*)$. A Hodge structure on the V_i is given by a morphism

$$h : \mathbb{S} \longrightarrow \prod GL(V_i).$$

Let G be the algebraic subgroup of the right hand side which fixes each of the s_j. The s_j are all of type $(0,0)$ if and only if h factors through G :

$$\mathbb{S} \longrightarrow G_{\mathbb{R}} \hookrightarrow \prod GL(V_i).$$

Starting with $h_0 : \mathbb{S} \longrightarrow G_{\mathbb{R}}$ we may produce a family of Hodge structures on the V_i simply by conjugating h_0 by an element of $G(\mathbb{R})$. Let X be the $G(\mathbb{R})$ conjugacy class of h_0. Let X^+ be a connected component of X. Motivated by the examples of families of abelian varieties, it is reasonable to ask that X^+ be a Hermitian symmetric domain and that it parametrizes a variation of polarized Hodge structures of constant weight.

Thus, Deligne only considers those X^+ so that the gradation by the weight is constant in the family (i.e. that for each i, the graded pieces of V_i (graded according to the weight) are independent of the $G(\mathbb{R})$ conjugate of h_0.) This is equivalent to requiring that $h_0(\mathbb{R}^\times)$ is contained in the center of $G(\mathbb{R})^+$. (In this case, we may write w_{X^+} for the weight morphism). Moreover, he asks that X^+ parametrize a variation of Hodge structure. (This means that the Hodge filtrations $\{F^p\}$ vary holomorphically in the family and that the associated bundles $\mathbf{F}^p$ satisfy Griffiths' transversality condition: the derivative of a section of $\mathbf{F}^p$ lies in $\mathbf{F}^{p-1}$. The reader is referred to the expository paper [GS] for precise definitions.) In addition, Deligne asks that the Hodge structures parametrized by X^+ be polarized in the following uniform sense: if V is a component of weight n of one of the V_i, there exists

$$\psi : V \otimes V \longrightarrow \mathbb{R}(-n)$$

which for *all* $h \in X$ is a polarisation of V.

The first condition, namely that w_{X^+} be constant in the family, can be expressed in terms of the adjoint representation

$$G(\mathbb{R}) \longrightarrow GL(\mathfrak{g})$$

which factors through the adjoint group $G^{\mathrm{ad}} = G/Z(G)$. Given a morphism

$$h_0 : \mathbb{S} \longrightarrow G_{\mathbb{R}},$$

composing it with the adjoint representation gives a map

$$h : \mathbb{S} \longrightarrow GL(\mathfrak{g})$$

and thus a Hodge structure on $\mathfrak{g}$. The condition that $h_0(\mathbb{R}^\times)$ be contained in the center of $G(\mathbb{R})$ means that the Hodge structure on $\mathfrak{g}$ is of weight 0.

Now, let W be a component of $\oplus V_i$ of a fixed weight. Then, there is a map of X^+ into a space of Grassmanians of $W_{\mathbb{C}}$ which sends the Hodge structure at $h \in X^+$ to the corresponding Hodge filtration:

$$\{ W^{p,q} \} \longmapsto \{ F^p \}.$$

This map is injective since the Hodge structure can be recovered from the filtration. It is also an embedding. There is then a unique complex structure on X^+ which ensures that the filtrations vary holomorphically, namely that inherited from the Grassmanians. (See [D2, 1.1.14(i)].)

In order that the $h \in X^+$ satisfy transversality, Deligne shows [D2, 1.1.14(ii)] that it is necessary and sufficient that the Hodge structure on $\mathfrak{g}$ be of type $\{(-1,1),(0,0),(1,-1)\}$. That is, if (p,q) is not one of these three ordered pairs, then the (p,q) piece of the Hodge structure is zero.

Let G_1 be the smallest algebraic subgroup of G through which all of the $h \in X^+$ factor. Let V be a component of V_i of weight n. For there to exist a map

$$\psi : V \times V \longrightarrow \mathbb{R}(-n)$$

which for *all* $h \in X^+$ is a polarisation of V, Deligne shows [D2, 1.1.14(iii)] that it is necessary and sufficient that G_1 be reductive and for all $h \in X^+$, the inner automorphism int $h(i)$ induce a Cartan involution on the adjoint group.

Under these conditions, Deligne shows [D2, 1.1.17] that X^+ is a Hermitian symmetric domain.

Note that from the above, it follows that the requirement that the family of Hodge structures given by the $h \in X$ be a variation of uniformly polarized Hodge structures is *independent* of the choice of faithful representations V_i of $G_{\mathbb{R}}$.

Now Deligne's axioms may be stated. Let G be a reductive algebraic group defined over $\mathbb{Q}$, and X a $G(\mathbb{R})$−conjugacy class of morphisms of $\mathbb{R}$−algebraic groups $\mathbb{S} \longrightarrow G_{\mathbb{R}}$. Suppose that (G, X) satisfies

(1) the Hodge structure on $\mathfrak{g}$ given by every $h \in X$ is of type

$$\{(-1,1),(0,0),(1,-1)\}.$$

(2) for all $h \in X$, the inner automorphism int $h(i)$ is a Cartan involution of the adjoint group $G_{\mathbb{R}}^{\mathrm{ad}}$.

(3) G^{ad} does not have a factor G' defined over $\mathbb{Q}$ on which h projects trivially.

X admits a unique complex structure such that for all faithful representations V of $G_{\mathbb{R}}$, the Hodge filtration of V varies holomorphically with h. For this complex structure, the connected components of X are Hermitian symmetric domains.

Also, if we decompose $G_{\mathbb{R}}^{\mathrm{ad}}$ into simple factors, then h projects trivially on the compact factors. (Indeed, if $G_{\mathbb{R}}$ is a compact group, int $h(i)$ acts trivially. This means that $h(i)$ is contained in the center of $G(\mathbb{R})$ and so the Hodge structure on $\mathfrak{g}$ is of type $(0,0)$. This means that $h = 1$.) Every connected component of X is a product of Hermitian symmetric spaces corresponding to the non-compact factors. The condition (3) may therefore be expressed as saying that G^{ad} does not have a factor G' defined over $\mathbb{Q}$ such that $G'(\mathbb{R})$ is compact.

Given (G, X) satisfying the above conditions, the Shimura variety associated to G, X and a compact open subgroup $K \subseteq G(\mathbb{A}_f)$ is

$$_K\mathrm{Sh}(G, X) = G(\mathbb{Q})\backslash(X \times G(\mathbb{A}_f)/K).$$

Associated to each $h \in X$, let

$$\mu_h : \mathbb{C}^\times \longrightarrow G(\mathbb{C})$$

be given by

$$\mu_h(z) = h(z, 1).$$

The set of cocharacters $\{\mu_h : h \in X\}$ lies in a single $G(\mathbb{C})$ conjugacy class, M_X (say). The image of any μ_h lies in a maximal torus of $G(\mathbb{C})$ and so some $G(\mathbb{C})$ conjugate of it has image contained in a maximal torus $T(\mathbb{C})$ of $G(\mathbb{C})$. This torus is defined and split over $\overline{\mathbb{Q}}$ and so any morphism $\mathbb{C}^\times \longrightarrow T(\mathbb{C})$ is defined over $\overline{\mathbb{Q}}$. Thus, M_X has a representative defined over $\overline{\mathbb{Q}}$. Also, note that if $\alpha, \beta : \overline{\mathbb{Q}}^\times \longrightarrow G(\overline{\mathbb{Q}})$ are $G(\mathbb{C})$ conjugate, then they must be $G(\overline{\mathbb{Q}})$ conjugate. Hence, any two elements of M_X which are defined over $\overline{\mathbb{Q}}$ lie in the same $G(\overline{\mathbb{Q}})$ conjugacy class. Thus we may define the *reflex field* $E(G, X)$ to be the field of definition of the $G(\overline{\mathbb{Q}})$−conjugacy class of this representative. In particular, $E(G, X)$ is contained in any field over which a maximal torus of G splits. Thus if G has a maximal split torus defined over $\mathbb{Q}$, then $E(G, X) = \mathbb{Q}$.

Example 1. Let $G = T$ be a torus. Then X consists of a single point and

$$_K\mathrm{Sh}(T, X) = T(\mathbb{Q})\backslash T(\mathbb{A}_f)/K$$

is a finite set of points. Let M be a number field and $T = R_{M/\mathbb{Q}}\mathbb{G}_m$. Thus T is a torus of dimension $[M : \mathbb{Q}]$ with $T(\mathbb{Q}) = M^\times$. Let $K = \prod_p(\mathcal{O}_M \otimes \mathbb{Z}_p)^\times$ where $\mathcal{O}_M$ denotes the ring of integers of M. Then

$$_K\mathrm{Sh}(T, X) = (M \otimes \mathbb{A}_f)^\times/M^\times K = M_\mathbb{A}^\times/M_\infty^\times M^\times K$$

which is the ideal class group of M. If M is imaginary quadratic, the reflex field $E(T, X)$ is M itself. In the general case, $E(T, X)$ depends on the choice of an $h : \mathbb{S} \longrightarrow T(\mathbb{R})$.

Example 2. Let $G = GL_2$ and consider the map

$$h_o : \mathbb{S} \longrightarrow GL_2$$

defined by

$$a + ib \longmapsto \begin{pmatrix} a & b \\ -b & a \end{pmatrix}.$$

In this case, X is the union of the upper half plane and the lower half plane and $E(G, X) = \mathbb{Q}$. Let $G(\mathbb{R})_+$ denote the inverse image of $G^{\mathrm{ad}}(\mathbb{R})^+ = PGL_2(\mathbb{R})^+$ in $G(\mathbb{R})$ and $G(\mathbb{Q})_+ = G(\mathbb{R})_+ \cap G(\mathbb{Q})$. Let K be the compact open subgroup of $G(\mathbb{A}_f)$ such that $K \cap G(\mathbb{Q})_+ = \Gamma(N)$ where $\Gamma(N)$ denotes the principal congruence subgroup of level N. Then

$$G(\mathbb{Q})_+ \backslash G(\mathbb{A}_f)/K \simeq (\mathbb{Z}/N)^\times.$$

We have

$$_K\mathrm{Sh}(G, X) = \cup_g \Gamma_g \backslash \mathfrak{h}$$

where the Γ_g are certain subgroups of $SL_2(\mathbb{Q})$ commensurate with $\Gamma(N)$. For a detailed treatment of this and similar cases, see the book of Shimura [Sh1].

In general, $_K\mathrm{Sh}(G, X)$ can be written as a finite union of arithmetic quotients of bounded symmetric domains as follows. Choose a set of representatives $\{g\}$ for the double cosets

$$G(\mathbb{Q})_+ \backslash G(\mathbb{A}_f)/K.$$

Let Γ_g denote the image of $gKg^{-1} \cap G(\mathbb{Q})_+$ in $G^{\mathrm{ad}}(\mathbb{R})^+$. Then Γ_g is an arithmetic subgroup. Let X^+ be a connected component of X. Then

$$_K\mathrm{Sh}(G, X) = \cup_g \Gamma_g \backslash X^+.$$

If K is sufficiently small, Γ_g is torsion-free. Thus, by the theorem of Baily and Borel, $_K\mathrm{Sh}(G, X)$ is a finite union of quasi-projective varieties.

Example 3. Let F be a real quadratic field and let $G = R_{F/\mathbb{Q}}GL_2$. Then

$$G(\mathbb{R}) = GL_2(F \otimes \mathbb{R}) \simeq GL_2(\mathbb{R}) \times GL_2(\mathbb{R})$$

and if we take for h the map

$$a + ib \longmapsto \left(\begin{pmatrix} a & b \\ -b & a \end{pmatrix}, \begin{pmatrix} a & b \\ -b & a \end{pmatrix} \right)$$

the space X can be identified with $\mathfrak{h} \times \mathfrak{h}$. Let

$$K = G(\hat{\mathbb{Z}}) = \prod GL_2(\mathcal{O}_F \otimes \mathbb{Z}_p).$$

Then $E(G,X) = \mathbb{Q}$ and

$$_K \mathrm{Sh}(G,X) = \cup\, \Gamma(\mathfrak{a})\backslash(\mathfrak{h} \times \mathfrak{h})$$

where $\mathfrak{a}$ ranges over a set of representatives for the ideal class group of F and

$$\Gamma(\mathfrak{a}) = \{\begin{pmatrix} a & b \\ c & d \end{pmatrix} \in SL_2(F) : a,d \in \mathcal{O}_F,\ c \in \mathfrak{a},\ b \in \mathfrak{a}^{-1}\}.$$

This is an example of a Hilbert-Blumenthal surface. For a detailed discussion of this case, see the excellent survey by Ramakrishnan [R].

If $K_1 \subseteq K_2$ we have a natural map

$$_{K_1}\mathrm{Sh}(G,X) \longrightarrow {}_{K_2}\mathrm{Sh}(G,X).$$

Taking the projective limit over all K let us set

$$\mathrm{Sh}(G,X) = \lim\,_K\mathrm{Sh}(G,X).$$

It comes equipped with a natural (right) $G(\mathbb{A}_f)$ action.

Example 4. Quaternionic Shimura Varieties : Let F be a totally real field and B a quaternion division algebra over F (that is, a central simple division algebra over F of dimension 4.) Let G be the linear algebraic group defined over $\mathbb{Q}$ such that $G(\mathbb{Q}) = B^\times$. Let $a,b \in \mathbb{Z}$ be such that $a \geq 1, b \geq 0$ and $a + b = d = [F : \mathbb{Q}]$. Suppose that

$$B \otimes \mathbb{R} = \prod B_v = M_2(\mathbb{R})^a \times \mathbb{H}^b$$

where v ranges over the archimedean primes of F and $\mathbb{H}$ denotes the Hamilton quaternions. Then

$$G(\mathbb{R}) = GL_2(\mathbb{R})^a \times (\mathbb{H}^\times)^b$$

acts on the a−fold product of the union of the upper and lower half planes: $(\mathfrak{h}^+ \cup \mathfrak{h}^-)^a$ through the first factor.

Consider the map

$$h_0 : \mathbb{S} \longrightarrow G(\mathbb{R})$$

given by

$$\alpha + \beta i \mapsto \left(\begin{pmatrix} \alpha & -\beta \\ \beta & \alpha \end{pmatrix}, \cdots, \begin{pmatrix} \alpha & -\beta \\ \beta & \alpha \end{pmatrix}, 1, \cdots, 1\right).$$

The $G(\mathbb{R})$ conjugacy class X of this map h_0 satisfies

$$X \simeq G(\mathbb{R})/K_\infty \simeq (\mathfrak{h}^+ \cup \mathfrak{h}^-)^a.$$

where

$$K_\infty = \{\begin{pmatrix} \alpha & -\beta \\ \beta & \alpha \end{pmatrix}\}^a \times \mathbb{H}^{\times b}.$$

Then for a compact open subgroup $K \subseteq G(\mathbb{A}_f)$,

$$_K \mathrm{Sh}(G, X)(\mathbb{C}) = G(\mathbb{Q})\backslash G(\mathbb{A})/K_\infty K.$$

This is an $a-$dimensional variety defined over the reflex field $E(G, X)$ which can be described as follows. Let $\tilde{F}$ be a Galois extension of F and let $W = \{v : B_v = M_2(\mathbb{R})\}$. Then,

$$E(G, X) = \mathbb{Q}(\sum_{\alpha \in W} \alpha a \ : \ a \in F)$$

and it is the fixed field of

$$\{\sigma \in \mathrm{Gal}(\tilde{F}/\mathbb{Q}) : \sigma W = W\}.$$

Let us write

$$F_{\mathbb{A}}^\times = \cup_{j=1}^h F^\times b_j F_\infty^+ (\det K)$$

where F_∞^+ denotes the totally positive elements in $F \otimes \mathbb{R}$. Then

$$G(\mathbb{A}) = \cup_{j=1}^h G(\mathbb{Q}) x_j G(\mathbb{R})^+ K$$

where $x_j = (x_{j,v}) \in G(\mathbb{A}_f)$ has reduced norm equal to b_j and

$$x_{j,v} = \begin{cases} 1 & \text{if } B \text{ is ramified at } v \\ \begin{pmatrix} b_{j,v} & 0 \\ 0 & 1 \end{pmatrix} & \text{otherwise} . \end{cases}$$

Let us set

$$\Gamma_j = G(\mathbb{Q}) \cap x_j G(\mathbb{R})^+ K x_j^{-1}.$$

Then

$$_K \mathrm{Sh}(G, X)(\mathbb{C}) = \cup_{j=1}^h \Gamma_j \backslash \mathfrak{h}^a.$$

If $b = 0$, then $E(G, X) = \mathbb{Q}$ and $\mathrm{Sh}(G, X)$ is a parameter space for abelian varieties of dimension d with multiplication by B.

A *model* of $\mathrm{Sh}(G, X)$ over a subfield E of $\mathbb{C}$ is a scheme S defined over E together with an $E-$rational action of $G(\mathbb{A}_f)$ such that there is a $G(\mathbb{A}_f)-$equivariant isomorphism

$$\mathrm{Sh}(G, X) \simeq S \otimes_E \mathbb{C}.$$

It is known that $\mathrm{Sh}(G, X)$ has a model over $E(G, X)$. By introducing the notions of special points and canonical models (which are central to the arithmetic theory of Shimura varieties) a much more precise statement is known by the fundamental work of Shimura with refinements by Deligne, Langlands and the work of Borovoi, Milne and Shih. (See [Mi] for details and references).

If the weight morphism w_X is defined over $\mathbb{Q}$, then $\text{Sh}(G, X)$ is expected to be a parameter space for a family of motives. For a brief discussion of this, see [Mi, ch. II, §3].

The determination of the zeta function of a Shimura variety and its expression in terms of $L-$functions of automorphic representations is viewed by Langlands as a higher dimensional version of Artin reciprocity. For a clear discussion of this problem and the difficulties involved, see [L1] and [LR] and the references therein.

REFERENCES

[AMRT] A. Ash, D. Mumford, M. Rapaport, Y. Tai, Smooth compactifications of locally symmetric varieties, Math. Sci Press, Brookline, 1975.

[D1] P. Deligne, Travaux de Shimura, Lecture notes in Mathematics 244 (1971), 123-165.

[D2] P. Deligne, Variétés de Shimura: interprétation modulaire et techniques de construction de modèles canoniques, Proc. Symp. Pure Math. AMS, 33(2)(1979),247-290.

[GS] P. Griffiths and W. Schmid, Recent developments in Hodge theory : a discussion of techniques and results, in: Discrete subgroups of Lie groups and applications to moduli, pp. 31-127, Tata Institute of Fundamental Research, Oxford, 1973.

[G-PS] M. Gromov and I. Piatetski-Shapiro, Non-arithmetic groups in Lobachevsky spaces, Publ. Math. IHES, 66(1988), 93-103.

[H] S. Helgason, Differential geometry, Lie groups and symmetric spaces, Academic Press, 1978.

[L1] R. Langlands, Some contemporary problems with origins in the Jugendtraum, Proc. Symp. Pure Math. AMS, 28(1976), 401-418.

[L2] R. Langlands, Automorphic representations, Shimura varieties, and motives: Ein Märchen, Proc. Symp. Pure Math. AMS, 33(2)(1979), 205-246.

[LR] The zeta functions of Picard modular surfaces, ed. R. Langlands and D. Ramakrishnan, Centre de Recherches Mathematiques, 1992.

[M] G. A. Margulis, Arithmetic properties of discrete subgroups, Uspekhi. Math. Nauk., 29(1974), 49-98.

[Mi] J. Milne, Canonical models of (mixed) Shimura varieties and automorphic vector bundles, in: Automorphic forms, Shimura varieties and $L-$functions, ed. L. Clozel and J. Milne, pp. 283-414, 1990, Academic Press.

[Mo] G. Mostow, Discrete subgroups of Lie groups, in: Elie Cartan et les mathematiques d' aujourd'hui, Asterisque, (1985), 289-309.

[P] G. Prasad, Semi-simple groups and arithmetic groups, Proc. ICM 1990, vol. 2, pp. 821-832, Springer-Verlag.

[PS] I. Piatetski-Shapiro, Geometry of classical domains and theory of automorphic functions, Gordon and Breach, New York, 1969.

[R] D. Ramakrishnan, Arithmetic of Hilbert-Blumenthal surfaces, in: Number Theory, ed. H. Kisilevsky and J. Labute, CMS Conf. Proc., 7(1987), 285-370.

[S] I. Satake, Algebraic structures of symmetric domains, Publ. Math. Soc. Japan, no. 14, Iwanami Shoten and Princeton, 1980.

[Sh1] G. Shimura, Arithmetic theory of automorphic functions, Publ. Math. Soc. Japan, no. 11, Iwanami Shoten and Princeton, 1971.

[Sh2] G. Shimura, On canonical models of arithmetic quotients of bounded symmetric domains, Annals of Math., I: 91(1970),141-222, II: 91(1970), 528-549.

[Sh-T] G. Shimura and Y. Taniyama, Complex multiplication of abelian varieties and its applications to number theory, Publ. Math. Soc. Japan, no. 6, Iwanami Shoten, 1961.

V. Kumar Murty
Department of Mathematics
University of Toronto
Toronto, CANADA, M5S 1A1
e-mail murty@math.toronto.edu

Galois Action on the Nilpotent Completion of the Fundamental Group of an Algebraic Curve

Takayuki Oda

Research Institute for Mathematical Science, Kyoto University

Introduction

This is a short report on our recent work on the Galois action on the nilpotent completion of the fundamental groups of algebraic curves ([14], [20], [21], [3]).

Let X be a normal geometrically irreducible algebraic variety over a field K. Then the fundamental group of X is defined by

$$\pi_1(X) = \lim_{Y_L \to X \ \text{étale}} \text{Gal}(L/K(X)).$$

Here $K(X)$ is the function field of X over K, and L runs over a finite Galois extension of $K(X)$ such that the canonical finite surjective homomorphism $Y_L \to X$ from the normal closure Y_L of X in L to X is étale.

We can define $\pi_1(X \otimes_K \overline{K})$ similarly for the separable closure $\overline{K}$ of K. Then we have an exact sequence

$$1 \to \pi_1(X \otimes_K \overline{K}) \to \pi_1(X) \to \text{Gal}(\overline{K}/K) \to 1.$$

For any element σ of $\text{Gal}(\overline{K}/K)$, we choose a pre-image $\tilde{\sigma}$ in $\pi_1(X)$. Then the transform by $\tilde{\sigma}$ induces an automorphism $Int(\tilde{\sigma})$ on $\pi_1(X \otimes_K \overline{K})$. For another choice $\tilde{\sigma}'$ of pre-image of σ, $Int(\tilde{\sigma}')$ differs from $Int(\tilde{\sigma})$ only by an inner automorphism of $\pi_1(X \otimes_K \overline{K})$. Thus we have a natural Galois representation

$$\text{Gal}(\overline{K}/K) \longrightarrow \text{Out } \pi_1(X \otimes_K \overline{K}).$$

Here Out means the outer automorphism group, i.e. the quotient of the continuous automorphism group Aut $\pi_1(X \otimes_K \overline{K})$ by the subgroup Inn $\pi_1(X \otimes_K \overline{K})$ of inner automorphisms.

Let l be a rational prime, and let $\pi_1(X \otimes_K \overline{K})_l$ be the maximal pro-l quotient of $\pi_1(X \otimes_K \overline{K})$. Then the kernel of the canonical surjection $\pi_1(X \otimes_K \overline{K}) \to \pi_1(X \otimes_K \overline{K})_l$ is a characteristic subgroup of $\pi_1(X \otimes_K \overline{K})$. Therefore we have a canonically induced homomorphism

$$\rho_X : \text{Gal}(\overline{K}/K) \longrightarrow \text{Out } \pi_1(X \otimes_K \overline{K})_l.$$

When U is the projective line P^1 minus three points $\{0, 1, \infty\}$ defined over $\mathbb{Q}$, Belyi (cf. [4]) shows that the natural homomorphism

$$\mathrm{Gal}(\overline{\mathbb{Q}}/\mathbb{Q}) \longrightarrow \mathrm{Out}\, \pi_1(U \otimes_{\mathbb{Q}} \overline{\mathbb{Q}})$$

is injective. Ihara [10] and Deligne [9] show that the homomorphism

$$\rho_U : \mathrm{Gal}(\overline{\mathbb{Q}}/\mathbb{Q}) \longrightarrow \mathrm{Out}\, \pi_1(U \otimes_{\mathbb{Q}} \overline{\mathbb{Q}})_l.$$

has very interesting arithmetical contents.

It is natural to attempt to investigate the corresponding representation for a curve of higher genus. Our purpose is to investigate this Galois representation using the lower central filtration on the pro-l fundamental group $\pi_1(X \otimes_K \overline{K})_l$, when X is a curve.

Here is the contents of this paper. In the first two sections, we discuss some results which are derived from the basic facts on commutator calculus of groups. In §1, we recall some basic facts on the central filtration, commutator calculus, the associated graded quotients, and the induced filtration on the automorphism groups. In §2, we consider the Galois action on the graded quotients associated to the lower central series of the pro-l fundamental group of an algebraic curve. If C is a complete curve, it is known ([14]) that as a Galois module, the graded Lie algebra associated to the l-adic completion of the fundamental group of the curve $C \times \overline{K}$ is isomorphic to a quotient of the free graded Lie algebra on the dual of $H^1_{\acute{e}t}(C \times \overline{K}, \mathbb{Z}_l)$ modulo the ideal generated by the images of the dual of the cup-product self-pairing

$$\overset{2}{\bigwedge} H^1_{\acute{e}t}(C \times \overline{K}, \mathbb{Z}_l) \to H^2_{\acute{e}t}(C \times \overline{K}, \mathbb{Z}_l).$$

Extending this, we give a similar but more complicated description, if C is not complete (Theorem 6).

In the latter two sections, we discuss the monodromy action when the curve is degenerating. In §3, we consider the action of the inertia group on the pro-l fundamental group of a curve defined over a discrete valuation ring with residual characteristic$\neq l$. We can show that this action of the inertia group on the pro-l fundamental group is trivial if and only if the curve has good reduction (Theorem 9).

It is natural to consider not only a single curve defined over a field, but also a family of curves with variable moduli. In §4, we discuss the analytical case where curves are defined over the complex number field $\mathbb{C}$. We consider locally universal families of degenerating curves, and give a combinatorial description of their local monodromy in terms of the dual graphs of degenerate curves. This local investigation in the "Betti realization" should be the corner stone of a further investigation. The results of §4 is a joint work with M. Asada and M. Matsumoto.

Acknowledgment

The author thanks Y. Ihara whose work is the starting point for him. He also thanks to M. Asada, M. Kaneko, M. Matsumoto, and H. Nakamura for discussions and various assists.

1 Filtration on groups

In this section, we fix basic notations for subsequent sections.

1.1 Central filtration and commutator calculus

The basic references of this section are Magnus-Karras-Solitar [17] and Bourbaki [5]. Let G be a discrete group, or a compact topological group. We consider only a finitely generated discrete group, or a compact topological group with finite number of topological generators. For subgroups H, K of G, we denote by $[H, K]$ the (closed) subgroup generated by commutators $[h, k] = hkh^{-1}k^{-1}$ ($h \in H, k \in K$).

Let us recall the notion of central filtration.

Definition 1. *Let $\{\varphi_m G\}_{m \geq 1}$ be a decreasing sequence of closed subgroups of G satisfying*

   (i) $\varphi_1 G = G$;

and

   (ii) $[\varphi_k G, \varphi_l G] \subset \varphi_{k+l} G$ *for any* $k, l \geq 1$.

Then $\{\varphi_m G\}_{m \geq 1}$ is called a central filtration *on G.*

In this case, $\varphi_m G$ is a normal (closed) subgroup of G for each $m \geq 1$, and the quotient group $gr_m^\varphi G = \varphi_m G / \varphi_{m+1} G$ is an abelian group. Moreover, when G is (topologically, resp.) finitely generated, then $gr_m^\varphi G$ is also (topologically, resp.) finitely generated.

For $\xi \in gr_k^\varphi G$ and $\eta \in gr_l^\varphi G$, choose representatives $x \in \varphi_k G$ modulo $\varphi_{k+1} G$ for ξ and $y \in \varphi_l G$ modulo $\varphi_{l+1} G$ for η. Put

$$[\xi, \eta] = (xyx^{-1}y^{-1} \text{ modulo } \varphi_{k+l} G) \text{ in } gr_{k+l}^\varphi G.$$

Then $[\xi, \eta]$ is determined independently of the choice of x, y. Obviously, we have $[\xi, \eta] = -[\eta, \xi]$. Let $\zeta \in gr_n^\varphi G$ be the third element. Then the Hall identity implies the Jacobi identity

$$[[\xi, \eta], \zeta] + [[\eta, \zeta], \xi] + [[\zeta, \xi], \eta] = 0.$$

Hence $gr_\bullet^\varphi G := \oplus_{m=1}^\infty gr_m^\varphi G$ has a canonical structure of graded Lie algebra.

One example of central filtration is the lower central series $\{\Gamma_m G\}_{m \geq 1}$ defined as follows.

Definition 2. *Set* $\Gamma_1 G = G$, *and inductively we put*

$$\Gamma_{m+1} G = \overline{[G, \Gamma_m G]} \quad \text{for each } m \geq 1.$$

Here $\overline{}$ *means the closure when G is a compact topological group.*

The lower central series is the most rapidly decreasing central filtration. (cf. Bourbaki [5], Chap. 2, §4).

Since the m-th higher commutator $\Gamma_m G$ is a characteristic subgroup of G for each m, any (topological) automorphism ρ of G induces canonically an automorphism on $\Gamma_m G, G/\Gamma_m G$, and $\Gamma_m G/\Gamma_{m+1} G$.

More generally, when a normal (closed) subgroup N of G is given, and when $\{\varphi_m G\}_{m \geq 1}$ is the most rapidly decreasing central filtration of G such that $\varphi_2 G \supset N$, then any automorphism of G which preserves N induces canonically an automorphism on $\varphi_m G, G/\varphi_m G$ and $gr_m^\varphi G$, respectively.

1.2   The induced filtration on automorphism groups

For a compact totally disconnected group G, topologically finitely generated, the group $\operatorname{Aut} G$ of continuous automorphisms is also a compact totally disconnected group, i.e. a profinite group.

Let G be a discrete group, or a compact topological group with a central filtration $\{\varphi_m G\}_{m \geq 1}$. Then the group $\tilde{\Gamma} = \operatorname{Aut} G$ of continuous automorphisms has a canonically induced filtration $\{\tilde{\Gamma}(m)\}_{m \geq 1}$ defined by

$$\tilde{\Gamma}(m) := \{\rho \in \tilde{\Gamma} \mid \text{ for any } k \geq 1, \text{and any } x \in \varphi_m G, \rho(x)x^{-1} \in \varphi_{m+k} G\}$$

Let $\Gamma = \operatorname{Out} G = \operatorname{Aut} G / \operatorname{Inn} G$ be the outer automorphism group of G, i.e. the quotient group of $\operatorname{Aut} G$ by the closed subgroup of inner automorphisms $\operatorname{Inn} G$. Then Γ has an induced filtration

$$\Gamma(m) = (\tilde{\Gamma}(m) \operatorname{Inn} G)/ \operatorname{Inn} G.$$

The following is well-known. (cf. Bourbaki,[5], Chapter 2, §4, Exercise 3)

Proposition 3. *(1)* $\Gamma(0) = \Gamma$

(2) $[\Gamma(k), \Gamma(l)] \subset \Gamma(k + l)$ *for any* $k, l \geq 0$.
In particular, $\Gamma(1)$ *has a central filtration* $\{\Gamma(m)\}_{m \geq 1}$.

2   Galois action on the graded quotients of the lower central series of the fundamental group

In this section, we recall some basic results of Kohno-Oda [14], enhancing them in some points.

2.1 Comparison theorem

Let C be a smooth geometrically irreducible curve over a field K. Let $\tilde{C}$ be the smooth projective compactification of C over K.

Let **L** be a set of rational primes. Then for a discrete group G, the pro-**L** completion of G is defined by

$$G_{\mathbf{L}} = \varprojlim G/H,$$

where H is a normal subgroup of finite index m in G such that the prime divisors of m are all contained in **L**. When $p =$ the characteristic of K, then we set $p' =$ all primes of $\mathbb{Z}-\{p\}$.

Let g be the genus of the compactification of $\tilde{C}$ and let n be the number of geometric points in $\tilde{C} - C$.

Let $\Pi_{g,n}$ be a discrete group with $2g + n$ generators

$$\alpha_1, \beta_1, \ldots \alpha_g, \beta_g, \gamma_1, \ldots \gamma_n$$

and a unique defining relation:

$$[\alpha_1, \beta_1] \ldots [\alpha_g, \beta_g]\gamma_1\gamma_2 \cdots \gamma_n = 1.$$

Here $[\alpha_i, \beta_i] = \alpha_i\beta_i\alpha_i^{-1}\beta_i^{-1}$. Then the following is a result of the comparison theorem on the tame fundamental group.

Theorem 4. (Grothendieck, Raynaud [22])

(1) $\pi_1(C \times \overline{K})_{p'} \cong (\Pi_{g,n})_{p'}$ as compact topological groups.

(2) In particular, if l is a prime number $l \neq$ ch K, then

$$\pi_1(C \times \overline{K})_l \cong (\Pi_{g,n})_l.$$

We have a similar result for $\tilde{C}$ replacing $\Pi_{g,n}$ by $\Pi_{g,0}$.

Let N be the kernel of the canonical surjective homomorphism $\pi_1(C \times \overline{K}) \to \pi_1(\tilde{C} \times \overline{K})$, and let $(N)_{p'}$ be the image of N in $\pi_1(C \times \overline{K})_{p'}$.

Definition 5. *Let $\{W_{-m}\pi_1(C \times \overline{K})_{p'}\}_{m \geq 1}$ be the most rapidly decreasing central filtration on $\pi_1(C \times \overline{K})_{p'}$ such that $(N)_{p'}$ is contained in $W_{-2}\pi_1(C \times \overline{K})_{p'}$. The filtration $\{W_{-m}\pi_1(C \times \overline{K})_{p'}\}_{m \geq 1}$ is called the* weight filtration. *(cf. Kaneko [12]).*

We can define similarly for $\pi_1(C \times \overline{K})_l$. We put

$$gr_m^W \pi_1(C \times \overline{K}) = \{W_{-m}\pi_1(C \times \overline{K})\}/\{W_{-m-1}\pi_1(C \times \overline{K})\}.$$

Remark. The minus sign for the index of the weight filtration is to make it compatible with the convention of Deligne. However for the associated graded modules, we take plus sign.

2.2 A variant of Witt formula

Using the standard results on the comparison of the filtration on discrete groups and their completion (cf. Lubotzky [16]), we have the following

Proposition 6. *(1) For each $m \geq 1$, $gr_m^W \pi_1(C \times \overline{K})_{p'}$ is a free $\mathbb{Z}_{p'}$-module of finite rank ρ_m.*

(2) Let t be a variable and set

$$\psi(t) = \prod_{m=1}^{\infty} (1 - t^m)^{-\rho_m}.$$

Then

$$\psi(t) = 1 - 2gt + (1 - n)t^2 \quad in \quad \mathbb{Z}[[t]].$$

Here $\mathbb{Z}[[t]]$ is the formal power series ring with integral coefficients.

Similarly $gr_m^W \pi_1(C \times \overline{K})_l$ is a free $\mathbb{Z}_l$-module of finite rank ρ_m for each m, and for ρ_m we have the same formula as (2).

Remark. When $n = 0$, Proposition 5 is a formula of Labute [15]. General case is due to Kaneko [12].

Since any finite nilpotent group is a direct product of its Sylow subgroups, we have an isomorphism of compact topological groups

$$\pi_1(C \times \overline{K})_{p'}/W_{-m}\pi_1(C \times \overline{K})_{p'} \cong \prod_{\substack{l \neq p \\ l \text{ primes}}} \pi_1(C \times \overline{K})_l/W_{-m}\pi_1(C \times \overline{K})_l.$$

Hence from now on, we consider only the pro-l completions.

2.3 Galois action on the graded quotients

In this section, we discuss another variant of Witt formula, which describe the Galois action on the graded quotients associated to the weight filtration on the fundamental groups of algebraic curves.

The Galois representations

$$\rho_C : \text{Gal}(\overline{K}/K) \longrightarrow \text{Out} \, \pi_1(C \times \overline{K})_l,$$

induces canonically a Galois representation

$$\psi_{C,m} : \text{Gal}(\overline{K}/K) \longrightarrow \text{Aut} \, gr_m^W \pi_1(C \times \overline{K})_l$$

for each $m \geq 1$. When $m = 1$, $\psi_{C,1}$ is nothing but the Galois representation on

$$\pi_1(C \times \overline{K})_l/(N)_l \cdot \overline{[\pi_1(C \times \overline{K})_l, \pi_1(C \times \overline{K})_l]}$$

$$\cong \pi_1(\tilde{C} \times \overline{K})_l / \overline{[\pi_1(\tilde{C} \times \overline{K})_l, \pi_1(\tilde{C} \times \overline{K})_l]},$$

that is the contragredient representation of the Galois representation

$$\mathrm{Gal}(\overline{K}/K) \longrightarrow \mathrm{Aut}\, H^1_{\acute{e}t}(\tilde{C} \times \overline{K}; \mathbb{Z}_l).$$

For each $m \geq 1$, let

$$\det(1 - \psi_{C,m}(\sigma) \cdot t^m)$$

be the characteristic polynomial of $\psi_{C,m}(\sigma)$ in $\mathrm{Aut}\, gr^W_m \pi_1(C \times \overline{K})_l$ with variable t^m for $\sigma \in \mathrm{Gal}(\overline{K}/K)$.

Put

$$\pi_1(C \times \overline{K})^{ab}_l = \pi_1(C \times \overline{K})_l / [\pi_1(C \times \overline{K})_l, \pi_1(C \times \overline{K})_l].$$

Theorem 7. *(1) When $n = 0$, i.e. $C = \tilde{C}$ is proper, as a $\mathrm{Gal}(\overline{K}/K)$-graded Lie algebra $gr^W_\bullet(\pi_1(C \times \overline{K})_l)$ is a quotient of the free graded Lie algebra $\mathcal{L}(\pi_1(C \times \overline{K})^{ab}_l)$ generated by $\pi_1(C \times \overline{K})^{ab}_l$ over $\mathbb{Z}_l$ by the ideal generated image of ' the canonical homomorphism*

$$H^2_{\acute{e}t}(C \times \overline{K}), \mathbb{Z}_l)^\vee \longrightarrow \overset{2}{\bigwedge} \pi_1(C \times \overline{K})^{ab}_l = \mathcal{L}(\pi_1(C \times \overline{K})^{ab}_l)_2$$

dual to the cup product

$$\overset{2}{\bigwedge} H^1_{\acute{e}t}(C \times \overline{K}, \mathbb{Z}_l) \longrightarrow H^2_{\acute{e}t}(C \times \overline{K}, \mathbb{Z}_l).$$

Here $H^2_{\acute{e}t}(C \times \overline{K}, \mathbb{Z}_l)^\vee$ is the dual $\mathbb{Z}_l$-module of $H^2_{\acute{e}t}(C \times \overline{K}, \mathbb{Z}_l)$ with the contragredient $\mathrm{Gal}(\overline{K}/K)$ action, and $\mathcal{L}(\pi_1(C \times \overline{K})^{ab}_l)_2$ is the homogeneous part of degree 2 in $\mathcal{L}(\pi_1(C \times \overline{K})^{ab}_l)$.

When $n > 0$, $gr^W_\bullet(\pi_1(C \times \overline{K})_l)$ is a graded $\mathrm{Gal}(\overline{K}/K)$ Lie algebra freely generated over the rank $2g + n - 1$ free $\mathbb{Z}_l$-module

$$\pi_1(\tilde{C} \times \overline{K})^{ab}_l \oplus \mathrm{Ker}(\pi_1(C \times \overline{K})^{ab}_l \longrightarrow \pi_1(\tilde{C} \times \overline{K})^{ab}_l)$$

with $\pi_1(\tilde{C} \times \overline{K})^{ab}_l$ as the degree 1 part and

$$\mathrm{Ker}(\pi_1(C \times \overline{K})^{ab}_l \longrightarrow \pi_1(\tilde{C} \times \overline{K})^{ab}_l)$$

as the degree 2 part.

(2) Let t be a variable and set

$$\psi(t; \sigma) = \prod_{m=1}^{\infty} \det(1 - \psi_{C,m}(\sigma)t^m)^{-1}$$

for each $\sigma \in \mathrm{Gal}(\overline{K}/K)$. Then

$$\psi(t; \sigma) = 1 - \mathrm{tr}(\sigma^{-1}|_{H^1_{\acute{e}t}(\tilde{C}, \mathbb{Z}_l)})t + (1 - n)\,\mathrm{tr}(\sigma^{-1}|_{H^2_{\acute{e}t}(\tilde{C}, \mathbb{Z}_l)})t^2$$

in $\mathbb{Z}_l[[t]]$.

Proof. When $n = 0$, the theorem is proved in [14], Proposition (6.1). We consider the case $n > 0$. By the results of Kaneko [12], $gr_\bullet^W(\pi_1(C \times \overline{K})_l)$ is a free graded Lie algebra with $2g + n - 1$ elements with $2g$ generators at degree 1, and $n - 1$ generators at degree $n - 1$. Thus, in order to show (1), it is enough to see that these constructions are compatible with Galois action, which is obvious.

To prove (2), we can extend the scalars from $\mathbb{Z}_l$ to $\mathbb{Q}_l$, and finally to $\overline{\mathbb{Q}}_l$. We may diagonalize the action $\psi_{C,1}(\sigma)$ of σ on $\pi_1(\tilde{C} \times K)_l^{ab}$, replacing it by its semisimplification if necessary. The action σ on $\mathrm{Ker}(\pi_1(C \times \overline{K})_l^{ab} \to \pi_1(\tilde{C} \times K)_l^{ab})$ is a sum of cyclotomic characters, hence a scalar multiple.

Thus the problem is reduced to the multi-degree version of the Witt formula for free Lie algebra. (cf. Bourbaki [5] chap.2, §3, Theorem 2), similarly as the case when $n = 0$ (cf. also [14], Lemma (6.3)).   (q.e.d.)

3 Bad reduction and the action of the inertia group

In this section, we consider a proper flat curve $\mathcal{C}$ defined over a discrete valuation ring R, whose generic fiber C is smooth and geometrically connected of genus g over the quotient field K. Let $\mathfrak{m}$ be the valuation ideal of R, and let J be the Picard variety of C. Then we have the following dichotomy:

(A) J has bad reduction modulo $\mathfrak{m}$;

(B) J has good reduction modulo $\mathfrak{m}$.

In the second case (B), the special fiber of of $\mathcal{C}$ is a stable curve of genus g by a fundamental result of Deligne-Mumford [7].

Assume that the residue field $R/\mathfrak{m}$ of R is an algebraically closed field of characteristic p, p being possibly 0. Let l be a rational prime distinct from p. Then we have a Galois representation of the inertia group $\mathrm{Gal}(\overline{K}/K)$:

$$\rho_C : \mathrm{Gal}(\overline{K}/K) \to \mathrm{Out}\,\pi_1(C \times \overline{K})_l.$$

We want to describe the relation of this homomorphism with the lower central filtration, improving the results of the previous papers [20], [21].

3.1 The case (A)

Consider the truncation of the representation ρ_C modulo the commutator $\Gamma_2\pi_1(C \times \overline{K})_l$ of $\pi_1(C \times \overline{K})_l$. Then we have a Galois representation

$$\rho_C^{ab} : \mathrm{Gal}(\overline{K}/K) \to \mathrm{Aut}\,\pi_1(C \times \overline{K})_l^{ab},$$

which is equivalent to the representation on on the Tate module $T_l(J)$ of the Jacobian variety J of C. Then by the criterion of Serre-Tate [23], ρ_C is not trivial.

By the stable reduction theorem, we can find a finite separable extension K' of K such that $J \otimes_K K'$, accordingly $C \otimes_K K'$ also, has stable reduction at $\mathfrak{m}'$. Here $\mathfrak{m}'$ is the maximal ideal of the integral closure R' of R in K'. We have a further dichotomy:

(A1) $J \otimes_K K'$ has good reduction modulo $\mathfrak{m}'$,

(A2) $J \otimes_K K'$ has stable bad reduction modulo $\mathfrak{m}'$.

In the first case, since $\rho_C^{ab}(\mathrm{Gal}(\overline{K}/K')) = \{1\}$, the image of ρ_C^{ab} is finite, and as K' we may take the invariant subfield of the separable closure of K under $\mathrm{Ker}(\rho_C^{ab})$.

Let us consider the second case (A2). We may assume that K is a finite Galois extension of K. Let

$$\rho_{C'}^{ab} = \mathrm{Gal}(\overline{K}/K') \to \mathrm{Aut}\, \pi_1(C \times \overline{K})_l^{ab}$$

be the restriction of ρ_C^{ab} to the normal subgroup $\mathrm{Gal}(\overline{K}/K')$ of $\mathrm{Gal}(\overline{K}/K)$. Since $C \otimes_K K'$ or $J \otimes_K K'$ has stable reduction at $\mathfrak{m}'$, the monodromy theory (cf. [8]) tells that the image of $\rho_{C'}^{ab}$ is a subgroup of a unipotent algebraic group of $\mathrm{Aut}\, \pi(C \times \overline{K})_l^{ab} \otimes_{\mathbb{Z}_l} \mathbb{Q}_l$. Hence the image of $\rho_{C'}^{ab}$ is a pro-l-group.

The ramification theory of Hilbert tells there exists an exact sequence

$$1 \to P \to \mathrm{Gal}(\overline{K}/K') \to \prod_{l' \neq p} \mathbb{Z}_{l'} \to 1.$$

Here P is the Sylow pro-p-group of $\mathrm{Gal}(\overline{K}/K')$. Thus $\rho_{C'}^{ab}(P) = \{1\}$, and $\mathrm{Im}\, \rho_{C'}^{ab}$ is an infinite quotient of $\mathbb{Z}_l$, because we cannot improve the reduction of $C \otimes_K K'$ replacing K' by a further extension K''. Since $\mathrm{Gal}(\overline{K}/K')_l \simeq \mathbb{Z}_l$ is a Hopfian group i.e. any continuous surjective endomorphism is an isomorphism, the map $\rho_{C'}^{ab}$ induces an injective homomorphism

$$\mathrm{Gal}(\overline{K}/K')_l \hookrightarrow \mathrm{Aut}\, \pi_1(C \times \overline{K})_l^{ab}.$$

Thus in the case (A2), the image of ρ_C^{ab} has a normal subgroup of finite index $\mathrm{Im}\, \rho_{C'}^{ab}$ isomorphic to $\mathbb{Z}_l$.

Let us recall some results of Asada-Kaneko [1] and Asada [2], on the induced filtration $\{\Gamma_{(m)}\}_{m \geq 0}$ on $\Gamma = \mathrm{Aut}\, \pi(C \times \overline{K})_l$. Among others they show that $\{\Gamma(m)/\Gamma(m+1)\}_{m \geq 1}$ is a free $\mathbb{Z}_l$-module of finite rank.

Let $\mathrm{Im}\, \rho_C(m) = \Gamma(m) \cap \mathrm{Im}\, \rho_C$ be the induced filtration on $\mathrm{Im}\, \rho_C$. Since $\mathrm{Im}\, \rho_{C'}^{ab} = \mathrm{Im}\rho_C'/\mathrm{Im}\rho_C' \cap \Gamma(1)$ is a pro-l-group, and $\mathrm{Im}\rho_C' \cap \Gamma(1)$ is a pro-l-group, $\mathrm{Im}\, \rho_{C'}$ itself is also a pro-l-group. Hence it is a quotient of $\mathrm{Gal}(\overline{K}/K')_l \simeq \mathbb{Z}_l$. Thus we see that the induced homomorphism $\mathrm{Im}\, \rho_{C'}' \to \mathrm{Im}\, \rho_{C'}^{ab}$ is an isomorphism. Hence $\mathrm{Im}\, \rho_C' \cap \Gamma(1) = \{1\}$. Therefore $\mathrm{Im}\, \rho_C \cap \Gamma(1)$ is a finite subgroup, that means $\mathrm{Im}\, \rho_C \cap \Gamma(1) = \{1\}$ because $\Gamma(1)$ is torsion-free. Thus $\mathrm{Im}\, \rho_C \to \mathrm{Im}\, \rho_C^{ab}$ is also an isomorphism. We have the following.

Theorem 8. *Let C be a curve over the quotient field K of a discrete valuation ring R with an algebraic closed residue field $R/\mathfrak{m}$. Assume that the Jacobian variety J of C has bad reduction at $\mathfrak{m}$. Let l be a rational prime distinct from the residual characteristic of $R/\mathfrak{m}$. Then for the representation $\rho_C : \mathrm{Gal}(\overline{K}/K) \to \mathrm{Out}\,\pi_1(C \times \overline{K})_l$, $\mathrm{Im}\,\rho_C$ is canonically isomorphic to $\mathrm{Im}\,\rho_C^{ab}$.*

3.2   The case (B)

Let us discuss the case (B), when $\mathrm{Gal}(\overline{K}/K)$ acts trivially on $\pi_1(C \times K)_l^{ab} \simeq T_l(J)$. Then the image $\mathrm{Im}\,\rho_C$ of $\rho_C : \mathrm{Gal}(\overline{K}/K) \to \mathrm{Out}\,\pi_1(C \times K)_l = \Gamma$ is included in $\Gamma(1)$, a pro-l-group. Therefore ρ_C factors through the maximal pro-l quotient $\mathrm{Gal}(\overline{K}/K)_l \simeq \mathbb{Z}_l$ of $\mathrm{Gal}(\overline{K}/K)$.

Theorem 9. *Let C be a smooth proper geometrically connected curve of genus $g \geq 2$ over the quotient field K of a discrete valuation ring R with an algebraically closed residue field $R/\mathfrak{m}$. Assume that the Jacobian variety J of C has good reduction at the valuation ideal $\mathfrak{m}$ of R. Let l be a rational prime distinct from the residual characteristic of R. Then*

(1) $\mathrm{Im}\,\rho_C$ is contained in $\Gamma(2)$;

(2) $\mathrm{Im}\,\rho_C \cap \Gamma(3) = \{1\}$;

(3) $\mathrm{Im}\,\rho_C \simeq \mathbb{Z}_l$.

Proof. We recall the results of a previous paper [21]. When the residual characteristic of $R/\mathfrak{m}$ is 0, then Theorem follows from a corresponding transcendental result (cf. Proposition (1.10) of [21]) by comparison of classical topology and étale topology. When R has unequal characteristic, i.e. ch $K = 0$ and ch $R/\mathfrak{m} = p > 0$, the argument to prove Theorem (2.2) of [21] works without substantial changes.

We have to discuss the case of equal characteristic ch $K = $ ch $R/\mathfrak{m} = p > 0$. We may assume that R is complete. Then by Cohen's structure theorem of a complete regular local ring, R is isomorphic to a formal power series ring $k[[T]]$ over the residue field $k = R/\mathfrak{m}$ with variable T.

Let $W(k)$ be the Witt ring for k, and let π be a prime element of $W(k)$. Then by an argument similar to that of Proposition (2.4) of [21], We can construct a flat proper curve $\tilde{C} \to \mathrm{Spec}\,W(k)[[T]]$ over a 2-dimensional scheme $\mathrm{Spec}\,W(k)[[T]]$ such that $\tilde{C} \otimes_{W(k)[[T]]} W(k)[[T]]\,(\frac{1}{T})$ is smooth over $W(k)[[T]]\,(\frac{1}{T})$ and $\tilde{C} \otimes_{W(k)[[T]]} W(k)[[T]]/(\pi)$ is isomorphic to the original curve $C \to \mathrm{Spec}\,k[[T]]$.

Put $A = W(k)[[T]]$, and let L be the quotient field of the strict henselization of A at T. Let V be the quotient field of $W(k)$. Put $W = \mathrm{Spec}\,A[1/T]$. Then Abhyankar's Lemma implies the natural isomorphisms:

$$\operatorname{Gal}(\bar{L}/L)_l \cong \pi_1(W)_l \cong \operatorname{Gal}(\bar{V}/V)_l$$

(cf. [22], Exposé XIII, §5). Also the specialization $A[1/T] \to R[1/T]$ induces an isomorphism

$$\pi_1(W)_l \cong \pi_1(\operatorname{Spec} R[1/T])_l = \operatorname{Gal}(\overline{K}/K)_l.$$

Since the image of ρ_C is contained in a pro-l group $\Gamma(1)$, the representation ρ_C factors through $\operatorname{Gal}(\overline{K}/K)_l \cong \mathbb{Z}_l$. The curve $\tilde{C}$ has good reduction with respect to the specialization $\operatorname{Spec} A[1/T] \to \operatorname{Spec} R[1/T]$. Hence we have an isomorphism of the fundamental groups:

$$\pi_1(C \times \overline{K})_l \cong \pi_1(\tilde{C} \otimes \bar{L})_l$$

compatible with the actions of the isomorphic inertia groups:

$$\operatorname{Gal}(\bar{L}/L)_l \cong \operatorname{Gal}(\overline{K}/K)_l.$$

Thus we have reduced the problem to the case of characteristic 0.   (q.e.d)

3.3   Application and remarks

The following is an immediate corollary of the previous two theorems.

Theorem 10. *Let C be a smooth proper geometrically connected curve of genus $g \geq 2$ over the quotient field K of a discrete valuation ring R with an algebraically closed residue field $R/\mathfrak{m}$. Let l be a rational prime distinct from the residual characteristic of R. Then the following two statements are equivalent*

(1) C has good reduction at $\mathfrak{m}$;

(2) the inertia group $\operatorname{Gal}(\overline{K}/K)$ acts trivially on $\pi_1(C \times \overline{K})_l$.

Remark. In the case of the characteristic 0, the transcendental version of the above theorem is found in Imayoshi [11], Theorem 1. Matsumoto and Montesinos-Amilibia [18] discuss the bijection between homotopy types of singular fibers and the deck transformations in Out π_1 in the transcendental case.

4   Degeneration of stable curves and local monodromy on the fundamental group

In this section, we review some results on the local monodromy representation on the fundamental groups for a universal degenerating family of punctured algebraic curves. In this section we consider analytic varieties. Complete proofs are found in [3].

A typical case is the following. We start with a stable curve C_0 of genus $g \geq 2$, which is most degenerate, i.e. a union of the projective lines. For such a curve, we associate the dual graph Y whose vertices correspond to the irreducible components of C_0 and edges to double points. Consider a local universal deformation $f : \mathcal{C} \to \mathcal{D}$ of C_0 over a small polydisk $\mathcal{D}$ of dimension $3g - 3$ in the category of stable curves. Let $\mathcal{D}^\circ$ be the open subset of $\mathcal{D}$, on which the fibers of f are smooth, which is the product of $3g - 3$ copies of the punctured disk. Let $\mathbf{t}$ be a point on $\mathcal{D}^\circ$. Then we obtain the monodromy map on the fundamental group $\pi_1(C_{\mathbf{t}}, b)$

$$\rho_{C_0} : \pi_1(\mathcal{D}^\circ, \mathbf{t}) \to \mathrm{Out}\ \pi_1(C_{\mathbf{t}}, b).$$

Here b is a base point in $C_{\mathbf{t}}$.

We consider the lower central series on the fundamental group of curves, which is preserved by the monodromy homomorphism, and describe the relation between the lower central series and the monodromy homomorphism for the local universal deformation of a most degenerate stable curve.

The main result for this special case is the following. Let I_Y be the image of the injective homomorphism ρ_{C_0} which is a free abelian group of rank $3g-3$, and let $\{I_Y^{(m)}\}_{m=0,1,2,\ldots}$ be the induced filtration on I_Y derived from the lower central filtration on $\pi_1(C_{\mathbf{t}}, b)$. Put

$$r_m(Y) = \mathrm{rank}_{\mathbb{Z}}\ I_Y^{(m)}/I_Y^{(m+1)} \text{ for all } m\ (m = 0, 1, 2, \ldots).$$

Then we have

$$r_m(Y) = 0, \text{ if } m \geq 3, \quad r_2(Y) = s_2(Y), \quad r_1(Y) = s_1(Y),$$

$$\text{and } r_0(Y) = 3g - 3 - s_1(Y) - s_2(Y).$$

Here $s_2(Y)$ is the number of bridges in the graph Y, and $s_1(Y)$ is also another geometric invariant of the graph Y (cf. §4.4). We also note here the equality $r_0(Y) = 3g - 3 - s_1(Y) - s_2(Y)$ is due to Brylinski [6].

The first motivation was to generalize the transcendental part of the previous paper [21], in which we discussed a similar problem when the base $\mathcal{D}$ is one-dimensional, and the graph of C_0 is a tree.

4.1 Stable n-pointed curves and their graph

Let us recall the definition of stable n-pointed curves Knusten [13], §1.

Definition 11. *A stable n-pointed curve (C, S) of genus g is a pair (C, S) of a proper connected curve C over the complex number field C and a subset of n-distinct smooth points on C satisfying the following conditions:*

(1) C *has only ordinary double points as singularities.* C_{sing} *denotes the locus of singularities. Let* $p : C^* \to C$ *be the normalization of* C*. Then we set* $C^*_{sing} = p^{-1}(C_{sing})$ *and identify* $p^{-1}(S)$ *with* S *via* p*.*

(2) (stability) On the normalization D^* *of each irreducible component* D *of* C *which is isomorphic to* P^1*, the sum of numbers of* $D^* \cap C_{sing}$ *and* $D^* \cap S$ *is at least 3.*

When $n = 0$, the above definition gives the notion of stable curves Deligne-Mumford [7].

A graph of a stable n-pointed curve
    For each stable n-pointed curve (C, S), we can associate the (dual) graph Y in the following manner [7], [19].

Definition 12. *(1) Each vertex* P *of the graph* Y *corresponds uniquely to an irreducible component* C_P *of* C*. Or equivalently, each vertex* P *corresponds uniquely to a connected component of the normalization* C^* *of* C*.*

(2) A pair $\{y, \bar{y}\}$ *of mutually inverse (oriented) edges of* Y *corresponds uniquely to a singular point* $q_{\{y,\bar{y}\}}$ *of* C*. If necessary, we refer to the pair* $\{y, \bar{y}\} = |y|$ *as a geometric edge associated with* y *or with* $\bar{y}$*. We also denote* $q_{\{y,\bar{y}\}}$ *by* q_y*,* $q_{\bar{y}}$*, or* $q_{|y|}$*. The set of geometric edges is denoted by* $\mathrm{Edge}(Y)_{geom}$*.*

(3) For each edge y*, its two extremities are given by the vertices* P_1*,* P_2 *so that*
$$q_y = C_{P_1} \cap C_{P_2} \qquad (if\ P_1 \neq P_2),$$
$$q_y = C_{P_1} \cap C_{sing} \qquad (if\ P_1 = P_2).$$

(4) There is a function
$$v : \mathrm{Vert}(Y) \to \mathbb{Z} \times \mathbb{Z}$$

from the set of vertices $\mathrm{Vert}(Y)$ *of* Y *to the product of the set of non-negative integers defined by* $v(P) = (g_P, n_P)$*. Here* g_P *is the genus of the normalization* C^*_P *of* C_P*, and* n_P *is the cardinality of the set* $S \cap C_P$*.*

The graph (Y, v) determines the homotopy type of $C - S$.

The most degenerate case

Definition 13. *A stable n-pointed curve* (C, S) *is called* most degenerate, *if it has no deformation with the same homotopy type.*

In this case, any irreducible component of C is of genus 0. Moreover the graph (Y, v) of (C, S) satisfies the following conditions.

Lemma 14. *If* (Y, v) *is the graph of a most degenerate stable n-pointed curve* (C, S) *of genus g. Then*

$$\#(\mathrm{Vert}(Y)) = 2(g - 1) + n; \quad \#(\mathrm{Edge}(Y)) = 3(g - 1) + n.$$

In particular,

$$h^1(Y) = \#(\mathrm{Edge}(Y)) - \#(\mathrm{Vert}(Y)) + 1 = g.$$

For each $P \in \mathrm{Vert}\, Y$,

$$n_P = 3 - \#\{\, y \in \mathrm{Edge}(Y) \mid t(y) = P \,\}.$$

Since Y *is connected,* $n_P = 0$, *1, or 2. When* $n_P = 2$, P *is a terminal point of* Y.

This is easy to prove and more or less well-known. Here the assigned number denotes n_P for each P if $n_P > 0$.

4.2 Weight filtration on the fundamental groups and induced filtration on the automorphism groups

A group isomorphic to the fundamental group of a compact Riemann surface of genus g is called a *surface group* of genus g. The fundamental group of an n-punctured Riemann surface is a free group, if $n > 0$. On these groups, similarly as pro-l case (*cf.* §2), we can define the weight filtration in the following way.

The weight filtration

Let π_1 be the surface group of genus g. Then the weight filtration $\{W_{-m}(\pi_1)\}_{m \geq 1}$ is the lower central series

$$W_{-m}(\pi_1) = \Gamma_m \pi_1 \text{ for each m} \geq 1.$$

Here $\Gamma_m \pi_1$ is the m-th higher commutator.

The case of the fundamental group of a punctured Riemann surface is slightly more complicated (*cf.* Kaneko [12]).

Let C be a compact Riemann surface of genus g, and S be a finite subset of C with cardinality n. Choose a base point $*$ in $C - S$. Let N be the kernel of the canonical surjection

$$\pi_1(C - S, *) \rightarrow \pi_1(C, *)$$

which is a normal subgroup of π_1 generated by the homotopy classes which correspond to the puncture.

We set $W_{-1}(\pi_1) = \pi_1$ the whole group, and $W_{-2}(\pi_1) = [\pi_1, \pi_1]N$. Then the weight filtration $\{W_{-n}(\pi_1)\}_{n \geq 1}$ is defined as the fastest decreasing central filtration.

Note that the quotient group $\pi_1(C - S, *)/W_{-1}(\pi_1)$ is isomorphic to the 1-st homology group $H_1(C, \mathbb{Z})$.

The induced filtration

Now we consider the induced filtration on the outer automorphism group of π_1 and its subgroup. For each point p in S, we choose an element γ_p in $\pi_1(C - S, *)$, which is free-homotopically equivalent to a small closed path that circles p one time counter-clockwise. Let $\mathrm{Aut}_S \ \pi_1$ be the subgroup of the automorphism group $\mathrm{Aut} \ \pi_1(C - S, *)$ consisting elements which map γ_p to its conjugate for each $p \in S$. Also by $\mathrm{Aut}_S^+ \ \pi_1$ the subgroup of $\mathrm{Aut}_S \ \pi_1$ given as the kernel of the composition of the canonical homomorphisms

$$\mathrm{Aut}_S \ \pi_1(C - S, *) \to \mathrm{Aut} \ \pi_1(C) \to \mathrm{Aut} \ H_2(\pi_1(C), Z).$$

When $g = 0$, $\mathrm{Aut}_S^+ \ \pi_1 = \mathrm{Aut}_S \ \pi_1$, and when $g > 0$, $\mathrm{Aut}_S^+ \ \pi_1$ is an index 2 subgroup of $\mathrm{Aut}_S \ \pi_1$.

Notation. We denote by $\tilde{\Gamma}_{g,n}$ the group $\mathrm{Aut}_S^+ \ \pi_1$, and by $\Gamma_{g,n}$ the group $\mathrm{Out}_S^+ \ \pi_1$.

By a classical theorem of Nielsen, $\Gamma_{g,n}$ is isomorphic to a *mapping class group* or a *Teichmüller group*.

The weight filtration on $\pi_1(C - S, *)$ canonically induces a filtration on $\tilde{\Gamma}_{g,n}$ by

$$\tilde{\Gamma}_{g,n}[k] = \{\sigma \in \tilde{\Gamma}_{g,n}| \text{ for any } l \geq 1,$$
$$\text{and any } x \in W_{-l}(\pi_1), \ \sigma(x)x^{-1} \in W_{-k-l}(\pi_1)\}.$$

Passing to the quotient $\Gamma_{g,n} = \tilde{\Gamma}_{g,n}/\mathrm{Inn}(\pi_1(C - S, *))$, we can define the induced filtration on $\Gamma_{g,n}$, by the image of the canonical homomorphism:

$$\Gamma_{g,n}[k] = \mathrm{Im}(\tilde{\Gamma}_{g,n}[k] \to \Gamma_{g,n})$$

for each k. Then we have the following

Proposition 15. *(1)* $\Gamma_{g,n}[0] = \Gamma_{g,n}$, *and*

$$[\Gamma_{g,n}[k], \Gamma_{g,n}[l]] \subset \Gamma_{g,n}[k + l] \text{ for any } k, \ l \geq 0;$$

(2) The quotient $\Gamma_{g,n}/\Gamma_{g,n}[1]$ *is isomorphic to the Siegel modular group* $Sp(g; \mathbb{Z})$;

(3) For any m *($m \geq 1$), the quotient group* $\Gamma_{g,n}[m]/\Gamma_{g,n}[m + 1]$ *is a free abelian group of finite rank.*

4.3   The non-abelian monodromy homomorphism

Let (C_0, S_0) be a stable n-pointed curve of genus g with the associated graph Y, which is most degenerate. Consider the local universal deformation of (C_0, S_0). For each edge $e = |y|$ ($y \in \text{Edge}(Y)$), let

$$u_e v_e = 0 \qquad (\text{in } (u_e, v_e) \in \mathbb{C}^2)$$

be the local defining equation of the singularity associated with e. Let

$$u_e v_e = t_e \qquad (\text{in } (u_e, v_e, t_e) \in \mathbb{C}^3)$$

be the local universal deformation of the above singularity (cf. [7], §1, and [13], §2).

For each e, we associate a small complex disk $\mathcal{D}_e = \{\, t_e \in \mathbb{C} \mid |t_e| < \varepsilon \,\}$. Then over the polydisk $\mathcal{D} = \prod_{e \in \text{Edge}(Y)} \mathcal{D}_e$, we have a local universal family

$$f : \mathcal{C} \to \mathcal{D}, \qquad \mathcal{S} : \{1, \ldots, n\} \times \mathcal{D} \to \mathcal{C}.$$

If $\mathbf{t} = (t_e)_{e \in \text{Edge}(Y)}$ satisfies $t_e \neq 0$ for any $e \in \text{Edge}(Y)$, the fiber $f^{-1}(\mathbf{t}) = C_{\mathbf{t}}$ is a smooth proper curve of genus g, and $\mathcal{S}(\mathbf{t}) = S_{\mathbf{t}}$ is a set of n distinct points on $C_{\mathbf{t}}$. Let $\mathcal{D}^0$ be the open subset of $\mathcal{D}$ consisting of such points. Choose such a point $\mathbf{t}_0$ in $\mathcal{D}^0$. Let

$$\pi_1(C_{\mathbf{t}_0} - S_{\mathbf{t}_0}, *)$$

be the fundamental group of the n-punctured Riemann surface $C_{\mathbf{t}_0} - S_{\mathbf{t}_0}$ with a base point $*$. Then we have the non-abelian monodromy homomorphism

$$\rho_{(C_0, S_0)} : \pi_1(\mathcal{D}^0, \mathbf{t}_0) \cong \mathbb{Z}^{3g-3+n} \to \text{Out } \pi_1(C_{\mathbf{t}_0} - S_{\mathbf{t}_0}, *).$$

By using a transcendental result, we can assure that the monodromy homomorphism $\rho_{(C_0, S_0)}$ is injective.

This monodromy homomorphism is compatible with the weight filtration.

Proposition 16. *The monodromy homomorphism $\rho_{(C_0, S_0)}$ preserves the weight filtration on $\pi_1(C_{\mathbf{t}_0} - S_{\mathbf{t}_0}, *)$. In particular, for any σ of $\pi_1(\mathcal{D}^0, \mathbf{t}_0)$, we have $\sigma(N) = N$, where N is the kernel of the canonical surjective homomorphism*

$$\pi_1(C_{\mathbf{t}_0} - S_{\mathbf{t}_0}, *) \to \pi_1(C_{\mathbf{t}_0}, *).$$

4.4 Invariants $s_1(Y)$, $s_2(Y)$ in a graph Y

In this subsection, we define some invariants of a graph.

Definition 17. *(1) An edge y is called a* bridge *, if the subgraph $Y - \{|y|\}$ is not connected.*

(2) A pair $\{|y_1|, |y_2|\}$ of geometric edges is called a cut pair, *if neither $|y_1|$ nor $|y_2|$ is a bridge, and the subgraph $Y - \{|y_1| \cup |y_2|\}$ is not connected.*

The following is easy to prove.

Lemma 18. *Let $\{|y_1|, |y_2|\}$ be a cut pair, and $\{|y_2|, |y_3|\}$ ($|y_3| \neq |y_1|$) be another cut pair. Then $\{|y_1|, |y_3|\}$ is also a cut pair.*

Definition 19. *We call a set E of geometric edges a* maximal cut system, *if*

(1) it contains at least two distinct geometric edges;

(2) any pair of two distinct geometric edges $|y|$, $|y'|$ in E is a cut pair;

(3) and no edge y'' outside E makes a cut pair with an edge in E.

Now we define two invariants of a graph Y which is used to describe the main result of this section.

Definition 20. *(1) Let $s_2(Y)$ be the number of bridges in the graph Y.*

(2) Put $s_1(Y) = \sum_E \{|E| - 1\}$, where E runs over the maximal cut systems in Y.

4.5 Main results

Let (Y, v) be a graph of a most degenerate n-pointed stable curve of genus g. Recall the monodromy homomorphism $\rho_{(C_0, S_0)}$ in Subsection (4.3).

Definition 21. *We denote by I_Y the image of the monodromy homomorphism*

$$\rho_{(C_0, S_0)} : \pi_1(\mathcal{D}^0, t_0) \cong \mathbb{Z}^{3g-3+n} \longrightarrow \text{Out } \pi_1(C_{t_0} - S_{t_0}, *).$$

in $\Gamma_{g,n} = \text{Out}_S^+(\pi_1)$.

Since $\rho_{(C_0,S_0)}$ is injective, I_Y is a free abelian subgroup of rank $3g - 3 + n$. Let $I_Y^{(m)} = I_Y \cap \Gamma_{g,n}[m]$ for each $m \geq 1$, and define the numbers $\{r_m(Y)\}_{m \geq 0}$ by

$$r_m(Y) = \mathrm{rank}_{\mathbb{Z}} I_Y^{(m)} / I_Y^{(m+1)} \qquad \text{for each } m \geq 0.$$

Note here each $I_Y^{(m)}/I_Y^{(m+1)} \subset \Gamma_{g,n}[m]/\Gamma_{g,n}[m+1]$ is a free abelian group of finite rank by Proposition 1.4, if $m \geq 1$. We will see later that $I_Y^{(0)}/I_Y^{(1)}$ is also a free $\mathbb{Z}$-module

Here is the main result of this section.

Theorem 22. *Let Y be an associated graph with a most degenerate stable pointed curve of type (g, n). Then*

(1) $r_0(Y) = 3g - 3 + n - s_1(Y) - s_2(Y)$;

(2) $r_1(Y) = s_1(Y)$;

(3) $r_2(Y) = s_2(Y)$;

(4) $I_Y^{(3)} = \{0\}$.

Remark. The first statement (1) is due to Brylinski [6], Proposition 5. The following is an immediate corollary of the above theorem.

Corollary. (1) When $n = 0$, the naturally induced homomorphism

$$\rho_{(C_0,S_0)}(mod\ 3) : \pi_1(\mathcal{D}^0, \mathbf{t}_0) \cong \mathbb{Z}^{3g-3+n} \to \mathrm{Out}(\pi_1(C_{\mathbf{t}_0} - S_{\mathbf{t}_0}, *)/W_{-4}\pi_1)$$

is injective;

(2) When $n > 0$, the homomorphism

$$\rho_{(C_0,S_0)}(mod\ 4) : \pi_1(\mathcal{D}^0, \mathbf{t}_0) \cong \mathbb{Z}^{3g-3+n} \to \mathrm{Out}(\pi_1(C_{\mathbf{t}_0} - S_{\mathbf{t}_0}, *)/W_{-5}\pi_1)$$

is injective.

The proof of the above theorem proceeds as follows. First, we describe the monodromy homomorphism in terms of *graph of groups* in the sense of Bass-Serre [24], i.e. a kind of homotopical version of the Picard-Lefschetz formula. This reduces our problem to a problem of combinatorial group theory. After this step, we can forget the geometric origin of the problem, and can use the inductive argument freely.

5 References

1. Asada, M. and Kaneko, M. (1987) On the automorphism group of some pro-*l* fundamental groups *Advanced Studies in pure math.*, **12**, 137-160

2. Asada, M. Two properties of the filtration of the outer automorphism groups of certain groups *Preprint*

3. Asada, M., Matsumoto, M. and Oda, T. Local monodromy on the fundamental groups of algebraic curves along a degenerate stable curve *Preprint* RIMS-850

4. Belyi, G. V. (1979) On Galois extensions of a maximal cyclotomic field *Izv. Acad. Nauk USSR*, **43-2**, = *Math. USSR Izv.*, **14-2**, 247-256

5. Bourbaki, N. (1960) *Groupes de algèbres de Lie*, Hermann, Paris

6. Brylinski, J.-L. (1979) Propriétés de ramification a l'infini du groupe modulaire de Teichmüller *Ann. Scient.Éc.Norm.Sup. 4e série*, **t.12**, 295-333

7. Deligne, P. and Mumford, D. (1969) The irreducibility of the space of curves of given genus *Publ. Math. I.H.E.S.*, **36**, 75–110

8. Deligne, P. (1973) La formule de Picard-Lefschetz, in "Groupes de monodromie en géométrie algébrique II" *Lecture Notes in Math.*, **340**, 165-196

9. Deligne, P. (1989) Le groupe fondamental de la droite projective moins trois points, in "Galois groups over ℚ" **Publ. MSRI, 16**, 79–297

10. Ihara, Y. (1986) Profinite braid groups, Galois representations, and complex multiplications *Ann. of Math.*, **123** , 43–106

11. Imayoshi, Y. (1981) Holomorphic families of Riemann Surfaces and Teichmüller spaces in " Riemann surfaces and related topics: Proceeding of the 1978 Stony Brook Conference" ed. by Irwin Kra and Bernard Maskit *Ann. of Math. Studies* **97**, Princeton University Press

12. Kaneko, M. (1989) Certain automorphism groups of pro-l fundamental groups of punctured Riemann surfaces. *J. Fac. Sci. Univ. Tokyo, Sect. IA, Math.*, **36**, 363-372

13. Knusten, F.F. (1983) The projectivity of the moduli space of stable curves II. *Math. Scad.*, **52**, 161l-199

14. Kohno, T. and Oda, T. (1987) The lower central series of the pure braid group of an algebraic curve *Advanced Studies in pure math.*, **12**, 201-219

15. Labute, J.P. (1970) On the descending central series of groups with a single defining relation *Journal of Algebra*, **14**, 16–23

16. Lubotzky, A. (1982) Combinatorial group theory for pro-p groups *Journal of pure and appl. Algebra*, **14**, 311-325

17. Magnus, W. Karras, A. and Solitar, D. (1966) *Combinatorial Group Theory* , Interscience

18. Matsumoto, Y. and Montesinos-Amilibia, J.-M. Pseudo-periodic maps and degeneration of Riemann surfaces. *In preparation*

19. Namikawa,Y. (1973) On the canonical holomorphic map from the moduli space of stable curves to the Igusa monoidal transform *Nagoya Math. J.*, **52** , 197–259

20. Oda, T.(1990) A note on ramification of the Galois representation on the fundamental group of an algebraic curve *J. of Number Theory*, **34**, 225–228

21. Oda, T. A note on the ramification of the fundamental group of an algebraic curve II. *To appear in J. of Number Theory*,

22. Raynaud, M. (1971) Proprietés cohomologiques des faisceaux d'ensembles et des faisceaux des groupes non commutatifs, in "Revêtements étales et groupe fondamental", *Lecture Notes in Math.*, **224**, 344-439

23. Serre, J.P. and Tate, J. (1968) Good reduction of abelian varieties *Annals of Math.*,**88**, 492–517

24. Serre, J.P (1977) Arbres, Amalgames, SL₂". *Astérisque*, **46**

Takayuki Oda
Research Institute for Mathematical Science
Kyoto University
Kyoto 060
Japan
e-mail oda@kurims.kyoto-u.ac.jp

Arithmetic of Subvarieties of Abelian and Semiabelian Varieties

Paul Vojta

University of California at Berkeley

We will discuss work of Faltings concerning rational points on subvarieties of abelian varieties. In the talk given at the conference, the present author announced a generalization of this theorem to the case of integral points on semiabelian varieties. The proof has some serious flaws, however, so the generalization still remains a conjecture.

Throughout this paper, unless otherwise specified, let X be a subvariety of an abelian variety A, and assume that both are defined over a number field k. Then Faltings' result is the following.

Theorem 1. *(Faltings, 1991 [2] and [12]) The set $X(k)$ of k-rational points on X is contained in a finite union $\bigcup B_i(k)$, where each B_i is a translated abelian subvariety of A contained in X.*

Corollary 2. *(Faltings, 1989 [1]) If $X \times_k \bar{k}$ does not contain any nontrivial translated abelian subvarieties of $A \times_k \bar{k}$, then $X(k)$ is finite.*

Note that, first of all, the theorem generalizes the corollary, and secondly that finiteness of $X(k)$ does not hold in this generality, since a nontrivial B_i may have infinitely many rational points.

This theorem was originally conjectured by S. Lang [4]. In that paper he also raised the question of whether the result can be generalized to subvarieties of semiabelian varieties. Recall that a semiabelian variety is a group variety A for which there exists an exact sequence of homomorphisms of group varieties

$$0 \to \mathbb{G}_m^\mu \to A \to A_0 \to 0,$$

where A_0 is an abelian variety. In general a base change may be needed before the first factor splits into a product of multiplicative groups, but it is no loss of generality to assume that such a base change has already been done. Thus, one may conjecture that Faltings' theorem 1 generalizes to the situation of semiabelian varieties as follows.

Conjecture 3. *Let k be a number field, with ring of integers R. Let S be a finite set of places of k, containing the set of archimedean places, and* let R_S be the localization of R away from (non-archimedean) places in S. Let X be a subvariety of a semiabelian variety A; assume both are defined over k. Let $\mathcal{X}$ be a model for X over $\operatorname{Spec} R_S$. Then the set $\mathcal{X}(R_S)$ of R_S-valued points in $\mathcal{X}$ is contained in a finite union $\bigcup \mathcal{B}_i(k)$, where each $\mathcal{B}_i$ is a subscheme of $\mathcal{X}$ whose generic fiber B_i is a translated semiabelian subvariety of A.

Of course, if $\mu = 0$ then A is an abelian variety, and integral points are the same as rational points. Thus, this conjecture (if proved) would generalize Theorem 1.

We will now discuss the proof of Theorem 1. A thirty minute talk does not give time for many details, so I will only give a brief overview.

The proof uses the methods of the Thue-Siegel-Roth theorem, as adapted in [11] to the case of rational points. That paper gave a new proof of the Mordell conjecture, which again follows from Corollary 2. (Recall that the Mordell conjecture, first proved by Faltings in 1983, asserts that if C is a projective curve of genus > 1, then $C(k)$ is finite. Embedding C into its Jacobian puts us into the situation of Corollary 2.)

1   Geometry of subvarieties of abelian varieties

For this section, until further notice, we shall assume that X and A are as above, except that they are defined over $\mathbb{C}$. We have two results in this situation concerning translated abelian subvarieties of A contained in X.

Definition 4. *Let $B(X)$ be the identity component of the algebraic group*

$$\{a \in A \mid a + X = X\}.$$

Then the restriction to X of the quotient map $A \to A/B(X)$ gives a fibration $X \to X/B(X)$ whose fibers are all isomorphic to $B(X)$. This fibration is called the **Ueno fibration** *of X. It is* **trivial** *when $B(X)$ is a point.*

Theorem 5. *(Ueno [9]Thm. 10.9) If $B(X)$ is trivial, then X is of general type (and conversely).*

In 1983, Bombieri [7] posed the question: given a variety X/k of general type, is $X(k)$ contained in a proper closed subset in the Zariski topology? Theorem 1 answers this question in the special case of a subvariety of an abelian variety. It also touches on conjectures in [6]Conj. 5.8 and [10].

Theorem 6. *(Kawamata structure theorem [3]Thm. 4) Let $Z(X)$ denote the union of all nontrivial translated abelian subvarieties of A contained in X. Then $Z(X)$ is a Zariski-closed subset, and each irreducible component of it has nontrivial Ueno fibration. In particular, $Z(X)$ is a proper subset of X if and only if the Ueno fibration of X is trivial.*

Returning to the situation over the number field k, it follows by a Galois theoretic argument that the sets $B(X)$ and $Z(X)$ are also defined over k. It suffices to prove Theorem 1 in the special case that X has trivial Ueno fibration. Otherwise, if $B(X)$ is nontrivial, then we can prove the result on $X/B(X)$ (which has trivial Ueno fibration), and then pull back the resulting B_i to X.

Since X has trivial Ueno fibration, it now suffices to show that $X \setminus Z(X)$ has only finitely many rational points. Indeed, this reduces the problem to considering only the rational points on $Z(X)$; this set can be handled by Noetherian induction.

2 The main argument

We start with some additional notation. Let L be a very ample divisor class on A. We also require L to be symmetric: an abelian variety A has a canonical involution given by taking the inverse under the group law; then L should be invariant under this involution. Also fix a projective embedding of A associated to L.

For rational points $P \in A(k)$ we can define a **height** relative to L. If $k = \mathbb{Q}$ this can be described explicitly as follows. Let $[x_0 : \ldots : x_N]$ be homogeneous projective coordinates such that $x_0, \ldots, x_N$ are integers with no common factor. Then $h(P) = \log \max |x_i|$. Thus the height measures the complexity of the point; there are only finitely many rational points with height less than any given bound. For more details covering the number field case see [5] or [8]. There is also a **canonical height** function, due to Néron and Tate, which has special properties relative to the group law on A. This is a function $\hat{h}_L$ on $A(k)$ which differs by a bounded function from the usual height function, and which is a quadratic form relative to the group law on A. This leads to an associated bilinear pairing on $A(k)$:

$$(P, Q)_L = \hat{h}_L(P + Q) - \hat{h}_L(P) - \hat{h}_L(Q).$$

This defines a Euclidean structure on the vector space $A(k) \otimes_{\mathbb{Z}} \mathbb{Q}$. By the Mordell-Weil theorem, this vector space is finite dimensional.

Generalizing the notion of the height of a *point*, there is also the definition of the height of a *subvariety*. This can be defined either as the height of the corresponding point on some associated Chow variety, or by arithmetic intersection theory as in Faltings [1]. For a subvariety Y of A, let $h(Y)$ denote its height.

The main argument of the proof of Theorem 1, then, is to assume that $X \setminus Z(X)$ has infinitely many rational points, and then derive a contradiction.

To begin, let

$$n = \dim X + 1.$$

We will be working with k-rational points $P_1, \ldots, P_n \in X \backslash Z(X)$. These will be chosen (later) so as to satisfy the following conditions, for certain constants $c_1, c_2 \geq 1$, and ϵ_1:

i. $h_L(P_1) \geq c_1$;

ii. $h_L(P_{i+1})/h_L(P_i) \geq c_2, \ i = 1, \ldots, n - 1$;

iii. $P_1, \ldots, P_n$ all point in roughly the same direction in $A(k) \otimes_{\mathbb{Z}} \mathbb{R}$:

$$(P_i, P_j)_L \geq (1 - \epsilon_1)\sqrt{(P_i, P_i)_L(P_j, P_j)_L}$$

for all i, j.

We call these conditions $C_P(c_1, c_2, \epsilon_1)$.

We also work with subvarieties $X_1, \ldots, X_n$ of X satisfying the following conditions, denoted $C_X(c_3, c_4, P_1, \ldots, P_n)$:

i. Each X_i contains P_i.

ii. The X_i are geometrically irreducible and defined over k.

iii. The degrees $\deg X_i$ satisfy $\deg X_i \leq c_3$.

iv. The heights $h(X_i)$ are bounded by the formula

$$\sum_{i=1}^{n} \frac{h(X_i)}{h_L(P_i)} < c_4 \sum_{i=1}^{n} \frac{1}{h_L(P_i)}.$$

The main step consists of finding subvarieties X_i' of X_i, not all equal, which also satisfy C_X (with different constants). This can be written explicitly as follows:

$\forall \ c_3, c_4$ and $\forall \ \delta_1, \ldots, \delta_n \in \mathbb{N}$

$\exists \ c_1, c_2, \epsilon_1, c_3', c_4'$ such that

$\quad \forall \ P_1, \ldots, P_n \in \big(X \backslash Z(X)\big)(k)$ satisfying $C_P(c_1, c_2, \epsilon_1)$ and

$\quad \forall \ X_1, \ldots, X_n \subseteq X$

$\quad\quad\quad$ satisfying $C_X(c_3, c_4, P_1, \ldots, P_n)$ and $\dim X_i = \delta_i \ \forall \ i$

$\quad\quad \exists \ X_1', \ldots, X_n'$ with $X_i' \subseteq X_i \ \forall \ i$ and $X_i' \neq X_i$ for some i,

$\quad\quad\quad$ and satisfying $C_X(c_3', c_4', P_1, \ldots, P_n)$.

Starting with $X_1 = \ldots = X_n = X$, we repeat the main step until some X_j has dimension zero. This implies that $X_j = P_j$, and in that case:

$$1 = \frac{h(X_j)}{h_L(P_j)} \leq \sum_{i=1}^{n} \frac{h(X_i)}{h_L(P_i)} \leq c_4 \sum_{i=1}^{n} \frac{1}{h_L(P_i)} \leq \frac{c_4 n}{h_L(P_1)}.$$

This leads to a contradiction if $c_1 > c_4 n$. Note that c_3 and c_4 will depend on the dimensions of $X_1, \ldots, X_n$, but that only finitely many such possibilities occur; we then choose c_1, c_2, and ϵ_1 to be the largest, largest, and smallest (respectively) of the corresponding quantities encountered in the above main step, and furthermore require that c_1 be larger than $c_4'n$ for all c_4' encountered.

For these values of c_1, c_2, and ϵ_1, it is possible to choose $P_1, \ldots, P_n$ satisfying C_P, deriving a contradiction.

3 The line sheaf

The main step outlined above involves the following object:

$$L_{-\epsilon,s} = \sum_{i<j} (s_i \cdot pr_i - s_j \cdot pr_j)^* L - \epsilon \sum_{i=1}^{n} s_i^2 \, pr_i^* \, L.$$

If $s = (s_1, \ldots, s_n)$ is a tuple of positive integers, then this is a divisor class on A^n. Here $pr_i \; A^n \to A$ is the projection onto the ith factor, and the multiplication and subtraction operations in parentheses in the first factor refer to the group law on A. By the theorem of the cube, this expression is homogeneous of degree two in $(s_1, \ldots, s_n)$, so we may extend the definition to $s \in \mathbb{Q}_{>0}^n$ by homogeneity; in that case $L_{-\epsilon,s}$ is a $\mathbb{Q}$-divisor class; *i.e.*, a divisor class with rational coefficients. Also, ϵ is a positive rational number.

To use this object, we let s_i be rational numbers close to $1/\sqrt{h_L(P_i)}$, and choose a suitable ϵ. If this ϵ is sufficiently small, then we can construct a small section of $\mathcal{O}(dL_{-\epsilon,s})$ on $X_1 \times \ldots \times X_n$, for some sufficiently divisible positive integer d. This section replaces the auxiliary polynomial in the more classical Thue-Siegel-Roth proof. As in that proof, arithmetic properties of the section imply that it must vanish in a certain way at $(P_1, \ldots, P_n)$. This leads to a construction of $X_1', \ldots, X_n'$.

One way to prove the semiabelian case might be to add another term to $L_{-\epsilon,s}$, similar to the first term, to deal with the $\mathbb{G}_m^\mu$ part of A. This leads to difficulties with obtaining a lower bound on h^0, however.

Briefly, the reason that this line sheaf works is that, over k, the first term is large enough to guarantee that there are a large number of global sections, while the second term has a minimal effect. However, looking at the point $(P_1, \ldots, P_n)$ over the ring of integers of k, it is the second term which dominates, since the P_i were chosen to minimize the effect of the first term. This interplay is what allows the proof to work.

References

1. G. Faltings, Diophantine approximation on abelian varieties, *Ann. Math.* **133** (1991), 549–576.

2. G. Faltings, The general case of S. Lang's conjecture (to appear).

3. Y. Kawamata, On Bloch's conjecture, *Invent. Math.* **57** (1980), 97–100.

4. S. Lang, Integral points on curves, *Publ. Math. IHES* **6** (1960), 27–43.

5. S. Lang, *Fundamentals of diophantine geometry*, Springer-Verlag, New York, 1983.

6. S. Lang, Hyperbolic and diophantine analysis, *Bull. AMS* **14** (1986), 159–205.

7. J. Noguchi, A higher dimensional analogue of Mordell's conjecture over function fields, *Math. Ann.* **258** (1981), 207–212.

8. J. H. Silverman, The theory of height functions. In: *Arithmetic geometry*, G. Cornell and J. H. Silverman, eds., Springer-Verlag, New York, 1986, 151–166.

9. K. Ueno, *Classification theory of algebraic varieties and compact complex spaces*, Lecture Notes in Mathematics 439, Springer-Verlag, Berlin-Heidelberg, 1975.

10. P. Vojta, *Diophantine approximations and value distribution theory*, Lecture Notes in Mathematics 1239, Springer-Verlag, New York, 1987.

11. P. Vojta, Siegel's theorem in the compact case, *Ann. Math.* **133** (1991), 509–548.

12. P. Vojta, Applications of arithmetic algebraic geometry to diophantine approximations, proceedings of CIME conference, Lecture Notes in Mathematics, Springer-Verlag, Heidelberg (to appear).

Paul Vojta
Department of Mathematics
University of California
Berkeley, CA 94720
USA
e-mail vojta@math.berkeley.edu

Projection from an Algebraic Quadratic Form
to Rational Quadratic Forms

Harvey Cohn
City College (CUNY), New York

DEDICATED TO PAULO RIBENBOIM

Abstract: If the prime $p \equiv 1 \mod 8$ then we can write $p = N(\pi)$ in $\mathbf{Q}(\sqrt{2})$ and we can decompose into integral squares $\pi = \xi^2 + \eta^2$, as is classically known. To go further, if (say) η is even, then $\sqrt{2}^t \| \eta$ with the maximum exponent t that satisfies $p = x^2 + 2 \cdot 2^t y^2 = u^2 + 4 \cdot 2^t v^2$ in $\mathbf{Z}$, (see *Math. Ann.* **265** (1983)). In effect, we can *project* from the representation of π to the representation of p, or we can *lift* in the opposite direction. Similar results are now found for any real field $\mathbf{Q}(\sqrt{D})$ of class number unity. Indeed other forms, e.g., $\xi^2 + 2\eta^2$ can also be used. (Such results for the cases $D = 2,3$ arose from modular forms in two variables taken on a diagonal, but the problem is treated here as purely algebraic, see *ibid*). The process creates a Diophantine analogue of the Dirichlet formula for the class number of a composite biquadratic field in terms of its quadratic subfields.

0. Introduction. Class field theory, in its historical origins, first centered on the study of conditions on a rational prime p which lead to its representation by a positive definite quadratic form in rational integers

$$(0.1) \qquad p = ax^2 + bxy + cy^2, \quad (d = b^2 - 4ac < 0),$$

where $(a,b,c) = 1$, but the discriminant d need not be fundamental. In particular, for a *principal* quadratic form, (where say $a = 1$), the condition for representability (0.1) in integral x,y is that p *split* a so-called "ring class field" (every polynomial generating an element of the field splits into linear factors $\mod p$). In the more elementary cases these are composite "genus" fields, $\mathbf{Q}(\sqrt{A}, \sqrt{B}, \cdots)$, but the more complicated cases defy such an algebraic generic description. This is a subject which is well-explored in the literature (see [13], [5], [12], [10]). It has a particular luster since Weber showed in all cases the ring class fields can be generated by values of the modular function $j(z)$, for example

$$(0.2) \qquad K = \mathbf{Q}(\sqrt{d}, j(\frac{d + \sqrt{d}}{2})).$$

The modular function theory and the ring class field theory are two independently conceived "free-standing" theories which happen to have as intersection a profound theory.

The same independence can not be claimed for forms over the ring of integers of a field, even in the simplest case of a quadratic base field. Often work on forms over quadratic fields has been motivated by generalizations of modular function theory from one variable to several variables and conversely ([8], [3]). There is, for example, a process by which modular functions of two variables reduce to modular functions of one variable taken on a diagonal. Parallel with this reduction process, quadratic forms over a quadratic ring of integers can be reduced to quadratic forms over rational integers. While this connection is severely limited in modular function theory to a few "known" cases, this process will be explored here as a separate study, purely algebraically.

In effect, for a real quadratic field $\mathbf{Q}(\sqrt{D})$ with square-free $D(> 0)$ and integer ring O_0, we factor the prime $p = \pi\pi'$, and consider the cases where π is represented by a form over O_0

$$(0.3) \qquad\qquad \pi = \xi^2 + A\eta^2,$$

with $A(> 0)$ a rational integer (not necessarily square-free). This relation will be seen to lead to simultaneous representations over $\mathbf{Z}$

$$(0.4) \qquad\qquad p = x^2 + Ay^2 = u^2 + ADv^2$$

by *projection* . The converse *lifting* from (0.4) to (0.3) is made possible by Dirichlet's biquadratic class number formula when $\mathbf{Q}(\sqrt{D})$ has class number one. This whole process must necessarily be related to relative-quadratic genus techniques (see [9]), but there are reasons for an independent approach. We have a special case easily accessible by elementary methods and we can deduce specific class field relations for nonfundamental forms.

1. Illustrative Example. Let us consider the pilot example (see [3]), the reduction of the form $\xi^2 + \eta^2$ over $O_0 = \mathbf{Z}[\sqrt{2}]$ to the forms $x^2 + y^2$ and $x^2 + 2y^2$ over $\mathbf{Z}$:

The ring O_0 is the integer ring of $k_0 = \mathbf{Q}(\sqrt{2})$, in which we consider odd primes π. They are the factors of primes p satisfying $(2/p) = 1$, (i.e., $p \equiv \pm 1 \bmod 8$), because the class number of k_0 is unity. The traditional question would have to be the representability of

$$(1.1) \qquad\qquad \pi = \xi^2 + \eta^2, \ (\xi, \eta \in O_0).$$

This imposes the further condition that $\pi > 0$ and $\pi' > 0$ (denoting the conjugate under $k_0/\mathbf{Q}$). There is also the condition that $\pi \in O_0^0$, where

$O_0^0 = \mathbf{Z}[2\sqrt{2}]$, a subring of O_0. Because of these conditions, $\pi\pi' = p$ where $\pi \equiv 1 \bmod 2\sqrt{2}$ and thus $p \equiv 1 \bmod 8$.

Next consider the field $k_4 = \mathbf{Q}(i, \sqrt{2})$. The odd rational primes p which split (completely) are those which split in each intermediate field, k_0, $k_1 = \mathbf{Q}(i)$, and $k_2 = \mathbf{Q}(\sqrt{-2})$. These, again, are $p \equiv 1 \bmod 8$. The class number of k_4 is again unity, so that these p are norms of integers of k_4. The generic integer of k_4 is

$$(1.2) \qquad O_4 = \{\, \Pi = \xi + \Lambda\eta \,\}, \text{ where } \Lambda = (1+i)/\sqrt{2}.$$

We should like to have $\Pi = \xi + i\eta$, but this is essentially true now. We can assume that ξ is odd and η is even, since otherwise we replace Π by $\Lambda\Pi/i$ (which has the same norm p). Then if we replace η by $\sqrt{2}\eta$,

$$(1.3) \qquad \pi = N_{k_4/k_0}(\Pi) = \xi^2 + \sqrt{2}\xi\eta + \eta^2.$$

With ξ odd, η must again have $\sqrt{2}$ as factor. Thus finally with $p \equiv 1 \bmod 8$, a factor π is reducible to the form $\xi^2 + \eta^2$, with a slight change of notation. (Some of the details were abridged here, but the references to the methods occur in Section 3 below).

Now to see the projection method at work, take the generic odd

$$(1.4) \qquad \Pi = \xi + i(\sqrt{2})^t\eta, \;\; \Pi^* = \xi - i(\sqrt{2})^t\eta,$$

$$\Pi' = \xi' + i(-\sqrt{2})^t\eta', \;\; \Pi'^* = \xi' - i(-\sqrt{2})^t\eta'$$

where ξ and η are both odd and $t \geq 1$. Here, of course,

$$(1.5a) \qquad \pi = \Pi\Pi^* = \xi^2 + 2^t\eta^2.$$

We introduce the abbreviations,

$$(1.5b) \qquad \xi\eta' = y + v\sqrt{2}, \;\; x = \xi\xi' + 2^t\eta\eta', \;\; y = \xi\xi' - 2^t\eta\eta'$$

(note y is odd). There are two new elements of norm p, namely

$$(1.5c) \qquad \Pi\Pi' = \Theta_1 \in k_1, \;\; \Pi\Pi'^* = \Theta_2 \in k_2$$

as follows (first for t even):

$$(1.5d) \qquad \Theta_1 = x + i2^{t/2}2y, \;\; \Theta_2 = u + \sqrt{-2} \cdot 2^{(t+1)/2}v$$

(and vice versa for t odd). Taking $p = N(\Theta_1) = N(\Theta_2)$ and making minor changes in notation (to accomodate t odd or even) we have

Pilot Example. *If $p \equiv 1 \mod 8$, then some factor π is expressible in O_0 as $\xi^2 + \eta^2$. For η even and $2^t || N(\eta)$ then t is determined as the maximum value for which, simultaneously, $p = x^2 + 2 \cdot 2^t y^2 = u^2 + 4 \cdot 2^t v^2$ in $\mathbf{Z}$.*

The process of going from one form in O_0 to two forms in $\mathbf{Z}$ is the *projection* process. Since the value of t is precisely determined in the projected forms, we also have a *lifting* from two forms in $\mathbf{Z}$ to one form in O_0. We shall generalize this example somewhat to find that the projection process is almost trivial but the lifting process goes somewhat deeply into class field theory.

2. Projection and Lifting Process. Consider the quadratic field and biquadratic extension parametrized by rational integers D and A,

$$(2.1a) \qquad \mathbf{Q} \subset \mathbf{Q}(\sqrt{D}) \, (= k_0) \subset \mathbf{Q}(\sqrt{D}, \sqrt{-A}) \, (= k_4)$$

Here we assume that the $D \, (> 1)$ is square-free but not necessarily so for $A \, (> 0)$. Of course k_4 has the additional subfields

$$(2.1b) \qquad k_1 = \mathbf{Q}(\sqrt{-A}), \quad k_2 = \mathbf{Q}(\sqrt{-AD}),$$

(and AD is not necessarily square-free either). Consider a prime p which factors into *principal positive* factors in $\mathbf{Q}(\sqrt{D})$, so

$$(2.2) \qquad p = \pi \pi', \, (\pi > 0, \, \pi' > 0).$$

Let us further consider the case where it is possible to properly choose π (among associates) so that for integers ξ, η in k_4

$$(2.3) \qquad \pi = \xi^2 + A\eta^2,$$

and likewise for π'. (To avoid ambiguity, if $A = 1$, we take ξ to be odd.) We are interested in the exponent m which satisfies

$$(2.4) \qquad 2^m \, || \, N(\eta)$$

for the norm N in $k_0/\mathbf{Q}$.

Projection Theorem. *Given the decomposition (2.3) of π, we have the simultaneous decomposition of p into two forms over rational integers:*

$$(2.5) \quad p = x^2 + Ay^2 = u^2 + ADv^2, \, (2|(y, v) \text{ for } D \equiv 2, 3 \mod 4).$$

(Again, to avoid ambiguity, if $A = 1$, we take x to be odd.)

From this point on, we ignore the case $D \equiv 1 \mod 8$, (where 2 has *two* distinct ideal prime divisors in k_0). For the remaining cases, we define g by

$$(2.6a) \qquad 2^g \, || \, (y, v)$$

and we set

$$(2.6b) \qquad e = \begin{cases} yv/2^{2g}, & D \text{ odd}, \\ y/2^g, & D \text{ even}. \end{cases}$$

Corollary to the Projection Theorem. *The values of* m *and* g *are related as follows:*

$$(2.7a) \qquad D \equiv 0 \mod 2, \quad m = \begin{cases} 2g - 2, & e \text{ odd,} \\ 2g - 1, & e \text{ even,} \end{cases}$$

$$(2.7b) \qquad D \equiv 3 \mod 4, \quad m = \begin{cases} 2g - 1, & e \text{ odd,} \\ 2g - 2, & e \text{ even,} \end{cases}$$

$$(2.7c) \qquad D \equiv 5 \mod 8, \quad m = \begin{cases} 2g, & e \text{ odd,} \\ 2g - 2, & e \text{ even.} \end{cases}$$

The proof of the Theorem and Corollary is based on the decomposition of (2.3) into conjugates in $k_4/\mathbf{Q}$,

$$(2.8a) \qquad \pi = \Pi\Pi^*, \quad \pi' = \Pi'\Pi'^*,$$

where the symbols Π^* and Π' denote conjugates in $k_1/\mathbf{Q}$ and $k_0/\mathbf{Q}$ respectively. Thus

(2.8b)
$$\Pi = \xi + \sqrt{-A}\eta, \ \Pi^* = \xi - \sqrt{-A}\eta, \ \Pi' = \xi' + \sqrt{-A}\eta', \ \Pi'^* = \xi' - \sqrt{-A}\eta'.$$

We can write

$$(2.9a) \qquad \Pi\Pi' = (\xi\xi' - A\eta\eta') + \sqrt{-A}(\xi'\eta + \eta'\xi) \in k_1,$$

$$(2.9b) \qquad \Pi\Pi'^* = (\xi\xi' + A\eta\eta') + \sqrt{-A}(\xi'\eta - \eta'\xi) \in k_2.$$

If, (as our hypothesis develops), we can write

$$(2.10a) \qquad \eta = 2^t \eta_0,$$

then the above products are represented in terms of the trace S in $k_0/\mathbf{Q}$ as

$$(2.10b) \qquad \Pi\Pi' = x + \sqrt{-A}2^t S(\xi'\eta_0) \in k_1,$$

$$(2.10c) \qquad \Pi\Pi'^* = u + \sqrt{-AD}2^t S(\xi'\eta_0/\sqrt{D}) \in k_2,$$

with the rational integers

(2.11) $$x = \xi\xi' - A\eta\eta', \quad u = \xi\xi' + A\eta\eta'.$$

From this point on, we again dismiss the case $D \equiv 1 \mod 8$ by verifiying (2.5) with $t = 0$. Each of the remaining three cases follows a rather uniform pattern:

(2.12a) $$D \equiv 0 \mod 2 : \xi'\eta_0 = y_0 + v_0\sqrt{D},$$
$$m \text{ even}, \ t = m/2, \ \xi'\eta_0 \text{ odd}, \ y_0 \text{ odd}$$
$$=> \ p = x^2 + 2^{m+2}Ay_0^2 = u^2 + 2^{m+2}ADv_0^2,$$
$$m \text{ odd}, \ t = (m-1)/2, \ \xi'\eta_0 \text{ even}, \ y_0 \text{ even}$$
$$=> \ p = x^2 + 2^{m+1}Ay_0^2 = u^2 + 2^{m+1}ADv_0^2;$$

(2.12b) $$D \equiv 3 \mod 4 : \xi'\eta_0 = y_0 + v_0\sqrt{D},$$
$$m \text{ even}, \ t = m/2, \ \xi'\eta_0 \text{ odd}, \ y_0 \not\equiv v_0 \mod 2$$
$$=> \ p = x^2 + 2^{m+2}Ay_0^2 = u^2 + 2^{m+2}ADv_0^2,$$
$$m \text{ odd}, \ t = (m-1)/2, \ \xi'\eta_0 \text{ even}, \ y_0v_0 \equiv 1 \mod 2$$
$$=> \ p = x^2 + 2^{m+1}Ay_0^2 = u^2 + 2^{m+1}ADv_0^2;$$

(2.12c) $$D \equiv 5 \mod 8 : \xi'\eta_0 = (y_0 + v_0\sqrt{D})/2, \ y_0 \equiv v_0 \mod 2,$$
$$m \text{ even } only, \ t = m/2, \ \xi'\eta_0 \text{ odd } only,$$
$$p = x^2 + 2^mAy_0^2 = u^2 + 2^mADv_0^2, \ y_0 \equiv v_0 \equiv 1 \mod 2$$
$$p = x^2 + 2^{m+2}Ay_1^2 = u^2 + 2^{m+2}ADv_1^2, \ (y_1 =) \ y_0/2 \not\equiv (v_1 =) \ v_0/2 \mod 2.$$

The completion of the proofs is now straightforward.

A much more challenging problem is the converse of the Projection Theorem:

Lifting Principle. *The lifting principle is said to be valid for D and A, when the representations (2.5) of p can be lifted to imply the representation (2.3) of π. (When the lifting principle holds, of course, the Projection Theorem and its Corollary also hold).*

This principle is far from universally true. In general it is false for fields k_0 which are not of class number unity (as the later discussion in Section 4 shall reveal). For instance for $D = 10$, the class number of $\mathbf{Q}(\sqrt{10})$ is 2, and indeed, with $A = 1$,

(2.13a) $$p = 41 = N(9 + 2\sqrt{10}) = 5^2 + 4^2 = 1^2 + 10 \cdot 2^2$$

whereas for π, (which is essentially unique in this context),

(2.13b) $$\pi = 9 + 2\sqrt{10} \neq \xi^2 + \eta^2.$$

Thus for the projection (2.13a) there is no lifting to (2.13b).

On the positive side, the lifting holds for the more elementary cases $D = 2, 3, 5, 6$ with $A = 1, 2$, as illustrated below in Table I (see the next section for details).

3. Biquadratic Field Theorems. We give a short review of theorems (see [7],[5],[11]) especially effective in proving the assertions in Table I.

We have the fields and the usual invariants:

(3.1)
$$k_0 = \mathbf{Q}(\sqrt{D}), \ k_1 = \mathbf{Q}(\sqrt{-A}), \ k_2 = \mathbf{Q}(\sqrt{-AD}), \ k_4 = \mathbf{Q}(\sqrt{D}, \sqrt{-A}),$$

O_j = integer ring of k_j, $j = 0, 1, 2, 4$,
$O_4^0 = [1, \sqrt{-A}]O_0 \ (\subseteq O_4)$,
O_j^* = corresponding ring of units,
$O_0^* = < -1, \varepsilon >$ (fundamental unit),
h_j = corresponding class numbers,
$U = |O_4^* : O_0^* \times O_1^* \times O_2^*|$ the unit index,
∂_j = different of $k_j/\mathbf{Q}$,
∂_{40} = different of k_4/k_0.

The differents can all be conveniently written as generators of ideals in k_4. The most special definition is the unit index U which in this context is (generically) 1 but it becomes 2 when a unit Ω exists such that (see [7])

(3.2) $\Omega \in O_4^*, \ \Omega \notin O_0^* \times O_1^* \times O_2^*, \ \Omega^2 \in O_0^* \times O_1^* \times O_2^*.$

For these definitions we have the panoply of formulas involving the differents (conveniently regarded as ideals in k_4):

(3.3a) ∂_j^2 = discriminant k_j, $j = 0, 1, 2$, ∂_4^4 = discriminant k_4,

(3.3b) $$\partial_{40} = \sqrt{\partial_1 \partial_2 / \partial_0}, \ = \partial_4 / \partial_0,$$

(3.3c) $$O_4 = [1, \Lambda]O_0, \text{ where } \Lambda - \Lambda^* = \partial_{40}.$$

This is the relative basis property, (valid, e.g., when $h_0 = 1$). Above all, we need the formula of Dirichlet (see [7])

(3.4) $$h_4 = h_0 \, h_1 \, h_2 \, U/2$$

Remark. *The above formulas prove the assertions in Table I, which is to be found in the next page.*

$$\mathbf{Q}(\sqrt{2})$$

$p \equiv 1 \bmod 8 \iff p = \pi\pi'$ where $\pi = \xi^2 + \eta^2$ (ξ odd), $2^m || N(\eta)$ ($m \geq 1$),

$$p = x^2 + 4y^2 = u^2 + 8v^2, \ 2^g || (y, v),$$

$y/2^g$ odd $\iff m = 2g$, $y/2^g$ even $\iff m = 2g + 1$.

$$\mathbf{Q}(\sqrt{3})$$

$p \equiv 1 \bmod 12 \iff p = \pi\pi'$ where $\pi = \xi^2 + \eta^2$ (ξ odd), $2^m || N(\eta)$ ($m \geq 1$),

$$p = x^2 + 4y^2 = u^2 + 12v^2, \ 2^g || (y, v),$$

$yv/2^{2g}$ odd $\iff m = 2g + 1$, $yv/2^{2g}$ even $\iff m = 2g$.

$p \equiv 1 \bmod 24 \iff p = \pi\pi'$ where $\pi = \xi^2 + 2\eta^2$, $2^m || N(\eta)$,

$$p = x^2 + 8y^2 = u^2 + 24v^2, \ 2^g || (y, v),$$

$yv/2^{2g}$ odd $\iff m = 2g + 1$, $yv/2^{2g}$ even $\iff m = 2g$.

$$\mathbf{Q}(\sqrt{5})$$

$p \equiv 1, 9 \bmod 20 \iff p = \pi\pi'$ where $\pi = \xi^2 + \eta^2$ (ξ odd), $2^m || N(\eta)$,

$$p = x^2 + y^2 = u^2 + 5v^2 \ (x \text{ odd}), \ 2^g || (y, v),$$

$yv/2^{2g}$ odd $\iff m = 2g$, $yv/2^{2g}$ even $\iff m = 2g - 2$.

$p \equiv 1, 9, 11, 19 \bmod 40 \iff p = \pi\pi'$ where $\pi = \xi^2 + 2\eta^2$, $2^m || N(\eta)$,

$$p = x^2 + 2y^2 = u^2 + 10v^2, \ 2^g || (y, v),$$

$yv/2^{2g}$ odd $\iff m = 2g$, $yv/2^{2g}$ even $\iff m = 2g - 2$.

$$\mathbf{Q}(\sqrt{6})$$

$p \equiv 1 \bmod 24 \iff p = \pi\pi'$ where $\pi = \xi^2 + \eta^2$ (ξ odd), $2^m || N(\eta)$ ($m \geq 1$),

$$p = x^2 + 4y^2 = u^2 + 24v^2, \ 2^g || (y, v),$$

$yv/2^{2g}$ odd $\iff m = 2g$, $yv/2^{2g}$ even $\iff m = 2g + 1$.

$p \equiv 1 \bmod 24 \iff p = \pi\pi'$ where $\pi = \xi^2 + 2\eta^2$, $2^m || N(\eta)$,

$$p = x^2 + 8y^2 = u^2 + 48v^2, \ 2^g || (y, v),$$

$yv/2^{2g}$ odd $\iff m = 2g$, $yv/2^{2g}$ even $\iff m = 2g + 1$.

Table I. Special Illustrations of Projections and Liftings.

These formulas show the reversability between the representation of primes π and p by quadratic forms. The value of π in $p = \pi\pi'$ requires a correctly chosen associate (not necessarily the same for $\pi = \xi^2 + \eta^2$ and for $\pi = \xi^2 + 2\eta^2$). For the case $\sqrt{5}$, it is possible that both ξ and η are odd in $\pi = \xi^2 + \eta^2$. Note for these cases, $A = 1$ and 2 only.

In Table II, the conditions on p for (2.3) and (2.5) happen to be linear congruences, (the same for both cases). The major steps are outlined in Table II; the details are standard exercises. Column 1 lists the fields k_4. Column 2 lists the character on (odd) p that it factor in k_0, (necessarily into principal factors). Column 3 lists the additional character for the condition that $\pi > 0$ and $\pi' > 0$, (needed only when $N(\varepsilon) = -1$). Column 4 lists the class number of k_4. Column 5, accordingly, lists the additional character (if any) for π to split in k_4 or to have principal factors as it splits. Column 6 shows the relative different ∂_{40}. Column 7 shows the values of Λ, indicating agreement of $(\Lambda - \Lambda^*)$ and ∂_{40}. For the cases where $D = 5$, $\partial_{40} = 1$ hence $O_4^0 = O_4$. In the other cases, $\Lambda = \Omega$ (and $U = 2$). We now have $\Pi \in O_4$, but our goal is to find an associate of Π for which $\Pi \in O_4^0$. This is possible, generally, if there are units $\Theta \notin O_0$ so that

$$(3.5) \qquad \Theta(\xi + \Lambda\eta) = \xi_0 + \Lambda\eta_0$$

produces a linear transformation of (ξ, η) into (ξ_0, η_0). (This process is most familiar in rational number theory as a means of going from the representation $p = x^2 - xy + y^2$ to $p = x^2 + 3y^2$ by multiplying $x + \frac{-1+\sqrt{-3}}{2}y$ by a suitable cube root of unity so as to make (x, y) become (x_0, y_0) with y_0 even). The only difficult case is $Q(\sqrt{6}, \sqrt{-2})$ where an additional character condition $(\frac{-1}{p})$ is required to ensure that $\Pi \in O_4^0$. The congruence characters for p which produce (2.3) also produce (2.5), hence the lifting is valid.

k_4	$p = \pm\pi\pi'$	$p = \pi\pi'$	h_4	$\pi = \Pi\Pi^*$	∂_{40}	Λ	
$Q(\sqrt{2}, \sqrt{-1})$	$(\frac{2}{p})$	$\cdots$	1	$(\frac{-1}{p})$	$\sqrt{2}$	$\frac{1+i}{\sqrt{2}}$	
$Q(\sqrt{3}, \sqrt{-1})$	$(\frac{3}{p})$	$(\frac{-1}{p})$	1	$\cdots$	1	$\frac{2+i+\sqrt{3}}{2}$	
$Q(\sqrt{5}, \sqrt{-1})$	$(\frac{5}{p})$	$\cdots$	1	$(\frac{-1}{p})$	2	i	$\cdots$
$Q(\sqrt{6}, \sqrt{-1})$	$(\frac{6}{p})$	$(\frac{-2}{p})$	2	$(\frac{-1}{p})$	$\sqrt{2}$	$\frac{1+i}{2+\sqrt{6}}$	
$Q(\sqrt{3}, \sqrt{-2})$	$(\frac{3}{p})$	$(\frac{-1}{p})$	2	$(\frac{-2}{p})$	2	$\frac{1+\sqrt{3}}{\sqrt{-2}}$	
$Q(\sqrt{5}, \sqrt{-2})$	$(\frac{5}{p})$	$\cdots$	1	$(\frac{-2}{p})$	$2\sqrt{2}$	$\sqrt{-2}$	
$Q(\sqrt{6}, \sqrt{-2})$	$(\frac{6}{p})$	$(\frac{-2}{p})$	1	$\cdots$	1	$\frac{\sqrt{-2}+\sqrt{6}}{2(2+\sqrt{6})}$	

Table II. Data for proof of Table I.

Note that the values of Λ are the special units Ω for fields with unit index $U = 2$. The exceptions, of course, are the general type cases, (with $D = 5$ and $U = 1$).

4. Proof of Lifting by Class Field Theory. Now standard theorems on density of ideals are used to prove a general result in the cases $A = 1$ and 2:

Main Theorem. *The lifting process is valid for $A = 1, 2$ and any square-free integer $D > 1$ for which $k_0 = Q(\sqrt{D})$ has class number unity.*

The special cases where $U = 2$ have already been treated in Table II. For the general case of the Main Theorem, $U = 1$. Once again we shall identify the primes satisfying (2.5) with those satisfying (2.3), but this can not be done as directly for the general case. We know by the Projection Theorem that there is an inclusion (which we shall make into an equality):

$$(4.1) \qquad \{p \text{ satisfying } (2.3)\} \subseteq \{p \text{ satisfying } (2.5)\}.$$

We now interpret this result in terms of ring class field theory. To review the definitions (suitably restricted), let $(0 <)B \in \mathbf{Z}$, and let us define

$$(4.2a) \qquad RCF(-4B) = \text{Ring Class Field for } x^2 + By^2$$

by the property that it is normal and that for a prime p with $(p, 2BD) = 1$,

$$(4.2b) \quad \{p \text{ splits in } RCF(-4B)/Q\} <=> \{p = x^2 + By^2, \, x, y \in \mathbf{Z}\}.$$

(The argument of RCF is the discriminant of the form). Then $RCF(-4B)$ has degree $2h(-4B)$, (the class number for forms of discriminant $-4B$). The term "split" will mean ideal factors of p are of degree one.

Analogously, starting with forms over O_0, for the same B, let us define

$$(4.3a) \qquad RCF[-4B] = \text{Ring Class Field for } \xi^2 + B\eta^2$$

by the property that it is normal and that for a prime $\pi | p$

$$(4.3b) \quad \{\pi \text{ splits in } RCF[-4B]/k_0\} <=> \{\pi = \xi^2 + B\eta^2, \, \xi, \eta \in O_0\}.$$

By the normality of the field, we can also say

$$(4.3c) \quad \{p \text{ splits in } RCF[-4B]/Q\} <=> \{\pi = \xi^2 + B\eta^2, \, \xi, \eta \in O_0\}.$$

The question of degree is not so simple now, but in the case where $O_4 = [1, \sqrt{-B}]O_0 \,(= O_4^0$ see Section 3), then $RCF[-4B]$ has degree $4h_4$.

Thus the set inclusion (4.1) asserts that

(4.4a) $\qquad RCF[-4A] \supseteq RCF(-4t^2A) \times RFC(-4t^2AD)$,

where $t = 1$ if $D \equiv 1 \bmod 4$ and $t = 2$ otherwise. In terms of degrees of fields,

(4.4b) $\qquad |RCF[-4A]| \geq \dfrac{|RCF(-4t^2A)|\,|RCF(-4t^2AD)|}{|RCF(-4t^2A) \cap RCF(-4t^2AD)|}.$

We need only prove equality to establish the main theorem. In other words, we want to prove

(4.4c) $\qquad RCF[-4A] = RCF(-4t^2A) \times RFC(-4t^2AD)$,

by proving

(4.4d) $\qquad |RCF[-4A]| = \dfrac{|RCF(-4t^2A)|\,|RCF(-4t^2AD)|}{|RCF(-4t^2A) \cap RCF(-4t^2AD)|}.$

It is easier to view the above relations in terms of (Dirichlet) density (see [5]). Let us define

(4.5a) $\qquad\qquad S_4 = \{p : p \text{ splits } RCF[-4A]/\mathbf{Q}\}$,

(4.5b) $\qquad\qquad S_1 = \{p : p \text{ splits } RCF(-4t^2A)/\mathbf{Q}\}$,

(4.5c) $\qquad\qquad S_2 = \{p : p \text{ splits } RCF(-4t^2AD)/\mathbf{Q}\}$.

Then in terms of

(4.6a) $\qquad\qquad d(S_i) = \text{(Dirichlet) density of } \{p \in S_i\}$

we are asserting in (4.1) that

(4.6b) $\qquad\qquad d(S_4) \leq d(S_1)\, d(S_2)\, /\, d(S_1 \cap S_2)$.

Our mission again is to establish equality.

According to classical theory, the primes which split in a normal extension K/Q have density $1/|K|$, but those which split into principal factors have density $1/|K|H$, with H the class number of K. In the case of a quadratic form of discriminant $-Bf^2$ the primes representable by that form are only a subset of those which split into principal factors in $k = \mathbf{Q}(\sqrt{-B})$.

For instance, if B is square-free and k has class number h, then (excluding $B = 1, 3$),

$$(4.7) \qquad d(\{p = x^2 + 2^{2t} By^2\}) = \begin{cases} 1/(2^{t+1}h) & \text{when } B \equiv 1, 2 \bmod 4 \\ 1/(2^{t+2}h) & \text{when } B \equiv 3 \bmod 4. \end{cases}$$

Finally, we need to refer to the lemma (see [3]) that the intersection of two ring class fields (over $\mathbf{Q}$) is abelian over each quadratic field, hence it is a genus field (determined by the characters of each form). In the present context of (4.4b) the characters will be chosen from

$$(4.8) \qquad\qquad \chi = \left(\frac{-1}{p}\right), \ \left(\frac{-2}{p}\right), \ \left(\frac{2}{p}\right).$$

The symbol χ shall denote all three characters.

A	$D \bmod 8$	t	$d(S_1)$	$d(S_2)$	χ	$d(S_1 \cap S_2)$	$d(S_4)$	∂_{40}
1	1, 5	1	$\frac{1}{2}$	$\frac{1}{2h_2}$	$\left(\frac{-1}{p}\right)$	$\frac{1}{2}$	$\frac{1}{4h_4}$	2
1	2	2	$\frac{1}{2}$	$\frac{1}{4h_2}$	$\left(\frac{-1}{p}\right)$	$\frac{1}{2}$	$\frac{1}{8h_4}$	$\sqrt{2}$
1	3, 7	2	$\frac{1}{2}$	$\frac{1}{8h_2}$	$\left(\frac{-1}{p}\right)$	$\frac{1}{2}$	$\frac{1}{16h_4}$	1
1	6	2	$\frac{1}{2}$	$\frac{1}{4h_2}$	$\left(\frac{-1}{p}\right)$	$\frac{1}{2}$	$\frac{1}{8h_4}$	$\sqrt{2}$
2	1, 5	1	$\frac{1}{2}$	$\frac{1}{2h_2}$	$\left(\frac{-2}{p}\right)$	$\frac{1}{2}$	$\frac{1}{4h_4}$	$2\sqrt{2}$
2	2	2	$\frac{1}{4}$	$\frac{1}{8h_2}$	$\left(\frac{2}{p}\right), \left(\frac{-1}{p}\right)$	$\frac{1}{4}$	$\frac{1}{16h_4}$	$\sqrt{2}$
2	3, 7	2	$\frac{1}{4}$	$\frac{1}{4h_2}$	$\left(\frac{2}{p}\right), \left(\frac{-1}{p}\right)$	$\frac{1}{4}$	$\frac{1}{8h_4}$	2
2	6	2	$\frac{1}{4}$	$\frac{1}{16h_2}$	$\left(\frac{2}{p}\right), \left(\frac{-1}{p}\right)$	$\frac{1}{4}$	$\frac{1}{32h_4}$	1

Table III. Details for proof of Main Theorem.

We are now ready to evaluate the right hand side of (4.6b). It is shown in Table III. Column 1 has the values $A = 1, 2$, Column 2 has the residue classes of D mod 8, and Column 3 has the values of t. Columns 4 and 5 have the densities calculated according to (4.7). Column 6 has generating Dirichlet characters common to the forms in S_1 and S_2 (which determine the density of primes for which such characters are unity, in Column 7). Thus in Column 8, we give the value of the right side of (4.6b), under the label "$d(S_4)$" (in anticipation of the equality we shall establish). The value

$$(4.9) \qquad\qquad h_4 = h_2/2$$

is used according to (3.4), since $h_0 = h_1 = 1$, $U = 1$.

To complete the identification of Column 8, we look at Column 9 which has ∂_{40}. If we define

$$(4.10a) \qquad J = (2\sqrt{A}/\partial_{40})^2,$$

then J is the discriminant ratio for O_4^0 over O_4. We note independently of (4.6b) that

$$(4.10b) \qquad d(S_4) = 1/(4h_4 J).$$

To see this start with the case $D \equiv 1, 5 \bmod 8$ where $J = 1$. Here $O_4 = O_4^0$ and the density of primes $p = N(\xi + \sqrt{-A}\eta)$ is $1/(4h_4)$. In general, if $2^r || J$, then a chain of modules

$$(4.11c) \qquad O_4^0 \subset O_4^1 \subset \cdots O_4^r (= O_4)$$

exists in which p is represented as norms. The density of such primes doubles at each step. Thus in (4.6b) the inequality is seen to be an inequality.

Corollary to the Main Theorem. *The lifting process for $A = 1$ and 2 is not generally valid if k_0 has class number $h_0 > 1$. Indeed the density of the primes for which it is valid is $1/h_0$.*

Actually, Table III is unchanged by the fact that $h_0 > 1$, but by (3.4), h_4 is now $h_0 h_2/2$ and $1/h_0$ is the ratio of the unequal sides of (4.6b).

5. Remarks on Specific Class Fields. There are many natural areas of further inquiry. The role of the base field k_0 may be broadened from quadratic to any 2-cyclic field (where the conjugate k_0' is uniquely determined as the involutory group element), and where the ideal 2 completely ramifies. For example, with minor modifications, the projection and lifting phenomenon is valid for an infinite tower of fields. We may consider the form $\xi^2 + \eta^2$ over integers in the sequence of fields (see [4]):

$$(5.1) \qquad \mathbf{Q}(\sqrt{2}), \ \mathbf{Q}(\sqrt{2 + \sqrt{2}}), \ \mathbf{Q}(\sqrt{2 + \sqrt{2 + \sqrt{2}}}), \cdots .$$

There is as yet no general analogue of the Main Theorem.

The process of lifting from two quadratic forms in Z to one form in $Z[\sqrt{2}]$ has a "simplifying" effect on the ring class fields. For instance, taking results for $x^2 + 4^m y^2$ (see [1], [2]), let us set

$$(5.2a) \qquad RCF(-4 \cdot 4^m) = K_m.$$

Then, with $\lambda = 2^{1/16}(1 + \sqrt{2})^{1/4}$ and $1 \leq m \leq 5$,

(5.2b)
$$K_1 = \mathbf{Q}(i), \ K_2 = K_1(\lambda^8), \ K_3 = K_2(\lambda^4), \ K_4 = K_3(\lambda^2), \ K_5 = K_4(\lambda).$$

Likewise, with reference to $x^2 + 2 \cdot 4^m y^2$, let us set

(5.3a) $RCF(-8 \cdot 4^m) = L_m.$

As before, with $\gamma = (2(1 + \sqrt{2}))^{1/8}$ and $0 \le m \le 4$,
(5.3b)
$$L_0 = \mathbf{Q}(\sqrt{-2}), \ L_1 = L_0(\gamma^8), \ L_2 = L_1(\gamma^4), \ L_3 = L_2(\gamma^2), \ L_4 = L_3(\gamma).$$

If we now refer to the Pilot Example (Section 1), with $\xi^2 + 2^m \eta^2$, we can define the new ring class fields F_m:

(5.4a) $RCF[-4 \cdot 2^m] \ (= RCF(-8 \cdot 2^m) \times RCF(-16 \cdot 2^m)) \ = \ F_m$ for $m \ge 1.$

Then with a range of m limited by (5.2ab) and (5.3ab), we obtain class fields which appear to have a simpler pattern:

$$F_1 = \mathbf{Q}(i, \sqrt{2}), \ F_2 = F_1(\sqrt{1 + \sqrt{2}}), \ F_3 = F_2(2^{1/4}), \ F_4 = F_3((1 + \sqrt{2})^{1/4}),$$

(5.4b) $F_5 = F_4(2^{1/8}), \ F_6 = F_5((1 + \sqrt{2})^{1/8}), \ F_7 = F_6(2^{1/16}).$

Incidentally, we can not expect F_8 (by "momentum") to contain $(1 + \sqrt{2})^{1/16}$; the reason is that class fields must be normal. Indeed, under the norm with $\sqrt{2} \to -\sqrt{2}$, F_8 would contain $\exp 2\pi/32$ (with the "three-story" radical $\sqrt{2 + \sqrt{2 + \sqrt{2}}}$). This would more than double the degree of F_7!

   If we consider the lifting from $\mathbf{Z}$ to $\mathbf{Z}[\sqrt{3}]$ the analogous simplification is less pronounced. We consider the ring class fields designated analogously with (4.3a) as

(5.5a) $R_m = RCF[\xi^2 + \eta^2]$, for $2^m | N(\eta), \ m \ge 1.$

(Note that we must use "$2^m | N(\eta)$" rather than "$2^m || N(\eta)$", since the ray concept is not based on *exact* divisibility). The projected ring class fields for forms over $\mathbf{Z}$ are those of (5.2a), (namely K_m), and

(5.5b) $M_m = RCF(-12 \cdot 4^m), \ m \ge 1.$

With $\mu = \sqrt{(1 + 2\sqrt{2} + \sqrt{3})\sqrt{(1 + \sqrt{3})/2}}$, the first few values [2] are

(5.5c) $M_1 = \mathbf{Q}(i, \sqrt{3})$, $M_2 = M_1(\mu^4)$, $M_3 = M_2(\mu^2)$, $M_4 = M_3(\mu)$.

We now find that for $m > 1$ the R_m are not quite the direct products of $RCF(\ )$ (as in (5.4a)). It is true that we start with

(5.5d) $\pi = \xi^2 + \eta^2,\ 2|N(\eta) \iff p = x^2 + 4y^2 = u^2 + 12v^2$,

but beyond this, from Table I, we have the parity-dependent relations
(5.5e)
$$\pi = \xi^2 + \eta^2,\ 2^{2t}|N(\eta) \iff p = x^2 + 4 \cdot 4^t y^2 = u^2 + 12 \cdot 4^t v^2$$

$$\pi = \xi^2 + \eta^2,\ 2^{2t+1}|N(\eta)$$

(5.5f) $\iff p = x^2 + 4 \cdot 4^t y^2 = u^2 + 12 \cdot 4^t v^2$ and $y \equiv v \bmod 2$.

We finally obtain as an analogue of the pattern of (5.4b):

$$R_1 = \mathbf{Q}(i, \sqrt{3}),\ R_2 = R_1(\sqrt{2}),\ R_3 = R_2(\sqrt{1 + \sqrt{3}}),$$

(5.5g)
$$R_4 = R_3(\sqrt{1 + \sqrt{3}}, 2^{1/4}),\ R_5 = R_4(\sqrt{(1 + 2\sqrt{2} + \sqrt{3})(1 + \sqrt{2})\sqrt{1 + \sqrt{3}}}).$$

Whatever simplification is created in these special cases should be ultimately explained by a general property of the lifting process.

REFERENCES

1. A. Aigner and H. Reichardt, *Steifenreihen und Potenzcharakter*, J. Reine Angew. Math. **184** (1940), 158–160.
2. H. Cohn, *Iterated Ring Class Fields and the Icosahedron*, Math. Annalen **255** (1981), 107–122.
3. H. Cohn, *Some examples of Weber-Hecke ring class field theory*, Math. Annalen **265** (1983), 83–100.
4. H. Cohn, *Representation of a prime as the sum of squares in a tower of fields*, Journ. reine angew. Math. **361** (1985), 135–144.
5. H. Cohn, *Introduction to the Construction of Class Fields*, Cambridge University Press, 1985.
6. H. Cohn and J. Deutsch, *Some singular moduli for* $\mathbf{Q}(\sqrt{3})$, Math. of Computation **58** (1992).
7. H. Hasse, *Über die Klassenzahl Abelscher Zahlkörper*, Akademie-Verlag Berlin, 1952.
8. E. Hecke, *Höhere Modulfunktionen und ihre Anwendung auf die Zahlentheorie*, Math. Annalen **71** (1912), 1–37.

9. D. Hilbert, *Über die Theorie des relativquadratischen Zahlkörpers*, Math. Annalen **51** (1899), 1–127.

10. K. Miyake, *The establishment of the Takagi-Artin class field theory*, Proc. History of Mathematics Symposium, Tokyo (1990); (to appear).

11. P. Ribenboim, *Gauss and the class number problem*, Symposia Gaussiana, (Proc. International Symp. on Math. and Theor. Phys. Guaruj/'a, Brazil) (1989); (to appear).

12. P. Ribenboim, *Algebraic Numbers*, Wiley-Interscience, 1972.

13. H. Weber, *Elliptische Funktionen und algebraische Zahlen*, Vieweg, Braunschweig, 1891.

Harvey Cohn
Department of Mathematics
City College, CUNY
New York, NY 10031
USA
e-mail hihcc@cunyvm.bitnet

Construction of $\tilde{S}_4$-fields
and modular forms of weight 1

Teresa Crespo

Universitat de Barcelona

We consider a Galois extension $L \mid \mathbf{Q}$ with Galois group $\tilde{S}_4$. Let $G_{\mathbf{Q}}$ denote the absolute Galois group of $\mathbf{Q}$. Using one of the two (conjugate) linear complex representations of $\tilde{S}_4$ of dimension 2, we obtain a Galois representation

$$\rho : G_{\mathbf{Q}} \to \mathrm{Gl}(2, \mathbf{C})$$

of octahedral type.

If ρ has odd determinant, we know by a theorem of Tunnell and a theorem of Weil-Langlands [7] that the Artin L-series of ρ:

$$L(s, \rho) = \sum_{n=1}^{\infty} a_n n^{-s}$$

gives rise to a cuspform $F(z) = \sum_{n=1}^{\infty} a_n q^n, q = e^{2\pi i z}$, newform of weight 1, level N equal to the Artin conductor of ρ and character ϵ, the quadratic character associated to $\det \rho$ [4].

In this paper, we build Galois extensions $\tilde{L} \mid \mathbf{Q}$ with Galois group $\tilde{S}_4$, by solving explicitly Galois embedding problems. This construction is applied to the computation of the coefficients of the modular form of weight 1, arising from the representation ρ, associated to $\tilde{L} \mid \mathbf{Q}$.

1 Construction of $\tilde{S}_4$-fields.

We explain here a particular case of the explicit resolution of embedding problems

$$\tilde{S}_n \to S_n \simeq \mathrm{Gal}(L \mid K)$$

for $n \geq 4$, $\tilde{S}_n$ the double cover of S_n in which transpositions lift to involutions and K a field of characteristic different from 2 (cf. [3]). The obstruction to the solvability of these embedding problems is given by a formula of Serre in terms of Hasse-Witt invariants (cf. [6]).

Let $f(X) \in \mathbf{Q}[X]$ be a polynomial realizing S_4 over $\mathbf{Q}$. Let L be the splitting field of f. We consider the embedding problem

$$\tilde{S}_4 \to S_4 \simeq \mathrm{Gal}(L \mid \mathbf{Q}).$$

Let x be a root of f, and let

$$E = \mathbf{Q}(x), \quad Q_E(X) = Tr_{E|\mathbf{Q}}(X^2), \quad d = \mathrm{disc}(E \mid \mathbf{Q}) \bmod \mathbf{Q}^{*2}.$$

According to Serre's formula, an equivalent condition to the solvability of the considered embedding problem is

$$w(Q_E) = (2, d) \in H^2(G_{\mathbf{Q}}, \{\pm 1\}),$$

where w denotes the Hasse-Witt invariant and $(\ ,\ )$ the Hilbert symbol. In this particular case, this condition is equivalent to

$$Q_E \sim_{\mathbf{Q}} Q^+ := X_1^2 + X_2^2 + 2X_3^2 + 2dX_4^2.$$

Let now R and M be the invertible matrices with entries in $\mathbf{Q}(\sqrt{d})$ and L, respectively, defined by:

$$R = \begin{pmatrix} 1 & 0 & 0 & 0 \\ 0 & 1 & 0 & 0 \\ 0 & 0 & 1/2 & 1/2 \\ 0 & 0 & 1/2\sqrt{d} & -1/2\sqrt{d} \end{pmatrix} \in \mathrm{Gl}_4(\mathbf{Q}(\sqrt{d})),$$

$$M = \left(x_i^j \right) \begin{matrix} 1 \le i \le 4 \\ 0 \le j \le 3 \end{matrix}$$

for $x = x_1, x_2, x_3, x_4$ the 4 roots of the polynomial f.

We obtain now the solutions to the embedding problem in terms of matrices:

Theorem 1. *If the embedding problem*

$$\tilde{S}_4 \to S_4 \simeq \mathrm{Gal}(L \mid \mathbf{Q})$$

is solvable, there exists a matrix P_0 in $\mathrm{Gl}_4(\mathbf{Q})$ such that

$$P_0^t(Q_E)P_0 = (Q^+)$$

and the element γ in L defined by

$$\gamma = \det(M P_0 R + I),$$

where I denotes the identity matrix, is not zero. Then $L(\sqrt{\gamma})$ is a solution to the embedding problem.

Remarks.

1) By developing the determinant, the element γ is obtained in terms of the 4 roots of the polynomial f.

2) All solutions of the embedding problem are the fields $L(\sqrt{r\gamma})$, with r running over $\mathbf{Q}^*/\mathbf{Q}^{*2}$.

Sketch of the proof. Let A_4 be the alternating group in 4 letters. Let $K := \mathbf{Q}(\sqrt{d}) = L^{A_4}$. We prove first that the element γ in the theorem is a solution to the embedding problem

$$\tilde{A}_4 \to A_4 \simeq \mathrm{Gal}(L \mid K),$$

where $\tilde{A}_4$ denotes the nontrivial double cover of A_4.

To this, we consider the two quadratic spaces (K^4, Q), $(E \otimes K, Q_E \otimes K)$, where $Q := X_1^2 + X_2^2 + X_3^2 + X_4^2$. We have the isomorphisms of quadratic spaces

$$f : L^4 \to E \otimes L,$$

associated to M^{-1}, which satisfies

$$(f^{-1} f^s)(e_i) = e_{s(i)},$$

for (e_1, e_2, e_3, e_4) the canonical basis of K^4, and

$$g : K^4 \to E \otimes K$$

associated to the matrix $P = P_0 R$.

We write down the element z in the Clifford algebra $C_L(Q_E)$ of $Q_E \otimes L$ defined by

$$z = \sum_{\epsilon_i = 0,1} v_1^{\epsilon_1} v_2^{\epsilon_2} v_3^{\epsilon_3} v_4^{\epsilon_4} w_4^{\epsilon_4} w_3^{\epsilon_3} w_2^{\epsilon_2} w_1^{\epsilon_1},$$

where $v_i = f(e_i)$, $w_i = g(e_i)$. We can choose P_0 such that z is invertible and by the spinorial interpretation of $\tilde{A}_4$ and the relation satisfied by f, we obtain that z satisfies

$$N(z)^s = b_s^2 N(z)$$

for s in A_4 and b_s in L^* such that

$$b_s b_t^s b_{st}^{-1} = a_{s,t},$$

for $a_{s,t}$ a 2-cocycle associated to $\tilde{A}_4$ (cf. [2]).

Now, $N(z) = 2^4 \det(MP + I)$ and so, $L(\sqrt{\gamma})$ is a solution to

Table 1.

pO_E	π_p	$\tilde{\pi}_p$	a_p
$P_1 P_1' P_1'' P_1'''$	1A	1A	2
		2A	-2
$P_2 P_2'$	2A	4A	0
$P_1 P_1' P_2$	2B	2B	0
$P_1 P_3$	3A	3A	-1
		6A	1
P_4	4A	8A	$i\sqrt{2}$
		8B	$-i\sqrt{2}$

$$\tilde{A}_4 \to A_4 \simeq \mathrm{Gal}(L \mid K).$$

It is easy to see that the element γ is invariant under the transposition $(3,4)$ of S_4, and then $L(\sqrt{\gamma})$ is a solution to

$$\tilde{S}_4 \to S_4 \simeq \mathrm{Gal}(L \mid \mathbf{Q}).$$

2   Modular forms

We go now to the problem of obtaining modular forms ([1,5]).

Coming back to the Galois representation ρ, we have that det ρ is associated to $K = \mathbf{Q}(\sqrt{d})$ and so, we get an odd ρ by taking $d < 0$.

For the level N, the prime factors are those of d and it can be computed by studying the higher ramification groups at the primes $p \mid d$.

For the coefficients a_n of F, as F is an eigenvector of the Hecke operators, they are determined by a_p, for p prime. For p non dividing N, we have

$$a_p = \mathrm{Tr}(\rho(\mathrm{Frob}_p)),$$

so we need to know the conjugation class of the Frobenius substitution at p in $\tilde{S}_4$. Let us denote this conjugation class by $\tilde{\pi}_p$ and let π_p be the conjugation class of the Frobenius at p in S_4. Now, π_p is determined by the reduction of the polynomial f modulo p and we know that the five conjugation classes of S_4 lift to eight conjugation classes in $\tilde{S}_4$. On the other hand, the character table of $\tilde{S}_4$ gives the value of a_p for each $\tilde{\pi}_p$. This information is gathered in table 1. In the first column, the subindex of each ideal denotes its residue degree.

We see then that $\tilde{\pi}_p$ and so a_p is determined if $\pi_p = $ 2A or 2B. Let us see now how to determine $\tilde{\pi}_p$ in the remaining cases. For $\pi_p = $ 1A, 3A, it is enough to know if the residue degree of $L(\sqrt{\gamma}) \mid L$ at a prime $\mathcal{P}$ in L over p is 1 or 2. We obtain then

Proposition 2. *If $\pi_p = $ 1A (resp. 3A), we have $a_p = $ 2 (resp -1) if and only if γ is a square mod $\mathcal{P}$.*

Let us note that this condition can be checked from the explicit expression of the element γ.

Now, for the case $\pi_p = $ 4A, the residue degree at $\mathcal{P}$ does not determine $\tilde{\pi}_p$. Nevertheless, it can be determined in the following way. We take an element s in the conjugation class 4A, namely $s= (1234)$ and let s_A in 8A, s_B in 8B be the two liftings of s in $\tilde{S}_4$. We have $\gamma^s = \gamma^{s_0} = b_{s_0}^2 \gamma$, for $s_0 = (123)$ (see the proof of theorem 1) and we can label things so that $s_A(\sqrt{\gamma}) = b_{s_0}\sqrt{\gamma}$. We obtain then

Proposition 3. *If $\pi_p = $ 4A, we have $a_p = i\sqrt{2}$ if and only if $\gamma^{\frac{p-1}{2}} \equiv b_{s_0}$ (mod $\mathcal{P}$).*

Now, by working with the elements z and z^{s_0} of the Clifford algebra $C_L(Q_E)$, we obtain the formula

$$b_{s_0} = -\frac{1}{2} + \frac{\gamma^{s^{-1}} - \gamma^s}{2\gamma}.$$

The element b_{s_0} can then be written down explicitly making it possible to check the condition in proposition 3.

We give now an example of this construction of modular forms.

Example

We consider the polynomial $f(X) = X^4 + 5X^3 + 6X^2 - 3$, having Galois group S_4 over $\mathbf{Q}$ and discriminant $-3^3 11^2$. Let $E = \mathbf{Q}(x)$, for x a root of f, and L the splitting field of f, as in section 1. The computation of the trace form of $E \mid \mathbf{Q}$ and its Hasse-Witt invariant gives that the embedding problem

$$\tilde{S}_4 \to S_4 \simeq \mathrm{Gal}(L \mid \mathbf{Q})$$

is solvable. Applying the formula in theorem 1, we obtain that a solution is the field $L(\sqrt{\gamma})$, with

$$\gamma = \quad -11(54x_1^3 x_2^2 + 222x_1^2 x_2^2 + 6x_1 x_2^2 - 336x_2^2 + 174x_1^3 x_2 + 798x_1^2 x_2$$

$$+ 357x_1 x_2 - 1152x_2 + 78x_1^3 + 537x_1^2 + 517x_1 - 1300),$$

for x_1, x_2, two roots of f.

Now the modular form associated to $\tilde{L}$ via one of the two linear complex representations of $\tilde{S}_4$ of dimension two has level $N = 3^3 11^2$. Its first coefficients are computed by applying the preceding results and given in the following table .

p	2	5	7	13	17	19	23
a_p	$i\sqrt{2}$	$i\sqrt{2}$	1	1	$i\sqrt{2}$	0	$-i\sqrt{2}$

p	29	31	37	41	43	47
a_p	0	1	0	$i\sqrt{2}$	0	$-i\sqrt{2}$

References

1. Bayer, P. and Frey, G. (1991). Galois representations of octahedral type and 2-coverings of elliptic curves. *Mathematische Zeitschrift*, **207**, 395-408.
2. Crespo, T. (1989). Explicit construction of $\tilde{A}_n$ type fields. *Journal of Algebra*, **127**, 452-461.
3. Crespo, T. (1990). Explicit construction of $2S_n$ Galois extensions. *Journal of Algebra*, **129**, 312-319.
4. Deligne, P. and Serre, J.-P. (1974). Formes modulaires de poids 1. *Ann. Sci. E.N.S.*, **7**, 507-530.
5. Lario, J. C. and Quer, J. (1988). De la solución de un problema de inmersión a la construcción de formas modulares de peso 1. *Actas de las XIII Jornadas hispanolusas de Matemáticas*, Universidad de Valladolid.
6. Serre. J.-P. (1984). L'invariant de Witt de la forme $\mathrm{Tr}(x^2)$. *Comment. Math. Helv.*, **59**, 651-676.
7. Tunnel, J. (1981). Artin's conjecture for representations of octahedral type. *Bull. Am. Math. Soc.*, **5**, 173-175.

Teresa Crespo
Department d'Àlgebra i Geometria
Universitat de Barcelona
Gran Via de les Corts Catalanes 585
08007 Barcelona
Spain

A Universal Format for Local/Global Theories

Paul Feit

University of Toledo

Abstract

The paper announces a proof of the existence of a category of global objects for any appropriate category of local structures. The proof is purely categorical, and does not require topological models. It applies equally well to examples whose present forms rely on topos theory.

1 Intuitions of Local and Global

A paper [3], to appear in the Memoirs of the A.M.S., will formulate a universal Existence Theorem which proves that a category of global objects exists for any reasonable notion of local structure. Global objects of number theory are typified by schemes, Tate's rigid analytic spaces and algebraic spaces. Analytic examples range from the basic manifold to Douady's espace banachique. In future work, we shall show that existence of all variants of orbifolds and, hopefully, of algebraic stacks are simple corollaries of the new machinery. The present article is intended to explain the desirability of such a universal format, and to outline our approach.

The result is foundational. It is intended to simplify existence and basic structure theory for a multitude of mathematical constructs. Typically (algebraic spaces or espaces banachiques are good examples), the crucial theorems of a local/global theory are demonstrated, while a myriad of routine details are left to the reader. The routine theory is complicated, not so much by the statements being proved, as by dependence on elaborate models. Heuristically, all these theories are the same, and the basic properties are tautological; in practice, the formalism is murky. Our thesis is (1) structure theory for all examples fits into one framework and (2) technical models, which have obscured the underlying unity, can be discarded.

Crudely, a theory with local/global behavior consists of a category $\mathcal{C}$ of all local objects and local morphisms, a corresponding completed category $\mathcal{C}^{gl}$ of all global objects, and a covariant canonical functor $\Phi : \mathcal{C} \longrightarrow \mathcal{C}^{gl}$. The archetypal example is the category of C^∞-manifolds. Here, manifolds make up $\mathcal{C}^{gl}$, $\mathcal{C}$ is the category of all open subsets of $\mathbb{R}^n$ (over all $n \in \mathbb{N}$) whose morphisms are C^∞-maps, and Φ is the subcategory embedding.

Figure 1. Consistency of Embeddings

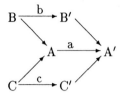

Traditional treatments often involved ingenious definitions for both $\mathcal{C}^{gl}$ and Φ. The category of schemes was another choice of $\mathcal{C}^{gl}$; the corresponding $\mathcal{C}$ was the opposite category of rings, and the functor Φ mapped each ring to its spectral sheaf. For Tate's rigid analytic spaces, $\mathcal{C}$ consisted of certain $\wp$-adic topological rings; after establishing the basics for $\mathcal{C}$, one still faced the sophisticated task of formulating rigorously what a $\mathcal{C}^{gl}$-object was. When $\mathcal{C}$ was the (opposite) category of rings with the étale topology, $\mathcal{C}^{gl}$ became the category of algebraic spaces; again, construction of $\mathcal{C}^{gl}$ was bedeviled by the need for a sufficiently rich class of objects to which the étale version of 'local' applied.

All examples, of local/global theory rely on a common intuition. The similarities are mathematical as well. Let $\Phi : \mathcal{C} \longrightarrow \mathcal{C}^{gl}$ be any of the preceding examples.

(1.1.a) In all cases, Φ preserves finite inverse limits.

(1.1.b) In all cases, for $A, B \in \mathcal{C}$, Φ determines a bijection

$$\mathrm{Mor}_{\mathcal{C}}(A, B) \longrightarrow \mathrm{Mor}_{\mathcal{C}^{gl}}(\Phi(A), \Phi(B)).$$

(1.1.c) In most examples, there is a cotangent bundle construction. In fact, there are entire theories of 'coherent' bundles.

(1.1.d) There are special morphisms, usually called *open embeddings* (or, for algebraic spaces, *étale* maps). These special morphisms exhibit certain properties which are independent of category. For example, consider the commutative diagram Figure 1 of $\mathcal{C}^{gl}$-objects. If a, b and c are open embeddings, and the corresponding products exist, then $b \times_a c : B \times_A C \longrightarrow B' \times_{A'} C'$ is an open embedding.

Classical methods of proof did not access these underlying themes. Instead, each case relied on a specific model. Grothendieck's definition of a scheme was based on the conversion from a ring to its spectral sheaf. For rigid analytic spaces, or even espaces banachiques, sheaves on topological bases were inadequate. Because each separate category was developed with its own, specific model, a universal proposition, of kind (1.1.a-d), required a different verification for each case. Demonstration of (1.1.a) for schemes was similar to the proof for algebraic spaces or manifolds; but, because the categories had radically different realizations, the three cases could not, previously, be handled by one argument.

The objectives of a universal framework are twofold.

(1.2.a) Given an arbitrary category $\mathcal{C}$ which supports a local/global theory, then $\mathcal{C}^{gl}$ is to be constructed in a manner independent of models. Indeed, $\mathcal{C}^{gl}$ is to be characterized by a universal property.

(1.2.b) Foundational propositions, like (1.1.a-d), must be established in complete generality.

We believe that both objectives have been achieved. The method is sketched below.

2 Grothendieck Topologies

Each classical theory with local/global behavior has some sort of topology. Grothendieck's formulation, with minor modification, will generalize all of these. In what follows, let $\mathcal{C}$ be a category.

Let b be a $\mathcal{C}$-morphism. We refer to the domain and codomain of b by dom b and cod b, respectively. The domain of a set-theoretic function f, regarded as a set rather than as an object, is denoted by $\mathrm{dom}(f)$. Certain categories of interest, such as the category of manifolds, are not closed under arbitrary pullback; a morphism b which has a pullback along any morphism to its codomain is called a *pullback base*.

A *universe of formal subsets* for $\mathcal{C}$ is a subclass $Sub \subseteq \mathrm{Mor}(\mathcal{C})$ for which

(2.1.a) every $\mathcal{C}$-isomorphism is in Sub,

(2.1.b) a composition of members of Sub is in Sub,

(2.1.c) each member of Sub is a pullback base, and every pullback of a member of Sub is again in Sub.

We say that Sub is a *universe of embeddings* if, in addition

(2.2) for $b \in \mathrm{Mor}(\mathcal{C})$ and $c \in Sub$ such that dom $c =$ cod b, $c \circ b \in Sub$ implies that $b \in Sub$.

If every member of Sub is a monomorphism, (2.2) holds.

Let θ be an indexed cone into $A \in \mathcal{C}$. A *subdivision function* for θ is a function which to each $j \in \mathrm{dom}(\theta)$ assigns a cone ϕ_j into $\mathrm{dom}\,\theta(j)$. Suppose ϕ is such a function. Put

$$\Lambda = \{(j, r) : j \in \mathrm{dom}(\theta),\ r \in \mathrm{dom}(\phi_j)\}.$$

The function $\tau : (j, r) \longmapsto \theta(j) \circ \phi_j(r)$ is called the *subdivision* of θ through ϕ.

Suppose Cov is a class of subsets of Sub such that

(2.3.a) each $S \in Cov$ is a non-empty cone,

(2.3.b) if $S \in Cov$, T is a subset of Sub such that T is a cone, and $S \subseteq T$, then $T \in Cov$,

(2.3.c) if b is a $\mathcal{C}$-isomorphism, then $\{b\} \in Cov$,

(2.3.d) if $S \in Cov$ and ϕ is a function which to each $s \in S$ assigns $\phi(s) \in Cov$ so that $\mathrm{cod}\,\phi(s) = \mathrm{dom}\,s$, then the subdivision of S through ϕ is in Cov,

(2.3.e) if $S \in Cov$, $B = \mathrm{cod}\,S$, $b : A \longrightarrow B$ is a $\mathcal{C}$-morphism and θ is a function which to each $s \in S$ assigns a pullback $\theta(s) = (b^{-1}s; \pi_A, \pi_s)$, then $\{(b^{-1}s, \pi_A) : s \in S\} \in Cov$,

(2.3.f) if $f \in Sub$ and π_1 and π_2 are projections of $f \times_{\mathrm{cod}\,f} f$ to $\mathrm{dom}\,f$, then $\{\pi_1\}, \{\pi_2\} \in Cov$.

We shall call Cov a *(Grothendieck) topology* for Sub; our usage is a variant of the standard term. We refer to $(\mathcal{C}, Sub, Cov)$ as a *topologized category*. A *covering morphism* is a $\mathcal{C}$-morphism b such that $\{b\} \in Cov$.

Suppose θ is a cone of $A \in \mathcal{C}$, indexed by a set J. Associate to θ a graph H, called its *canopy*, which is indexed by $J \amalg J^2$ and such that

$$H(j) = \mathrm{dom}\,\theta(j) \quad \text{for } j \in J, \tag{2.4}$$
$$H(j, k) = \mathrm{dom}\,\{\theta(j) \times_A \theta(k)\} \quad \text{for } j, k \in J,$$

and the morphisms of H are all canonical projections $H(j, k) \longrightarrow H(j)$ and $H(j, k) \longrightarrow H(k)$. Each object of H has a canonical morphism into A, which yields a cone $\alpha : H \longrightarrow A$. We say θ is *intrinsic* if α is a colimit. A Grothendieck topology Cov is called *intrinsic* if each cone associated to an indexed cover is intrinsic. Note that if Cov is intrinsic, then each cone of a cover will actually be an absolute colimit.

The first issues in a universal theory are

(2.5.a) Find a categorical hypothesis under which $\mathcal{C}$ serves as the basis of a local/global theory. Such a $\mathcal{C}$ is called a *local structure*. (Having an intrinsic topology is part of this hypothesis.)

(2.5.b) Given $\mathcal{C}$, we wish to construct a new topologized category $\mathcal{C}^{gl}$, which is to be interpreted as the category of all global structure which are locally in $\mathcal{C}$. Obviously, $\mathcal{C}^{gl}$ should be another local structure. But it must also have some extra property, the categorical version of being an 'entire' category of global objects.

The precise definition of local structure is omitted. Of greater importance is difference between a local structure and a *global structure*—that is, between an initial category of local objects and a full category of global objects.

Recall the classical examples. The difference between local objects and global objects is that the category of the latter is closed under *descent*, also referred to as *cut-and-paste*. For manifolds, the descent theorem is as follows:

(2.6) Let X be a topological space and let $\{U_i\}_{i \in I}$ be an indexed cover of X by open subsets. For $i, j \in I$, put $U_{i,j} = U_i \cap U_j$. Suppose U_i is assigned a differential structure for each i. Then, provided the structures are consistent on $U_{i,j}$ for all indices i and j, we conclude that X has a unique differential structure whose restriction to each U_i is the given structure.

Generally, a descent theorem for a category $\mathcal{C}$ states that given a family of objects $\{U_i \in \mathcal{C}\}_{i \in I}$ along with some notion of 'intersections' $U_{i,j}$, then, under a consistency hypothesis, the objects U_i somehow combine to form a new $\mathcal{C}$-object. A categorical formulation, independent of type of object involved, is needed.

Fix formal symbols ρ_1 and ρ_2. Let J be a be a non-empty set. Define a graph $\text{Int}(J)$ as the set $S(J) = J \amalg J^2$ with the set of morphisms

$$M(J) = \{((i,j), \rho_1, i) : i, j \in J\} \cup \{((i,j), \rho_2, j) : i, j \in J\}$$

Each triple $(A, \rho, B) \in M(J)$ has domain $A \in S(J)$ and codomain B. We refer to $\text{Int}(J)$ as the *intersection graph* on J. Although the above definition is meaningful even when J is empty, for our purposes *an intersection graph is never indexed by the empty set*. A graph F of $\mathcal{C}$-objects is called an intersection graph (over J) if it is of this type. For such an F, denote J by $\Lambda(F)$. For $i, j \in J$, we frequently denote $F[(i,j), \rho_1, i](\rho_1)$ by ρ_1 or ρ_i, and $F[(i,j), \rho_2, j](\rho_2)$ by ρ_2 or ρ_j. A cone θ of F is uniquely determined by its values on J, and is often treated as a function on J.

Let F be an intersection graph over $\mathcal{C}$. For $i \in \Lambda(F)$, a morphism $\delta : F(i) \longrightarrow F(i, i)$ such that $\rho_1 \circ \delta = \rho_2 \circ \delta = 1_{F(i)}$ is called a *diagonal at* i (for F). Now let $(i, j, k) \in \Lambda(F)^3$, and suppose

$$(P; \pi_i, \pi_k) \text{ is a fibered product for } (F(i,j), \rho_j) \times_{F(j)} (F(j,k), \rho_j).$$

A morphism $\omega : P \longrightarrow F(i, k)$ is called a *transition morphism* (for (i, j, k) and F) if

(2.7.a) $(P; \omega, \pi_i)$ is a fibered product for $(F(i,k), \rho_i) \times_{F(i)} (F(i,j), \rho_i)$,

(2.7.b) $(P; \omega, \pi_k)$ is a fibered product for $(F(i,k), \rho_k) \times_{F(k)} (F(j,k), \rho_k)$.

Definition 1. *An intersection graph F is a* (formal) *canopy with respect to a topological category $(\mathcal{C}, Sub, Cov)$ if the following conditions hold:*

(2.8.a) *For $i, j \in \Lambda(F)$, $\rho_i : F(i,j) \longrightarrow F(i)$ and $\rho_j : F(i,j) \longrightarrow F(j)$ are formal subsets.*

(2.8.b) *For $i, j \in \Lambda(F)$, $X \in \mathcal{C}$ and $f, g \in Mor_{\mathcal{C}}(X, F(i,j))$, if $\rho_i \circ f = \rho_i \circ g$ and $\rho_j \circ f = \rho_j \circ g$, then $f = g$.*

(2.8.c) *For $i \in \Lambda(F)$, there is a diagonal at i for F.*

(2.8.d) *For $i, j \in \Lambda(F)$, there is a $\mathcal{C}$-morphism $\phi : F(i,j) \longrightarrow F(j,i)$ such that $\rho_i \circ \phi = \rho_i$ and $\rho_j \circ \phi = \rho_j$.*

(2.8.e) *There exists a transition morphism for each $(i,j,k) \in \Lambda(F)^3$.*

The five conditions of Definition 1 imply stronger constraints. Condition (2.8.b) assures there is a unique diagonal at each $i \in \Lambda(F)$. The morphism ϕ of (2.8.d) is both uniquely defined and an isomorphism. Condition (2.8.a) assures existence of the fibered product upon which the transition morphism of (2.8.e) is to be defined; again, the transition morphism of a triple (i, j, k) is unique. For $i \in \Lambda(F)$, elementary theory implies that each of the morphisms ρ_1 and ρ_2 on $F(i,i)$ is a covering morphism.

The canopy of an indexed family of formal subsets is a formal canopy in the above sense. Intuitively, a formal canopy F is a 'new' object for $\mathcal{C}$ with an assigned cover $\{\iota_j : F(j) \longrightarrow F\}_{j \in \Lambda(F)}$. Each triple $(F(j,k), \rho_j, \rho_k)$ should be the fibered product $F(j) \times_F F(k)$.

Definition 2. *A global structure is a local structure in which each formal canopy is the canopy of a Grothendieck cover for some object (in the sense of (2.4)).*

Let $\mathcal{C}$ be a category. Define a globalization of $\mathcal{C}$ to be a category $\mathcal{D}$ with a covariant functor $\Phi : \mathcal{C} \longrightarrow \mathcal{D}$ such that

(2.9.a) *$\mathcal{D}$ is a global structure,*

(2.9.b) *Φ is continuous (in some technical sense), and*

(2.9.c) *$\Phi : \mathcal{C} \longrightarrow \mathcal{D}$ is universal (i.e., up to functorial equivalence) among functors which satisfy (2.9.a,b).*

3 Methods and Results

Formal canopies allow us to phrase descent universally. More than that, they suggest how a globalization could be built without a concrete model.

Let $\mathcal{C}$ be a local structure, and let $\mathcal{C}^+$ be the class of all formal canopies of $\mathcal{C}$-objects. There is a canonical embedding $\mathrm{Obj}(\mathcal{C}) \longrightarrow \mathcal{C}^+$ given by sending an object A to the canopy of the cover $\{1_A\}$. Intuitively, each graph $\Gamma \in \mathcal{C}$ represents a global object paired with a particular Grothendieck cover. We can formalize this interpretation by turning $\mathcal{C}^+$ into a local structure! This requires a definition for a morphism between two formal canopies, and also the selection of a 'canonical' Grothendieck topology. The details are not trivial, and technicalities shape the axioms required of a local structure. But, it is possible to make $\mathcal{C}^+$ a local structure without reference to the kind of object in $\mathcal{C}$. The result is an intrinsic construction which applies regardless to manifolds, schemes, orbifolds or algebraic spaces.

Theorem 3. *Let $\mathcal{C}$ be a local structure. Let $\mathcal{C}^+$ be the class of formal canopies of $\mathcal{C}$-objects, and let ι be the canonical assignment $\mathrm{Obj}(\mathcal{C}) \longrightarrow \mathcal{C}^+$. Then there is a canonical definition of morphism between $\mathcal{C}^+$ objects, a canonical Grothendieck topology, and a canonical lifting of ι to morphisms under which $\mathcal{C}^+$ becomes a local structure and the functor $\iota : \mathcal{C} \longrightarrow \mathcal{C}^+$ is characterized by a universal factoring property.*

Again, let $\mathcal{C}$ be an arbitrary local structure. Although $\mathcal{C}^+$ might not be a global structure, the second iterate of the process

$$\mathcal{C} \longrightarrow \mathcal{C}^+ \longrightarrow (\mathcal{C}^+)^+$$

must be a globalization.

Theorem 3 does not end the project. There are obvious questions:

(3.1.a) We began with many explicit local/global theories $\Phi : \mathcal{C} \longrightarrow \mathcal{C}^{gl}$. Are the classical examples really globalizations in the sense of Definition 2? If so, is this hard to check?

(3.1.b) What does the universal property mean? That is, do we get anything from the mere statement that $\mathcal{C}$ sits inside a category closed under descent? Structural lemmas (1.1.a-d) do not appear to have anything to do with descent.

(3.1.c) Sheaf cohomology applies to objects modeled by sheaves of abelian groups. Do cohomological tools lift to the abstract globalization?

(3.1.d) In short, can an abstract model be manipulated as facilely as a concrete one?

The author offers several points.

Consistency with Existing Theories Let $\Phi : \mathcal{C} \longrightarrow \mathcal{C}^{gl}$ be a classical local/global theory. Let $\Phi^* : \mathcal{C} \longrightarrow \mathcal{C}^*$ be an abstract globalization. In each classical case, it is known that C^{gl} is closed under descent. Thus, there is a continuous functor $\gamma : C^* \longrightarrow C^{gl}$ which factors Φ through Φ^*. Now Φ is fully faithful, and (in some technical sense) the topology of C is the pullback of the topology of C^{gl}. (Although not presented in this language, the properties involved are established in every classical theory.) We have a universal theorem which states that the factoring of such a functor exhibits the same properties. That is, γ is automatically fully faithful and pulls back topology. Since every member of C^{gl} is covered by C-objects, it is immediate that γ is both a functorial equivalence and an equivalence of topologies.

Universality The last paragraph stresses that abstract globalization has more than one universal factoring property. Some deal with non-continuous functors. A functor from a topological category C to an arbitrary category D is called a *functor of sections* or a (topos-theoretic) *sheaf* if it maps the cone of any C-cover's canopy to an absolute colimit in D. In the canonical example, let D be the category of all algebras of a fixed type, let C be the category of sheaves of such algebras (on topological bases), and then the global section functor $C \longrightarrow D$ has this property. In the new format, it is proven that a functor of sections on a local structure C has a unique (up to functorial equivalence) extension to C^*. For example, the global section functor on schemes can be interpreted as the unique extension of the identity functor on the opposite category of rings, rather than as a consequence of the spectral model. (Here, the domain category is treated as a local structure while the codomain is not assigned a topology.)

For cohomological methods, an explicit sheaf model of global spaces provides

(3.2.a) many subcategories which are abelian, and

(3.2.b) a global section functor.

In the context of a Grothendieck topology, a functor (3.2.b) is sufficient to define Čech cohomology. We stated above that the new format accounts for functors like (3.2.b). Below, we comment on (3.2.a).

Ease of Manipulation Consider an alternative definition for the category of coherent bundles over schemes. Let **Ring** be the category of commutative rings, and let **Ring**^{o} be its opposite, with the standard (i.e., not étale) topology. Let **Mod** be the class of pairs (R, M) where $R \in$ **Ring** and M is an R-module. To make **Mod** a category, define a morphism between members $(R, M) \longrightarrow (R', M')$ to be a pair $(\rho : R \longrightarrow R', \mu : M \longrightarrow M')$ where ρ is a ring homomorphism, μ is an additive homomorphism, and

$$\mu(b \cdot x) = \rho(b) \cdot \mu(x) \text{ for all } b \in R \text{ and } x \in M.$$

Let **Mod**^{o} be the opposite of **Mod**. Let $\ell :$ **Mod**$^o \longrightarrow$ **Ring**^{o} be the obvious functor $(R, M) \longmapsto R$; it is 'passing to the base space'. Define a formal subset of $\mathbf{Mod}^o$ to be any pair $(\rho, \mu) : (R, M) \longrightarrow (R', M')$ where ρ is an open embedding of $\mathbf{Ring}^o$ and μ is a tensor product $M \longrightarrow R' \otimes_R M$. Clearly, the $\mathbf{Ring}^o$ topology on base spaces induces a topology for $\mathbf{Mod}^o$. The category of schemes $\mathbf{Sch}$ is a globalization of $\mathbf{Ring}^o$, the category $\mathbf{Coh}$ of coherent bundles of modules is a globalization of $\mathbf{Mod}^o$, and the universal property assures that ℓ has a canonical extension $\ell^* : \mathbf{Coh} \longrightarrow \mathbf{Sch}$.

The question is whether the above characterization of $\mathbf{Coh}$ is as useful as the traditional, concrete, model of coherent bundles as sheaves over a topological base. Many constructions for $\mathbf{Coh}$ are motivated by the intuition of a base space; but, the issue is how intuition can be justified formally. In the new framework, rigorous demonstration or definition for a functorial process often becomes a trivial application of an established universal property. As already mentioned, there are several functorial mapping properties. For example, the functor $(R, M) \longrightarrow M$ has a canonical extension to a functor of sections from $\mathbf{Mod}^o$ to the category of abelian groups. This observation recaptures the global section functor. Similarly, the functor which sends a ring R to (R, Ω_R), where Ω_R is the differential module of R, extends canonically to the cotangent bundle functor $\mathbf{Sch} \longrightarrow \mathbf{Coh}$.

Functorial properties do not describe internal structure of $\mathbf{Mod}^o$. However, creation of the abstract globalization involves structural propositions, including (1.1.a-d). Behavior which, traditionally, is based on an explicit topological model, can be asserted in complete generality.

For $R \in \mathbf{Ring}^o$, the inverse image of R under ℓ is an abelian category. Moreover, the additive structure is preserved by pullback along $\mathbf{Ring}^o$-morphisms. Although the topic is not pursued in [3], proof that this relationship lifts to ℓ^* is known to the author. The advantage of the present language is that the analogous claim for coherent bundles in analytic and rigid analytic cases is established at the same stroke.

References

1. Artin, M. (1970). Algebraization of formal moduli: I, *Global Analysis, Papers in Honor of K. Kodaira.* University of Tokyo Press and Princeton University Press

2. Artin, M. (1970). Algebraization of formal moduli: II - Existence of modifications. *Ann. Math.*, **91**, 88-135

3. Feit, P. Axiomization of Passage from 'Local' Structure to 'Global' Object. To appear in *Memoirs of A.M.S.*

4. Grothendieck A. (1971). Revêtements étales et groupe fondamentale. *Lect. Notes in Math 224.*, Springer-Verlag: New York

5. Knutson D. (1971) Algebraic Spaces. *Lect. Notes in Math. 203.*, Springer-Verlag: New York

6. Milne, J.S. (1977) *Étale Cohomology.* Princeton University Press: New York

7. Satake, I. (1956). On a generalization of the notion of a manifold. *Proc. Nat. Acad. of Science USA,* **42**, 359-363

8. Thurston, W. *The Geometry and Topology of Three Manifolds.* under active revision

Paul Feit
Department of Mathematics
University of Toledo
Toledo, OH 43606
USA
e-mail fac3551@uoft01.bitnet

Shimura Reciprocity for Modular Functions
with Rational Fourier Coefficients

Alan J. Laing

University of Maryland, College Park

. Introduction

We give an exposition of the computational aspects of Shimura's explicit reciprocity law. First we discuss the classical formulation as in Shimura [4], Thm 6.31, and Lang [2], 11 §1 Thm 1. Then we consider Robert's formulation in the case that we have rational Fourier coefficients. Finally, we interpret Robert's formulation in terms of matrices and give an example.

. Shimura's Explicit Reciprocity Law

Class field theory provides a reciprocity law, Artin Reciprocity, describing the Galois group $\mathrm{Gal}(k^{\mathrm{ab}}/k)$ for any number field k. This description is given by the Artin Symbol $(\cdot, k)$ on the idèle group $\mathbb{J}_k$. We have the following exact sequence.

$$0 \longrightarrow D_k \longrightarrow C_k \xrightarrow{(\cdot,k)} \mathrm{Gal}(k^{\mathrm{ab}}/k) \longrightarrow 0,$$

where D_k is the connected component of $C_k = \mathbb{J}_k/k^{\times}$. It is easily shown that

$$D_k = \bigcap_{L/k,\text{abel.}} N_{L/k} C_L.$$

Unfortunately, Artin reciprocity is not explicit except in the simplest cases.

Shimura's explicit reciprocity law gives an explicit calculation of the Artin map for imaginary quadratic extensions k of $\mathbb{Q}$ in terms of rational matrices and permutations of various singular values. This effectively describes the Galois group of certain generalized dihedral extensions (abelian extensions of imaginary quadratic extensions k) of $\mathbb{Q}$ by relating the idèles of k to the adèlized $\mathrm{Gl}_2(\mathbb{Q})$. All that is necessary is to determine the action of complex conjugation on $\mathbb{J}_k$.

Using the theory of complex multiplication and elliptic curves, Shimura realizes abelian extensions of k as specializations of the various subfields of the field of modular functions over the rationals. From this he is able to describe the Galois group $\mathrm{Gal}(k^{\mathrm{ab}}/k)$ in terms of the automorphisms of the field of modular functions.

The field $\mathcal{F}$ of modular functions over $\mathbb{Q}$ is the compositum over all N of the fields $\mathcal{F}_N$ of modular functions of level N over $\mathbb{Q}$. A complex function is said to be modular of level N if it is a meromorphic function on $\mathfrak{H}$, the upper half plane, which induces a meromorphic function on the modular curve of level N

$$X(N) = \overline{\Gamma(N)\backslash\mathfrak{H}},$$

where $\Gamma(N)$ is the congruence subgroup of $\mathrm{Sl}_2(\mathbb{Z})$ defined as the kernel of the reduction mod N map.

The field of modular functions of level N over $\mathbb{C}$ is generated over $\mathbb{C}$ by the j-invariant

$$j(\tau) = j(\mathbb{C}/(1,\tau))$$

and the Fricke functions

$$f_{a,b}(\tau) = -2^7 3^5 \frac{g_2(\tau)g_3(\tau)}{\Delta(\tau)} \wp(\frac{a\tau + b}{N}; \tau)$$

for $a, b \pmod{N}$ not both 0. We define $\mathcal{F}_N$ to be the extension of $\mathbb{Q}$ by these same functions. Every function in $\mathcal{F}_N$, being meromorphic on $X(N)$, has a Laurent expansion at each of the cusps of $X(N)$ (translates of $\infty \in \overline{\mathrm{Sl}_2(\mathbb{Z})\backslash\mathfrak{H}}$ by $\mathrm{Sl}_2(\mathbb{Z})/\Gamma(N)$). The local parameter at each cusp is a conjugate of the parameter at infinity, $q_N = e^{2\pi i z/N}$. The "Fourier coefficients" of a modular function are the coefficients of the expansion at infinity in the local parameter. For the Fricke functions of level N it can be shown that their Fourier coefficients are elements of $\mathbb{Q}(\zeta_N)$. Moreover, $\mathcal{F}_N \cap \overline{\mathbb{Q}} = \mathbb{Q}(\zeta_N)$. We will be concerned with those modular functions whose Fourier coefficients are actually in $\mathbb{Q}$.

The automorphisms of the field of modular functions of level N over $\mathbb{Q}$ include a copy of $\mathrm{Gl}_2(\mathbb{Z}/N\mathbb{Z})$, whose action is to permute the Fricke functions according to their indices. In the limit, this gives an action of

$$U = \varprojlim \mathrm{Gl}_2(\mathbb{Z}/N\mathbb{Z}) = \prod_p \mathrm{Gl}_2(\mathbb{Z}_p)$$

on $\mathcal{F}$.

The set of rational matrices with positive determinant, denoted $\mathrm{Gl}_2(\mathbb{Q})^+$ also acts on $\mathcal{F}$ by composition, though in general does not preserve level. Using the Shimura exact sequence

$$0 \longrightarrow \mathbb{Q}^\times \longrightarrow \mathrm{Gl}_2(\mathbb{A}_f) \overset{\sigma}{\longrightarrow} \mathrm{Aut}(\mathcal{F}) \longrightarrow 0$$

where $\mathbb{Q}^\times$ embeds via diagonal matrices into $\mathrm{Gl}_2(\mathbb{A}_f)$ (cf. Lang [2] 7§3 Thm. 6) and the decomposition of the adèlized $\mathrm{Gl}_2(\mathbb{Q})$,

(1) $$\mathrm{Gl}_2(\mathbb{A}_f) = U \cdot \mathrm{Gl}_2(\mathbb{Q})^+$$

(cf. Lang [2] 7§2 Thm. 1), we have determined $\mathrm{Aut}(\mathcal{F})$.

The Shimura reciprocity law provides the computational tool necessary to determine the Galois action induced when we specialize to a point $z \in \mathfrak{H} \cap k$, where k is imaginary quadratic. In order to do this,

we need an embedding of the idèles $\mathbb{J}_k$ into $\mathrm{Gl}_2(\mathbb{A}_f)$. This embedding is given locally by the matrix $q_{z,p} \in \mathrm{Gl}_2(\mathbb{Q}_p)$ which gives multiplication in $k_p = k^\times \otimes \mathbb{Q}_p^\times$ under the $\mathbb{Q}_p$ basis $(1, z)$ of k_p. Specifically, the identity

$$q_{z,p}(\xi) \begin{bmatrix} z \\ 1 \end{bmatrix} = \begin{bmatrix} \xi z \\ \xi \end{bmatrix},$$

for $\xi = \alpha + \beta z \in k_p^\times$ defines $q_{z,p}$ as

$$q_{z,p}(\xi) = \begin{pmatrix} \alpha + \beta \, \mathrm{Tr}_{k/\mathbb{Q}}(z) & -\beta \, \mathrm{N}_{k/\mathbb{Q}}(z) \\ \beta & \alpha \end{pmatrix}.$$

Thus for $t = (\dots, t_p, \dots) \in \mathbb{J}_k$ we define $q_z(t) = (\dots, q_{z,p}, \dots) \in \mathrm{Gl}_2(\mathbb{A}_f)$, ignoring the infinite components. The decomposition (1) allows us to write

$$q_z(t) = u \cdot A,$$

with

$$A \in \mathrm{Gl}_2(\mathbb{Q})^+,$$

and

$$u = (\dots, u_p, \dots) \in \prod_p \mathrm{Gl}_2(\mathbb{Z}_p).$$

Note that this decomposition is unique up to multiplication of A by an element $B \in \mathrm{Sl}_2(\mathbb{Z})$. The corresponding u is then uB^{-1}, where $\mathrm{Sl}_2(\mathbb{Z})$ embeds into the diagonal in $\prod_p \mathrm{Gl}_2(\mathbb{Z}_p)$.

We are now ready to state Shimura's Reciprocity Law:

Theorem 1. (Shimura) *Let k be an imaginary quadratic field and $z \in k \cap \mathfrak{H}$. Then for every modular function f defined at z and every idèle s of k, we have*

$$f(z)^{(s,k)} = f^\sigma(z),$$

where $\sigma = \sigma(q_z(s^{-1})) = \sigma(u) \cdot \sigma(A)$.

Here $\sigma(u)$ is the automorphism induced by permuting the Fricke functions and $\sigma(A)$ is the composition $f \longmapsto f \circ A$.

§3. ROBERT'S FORMULATION

For modular functions with rational Fourier coefficients, Robert achieves an equivalent statement of Shimura's explicit reciprocity law which is very similar to the results from complex multiplication for the $\mathrm{Gal}(k^{\mathrm{ab}}/k)$-action on the j-invariant when viewed as a function on lattices in $\mathbb{C}$:

$$j(\langle 1, z \rangle)^{(s,k)} = j(s^{-1}\langle 1, z \rangle).$$

Here "multiplication" of a lattice L by an idèle t is defined as the intersection over all p of the local lattices $L \otimes \mathbb{Z}_p$ multiplied by the corresponding local component t_p of t. In our case,

$$s^{-1}\langle 1, z \rangle = \bigcap_p \left(k \cap \langle s_p^{-1}, s_p^{-1} z \rangle \right),$$

where $\langle s_p^{-1}, s_p^{-1} z \rangle$ is a lattice in $k \otimes \mathbb{Z}_p \cong \bigoplus_{\mathfrak{p}|p} k_{\mathfrak{p}}$. It becomes clear in this case that $j(s^{-1}\langle 1, z \rangle)$ depends only on the $\mathrm{Gl}_2(\mathbb{Q})^+$-component of q_z, the $\mathrm{Gl}_2(\mathbb{Z}_p)$ terms serving only to change bases of the local lattices

$$\langle s_p^{-1}, s_p^{-1} z \rangle \otimes \mathbb{Z}_p = u_p \langle \omega_1, \omega_2 \rangle \otimes \mathbb{Z}_p.$$

This formulation cannot hold in general though, since the values of modular functions of higher level are not fixed under the different decompositions of q_z.

However, Robert proves that this type of formulation holds more generally for any modular function with rational Fourier coefficients if we consider more data.

Let f be a modular function of level N. Then f induces a function $\tilde{f}$ on triples $(L, L_N, \dot{\omega})$, where L and L_N are complex lattices $L \subset L_N$ with cyclic quotient of order N, and $\dot{\omega}$ is a coset of L/NL whose class modulo NL_N generates the cyclic group L/NL_N of order N. More precisely,

$$\tilde{f}(L, L_N, \dot{\omega})) = f(\tau),$$

with $\tau \in \mathfrak{H}$ given by

$$\tau = \frac{\omega_2}{\omega_1},$$

where

$$L = \langle \omega_1, \omega_2 \rangle, \qquad L_N = \langle \frac{\omega_1}{N}, \omega_2 \rangle, \qquad \dot{\omega} = \omega_2 + NL.$$

The fact that such a basis of L exists is verified by Robert [3]($\S$1 Lemma 1).

Conversely, for any $\tau \in \mathfrak{H}$, we set

$$L = \langle 1, \tau \rangle, \qquad L_N = \langle \frac{1}{N}, \tau \rangle, \qquad \dot{\omega} = \tau + N\langle 1, \tau \rangle,$$

so that $\tilde{f}(L, L_N, \dot{\omega}) = f(\tau)$.

Robert expresses Shimura's explicit reciprocity law in the case that f has rational Fourier coefficients by the following:

Theorem 2. (Robert) *Let f be a modular function of level N with rational Fourier coefficients and let $\tilde{f}$ be the induced function on triples. Let k be an imaginary quadratic field, let $\tau \in \mathfrak{H} \cap k$ and define*

$$L = \langle 1, \tau \rangle, \qquad L_N = \langle \frac{1}{N}, \tau \rangle, \qquad \dot{\omega} = \tau + N \langle 1, \tau \rangle.$$

Then

$$f(\tau)^{(s,k)} = \tilde{f}(L, L_N, \dot{\omega})^{(s,k)}$$
$$= \tilde{f}(s^{-1}L, s^{-1}L_N, s^{-1}\dot{\omega}).$$

The last equality depends on the fact that f has rational Fourier coefficients. In fact, see [3],§2 Theorem 1 for a converse.

§4. INTERPRETATION BY MATRICES

Robert's theorem becomes a very useful computational aid when we interpret his results in terms of rational matrices. It is especially convenient for functions invariant under the subgroups $\Gamma_0(N)$, $\Gamma_1(N)$, or $\Gamma^1(N)$ of $\Gamma(N)$.

Theorem 3. *Given a modular function f of level N having rational Fourier coefficients and defined at $z \in \mathfrak{H} \cap k$, and an idèle $s = (\dots, \alpha_p + \beta_p z, \dots)$, there exists a rational matrix $A \in Gl_2(\mathbb{Q})^+$ so that*

$$f(z)^{(s,k)} = f(Az).$$

$A = \begin{pmatrix} a & b \\ c & d \end{pmatrix}$ *is determined up to multiplication by elements of $Sl_2(\mathbb{Z})$ by the condition*

(2) $$u = qA^{-1} \in U (= \prod_p Gl_2(\mathbb{Z}_p)),$$

where $q = q_z(s^{-1})$ is the "multiplication by s^{-1}" matrix defined in §2.
   A is determined up to multiplication by elements in $\Gamma_0(N)$ by the congruence condition

$$c\bar{\alpha} \equiv d\bar{\beta} \quad (\mathrm{mod}\ N \cdot \prod_{p|N} p^{\mathrm{ord}_p(\det(\bar{q}))}),$$

where $\bar{q} = \begin{pmatrix} \bar{\alpha} + \bar{\beta} \cdot \mathrm{Tr}(z) & -\bar{\beta} N(z) \\ \bar{\beta} & \bar{\alpha} \end{pmatrix}$ is the image of q mod N. Alternatively, this can be stated as

(3) $$u = qA^{-1} \in U \cap \begin{pmatrix} 1 & 0 \\ 0 & N \end{pmatrix} U \begin{pmatrix} 1 & 0 \\ 0 & \frac{1}{N} \end{pmatrix}$$

If f is $\Gamma_0(N)$-invariant, this is sufficient to determine $f(z)^{(s,k)}$. If not we need to specify the coset of A mod $\Gamma_1(N)$ (or $\Gamma^1(N)$) by the condition that

$$\bar{u} \equiv \begin{pmatrix} \bar{x} & \bar{y} \\ \cdot & \cdot \end{pmatrix} \quad (\text{mod } N)$$

with $\bar{x} \equiv 1 \ (\text{mod } N)$, $(\bar{x} \equiv 1 \text{ and } \bar{y} \equiv 0 \ (\text{mod } N))$, respectively.

Proof. Let

$$q = q_z(s^{-1}) = \left(\dots, \begin{pmatrix} \alpha_p + \beta_p \operatorname{Tr}_{k/\mathbb{Q}}(z) & -\beta_p \operatorname{N}_{k/\mathbb{Q}}(z) \\ \beta_p & \alpha_p \end{pmatrix}, \dots \right)$$

and let $q = u \cdot A$ be a decomposition of q as in (1). Then

$$L' = s^{-1}L = \bigcap (k \cap u_p A \cdot \langle 1, z \rangle \otimes \mathbb{Z}_p)$$
$$= \bigcap_p (k \cap A \langle 1, z \rangle \otimes \mathbb{Z}_p)$$
$$= A \langle 1, z \rangle.$$

According to Robert's formulation, $f^{(s,k)} = f(\frac{\omega_2}{\omega_1})$, for some basis $\{\omega_1, \omega_2\}$ of L'. Since any two bases of L' differ by an element of $\mathrm{Sl}_2(\mathbb{Z})$, A is determined up to $\mathrm{Sl}_2(\mathbb{Z})$.

Note that if f were modular of level 1, i.e. a rational function of j, then this would be enough to specify $f(z)^{(s,k)}$.

To specify A up to multiplication by $\Gamma_0(N)$, we need to consider the second term of the triple, $L'_N = s^{-1} L_N$. Similar to the way we defined the matrix q we can define a "multiplication by s^{-1}" matrix q^N locally on each k_p using the basis $\{\frac{1}{N}, z\}$. q^N is related to q by conjugation by $\begin{pmatrix} 1 & 0 \\ 0 & N \end{pmatrix}$ in each component (hence the notation: $x^N = $ conjugation by $\begin{pmatrix} 1 & 0 \\ 0 & N \end{pmatrix}$).

Again we have a decomposition $q^N = u'A'$ of q^N into a rational matrix and a product of local change of bases. Since conjugation distributes, we can write

$$u'A' = q^N = \begin{pmatrix} 1 & 0 \\ 0 & \frac{1}{N} \end{pmatrix} uA \begin{pmatrix} 1 & 0 \\ 0 & N \end{pmatrix}$$
$$= \begin{pmatrix} 1 & 0 \\ 0 & \frac{1}{N} \end{pmatrix} uB^{-1} \begin{pmatrix} 1 & 0 \\ 0 & N \end{pmatrix} \begin{pmatrix} 1 & 0 \\ 0 & \frac{1}{N} \end{pmatrix} BA \begin{pmatrix} 1 & 0 \\ 0 & N \end{pmatrix},$$

for some $B \in \mathrm{Sl}_2(\mathbb{Z})$. We want the decomposition of $q^N = u'A'$ so that

$$u' = (\dots, (u_p B^{-1})^N, \dots) \in \prod \mathrm{Gl}_2(\mathbb{Z}_p).$$

This condition determines B mod N up to multiplication by an element of $\Gamma_0(N)$. By changing the decomposition of $q = uA$ to $q = (uB^{-1})(BA)$, we have obtained a basis $BA \begin{bmatrix} z \\ 1 \end{bmatrix} = \begin{bmatrix} \omega_2 \\ \omega_1 \end{bmatrix}$ of $s^{-1}L$ for which $\begin{bmatrix} \omega_2 \\ \frac{\omega_1}{N} \end{bmatrix}$ is a basis of $s^{-1}L_N$.

To determine the correct A up to multiplication by an element of $\Gamma_1(N)$ or $\Gamma^1(N)$ we need to consider the third element of the triple. $\dot{\omega}' = s^{-1}\dot{\omega}$ is defined to be a coset of L'/NL' whose coset mod NL'_N generates the cyclic group L'/NL'_N of order N. In our case,

$$\dot{\omega}' = s^{-1}z + NA\langle 1, z \rangle.$$

Embedding this into the adèles, $\mathbb{A}_k$, we have

$$(\ldots, s_p^{-1}z + N\omega_2\mathbb{Z}_p \oplus N\omega_1\mathbb{Z}_p, \ldots).$$

Since $uA \begin{bmatrix} z \\ 1 \end{bmatrix} = \begin{bmatrix} s^{-1}z \\ s^{-1} \end{bmatrix}$ and $A \begin{bmatrix} z \\ 1 \end{bmatrix} = \begin{bmatrix} \omega_2 \\ \omega_1 \end{bmatrix}$, we have $u \begin{bmatrix} \omega_2 \\ \omega_1 \end{bmatrix} = \begin{bmatrix} s^{-1}z \\ s^{-1} \end{bmatrix}$. Writing

$$u = \left(\ldots, \begin{pmatrix} x_p & y_p \\ \cdot & \cdot \end{pmatrix}, \ldots \right) \in U$$

we have

$$\dot{\omega}' = (\ldots, x_p\omega_1 + y_p\omega_2 + N\omega_2\mathbb{Z}_p \oplus N\omega_1\mathbb{Z}_p, \ldots).$$

Now, for p prime to N, $N\mathbb{Z}_p = \mathbb{Z}_p$, so we can ignore the terms away from N. If we approximate $\bar{x} \equiv x_p$ and $\bar{y} \equiv y_p$ (mod $p^{\mathrm{ord}_p(N)}$) for all $p|N$, we obtain

$$\dot{\omega}' = (\ldots, \bar{x}\omega_2 + \bar{y}\omega_1 + N\omega_2\mathbb{Z}_p \oplus N\omega_1\mathbb{Z}_p, \ldots).$$

Intersecting to obtain a coset of the k-lattice, NL', we get

$$\dot{\omega}' = \bar{x}\omega_2 + \bar{y}\omega_1 + NL'.$$

Now, the pullback of the triple $(L', L'_N, \dot{\omega}')$ to ordered bases of $\mathbb{C}$-lattices is such that $\dot{\omega} = \omega_2 + NL' \in L'/NL'$. This requires that $\bar{x} \equiv 1 \pmod{N}$ and $\bar{y} \equiv 0 \pmod{N}$. If only the first of these conditions holds, then we still have $[\dot{\omega}] = \omega_2 + NL'_N \in L'/NL'_N$. According to the Corollary of [3], §1 Lemma 1, this is enough to determine A up to multiplication by elements of $\Gamma_1(N)$.

§5. AN EXAMPLE

For an example of the usefulness of the formulation, consider the following Weber function

$$\mathfrak{f}(z) = \zeta_{48}^{-1} \frac{\eta\left(\frac{z+1}{2}\right)}{\eta(z)}.$$

This is a modular function of level 48 and has rational Fourier coefficients, as is clear from the product expansion,

$$\mathfrak{f}(z) = q_{48} \prod_{m \text{ odd}} (1 + q_{48}^{24m})$$

(here $q_{48} = e^{2\pi i z/48}$ is the parameter at infinity for the modular group $X(48)$). The transformation law for $\mathfrak{f}$ can be determined from the transformation law for η

$$\eta(z + 1) = \zeta_{24}\eta(z);$$
$$\eta(\frac{-1}{z}) = \sqrt{-iz}\,\eta,$$

and the fact that $\mathfrak{f}^{24}(z) = \frac{\Delta(\frac{z+1}{2})}{\Delta(z)}$ is $\Gamma(2) \cup \begin{pmatrix} 0 & -1 \\ 1 & 0 \end{pmatrix} \Gamma(2)$- invariant. In particular, $\mathfrak{f}$ is invariant under the inversion $z \longmapsto \frac{-1}{z}$, and $\mathfrak{f}(z + 2) = \zeta_{24}^{-1}\mathfrak{f}(z)$.

Our theorem then specifies that for $s \in \mathbb{J}_k$ and $z \in \mathfrak{H} \cap k$, where $\mathfrak{f}(z) < \infty$, we have

$$\mathfrak{f}(z)^{(s,k)} = \mathfrak{f}\left(\frac{az + b}{cz + d}\right),$$

where $A = \begin{pmatrix} a & b \\ c & d \end{pmatrix}$ satisfies

$$u = q_z(s^{-1})A^{-1} \in U \cap \begin{pmatrix} 1 & 0 \\ 0 & \frac{1}{N} \end{pmatrix} U \begin{pmatrix} 1 & 0 \\ 0 & N \end{pmatrix},$$

and in fact, $\bar{u} \equiv u \pmod{48}$ satisfies

(4) $$\bar{u} \equiv \begin{pmatrix} 1 & 0 \\ 0 & \cdot \end{pmatrix} \pmod{48}.$$

Alternatively, if A' is chosen (easily enough) so that

$$q = q_z(s^{-1})A'^{-1} \in U,$$

we can choose $A = BA'$, with $B \in \mathrm{Sl}_2(\mathbb{Z})$ so that $A = BA'$ and $u = qA^{-1}$ satisfies (4).

To illustrate this, consider $k = \mathbb{Q}(\sqrt{-31})$. It can be shown by other methods (cf. [1] for example) that the extension of k by $\frac{\mathfrak{f}(\sqrt{-31})}{\sqrt{2}}$ is the maximal unramified abelian extension of k, i.e. the Hilbert class field of k. By considering the classes of binary quadratic forms of discriminant $-4 \cdot 31$, we see that the degree of this extension is three. In fact, we have the reduced forms $\{1, 0, 31\}, \{5, \pm 4, 7\}$ for representatives of the classes of binary quadratic forms of discriminant $-4 \cdot 31$. This gives the principal root $z = \sqrt{-31} \in \mathfrak{H} \cap k$ and a generator, $\langle 5, -2 + \sqrt{-31} \rangle$ of the ideal class group. Associated to this generator is the idèle $s = (1, 1, -2 + z, 1, \dots)$. We compute

$$s^{-1} = (1, 1, \frac{2 + \sqrt{-31}}{35}, 1, \dots)$$

and

$$q = q_z(s^{-1}) = \left(\mathbb{1}_2, \mathbb{1}_2, \begin{pmatrix} \frac{2}{35} & \frac{-31}{35} \\ \frac{1}{35} & \frac{2}{35} \end{pmatrix}, \mathbb{1}_2 \dots \right).$$

Now, a first approximation for a decomposition of q is $q = u_1 A_1$, where

$$A_1 = \begin{pmatrix} \frac{1}{5} & \frac{2}{5} \\ 0 & 1 \end{pmatrix},$$

and

$$u_1 = q A_1^{-1} = \left(\begin{pmatrix} 5 & -2 \\ 0 & 1 \end{pmatrix}, \begin{pmatrix} 5 & -2 \\ 0 & 1 \end{pmatrix}, \begin{pmatrix} \frac{2}{7} & -1 \\ \frac{1}{7} & 0 \end{pmatrix}, \begin{pmatrix} 5 & -2 \\ 0 & 1 \end{pmatrix}, \dots \right).$$

This decomposition clearly satisfies (2) with $A = A_1$ and $u = u_1$. Therefore $A = B A_1$ for some $B \in \mathrm{Sl}_2(\mathbb{Z})$.

Writing $B^{-1} = \begin{pmatrix} v & w \\ x & y \end{pmatrix}$, the condition (4) for $\bar{u} \equiv u_1 B^{-1} \pmod{48}$ is satisfied if

$$x \equiv 0 \pmod{48},$$
$$5v \equiv 1 \pmod{48},$$

and

$$5w - 2v \equiv 0 \pmod{48}.$$

Notice that $(5, 48) = 1 = (-19)5 + 2 \cdot 48$ so that the matrix

$$B_1^{-1} = \begin{pmatrix} -19 & 2 \\ -48 & 5 \end{pmatrix}$$

satisfies the congruence condition on the first column of $\bar{u}$. In fact

$$u_2 = u_1 B_1^{-1}$$
$$= \left(\begin{pmatrix} 1 & 0 \\ -48 & 5 \end{pmatrix}, \begin{pmatrix} 1 & 0 \\ -48 & 5 \end{pmatrix}, \begin{pmatrix} \frac{298}{7} & \frac{-31}{7} \\ \frac{-19}{7} & \frac{2}{7} \end{pmatrix}, \begin{pmatrix} 1 & 0 \\ -48 & 5 \end{pmatrix}, \dots \right)$$

whose image mod 48 is $\bar{u} \equiv \begin{pmatrix} 1 & 0 \\ 0 & 5 \end{pmatrix}$.

Therefore,

$$A = B_1 A_1 = \begin{pmatrix} 5 & -2 \\ 48 & -19 \end{pmatrix} \begin{pmatrix} \frac{1}{5} & \frac{2}{5} \\ 0 & 1 \end{pmatrix} = \begin{pmatrix} 1 & 0 \\ \frac{48}{5} & \frac{1}{5} \end{pmatrix}$$

so that

$$\mathfrak{f}(\sqrt{-31})^{(s,k)} = \mathfrak{f}\left(\frac{5\sqrt{-31}}{48\sqrt{-31}+1}\right).$$

We can verify this numerically by evaluating the reduced class equation of discriminant $-4 \cdot 31$, given by $x^3 - x^2 - 1$, at the conjugated class invariant $\left(\frac{\mathfrak{f}(\sqrt{-31})}{\sqrt{2}}\right)^{(s,k)}$. See [1] for an alternate construction of the reduced class equations.

References

1. Erich Kaltofen and Noriko Yui, *Explicit Construction of the Hilbert Class Fields of Imaginary Quadratic Fields by Integer Lattice Reduction*, New York Number Theory Seminar: Lecture Notes in Mathematics (1990), Springer Verlag.
2. Serge Lang, *Elliptic Functions*, Addison–Wesley Publishing Company, Inc., 1973.
3. Gilles Robert, *Multiplication Complexe et Lois de Réciprocité: Le Cas des Fonctions Modulaires Elliptiques*, Max–Planck–Institut für Mathematik, Bonn (1990).
4. Goro Shimura, *Introduction to Arithmetic Theory of Automorphic Forms*, Iwanami Shoten, Publ. and Princeton University Press, 1971.

Alan J. Laing
Department of Mathematics
University of Maryland
College Park
Maryland 20742
USA

On Serre's conjecture (3.2.4?) and vertical Weil curves

Joan-C. Lario

Universitat Politècnica de Catalunya, Barcelona

1 Introduction

Let $G_{\mathbb{Q}} = \mathrm{Gal}\,(\overline{\mathbb{Q}}/\mathbb{Q})$ and let

$$\rho : G_{\mathbb{Q}} \to \mathrm{GL}_2(\overline{\mathbb{F}}_p)$$

be a continuous, irreducible and odd representation. In [1], by means of a recipe, Serre attaches to ρ three invariants:

$$
\begin{array}{lll}
\text{the conductor} & N_\rho \in \mathbb{Z} - p\mathbb{Z}\,, \\
\text{the weight} & k_\rho \in \mathbb{Z} \cap [2, p^2 - 1]\,, \\
\text{the character} & \varepsilon_\rho \; : (\mathbb{Z}/N_\rho)^* \to \overline{\mathbb{F}}_p^*\,.
\end{array}
$$

He states there the conjecture (3.2.4?): such a ρ comes from a Hecke cusp form $(\bmod\, p)$ of type $(N_\rho, k_\rho, \varepsilon_\rho)$.

The first purpose of this contribution is to obtain a general criterion in order to verify this conjecture. In a joint work with P.Bayer [2], we verify (3.2.4?) for the Galois representations defined by the p-torsion points of p-vertical Weil curves (i.e., modular elliptic curves having bad but potentially good ordinary reduction at p), for $p > 7$. Here, we obtain another proof for this case as an application of our general criterion.

2 A numerical experiment

Let us first consider the following example: Let E be the elliptic curve, defined over $\mathbb{Q}$, given by the Weierstraß model

$$y^2 + xy + y = x^3 + 7x - 15\,;$$

this elliptic curve can be found in [3]. Its discriminant is $\Delta_E = -359^2$, and its conductor is $N_E = 359^2$. Using Tate's algorithm [4], we find that

---

Partially supported by a Grant from D.G.I.C.Y.T., PB89-0215-C02-01 and UPC, PRE89

Table 1. Coefficients of $L(E, s)$

n	1	2	3	4	5	6	7	8	9	10	11	12	13	14	15
A_n	1	1	-2	-1	1	-2	-4	-3	1	1	-2	2	3	-4	-2

the special fibre of its Néron model over the local ring $\mathbb{Z}_{359}$ has Kodaira symbol II. On the other hand, since $359 \equiv 2 \,(\mathrm{mod}\,3)$, E has potentially good supersingular reduction at 359.

The representation of the absolute Galois group

$$\rho : G_{\mathbb{Q}} \longrightarrow \mathrm{Aut}\,(E_{359})\,,$$

defined by the action of $G_{\mathbb{Q}}$ on the 359-torsion points of E, is continuous, odd and absolutely irreducible (since $359 > 163$, cf. [1]). Serre's invariants attached to ρ (cf. [2]) are:

$$N_\rho = 1\,, \qquad k_\rho = 21840\,, \qquad \varepsilon_\rho = 1\,.$$

Thus, Serre's conjecture predicts the existence of a Hecke cusp form (mod 359)

$$f \in S_{21840}(1)\,,$$

with Fourier coefficients a_ℓ congruent $(\mathrm{mod}\,359)$, for any prime $\ell \neq 359$, to the coefficients of the Hasse-Weil L-function

$$L(E, s) = \sum_{n=1}^{\infty} \frac{A_n}{n^s}\,.$$

Note that the dimension of the space of cusp forms, where we must find f, is

$$\dim S_{21840}(1) = 1821\,.$$

This dimension is excessive for calculus on a medium size computer. Anyway, we are going to consider the twisted representation

$$\rho(60) := \rho \otimes \chi^{-60}\,,$$

where χ is the pth cyclotomic character of $G_{\mathbb{Q}}$. Following Serre's recipe, we obtain the invariants for $\rho(60)$:

$$N_{\rho(60)} = 1\,, \qquad k_{\rho(60)} = 240\,, \qquad \varepsilon_{\rho(60)} = 1\,.$$

Now $\dim S_{240}(1) = 20$ and a base of this space is given by

$$S_{240}(1) = < f_i >_{i=1,20}\,, \quad f_i = \Delta^i\, E_6^{40-2i}\,,$$

where

$$\Delta(q) \;=\; q \prod_{n=1}^{\infty} (1 - q^n)^{24} \in S_{12}(1)\,,$$

$$E_6(q) = 1 - 504 \sum_{n=1}^{\infty} \sigma_5(n)q^n \in M_6(1)$$

are the discriminant invariant and the Eisenstein series of weight 6, respectively.

We have implemented a program, written in **FORTRAN**, which obtains the eigenvectors for the Hecke operator T_2 acting on $S_k(1)$. For $k = 240$, in terms of the preceding base, we get the eigenform

$$G = (1, 200, 8, 89, 210, 347, 117, 284, 158, 24, 242, 6,$$
$$264, 238, 190, 233, 238, 320, 303, 281).$$

The first Fourier coefficients on the q-expansion of G are:

$$
\begin{aligned}
& q & - \\
& 18976\,q^2 & + \\
& 174762740\,q^3 & - \\
& 1040444733543\,q^4 & + \\
& 4499651242530976\,q^5 & - \\
& 15060060147344832709\,q^6 & + \\
& 40580143460545124102245\,q^7 & - \\
& 90415252923989878420327980\,q^8 & + \\
& 169773777843010223881644216947\,q^9 & - \\
& 272439902993476545159400919943020\,q^{10} & + \\
& 377555704481308893617379138488273600\,q^{11} & - \\
& 455414178223644406628199528768738304222\,q^{12} & + \\
& 480912466381700331716014103399776614240292\,q^{13} & - \\
& 446423056402027970479412561736757701070300244\,q^{14} & + \\
& 365253275569640340540625846681467784312619400533\,q^{15} & -
\end{aligned}
$$

$$\cdots$$

By reducing $G \pmod{359}$ we get the cusp form $g(q) = \sum b_n q^n$. The 60th power of Ramanujan-Katz operator yields

$$f(q) := \theta^{60} g(q) = \sum_{n=1}^{\infty} a_n q^n.$$

Table 2. Fourier coefficients of g

n	1	2	3	4	5	6	7	8	9	10	11	12	13	14	15
b_n	1	-308	104	-88	47	-81	140	-181	191	-116	279	177	184	-40	221

Table 3. Fourier coefficients of f

n	1	2	3	4	5	6	7	8	9	10	11	12	13	14	15
a_n	1	-358	357	-1	1	-2	355	-3	1	-358	357	-357	3	-4	357

Note the congruences $(\mathrm{mod}\,359)$ between this Fourier coefficients and those of $L(E, s)$. One can then check that the existence of the cusp form f verifies Serre's conjecture for ρ.

3   Minimal and companion representations

Let

$$\rho \,:\, G_{\mathbb{Q}} \to \mathrm{GL}_2(\overline{\mathbb{F}}_p)$$

be a continuous, irreducible and odd representation . Let N_ρ, k_ρ, ε_ρ be the invariants attached to ρ.

For each $i \in \mathbb{Z}$, we consider the representation

$$\rho(i) \,:\, G_{\mathbb{Q}} \to \mathrm{GL}_2(\overline{\mathbb{F}}_p)\,,$$

twisted of ρ by the inverse of the ith power of the cyclotomic character $\chi \,:\, G_{\mathbb{Q}} \to \mathbb{F}_p^*$; i.e.,

$$\rho(i) := \rho \otimes \chi^{-i}\,.$$

Remarks. i) Actually, we only have $p - 1$ representations $\rho(i)$ since $\rho(i)$ depends on the residual class of i modulo $p - 1$. Obviously, $\rho(0) = \rho$.

ii) The representations $\rho(i)$ are also continuous, irreducible and odd. Note that $\det \rho(i) = \chi^{-2i} \det \rho$.

Let $N_{\rho(i)}$, $k_{\rho(i)}$, $\varepsilon_{\rho(i)}$ be Serre's invariants for the twisted representation $\rho(i)$. Observe that, since χ is a character unramified outside p, we have

$$N_{\rho(i)} = N_\rho \quad \text{and} \quad \varepsilon_{\rho(i)} = \varepsilon_\rho\,.$$

Nevertheless, the weights $k_{\rho(i)}$ do not remain unchanged by twisting; we know that

$$k_{\rho(i)} \in [2, p^2 - 1]\,.$$

Now we give

Definition 1. *We define the minimal weight for the representation ρ as the minimal value of the weights $k_{\rho(i)}$ attached to the twisted representations $\rho(i)$. If $k_{\rho(m)}$ is the minimal weight for ρ, we say that $\rho(m)$ is the minimal representation for ρ.*

Let $D_p \supset I_p$ be a decomposition group and the inertia group for p; denote by I_t the tame inertia group. Let

$$a = \text{inv}(\phi)$$
$$b = \text{inv}(\phi'),$$

the invariants of the characters ϕ, ϕ' (cf. [1]), where $\rho^{ss}|_{I_t} = \begin{pmatrix} \phi & 0 \\ 0 & \phi' \end{pmatrix}$.

It is easy to obtain

Proposition 2. i) *There exist an integer n such that the twisted representation $\rho(n)$ for ρ verifies $k_{\rho(n)} \leq p+1$. In particular, the minimal weight $k_{\rho(m)}$ for ρ is $\leq p+1$.*

ii) *There are, at the most, two classes $n \pmod{p-1}$ giving a twisted representation with weight $\leq p+1$; one of them is the minimal for ρ.*

The table yields the values of $n \pmod{p-1}$ which satisfy $k_{\rho(n)} \leq p+1$.

$\rho\vert_{D_p}$	conditions	weight k	n	weight $k_{\rho(n)}$
			a	$1+b-a$
irreducible	$0 \leq a < b \leq p-1$	$1+pa+b$		
			$b-1$	$2+p+a-b$
			a	$1+b-a$
completely	$0 \leq a \leq b \leq p-2$	$1+pa+b$		
			b	$p+a-b$
reducible				
	$a=b=0$	p	0	p
	$\beta > \alpha$			$1+b-a$
$\begin{pmatrix} \chi^\beta & * \\ 0 & \chi^\alpha \end{pmatrix}$	$\beta \neq \alpha+1$	$1+pa+b$	α	
	$\beta \leq \alpha$			$p+a-b$
$a = \min\{\alpha,\beta\}$	T.R.	$2+\alpha(p+1)$		2
	$\beta = \alpha+1$		α	
$b = \max\{\alpha,\beta\}$	P.R.	$(\alpha+1)(p+1)$		$p+1$

Before giving the proof, let us clarify the table. Here, $* \neq 0$ which means $\rho\vert_{D_p}$ reduces non-completely. When $\rho\vert_{D_p}$ reduces non-completely, the twisted representation $\rho(m)$ which gives a weight $\leq p+1$ is unique and, therefore, yields the minimal weight $k_{\rho(m)}$.

When $\rho\vert_{D_p}$ is irreducible or completely reducible, the ambiguity to decide the minimal weight can be solved depending on the values of a and b:

If $\rho|_{D_p}$ is irreducible,

$$k_{\rho(m)} = \begin{cases} 1 + b - a & \text{if } b - a \le \frac{p+1}{2}, \\ 2 + p + a - b & \text{if } b - a > \frac{p+1}{2}. \end{cases}$$

If $\rho|_{D_p}$ is completely reducible,

$$k_{\rho(m)} = \begin{cases} 1 + b - a & \text{if } b - a \le \frac{p-1}{2}, \\ p + a - b & \text{if } b - a > \frac{p-1}{2}. \end{cases}$$

Proof. It is enough to see that the weight, attached to a representation ρ, is $\le p + 1$ if and only if one of the following conditions holds:

$\rho|_{D_p}$ irreducible and $\rho^{ss}|_{I_t} = \begin{pmatrix} \vartheta^b_{p^2-1} & 0 \\ 0 & \vartheta^{pb}_{p^2-1} \end{pmatrix}$ with $1 \le b \le p - 1$,

or,

$$\rho|_{I_p} = \begin{pmatrix} \vartheta^b_{p-1} & 0 \\ 0 & 1 \end{pmatrix} \quad \text{with } 0 \le b \le p - 2,$$

or,

$$\rho|_{I_p} = \begin{pmatrix} \vartheta^\beta_{p-1} & * \\ 0 & 1 \end{pmatrix} \quad \text{with } 1 \le \beta \le p - 1.$$

Thus, taking into account the recipe in [1] we are done.

Definition 3. *Let ρ as above. If there exist n_1 and n_2 two different classes* (mod $p - 1$) *such that the twisted representations $\rho(n_1)$ and $\rho(n_2)$ have weights $k_{\rho(n_1)}$ and $k_{\rho(n_2)}$ less or equal than $p + 1$, we say that $\rho(n_1)$ and $\rho(n_2)$ are two companion representations for ρ.*

Before the next theorem, we need a technical result.

We fix the embedding $\overline{\mathbb{Q}} \subset \overline{\mathbb{Q}}_p$. Let $\psi : (\mathbb{Z}/p\mathbb{Z})^* \to \overline{\mathbb{Q}}^*$ be the Dirichlet character associated to the inverse of the Teichmüller character. Recall that ψ satisfies

$$\psi(n)n \equiv 1 \pmod{\mathfrak{P}},$$

being $\mathfrak{P}$ the place of $\overline{\mathbb{Q}}$ lying over p fixed by our embedding.

Consider the Eisenstein series

$$E_{1,\psi} = 1 + c_\psi \sum_{m>0,\, m_1>0} \psi(m) e^{2\pi i m\, m_1 z}.$$

We know that $E_{1,\psi}$ is a modular form of type $(p, 1, \psi)$, defined over $\overline{\mathbb{Q}}$.

Proposition 4. (cf. [2]). *Let $G(q) = \sum B_n q^n \in S_k(Np, \phi\psi^{-t})$ be a newform with $1 \le t < p - 1$ and ϕ a Dirichlet character modulo N, $p \nmid N$. Let d be a positive integer such that $d \equiv t \pmod{p - 1}$. Then*
i) $\mathrm{Tr}\,(\mathrm{GE}^d_{1,\psi}) \in S_{k+d}(N, \phi)$,
ii) $\mathrm{Tr}\,(\mathrm{GE}^d_{1,\psi}) \equiv G \pmod{\mathfrak{P}} \iff 2 - k + \frac{d-t}{p-1} + v_p(\overline{B}_p) > 0$.

Here Tr is the p-trace operator between modular forms. Now, we obtain the following

Theorem 5. *Let*

$$\rho : G_{\mathbb{Q}} \to \mathrm{GL}_2(\overline{\mathbb{F}}_p)$$

be a continuous, irreducible and odd representation.
i) *Suppose that its minimal representation $\rho(m)$ verifies Serre's conjecture; that is, there is a Hecke cusp form (mod p) g of type $(N_\rho, k_{\rho(m)}, \varepsilon_\rho)$, such that $\rho(m) \simeq \rho_g$. Then ρ verifies Serre's conjecture.*
ii) *Suppose ρ admits companion representations $\rho(n_1)$ and $\rho(n_2)$. Then:*

$\rho(n_1)$ *verifies Serre's conjecture* $\Longleftrightarrow$ $\rho(n_2)$ *verifies Serre's conjecture.*

Proof. (cf. [5]). To see i), we look for a Hecke cusp form (mod p) $f(q) = \sum a_n q^n$, of type $(N_\rho, k_\rho, \varepsilon_\rho)$, such that $\rho_f \simeq \rho$.
We distinguish several cases depending on the class of the integer m which gives the minimal representation.
First case. Suppose $m = a$ and that $\rho(a) = \rho \otimes \chi^{-a}$ verifies Serre's conjecture. Then, there exists a Hecke cusp form (mod p) $g(q) = \sum b_n q^n$, of type $(N_\rho, k_{\rho(a)}, \varepsilon_\rho)$, such that $\rho_g \simeq \rho \otimes \chi^{-a}$. If θ denotes the Ramanujan-Katz operator, one has

$$\rho_{\theta^a g} \simeq \rho,$$

and, moreover, $\theta^a g \in S_{a(p+1)+k_{\rho(a)}}(N_\rho, \varepsilon_\rho)$. Since Proposition 2 tells us

$$k_\rho = a(p+1) + k_{\rho(a)},$$

we are done by taking $f(q) := \theta^a g(q)$.
Second case. Now, suppose $m = b$ and that $\rho(b) = \rho \otimes \chi^{-b}$ verifies Serre's conjecture. Then, there exists a Hecke cusp form (mod p) $g(q) = \sum b_n q^n$, of type $(N_\rho, k_{\rho(b)}, \varepsilon_\rho)$, such that $\rho_g \simeq \rho \otimes \chi^{-b}$.
We can assume (cf. [6]) that g is new, its filtration is $k_{\rho(b)}$ and, moreover, is a p-ordinary form (i.e., $b_p \neq 0$).
If $k_{\rho(b)} = 2$, note that $b - a = p - 2$ and, therefore, we can take $f(q) := \theta^b g(q)$. Indeed, $f(q) = \theta^{a-1}\theta^{b-a+1}g(q) = \theta^{a-1}\theta^{p-1}g(q)$ and, since $\theta^{p-1}g(q)$ has filtration $2p$, then the weight of f is $2p+(a-1)(p+1) = 1+pa+b = k_\rho$. Also $\rho_f \simeq \rho$.
If $k_{\rho(b)} \geq 3$, consider the newform G in $S_2(N_\rho p, \varepsilon \chi_p{}^{k_{\rho(b)}-2})$ which is a lifting of $g(q) = \sum b_n q^n$ in characteristic zero as in [6]. Thus,

$$G(q) = \sum B_n q^n,$$

with $B_n \in \overline{\mathbb{Z}}_p$ and $B_n \equiv b_n(\mathrm{mod}\ \mathfrak{P})$ for all n. (Here, ε is the multiplicative lifting of ε_ρ). Since $b_p \neq 0$, we have $|B_p| = \sqrt{p}$ (cf. [7-8-9]). Therefore we get $v_p(B_p) = 0$, $v_p(\overline{B}_p) = 1$, being v_p the p-adic normalized valuation of $\mathbb{Q}_p$ defined by $\mathfrak{P}$.

Let

$$G'(q) = \sum B'_n q^n \in S_2(N_\rho p, \varepsilon \overline{\chi}_p^{\,k_{\rho(b)}-2}),$$

be the normalized newform which differs on a scalar multiple of $G|\omega_\zeta$, where ω_ζ is the Weil involution. We know (cf. [6])

$$\begin{aligned}
B'_p \cdot B_p &= p\varepsilon(p) \\
B'_n &= B_n \cdot \psi^{a-b+p-2}(n) \quad \text{if } n \text{ is prime to } p
\end{aligned} \qquad (3.1)$$

We are ready to apply Proposition 4 to the newform G'. By normalizing the exponent of its character, we get

$$\overline{\chi}_p^{\,k_{\rho(b)}-2} = \psi^{-(2-k_{\rho(b)})} = \psi^{-(2-k_{\rho(b)}+p-1)}.$$

Since $v_p(\overline{B}'_p) = 0$ and G' is of weight 2, the minimal integer d such that

$$\mathrm{Tr}\,(G' E^d_{1,\psi}) \equiv G' (\mathrm{mod}\ \mathfrak{P})$$

is $d = 2p - k_{\rho(b)}$. Note that

$$\mathrm{Tr}\,(G' E^{2p-k_{\rho(b)}}_{1,\psi}) \in S_{2(p+1)-k_{\rho(b)}}(N_\rho, \varepsilon).$$

Then, the form

$$f := \theta^{a-1}\mathrm{Tr}\,(G' E^{2p-k_{\rho(b)}}_{1,\psi})$$

has weight

$$2(p+1) - k_{\rho(b)} + (a-1)(p+1) = 1 + pa + b = k_\rho,$$

and by (3.1), the residual representation

$$\rho \otimes \chi^{-b} \otimes \chi^{-(a-b+p-2)} = \rho \otimes \chi^{-a-p+2} = \rho \otimes \chi^{-a+1} = \rho \otimes \chi^{-(a-1)} = \rho(a-1)$$

is the reduction of the representation $\rho_{G'}$, attached to G'. Thus, ρ arises from f.

Third case. Suppose $m = b - 1$. Let $g(q) = \sum b_n q^n$ be the Hecke cusp form $(\mathrm{mod}\ p)$, of type $(N_\rho, k_{\rho(b-1)}, \varepsilon_\rho)$, such that $\rho_g \simeq \rho(b-1)$.

Now, g is a p-supersingular newform. Observe that $k_{\rho(b-1)} = 2+p+a-b$ satisfies $3 \le k_{\rho(b-1)} \le (p+3)/2$ and, therefore, θ^{b-1-a} has filtration $1+b-a$ (cf. [10]). Then, the form $f(q) := \theta^{b-1-a}\theta^a g(q)$ satisfies Serre's conjecture in this case.

To see ii), we use the same kind of arguments plus the deep result stated in [6] concerning companion forms. One can assume that $\rho(n_1) = \rho(m)$ is the minimal representation for ρ. If $\rho(n_1)$ verifies Serre's conjecture, so does $\rho(n_2)$ because of i). To see the reciprocal, we distinguish two cases: when $\rho|_{D_p}$ is irreducible, we are done by using filtrations as above; when $\rho|_{D_p}$ is completely reducible, Theorem 13.10 in [6] asserts that $\rho(n_1)$ verifies Serre's conjecture.

As a consequence we get the following criterion which reduces the verification of Serre's conjecture to representations having weight smaller than $p+1$.

Corollary 6. *Let ρ as above. Let $\rho(\hat{m})$ be a twisted representation for ρ having weight $\leq p+1$. If $\rho(\hat{m})$ verifies Serre's conjecture, so does ρ.*

4 Application to vertical Weil curves

Now, we have

Theorem 7. (cf. [2]). *Serre's conjecture $(3.2.4_?)$ is true for the irreducible Galois representation*

$$\rho \,:\, G_{\mathbb{Q}} \longrightarrow \mathrm{Aut}\,(E_p)\,,$$

provided that E is a vertical Weil curve at $p > 7$.

Proof. For this kind of representations we have always a weight $k_\rho > p+1$. The idea of the proof is to check Serre's conjecture for the minimal or the companion representation $\rho(\hat{m})$ for ρ.

First step. Denote by $p-\mathrm{type}(E)$ the Kodaira symbol for the special fibre at p of the Néron model of E. The Galois representations ρ for the cases with p-type$(E) = II^*, III^*, IV^*$ are related to the cases without asterisk by the twist $\rho(\frac{p-1}{2})$. Observe that the minimal or companion representation in both cases does not change.

Therefore, we can reduce our job working with the Weil curves E such that

$$p - \mathrm{type}(E) = II, III, IV\,.$$

Second step. Now, we compute $\hat{m}$ such that $\rho(\hat{m})$ has weight $\leq p+1$. Under our assumptions we find:

$$\hat{m} \;=\; \frac{p-1}{e}\,,$$

$$\rho(\hat{m})_{|I_p} \;=\; \begin{pmatrix} \chi^{p-1+e} & * \\ 0 & 1 \end{pmatrix}$$

$$k_{\rho(\hat{m})} \;=\; p+1-2e$$

where $e = 6, 4, 3$ depending on p-type$(E) = II, III, IV$, respectively.

Third step. Let Np^2 be the geometric conductor of the elliptic curve E. Denote $F(q) = \sum A_n q^n \in S_2(Np^2)$ the newform attached to E by the Eichler-Shimura congruences. If ψ is the Dirichlet character as above, we can ensure that

$$F \otimes \psi^{\frac{p-1}{e}} \in S_2(Np^2, \psi^{2\frac{p-1}{e}})$$

is a Hecke cusp form which is not a new one. Let $G(q) = \sum B_n q^n$ be the newform with the same eigenvalue system as $F \otimes \psi^{\frac{p-1}{e}}$ (theorem of multiplicity one, cf. [11]). After Carayol's work [12], we know how to compute the level of G. It coincides with Artin's conductor of the ℓ-adic representation attached to $F \otimes \psi^{\frac{p-1}{e}}$ ($\ell \nmid Np^2$). Since E is a vertical Weil curve at p, we get

$$G(q) = \sum B_n q^n \in S_2(Np, \psi^{2\frac{p-1}{e}}) \quad \text{(cf. [2])}.$$

Fourth step. Finally, in order to apply Proposition 4, we must be careful of the "hidden" coefficient B_p of the newform G.

Let A over $\mathbb{Q}$ be the abelian variety attached to G, which is a factor of the jacobian variety $J_1(Np)$ of the modular curve $X_1(Np)$. Since $B_p \neq 0$, by a theorem of Langlands (cf. [13], Thm. 7.1, 7.4), A has potentially good reduction at p. Assume that A has good reduction over $L \subset \overline{\mathbb{Q}}_p$.

If p-type$(E) = II, III, IV$, looking at the representation $\rho(\hat{m})$, we see that the p-divisible group $A_{/L}[\mathfrak{P}^\infty]$ is ordinary. That means that the newform G is $\mathfrak{P}$-ordinary (cf. [14]). Then, since $|B_p| = \sqrt{p}$, we obtain $v_p(\overline{B}_p) = 1$.

Just applying Proposition 4, we show that $\rho(\hat{m})$ arises from

$$g(q) = \mathrm{Tr}(G\widetilde{E_{1,\psi}^{\hat{m}(e-2)}}) \in S_{k_{\rho(\hat{m})}}(N).$$

If necessary, we can use the results of Livné in [15] to adjust the right level.

5   References

1. Serre, J.-P. (1987). Sur les représentations modulaires de degré 2 de Gal$(\overline{\mathbb{Q}}/\mathbb{Q})$. *Duke Math. Journal*, **54**, 179-230.

2. Bayer, P; Lario, J-C. (1991). Galois representations defined by torsion points of modular elliptic curves. To appear in *Compositio Math.*

3. Edixhoven , B.; de Groot, A.; Top, J. (1990). Elliptic curves over the rationals with bad reduction at only one prime. *Math. of Computation*, **54**, 413-9.

4. Tate, J. (1975). Algorithm for determining the type of a singular fibre in an elliptic pencil. *Springer LN in Math.*, **476**, 33-52.

5. Lario, J.C. (1991). Representacions de Galois i corbes el·líptiques. *Universitat de Barcelona*. Thesis.

6. Gross, B.H. (1990). A tameness criterion for Galois representations associated to modular forms. *Duke Math. Journal*, **61**, 445-517.

7. Asai, T. (1976). On the Fourier coefficients of automorphic forms at various cusps and some applications to Rankin's convolution. *J. Math. Soc. Japan*, **28**, 48-61.

8. Li, W. (1975). Newforms and functional equations. *Math. Ann.*, **212**, 285-315.

9. Ogg, A.P. (1969). On the eigenvalues of Hecke operators. *Math. Ann.*, **179**, 1-21.

10. Jochnowitz, N. (1982). The local components of the Hecke algebra modℓ. *Trans. of the AMS*, **270**, 253-267.

11. Atkin, A.O.L. (1978). Twists of newforms and pseudoeigenvalues of W-operators. *Invent. Math.*. **48**, 221-244.

12. Carayol, H. (1986). Sur les représentations ℓ-adiques associées aux formes modulaires de Hilbert. *Ann. Sci. École Nor. Sup.*, **19**, 409-468.

13. Langlands, R.P. (1973). Modular forms and ℓ-adic representations. *Springer LN in Math.*, **349**, 361-500.

14. Mazur, B.-Tilouine, J. (1990). Représentations galoisiennes, différentielles de Kähler et "conjectures principales". *Publ. Math. IHES*, **71**, 65-103.

15. Jordan, B.-Livné. (1989). Conjecture "epsilon" for weight $k > 2$. *Bull. of the AMS*, **21**, 51-56.

Joan-C. Lario
Departament de Mathemàtica Applicada II
Facultat d'Informatica de Barcelona
Universitat Polytècnica de Catalunya
Pau Gargallo, 5
E-08028 Barcelona
Spain
e-mail lario@ma2.upc.es

Units in Number Fields and Elliptic Curves

Odile Lecacheux

Université Paris 6

Introduction

Determination of units in number fields is often the first stage when we study their arithmetic properties. It is indeed natural to be interested in number fields in which some units are obvious; let us give some examples.

If n is an element of $\mathbb{Z}$, let's consider fields generated by x, one of the roots of following polynomials

$$X^2 - nX - 1 \tag{1}$$

$$X^3 - nX^2 + (n-3)X + 1 \tag{2}$$

$$X^4 - nX^3 - 6X^2 + nX + 1 \tag{3}$$

$$X^4 - nX^3 + 2X^2 + nX + 1 \tag{4}$$

and if $n \equiv 2 \bmod 4$

$$X^6 - \frac{(n-6)}{2}X^5 - \frac{5(n+6)}{4}X^4 - 20X^3 + \frac{5(n-6)}{4}X^2 + \frac{(n+6)}{2}X + 1 \tag{5}$$

All these extensions are abelian and it is generally possible to give generators of their unit groups.

In the first part of this paper we recall how to construct the two first examples and others with the modular covering $X_1(N) \to X_0(N)$. By specializing modular units give units in number fields. An estimate of their regulators is given in part 2.

When $N=16$ and $N=18$, it is possible to construct other coverings which give the two examples of degree 4 extensions, and another example of a sextic field with 4 independent units.

In the last part of this paper, we give equations of two families of elliptic curves. They are defined over $\mathbb{Q}(n)$, have a point of order 7 (resp. 5) in an abelian extension of degree 6 (resp. 4) which corresponds to example (5) (resp. (3)). On the first family of curves, we construct two non-torsion points, one in $\mathbb{Q}(n)$, the other in $\mathbb{Q}((-7)^{1/2}n)$. By specialising we obtain some curves with $\mathbb{Q}$-rank ≥ 2 and two with $\mathbb{Q}$-rank ≥ 4.

1 Modular construction of units

If E is an elliptic curve defined over $\mathbb{Q}$, given by a Weierstrass equation, and P a N-torsion point on E, if x_i and y_i are the coordinates of iP, the quotients $(x_i - x_j)/(x_i - x_k)$ are independent of the equation and could be considered as modular functions. Recall that there exists a projective curve $X_1(N)/\mathbb{Q}$, such that points of $X_1(N)(K)$, correspond to equivalence classes of pairs (E, P), and each equivalence class contains a pair such that E is on K and P on $E(K)$. As above, there exists a curve, $X_0(N)/\mathbb{Q}$ corresponding to pairs $(E, C_N = < P >)$, and each equivalence class contains a pair such that E is over K and C_N is mapped to itself by $\mathrm{Gal}(\overline{K}/K)$. We have then an abelian covering, namely $X_1(N) \to X_0(N)$ of degree $\phi(N)/2$. The action of the Galois group corresponds to automorphisms of $X_1(N)$ represented by $(E, P) \to (E, iP)$. Moreover, choose N so that $X_0(N)$ is of genus 0. This occurs when $1 < N < 11, N = 12, 13, 16, 18$, and 25. Quotients $f_{i,j,k} = (x_i - x_j)/(x_i - x_k)$ are functions on $X_1(N)$. In these cases we construct a function n generating the function field of $X_0(N)$. For some values of i, j, k the functions $f_{i,j,k}$ are units on $\mathbb{Z}[n]$.

By specializing n with $n \in \mathbb{Z}$, we obtain units in number fields. Equations and constructions appear in details for $N = 13, 25, 16$ in [11, 12, 13]. We obtain extensions of degree 6, 10 and 4 with small regulators. We give others examples and make precise the function n.

If N is prime or a power of a prime number not equal to two, we can take for n the function $N^a(\eta(N\tau)/\eta(\tau))^b$ with a and b such that the q-development of $n|w_N$ begins with $1/q+\dots$ ($q = \exp(2i\pi)$, and w_N is the involution of the modular curve).

The example of the cubic field given in the introduction, often called the "simplest cubic field," can be constructed with $N = 7, n = 7^2(\eta(7\tau)/\eta(\tau))^4$, and $X = f_{1,2,3}$. We obtain the equation $X^3 - (n+8)X^2 + (n+5)X + 1 = 0$. But it can be also constructed with $N = 9, n = 3^3(\eta(3\tau)/\eta(\tau))^3$ and $X = f_{1,2,4}$. We obtain the equation $X^3 + (n-3)X^2 - nX + 1 = 0$.

If $N = 16$, $n = 2\eta^6(2\tau)\eta^{-2}(4\tau)\eta(\tau)^{-4} = 2e^{-i\pi/12}\eta^{-2}(\tau)\eta^2(\tau + 1/2)$, and $X = f_{2,3,7}f_{6,1,3}$. The relation between functions X and n is a model for $X_1(16)$:

$$X^4 - n^2 X^3 - X^2(n^3 + 2n^2 + 4n + 2) - n^2 X + 1 = 0.$$

So we obtain a family of cyclic fields of degree 4. The action of Galois group is given by $X \longrightarrow (X - m)/(-Xm + 1)$ where $m^2 - nm - 1 = 0$.

For $N = 18$, we choose $n = \eta^3(\tau)\eta(6\tau)\eta^{-3}(2\tau)\eta^{-1}(3\tau)$ and $X = f_{9,2,8}f_{6,2,4}$. The relation between functions X and n gives an equation for the modular curve $X_1(18)$ namely

$$X^3 + (n^3 - n^2 + 2n + 1)X^2 - (n - 1)^2 X - 1 = 0.$$

The action of Galois group is given by $X \mapsto (X^2 - 1 + Xn)/(X^2 n)$. The discriminant of this polynomial is equal to $n^2(n^2 + n + 1)^2(n^2 - 2n + 4)^2$ and n is not ramified.

In general, these last cubic fields are not "simplest cubic fields". By specializing n with $n \in \mathbb{Z}$, if x and x', roots of the polynomial generate the units group modulo ± 1, the ring of integers is not $\mathbb{Z}[\alpha]$ with some α. For this we use a property of M. N. Gras, namely $\text{Trace}(x) + 1$ and $\text{Trace}(1/x) + 1$ must be coprime, here n divides these integers [3]. If $n \not\equiv 1 \bmod 3$ and $d = (n^2 + n + 1)(n^2 - 2n + 4)$ without square factors, we can show that the trace of the d-th root of unity to this field is the sum of an integer $(n(n-2)/3))$ and a unit (namely $-1/X$). The only fields with this property and prime conductor are "simplest cubic fields" [18].

2 Regulators

Using regulators we can calculate index of unit groups (see [24] p 41). Using cusps of $X_1(N)$ gives results on the regulator of units. If we consider the q-developments of the modular functions

$$\frac{\wp(i/N; \tau, 1) - \wp(j/N; \tau, 1)}{\wp(i/N; \tau, 1) - \wp(k/N; \tau, 1),}$$

we show they have at cusp r/s, poles or zeros equal or superior to $\text{Inf}\{[is], [js]\} - \text{Inf}\{[is], [ks]\}$ where $[a]$ is the distance from a to $\mathbb{Z}$ (See [17]).

Let S be the set of poles of n. By the choice of n, poles are simple and S is included in the set of rational cusps. The function $1/n$ could be chosen as a local parameter for $A \in S$. Functions which give units have their poles and zeros in S (it's necessary for existence of parameterized units).(See [6]).

For example, with $N = 13$, we show how to evaluate regulators. We construct a function $X = f_{5,1,3} f_{1,5,3} f_{2,3,1} f_{2,3,5} f_{4,5,6}$ ($H - h$ in [11]) . If T is the automorphism corresponding to $\tau \mapsto (-6\tau + 1)/(-13\tau + 2)$, or $(E, P) \mapsto (E, 2P)$, the functions X and $X|T^i$ with $1 \leq i \leq 5$ generate the group of functions on $X_1(13)$, whose supports of divisors are rational cusps. Moreover $\text{Pic}^S(X_1(13))$ is finite and of order 19. The determinant of the 5×5 matrix whose coefficients are orders of poles and zeros of $X|T^i$ is equal to 19. Each function $X|T^i$ has an expansion in $A \in S$ such as $n^{-a_{i,A}} + ..$ where $a_{i,A}$ is the order of poles or zeros of the function in A. Using these estimations of $\log|X|$ and $\log|X|T^i|$, we prove that the regulator of such units is small in an asymptotic sense, namely is equal to $19 \log^5(|n|) + o(\log^5(|n|))$.

If N is prime recall that the order of the group generated by rational cusps is equal to $\mathcal{B} = N \prod_{\chi \neq \chi_0} (1/4 B_{2,\chi})$, where χ is an even character modulo N and

$$B_{2,\chi} = \frac{1}{N} \sum_{a \bmod N} \chi(a)(a^2 - Na).$$

[8]. We have constructed units with a regulator equivalent to

$$\mathcal{B} \log^{\phi(N)/2-1}(n).$$

If N is not prime we have an analogous formula for $\mathcal{B}$. We have construct units with a regulator equivalent to $a(N)\log^{\phi(N)/2-1}(n)$, but the choice of n involves that $a(N)$ divides $\mathcal{B}$. For example if N equals 18, X and its conjugate have a regulator equal to $7\log^2(|n|)+o(\log^2(|n|))$. It is this property which allows to show that, with some conditions on the discriminant of the field, these units generate all of the group of units [11,18] : for example for $N = 18$, we use the following inequality :

$$R \geq \frac{1}{16}\log^2(\frac{D}{4})$$

where D is the discriminant of the number field and R its regulator. (see [2 or 1]). So, if $n^2 + n + 1$ and $n^2 - 2n + 4$ have no square factors and if n is odd, and big enough, X and its conjugate generate the unit group.

3 The modular curve $X_1(16)$

The last three examples given in the introduction have not been obtained with the modular construction given in the first paragraph. In fact such a construction would give, for degree 4, a curve of genus two. Nevertheless we keep the idea to construct extensions using a curve and an automorphism of this curve, all of them being defined on $\mathbb{Q}$.

The covering $X_1(16) \to X_0(16)$ is of degree 4 and the Galois group is isomorphic to $(\mathbb{Z}/16\mathbb{Z})^*/ \pm 1$. This group contains an order two subgroup corresponding to the intermediate covering X_G, where G is the group $\Gamma_1(8) \cap \Gamma_0(16)$. The group G is normal in $\Gamma_0(8)$ and an other automorphism of order 2, defined on $\mathbb{Q}$, could be constructed as follow : a point on X_G is represented by $(E, 2P, < P >)$ where E is an elliptic curve, $2P$ is of order 8 and $< P >$ is the group generated by P. We associate to this $(E', \phi(P), < \phi(P_1) >)$, where $E' = E/ < 4P + B >, 2B = 0$ and $B \neq 8P$, ϕ the isogeny $E \longrightarrow E'$ and $2P_1 = P$. By composition, we construct an automorphism of order 4, named M. The quotient of X_G by M called X_{G_M} corresponds to the first family of number fields of degree 4 in the introduction. In the same way the automorphism M^2 and another element of the normalizer of G, corresponds to the last family of quartic fields given in the introduction. The covering $X_1(16) \longrightarrow X_{G_M}$ gives by specialisation a family of fields of degree 8 studied in [13,14].

With the same idea we can also construct another family of sextic number field with the modular curve $X_1(18)$ and the involution w_2 which is defined on $\mathbb{Q}$. In this case it is only possible to find 4 independent units, roots of the polynomial

$$X^6 + X^5(-a+2) + X^4(-a^2 - a - 9) + X^3(a^3 - 2a^2 + 8a - 14)+$$

$$X^2(a^3 + a^2 + 7a + 10) + X(a^2 + 8) + 1$$

These fields are composed of quadratic and cubic extensions defined by the two polynomials $X^2 - aX - 2$ and $X^3 - (a+1)X^2 + (a-2)X + 1$.

4 Elliptic curves and units

If $(E, C_N =< P >)$ is an elliptic curve with an N-isogeny defined over $\mathbb{Q}$, the two coordinates of P generate an abelian extension K' of degree dividing $\phi(N)$. To such a couple corresponds a point on $X_0(N)(\mathbb{Q})$. If $N = 7$, and n takes on this point an integer value, recall that the polynomial $X^3 - (n+8)X^2 + (n+5)X + 1$ is irreducible and so the degree of K' is 3 or 6. The field K' contains a root x of this polynomial. Then the element (E, P) corresponds to a point of $X_1(7)(\mathbb{Q}(x))$ and for this the action of $\text{Gal}(K'/\mathbb{Q}(x))$ is an automorphism of (E, P). This automorphism could be only $(E, P) \to (E, -P)$, because otherwise E has complex multiplication, which only happens when $n = \pm 7^a$ with $a=0,1,2$. Thus the abscissa of P is in $\mathbb{Q}(x)$.

Conversely to each point of $X_0(7)(\mathbb{Q})$ corresponds an infinity of elliptic curves twisted by a quadratic character. One of them allows us to construct a sextic extension with parametrized units.

Theorem 1 : *If z is a root of the polynomial*

$$X^6 - X^5(10+2n) - X^4(40+5n) - 20X^3 + (25+5n)X^2 + (16+2n)X + 1 = 0,$$

x is a root of
$$X^3 - (n+8)X^2 + (n+5)X + 1 = 0,$$

and
$$d = n^2 + 13n + 49,$$

then the elliptic curve

$$E(n) : Y^2 - \frac{dX^2}{4} = X^3 - X^2(n+8) + X(n+5) + 1$$

has a 7-rational isogeny whose kernel is generated by $P = (x, xd^{1/2}/2)$ and is defined over $\mathbb{Q}(z)$.

The discriminant of this curve is nd^2 and d^2 is the discriminant of the cubic polynomial.

If

$$s = \frac{z(2+z)(1+z+z^2)}{(z^2-1)(1+2z)},$$

the change of coordinates

$$X - x = s^2 \mathcal{X}, \quad Y - \frac{Xd^{1/2}}{2} = s^3(\mathcal{Y} + \frac{x-1}{x^2}\mathcal{X})$$

gives an isomorphism between the curve

$$\mathcal{Y}^2 + \mathcal{X}\mathcal{Y}\frac{x^2+x-1}{x^2} + \mathcal{Y}\frac{(x-1)^2}{x^3} = \mathcal{X}^3 + \frac{(1-x)^2}{x^3}\mathcal{X}^2$$

and $E(n)$, under which the 7-torsion point corresponds to $(\mathcal{X}=0, \mathcal{Y}=0)$.

Proof. The sextic field generated by z has been studied by M.N.Gras in [4]. If d has no square factors then the group of relative units is generated by $z(2z+1)/(z+2)$ and its conjugates, except for two cases, and if the quadratic units are known, all the unit group is known.   □

Theorem 2 : *The rank of $E(n)$ on $\mathbb{Q}(n)$ is greater than 1. The point $Q = (0,1)$ is not a torsion point. The two points $Q = (0,1)$ and Q' given by*

$$X = \frac{-128n^2}{7^3} - \frac{32n}{7} - 16,$$

$$Y = i\frac{832n^3 + 18032n^2 + 134456n + 352947}{7^{9/2}}$$

are independent. The two points M_1 with coordinates

$$X = \frac{13 + i7^{1/2}}{32}, \quad Y = 1 - \frac{7}{4}X + \frac{(n+8)}{2}X$$

and its conjugate M_2 generate $E(\overline{\mathbb{Q}}(n))$, and we have $-Q = M_1 + M_2$, $M_1 - M_2 = Q'$.

Proof. If n is an integer, the point Q could not be an s-torsion point for almost every n, because there are only a finite number of elliptic curves with a $7s$-rational isogeny, and in this case $s = 2$ or 3.

The point Q' can be constructed by considering the quotient curve $E' = E(n)/<P>$. Using the formulas in [23] to write E', we assert that E' is a twist of the curve $E(7^2/n)$ by $-7^{1/2}$ (recall that the involution w_7

changes n to $7^2/n$). The image of Q gives Q'. We may now conclude using [7]. □

Examples

If n takes an integer value we can find curves with "small" conductor and rank at least 2.

1) $n = -16$. The curve $Y^2 - 11XY = X^3 + 2X^2 - 11X + 1$ has conductor $18,818 = 2.97^2$. The following integer points are on the curve : $(-2, -23)$, $(0, 1)$, $(12, 145)$, $(30, 401)$, $(232, -2495)$. The points $(12, 145)$ and $(30, 401)$ are independent.

2) $n = -11$. The curve $Y^2 = X^3 + 39X^2 - 96X + 64$ has conductor $14,256 = 11. \, 2^4. \, 3^4$. The following integer points are on the curve : $(-40, 48)$, $(-22, 102)$, $(-7, 48)$, $(-3, 26)$, $(0, 8)$, $(2, 6)$, $(8, 48)$, $(32, 264)$, $(90, 1018)$. The points $(0, 8)$ and $(2, 6)$ are independent.

3) $n = 23$. The rank of the curve $Y^2 + 31XY = X^3 - 52X^2 + 28X + 1$ is at least 4. Its conductor is $17,689,967 = 23.877^2$. The points $A = (0, 1)$, $B = (3, -4)$, $C = (-21, 55)$, and $D = (-72, 341)$ are independent and the points $B - A = (19, -20) = E$, $-C + B = (-80, 409) = F$, $E - D = (-105, 671) = H$, $C - D = (325, 2326) = I$, $2A + B = (825, 13474)$, and $F + H = (5109, 292655)$ have integer coordinates.

4) $n = 44$. The curve $Y^2 + 49XY = X^3 - 13X^2 + 49X + 1$ has conductor $143,841,478 = 2.11.2557^2$. Its rank is at least 4. The points $A = (0, 1)$, $B = (-240, 4471)$, $C = (-32, 753)$, and $D = (132, 3541)$ are independent ; the point $(7150, 628926)$ is equal to $-3A - B + C$. The height regulator is 2.64183.

For N=5, we have a similar result.

Theorem 3 : *The elliptic curve*

$$Y^2 - \frac{(n^2 - 22n + 125)}{4}(X^2 + 1) = (X^2 + (n - 11)X - 1)(X + \frac{n - 11}{2})$$

has a 5-isogeny, with kernel is generated by $P = (a, b)$, *where*

$$a^2 + (n - 11)a - 1 = 0,$$

$$\frac{b = (a + 1/a)(x + 1/x)}{2,}$$

and x *satisfies*

$$x^4 - 2(n - 11)x^3 - 6x^2 + 2(n - 11)x + 1 = 0.$$

The last quartic field have been studied by M. N. Gras and A. Lazarus. We also notice that the point $((11 - n)/2, (n^2 - 22n + 125)/4)$ is, for almost n, a non-torsion point.

References

1. Bergé, A-M. and Martinet, J. (1990-1991). *Sur les minorations géométriques des régulateurs.* Séminaire de Théorie des Nombres, Birkhauser.

2. Cusik, T. W. (1984). *Lower bounds for regulators.* Number Theory, Noordwijkerhout 1983. Proceedings of the Journées Arithmétiques.Lecture Notes 1068.

3. Gras, M-N. (1976). *Sur le lien entre le groupe des unités et la monogénéité des corps cubiques cycliques.* Séminaire de Théorie des Nombres 1975-1976 Publ. Math. Fac. Sci. Besançon. Fasc **1**.

4. Gras, M-N.(1987). *Special units in real cyclic sextic fields.* Math. Comp., **48**, 179-182.

5. Gross, B. (1987). *Heegner points and the modular curve of prime level.* J. Math. Soc. Japan, **32**, 347-362.

6. Hellegouarch, H. Mc Quillan, D.L. Paysant-Leroux, R. (1987). *Unités de certains sous-anneaux des corps de fonctions algébriques.* Acta Arithmetica, **49**.

7. Kuwata, M. (1990). *The canonical height and elliptic surfaces.* Jour. of Number Theory, **36**, 201-211.

8. Kubert, D.S.Lang, S. (1981). *Modular units.* Springer-Velag. Comprehensive Studies in Mathematics **244**.

9. Lazarus, A. (1991). *On the class number and unit index of simplest quartic fields.* Nagoya Math. J., **121**, 1-13.

11. Lecacheux, O. (1989). *Unités d'une famille de corps cycliques réels de degré 6 liés à la courbe modulaire $X_1(13)$.* J. Number Theory, **31**, 54-63.

12. Lecacheux, O. (1990). *Unités d'une famille de corps liés à la courbe $X_1(25)$.* Annales de l'Institut Fourier, **40**, 237-253.

13. Lecacheux, O. (1991) *Familles de corps de degré 4 et 8 liés à la courbe modulaire $X_1(16)$.* Séminaire de Théorie des Nombres, Birkhauser.

14. Lehmer, E.(1988). *Connection between Gaussian periods and cyclic units.* Math. Comp. **30**, 535-541.

15. Levi, B. (1908). *Saggio per una teoria arithmetica delle forme cubiche ternarie.* Atti della R. Academia delle Scienze di Torino, **43**, 99-120.

16. Mestre, J.F. (1981). *Corps Euclidiens, Unités exceptionnelles et courbes elliptiques.* J. of Number Theory, **13**, 123-137.

17. Ogg, A. P. (1973). *Rational points on certain elliptic modular curves.* Proc. Symposia Pure Math. **24** Amer. Math. Soc. Providence,R.I., 221-231.

18. Schoof, R. and Washington, L. (1988). *Quintic polynomials and real cyclotomic fields with large class numbers.* Math. Comp. **50**, 182 542-555.

19. Seah, E., Washington, L. C., Williams, H. C. (1983). *The calculation of a large cubic class number with an application to real cyclotomic fields.* Math. Comp. **41**, 303-305.

20. Shanks, D. (1974). *The simplest cubic fields.* Math. Comp. **28**,1137-1152.

21. Shioda, T. (1972). *On elliptic modular surface.* J. Math. Soc. Japan **24** 20-59.

22. Silverberg, A. *Universal families of abelian varieties.* Number theory and cryptography, London Mathematical Society. Lecture Note Series, **154** Cambridge Press. Edited by J.H.Loxton.

23. Vélu, J.(1971). *Isogénies entre courbes elliptiques.* C.R. Acad Sc Paris Série **273**, 238-241.

24. Washington, L. C. *Introduction to Cyclotomic Fields.* Springer.1982.

25. Washington, L. C. (1991). *A family of cyclic quartic fields arising from modular curves.* Math. of Computation. **57**, 763-775.

26. Washington, L. C. (1987). *Class numbers of the simplest cubic fields.* Math. of Comput. **48**, 371-384.

Odile Lecacheux
Université Paris 6
4 Place Jussieu
F-75252 PARIS Cedex 05
France
e-mail lecacheu@FRCIRP81.bitnet

Descent by 3-isogeny and 3-rank of quadratic fields

Jaap Top

Erasmus University Rotterdam

*To Paulo and Huguette Ribenboim,
who were during our year in Kingston
not just friends but also spare grand-
parents of our daughters.*

Abstract

In this paper families of elliptic curves admitting a rational isogeny of degree 3
are studied. It is known that the 3-torsion in the class group of the field defined
by the points in the kernel of such an isogeny is related to the rank of the elliptic
curve. Families in which almost all the curves have rank at least 3 are constructed.
In some cases this provides lower bounds for the number of quadratic fields which
have a class number divisible by 3.

1   Introduction

By the 3-rank $r_3(K)$ of a number field K we will mean the dimension (as
a vector space over the field with 3 elements $\mathbf{F}_3$) of the 3-torsion in the
class group of K. As usual one writes $r_3(d) := r_3(\mathbf{Q}(\sqrt{d}))$. A remarkable
classical result of Scholz says that the numbers $r_3(d)$ and $r_3(-3d)$ differ by
at most 1. Most other literature on the function $r_3(d)$ concerns its possible
values.

So far, the record seems to be $r_3(d) = 6$. In his thesis written in 1987,
Quer exhibited 3 negative integers d for which this value is taken [8],[9].
The method exploited is that one can show that under certain conditions
$2r_3(d) + 1$ is an upper bound for the rank of the group of $\mathbf{Q}$-rational points
on the elliptic curve given by $y^2 = x^3 + d$. Given d such that this curve
has rank 12, one obtains $r_3(d) \geq 6$.

By class field theory the 3-torsion in the class group of $\mathbf{Q}(\sqrt{d})$ corresponds to degree 3 Galois extensions of $\mathbf{Q}(\sqrt{d})$ which are unramified at all finite primes. Unramified extensions with –more generally– Galois group the alternating group A_n have been constructed by Fröhlich and by Uchida [15]. An example in case $n = 3$ (with in fact even more special properties) is given by the fields $\mathbf{Q}(\sqrt{-27m^2 - 4m})$, for any integer m which cannot be written in the form $n^3 - n^2$. This last example appears in a recent paper of Brinkhuis [2].

From the example given above it follows easily that infinitely many fields $\mathbf{Q}(\sqrt{d})$ exist for which $r_3(d) \geq 1$. The same holds for $r_3(d) \geq 2$ (Shanks, [11]) and even $r_3(d) \geq 4$ (Craig, [4]). We will discuss the weakest of such results, namely $r_3(d) \geq 1$. The reason to do this is that we hope it illustrates a general method, which is probably also of interest in other situations.

The method consists of studying the relation between the rank of certain elliptic curves and the 3-rank. Firstly it is shown that Quer's technique hinted at above applies more generally to any elliptic curve admitting a rational 3-isogeny. This puts classical work of Selmer and of Cassels and more recently of Satgé [10] and Nekovář [6] on rank calculations for elliptic curves with j-invariant 0 into a more general framework.

Next we construct families of elliptic curves for which this relation is valid. The rational points on the generic curve in such a family yield a group equipped with a well-understood quadratic form. Our examples will illustrate how the theory of these so-called Mordell-Weil lattices as developed by Shioda immediately gives a lot of useful information about the families under consideration.

Specializing the parameter in our families to a rational number now gives elliptic curves for which our theory works. Hence some lower bound (which may well be zero) for r_3 of the fields involved holds. The d's obtained in this way are up to a square the values of some binary form. Using estimates of the number of square free values taken by binary forms (compare [14]) one can hope to obtain a density result for the number of quadratic fields with positive 3-rank. Unfortunately it seems that one has to put rather restrictive conditions on a family to obtain non-trivial bounds.

2 Results

To state the results of this paper some notation is needed. Let $E/\mathbf{C}(t)$ be an elliptic curve which is not isomorphic over $\mathbf{C}(t)$ to a curve already defined over $\mathbf{C}$. As in [12], 8.5 and 8.6, the finitely generated abelian group $E(\mathbf{C}(t))$ is equipped with a height pairing.

By A_2^* the lattice $\frac{1}{3}\sqrt{-3}\mathbf{Z}[-\frac{1}{2} + \frac{1}{2}\sqrt{-3}]$, with the quadratic form $\mathrm{tr}\,(x\bar{y})$ is denoted. Alternatively one can describe it as the dual of the root lattice of the Lie group A_2. Similarly one has A_1^* which is just **Z** with the quadratic form $x^2/2$.

Theorem 1. *Let E be an elliptic curve given by $y^2 = x^3 + a(t)(x - b)^2$, in which $b \in \overline{\mathbf{Q}}^*$ and in which $a(t) \in \overline{\mathbf{Q}}[t]$ is a polynomial of degree 2.*

Assume that neither $a(t)$ nor $4a(t) + 27b$ is a square.

Then the Mordell-Weil lattice $E(\overline{\mathbf{Q}}(t))$ equipped with the height pairing is isomorphic to $A_1^* \oplus A_2^*$. The points with x-coordinate $x = b$ generate the A_1^*; the remaining 6 points with constant x-coordinate correspond to the vectors with minimal norm in A_2^* (and hence they generate, too).

Corollary. With the notation from Theorem 1, let $a(t) = t^2 - (\beta^2 + \beta + 1)^3$ and $b = (\beta^2 + \beta)^2$, for a $\beta \in \mathbf{Q}$ with $\beta \neq 0, -1$.

Then $E(\overline{\mathbf{Q}}(t)) = E(\mathbf{Q}(t)) \cong \mathbf{Z}^3$.

Concerning the relation between ranks of certain elliptic curves over **Q** and 3-rank of quadratic fields the following general statement holds.

Theorem 2. *Denote by $E_{a,b}/\mathbf{Q}$ the elliptic curve which is given by the equation $y^2 = x^3 + a(x - b)^2$ with $a, b \in \mathbf{Z}, a \neq 0, b \neq 0, 4a + 27b \neq 0$. Assume that*

1. $a \equiv 3 \bmod 4$ and $b \equiv 1 \bmod 2$ (or $a \equiv 1 \bmod 2$ and $b \equiv 2 \bmod 4$);

2. $a \equiv 2 \bmod 3$;

3. a is square free.

Write s for the number of primes $p \geq 5$ such that $p | b$ and the Legendre symbol $\left(\frac{a}{p}\right) = 1$. Similarly let t denote the number of primes $p \geq 5$ for which $p | 4a + 27b$ and $\left(\frac{-3a}{p}\right) = 1$.

Then

$$\operatorname{rank} E_{a,b}(\mathbf{Q}) \leq r_3(a) + r_3(-3a) + s + t + 1.$$

3 Rational 3-isogenies

We recall some general facts about 3-isogenies. For the basic theory of elliptic curves Silverman's book [13] is an excellent reference.

Let K be a field of characteristic different from $2, 3$. Suppose E/K is an elliptic curve and $T \subset E(\overline{K})$ is a subgroup of order 3 which is stable under the action of $\operatorname{Gal}(K^{\mathrm{sep}}/K)$. We can give E/K by an equation $y^2 = f(x)$ with f of degree 3. In these coordinates T consists of the point at infinity (the standard convention to take this point as the zero for the group law on E is used), plus two other points $P = (\alpha, \beta)$ and $2P = -P = (\alpha, -\beta)$. The Galois invariance implies $\alpha \in K$ and $\beta^2 \in K$. By a change of coordinates we can assume $\alpha = 0$. The curve is now given by an equation $y^2 = x^3 + ax^2 + cx + d$, and the point $(0, \sqrt{d})$ on this curve has order 3 precisely when $c^2 = 4ad$.

In case $c = 0$ our equation is $y^2 = x^3 + d$. In the other case $c \neq 0$ the equation can be written as $y^2 = x^3 + a(x - b)^2$.

Dividing out by the subgroup T yields another elliptic curve, which is again equipped with a rational subgroup of order 3. The function field of this curve is the subfield of $K(x, y)$ generated by the functions $x + \xi + \xi'$ and $y + \eta + \eta'$. Here (ξ, η) and (ξ', η') are the functions describing translation over the points $(0, \sqrt{d})$ and $(0, -\sqrt{d})$ respectively:

$$(\xi, \eta) = (x, y) + (0, \sqrt{d})$$

where '+' denotes addition in the group law on E.

One can choose coordinates on the new curve such that it is given in the same way as the original curve E. In particular, the rational subgroup on it (which corresponds to the 3-torsion points on E modulo T) is again given by points with first coordinate 0. A routine calculation reveals:

Case E : $y^2 = x^3 + a(x - b)^2$: The quotient curve E/T is given by the equation

$$\eta^2 = \xi^3 - 27a(\xi - 4a - 27b)^2$$

and the quotient map by

$$\xi = 9(2y^2 + 2ab^2 - x^3 - \frac{2}{3}ax^2)x^{-2}$$

and

$$\eta = 27y(-4abx + 8ab^2 - x^3)x^{-3}.$$

Case E : $y^2 = x^3 + d$: The quotient curve E/T is given by the equation

$$\eta^2 = \xi^3 - 27d$$

and the quotient map by

$$\xi = (y^2 + 3d)x^{-2} \text{ and } \eta = y(x^3 - 8d)x^{-3}.$$

Repeating the process, i.e., taking the quotient by the new subgroup on the new curve, corresponds to taking the quotient by all 3-torsion on the original curve; this is just multiplication by 3.

Explicit formulas for isogenies as the one above can be obtained more generally (using the same ideas) from a paper of Vélu [16].

4 The Kummer sequence of a rational 3-isogeny

In this section the procedure for relating the rank of elliptic curves admitting a 3-isogeny as above to the 3-rank of certain quadratic fields is discussed. For the little bits of Galois cohomology needed, any textbook on the topic will suffice; we need nothing beyond the relevant Appendix in Silverman's book [13].

If one has to compute the rank of an elliptic curve E over a number field K, the usual strategy is to bound this rank from above by embedding a quotient $E(K)/nE(K)$ into a more understandable finite group. The relatively easy case where $K = \mathbf{Q}, n = 2$ and multiplication by 2 can be factored as a product of isogenies of degree 2 is well known; the first written account of it seems to be the Haverford Lectures by Tate which form the basis of Husemöller's book on elliptic curves. The next simplest case is the one to be discussed here. In Tate's situation the target group is a finite subgroup of $\mathbf{Q}^*/\mathbf{Q}^{*2}$. Here following Quer we land in a finite subgroup of

$$\text{Kernel } K^*/K^{*3} \overset{\text{Norm}}{\longrightarrow} \mathbf{Q}^*/\mathbf{Q}^{*3},$$

in which $K/\mathbf{Q}$ is a quadratic extension. We start by introducing these subgroups and showing their relation to the class group of K.

Let $K = \mathbf{Q}(\sqrt{d})$ be a quadratic extension of $\mathbf{Q}$. The Norm homomorphism $N : K^* \to \mathbf{Q}^*$ given by $N(a + b\sqrt{d}) = a^2 - b^2d$ (for $a, b \in \mathbf{Q}$) induces a homomorphism $K^*/K^{*3} \to \mathbf{Q}^*/\mathbf{Q}^{*3}$ which will also be denoted N.

For a finite set of primes $p_1, \ldots, p_t \in \mathbf{Z}$ which all split in K, write

$$H(p_1, \ldots, p_t) := \{x \in \text{Ker}\,(N); v_\wp(x) \bmod 3 = 0 \; \forall \wp \in \mathcal{S} \text{ with } p_1 \cdots p_t \notin \wp\}$$

in which $\mathcal{S}$ denotes the set of all split primes of $K/\mathbf{Q}$. Decomposing the ideal of an $x \in K^*$ whose norm is a cube into prime ideals, and using that $v_\wp(x) + v_{\overline{\wp}}(x) \equiv 0 \bmod 3$, one obtains

Lemma 3.

$$\dim_{\mathbf{F}_3} H(p_1, \ldots, p_t) \leq r_3(d) + t + \dim_{\mathbf{F}_3} U/U^3.$$

Here U denotes the units in the ring of integers of K, hence the dimension of U/U^3 is 1 if d is positive or $K = \mathbf{Q}(\sqrt{-3})$ and 0 otherwise.

Suppose $E/\mathbf{Q}$ is an elliptic curve admitting a rational 3-isogeny $\psi : E \to E'$. Write ψ' for the dual isogeny; so $\psi'\psi$ is multiplication by 3 on E. If one assumes that E has no point of order 3 defined over $\mathbf{Q}$, or equivalently that both the kernel of ψ and the kernel of ψ contain no non-trivial rational point, then

$$\operatorname{rank} E(\mathbf{Q}) = \dim_{\mathbf{F}_3} E(\mathbf{Q})/3E(\mathbf{Q}).$$

Moreover under our assumptions the group written on the right fits in the exact sequence

$$0 \to E'(\mathbf{Q})/\psi E(\mathbf{Q}) \xrightarrow{\psi'} E(\mathbf{Q})/3E(\mathbf{Q}) \longrightarrow E(\mathbf{Q})/\psi' E'(\mathbf{Q}) \to 0.$$

What remains to be done is estimating the dimensions on the left and the right in the above sequence. Here Galois cohomology comes in. Write $G_K = \operatorname{Gal}(\overline{\mathbf{Q}}/K)$. In the preceding section it was shown that the kernel T of ψ consists of points which are rational over a field $\mathbf{Q}(\sqrt{d})$. The 'Kummer sequence of our 3-isogeny'

$$0 \to T \to E(\overline{\mathbf{Q}}) \to E'(\overline{\mathbf{Q}}) \to 0$$

yields a long exact sequence

$$\ldots \to E(\mathbf{Q}) \to E'(\mathbf{Q}) \to H^1(G_{\mathbf{Q}}, T) \to \ldots$$

and therefore an injection

$$E'(\mathbf{Q})/\psi E(\mathbf{Q}) \longrightarrow H^1(G_{\mathbf{Q}}, T).$$

Let K be any quadratic extension of $\mathbf{Q}$ and $\langle \tau \rangle = \operatorname{Gal}(K/\mathbf{Q})$. From the inflation-restriction sequence one obtains an injection (in fact an isomorphism)

$$H^1(G_{\mathbf{Q}}, T) \longrightarrow H^1(G_K, T)^{\langle \tau \rangle}.$$

The action of $\langle \tau \rangle$ on $H^1(G_K, T)$ here is given on cocycles as ${}^\tau \xi(\sigma) = \tilde{\tau}(\xi(\tilde{\tau}\sigma\tilde{\tau}^{-1}))$ in which $\tilde{\tau}$ is any element of $G_{\mathbf{Q}}$ which acts on K by τ.

In the situation we consider it seems very natural to take the restriction to $G_{\mathbf{Q}(\sqrt{d})}$. This is what Satgé [10] does. The group $G_{\mathbf{Q}(\sqrt{d})}$ acts trivially on T, hence the H^1 consists of homomorphisms from $G_{\mathbf{Q}(\sqrt{d})}$ to $\mathbf{Z}/3\mathbf{Z}$, or equivalently of cubic cyclic extensions of $\mathbf{Q}(\sqrt{d})$. There is however a second restriction which appears helpful, namely the one to $G_{\mathbf{Q}(\sqrt{-3d})}$. We will use the latter restriction; it is also considered by Quer [8].

One has $T \cong \mu_3$ as $G_{\mathbf{Q}(\sqrt{-3d})}$-modules, hence using Hilbert 90 one obtains

$$H^1(G_{\mathbf{Q}(\sqrt{-3d})}, T) \cong H^1(G_{\mathbf{Q}(\sqrt{-3d})}, \mu_3) \cong \mathbf{Q}(\sqrt{-3d})^* / \mathbf{Q}(\sqrt{-3d})^{*^3}.$$

Note that we deal with the $\langle \tau \rangle$-invariants of the first group. It is clear that a lifting $\tilde{\tau}$ fixes all of T precisely when it acts by inversion on μ_3. Hence the action induced on $\mathbf{Q}(\sqrt{-3d})^* / \mathbf{Q}(\sqrt{-3d})^{*^3}$ is not the natural one, but the one given by $^\tau x = \tau(x)^{-1}$. In particular the invariants are given as the kernel of the norm map.

Summarizing, one obtains an injective homomorphism

$$E'(\mathbf{Q})/\psi E(\mathbf{Q}) \longrightarrow \mathrm{Ker}\,(\ N\ :\ \mathbf{Q}(\sqrt{-3d})^*/\mathbf{Q}(\sqrt{-3d})^{*^3} \rightarrow \mathbf{Q}^*/\mathbf{Q}^{*^3}\,).$$

Using the explicit description of the morphism ψ as given in the previous section, and the definition of the various maps in cohomology it turns out that this injection can be given (up to a choice of an isomorphism $T \cong \mu_3$) as

$$(x,y) \mapsto (y + \sqrt{-27k}) \cdot \mathbf{Q}(\sqrt{-3k})^{*^3}$$

in case E' is given by $y^2 = x^3 - 27k$, and

$$(x,y) \mapsto (y + (x - 4a - 27b)\sqrt{-27a}) \cdot \mathbf{Q}(\sqrt{-3a})^{*^3}$$

in case E' is defined by an equation $y^2 = x^3 - 27a(x - 4a - 27b)^2$. However this explicit description will not be used.

The next task will be to bound the image of this injection. This is done by showing it is contained in a group $H(p_1, \ldots, p_t)$ as defined above. Let $p \in \mathbf{Z}$ be a prime which splits in $K = \mathbf{Q}(\sqrt{-3d})$. Take $\wp|p$ in K. Assume $p \neq 3$; then for $x \in K^*/K^{*3}$ to have $v_\wp(x) \bmod 3 = 0$ precisely means that x maps to 1 under the composition

$$K^*/K^{*3} \longrightarrow \mathbf{Q}_p^*/\mathbf{Q}_p^{*3} \longrightarrow \mathbf{Q}_p^{\mathrm{un}*}/\mathbf{Q}_p^{\mathrm{un}*3}.$$

Here the first map is defined using $\wp$ and $\mathbf{Q}_p^{\mathrm{un}}$ denotes the maximal unramified extension of $\mathbf{Q}_p$. In virtue of the commutative diagram

$$
\begin{array}{ccc}
E'(\mathbf{Q})/\psi E(\mathbf{Q}) & \longrightarrow & K^*/K^{*3} \\
\downarrow & & \downarrow \\
E'(\mathbf{Q}_p)/\psi E(\mathbf{Q}_p) & \longrightarrow & \mathbf{Q}_p^*/\mathbf{Q}_p^{*3} \\
\downarrow & & \downarrow \\
E'(\mathbf{Q}_p^{\mathrm{un}})/\psi E(\mathbf{Q}_p^{\mathrm{un}}) & \longrightarrow & \mathbf{Q}_p^{\mathrm{un}*}/\mathbf{Q}_p^{\mathrm{un}*3}
\end{array}
$$

(in which all the horizontal arrows are injections) this means that the only primes $p \neq 3$ we need to consider are the ones that split in K, and moreover the map

$$E'(\mathbf{Q}_p)/\psi E(\mathbf{Q}_p) \longrightarrow E'(\mathbf{Q}_p^{\mathrm{un}})/\psi E(\mathbf{Q}_p^{\mathrm{un}})$$

is not the zero map.

Note that the prime $p = 3$ of course needs special attention. However, by assuming 3 does not split in K this can be completely avoided. Let $p \neq 3$ be given; we assume that the equations defining E and E' are both minimal at p and that both curves have either good reduction at p or reduction of type II or I_ν^* for some $\nu \geq 0$. For $p = 2$ this can be achieved with the curve given by $y^2 = x^3 + a(x-b)^2$ by demanding $a \equiv 3 \bmod 4, b \equiv 1 \bmod 2$, or $a \equiv 1 \bmod 2, b \equiv 2 \bmod 4$. In both these situations E and E' have reduction of type II. For $p \geq 5$ one has good reduction if $p \nmid ab(4a + 27b)$; type II reduction if $p||a, p \nmid b(4a + 27b)$ and type I_ν^* reduction if $p||a, p|b$. Recall the standard notation [13], Chapter VII, or [1], pp. 41–46, $E_0(\mathbf{Q}_p)$ is the group of points which reduce mod p to points in the smooth part $E_0(\mathbf{F}_p)$ of $E(\mathbf{F}_p)$, and $E_1(\mathbf{Q}_p)$ is the kernel of the reduction map $E_0(\mathbf{Q}_p) \to E_0(\mathbf{F}_p)$.

By the minimality assumption ψ maps E_i to E'_i. Moreover the assumption on the reduction implies that the quotients $E(\mathbf{Q}_p)/E_0(\mathbf{Q}_p)$ and $E'(\mathbf{Q}_p)/E'_0(\mathbf{Q}_p)$ are isomorphic to groups of order $1, 2$ or 4. Hence the Kernel-Cokernel sequence of the commutative diagram

$$
\begin{array}{ccccccccc}
0 & \to & E_0(\mathbf{Q}_p) & \longrightarrow & E(\mathbf{Q}_p) & \longrightarrow & (*) & \to & 0 \\
  &   & \downarrow & & \downarrow & & \downarrow & & \\
0 & \to & E'_0(\mathbf{Q}_p) & \longrightarrow & E'(\mathbf{Q}_p) & \longrightarrow & (*) & \to & 0
\end{array}
$$

yields $E'(\mathbf{Q}_p)/\psi E(\mathbf{Q}_p) \cong E'_0(\mathbf{Q}_p)/\psi E'_0(\mathbf{Q}_p)$.

To compute the latter quotient (or rather its image over $\mathbf{Q}_p^{un}$) one uses the same technique applied to the diagram

$$
\begin{array}{ccccccccc}
0 & \to & E_1(\mathbf{Q}_p^{un}) & \longrightarrow & E_0(\mathbf{Q}_p^{un}) & \longrightarrow & E_0(\overline{\mathbf{F}}_p) & \to & 0 \\
  &   & \downarrow & & \downarrow & & \downarrow & & \\
0 & \to & E'_1(\mathbf{Q}_p^{un}) & \longrightarrow & E'_0(\mathbf{Q}_p^{un}) & \longrightarrow & E'_0(\overline{\mathbf{F}}_p) & \to & 0.
\end{array}
$$

The vertical arrow on the left is an isomorphism since multiplication by 3 defines an isomorphism on both E_1 and E'_1 ([13], IV.2.3 and VII.2.2). The vertical arrow on the right is surjective, hence it follows that the image of $E'(\mathbf{Q}_p)/\psi E(\mathbf{Q}_p)$ in $E'(\mathbf{Q}_p^{un})/\psi E(\mathbf{Q}_p^{un})$ is zero.

Note that in fact the above argument is only needed in case $p = 2$ and in case $p \geq 5$ such that E has good reduction at p. In the remaining cases p does not split in K, hence we can ignore it. Note also that the argument applies verbatim to the dual isogeny ψ'.

The proof of Theorem 2 is now almost complete. Assume the conditions mentioned in 2. By the argument given above, $E_{a,b}(\mathbf{Q})/\psi' E_{a,b}'(\mathbf{Q})$ injects into $H(p_1, \ldots, p_t)$, where the p_i are all primes such that $\left(\frac{a}{p_i}\right) = 1$ and $E_{a,b}$ has multiplicative reduction at p_i. Assume p is such a prime, then working over $\mathbf{Q}_p^{un}$ we have two cases:

1. If $p^n || 4a + 27b$ and $n > 0$ then one obtains a commutative diagram with exact rows and columns

$$
\begin{array}{ccccccc}
0 \to & 0 & \to & 0 & \to & 0 \\
& \downarrow & & \downarrow & & \downarrow \\
0 \to & 0 & \to & E'_0(\mathbf{Q}_p^{un}) & \to & E_0(\mathbf{Q}_p^{un}) \\
& \downarrow & & \downarrow & & \downarrow \\
0 \to & \mathbf{Z}/3\mathbf{Z} & \to & E'(\mathbf{Q}_p^{un}) & \to & E(\mathbf{Q}_p^{un}) \\
& \downarrow & & \downarrow & & \downarrow \\
0 \to & \mathbf{Z}/3\mathbf{Z} & \to & \mathbf{Z}/3n\mathbf{Z} & \to & \mathbf{Z}/n\mathbf{Z}.
\end{array}
$$

An isomorphism $E(\mathbf{Q}_p^{un})/\psi' E'(\mathbf{Q}_p^{un}) \cong E_0(\mathbf{Q}_p^{un})/\psi' E'_0(\mathbf{Q}_p^{un})$ follows. Since ψ' induces isomorphisms on the first and third group in the sequence

$$
E'_1(\mathbf{Q}_p^{un}) \longrightarrow E'_0(\mathbf{Q}_p^{un}) \longrightarrow \overline{\mathbf{F}_p}^*
$$

we conclude that the quotient we are computing is zero in this case.

2. The remaining case is $p|b$. Here the analogous argument shows that the quotient over $\mathbf{Q}_p^{un}$ is cyclic of order 3, and the quotient one obtains over $\mathbf{Q}_p$ surjects onto it. Hence these primes can not be avoided in general.

A similar argument shows that for the dual isogeny precisely the primes p of multiplicative reduction and $p|4a + 27b$ and $\left(\frac{b}{p}\right) = 1$ have to be taken. Since $\left(\frac{-3a}{p}\right) = \left(\frac{b}{p}\right)$ for $p|4a + 27b$ this finishes the proof of Theorem 2. $\square$

5   Some applications of the rank estimate

We will briefly discuss an attempt to prove the existence of many quadratic fields with a high 3-rank. First, note that e.g. Brinkhuis's example mentioned in the introduction already yields using the sieving results from [14]

$$
\# \{d \in \mathbf{Z} \; ; \; |d| < T \ \& \ r_3(d) \neq 0\} \geq C \cdot T^{1/2}
$$

for a positive constant C.

The simplest way to obtain positive 3-rank via our approach seems to be to look for elliptic curves of moderately high rank, such that by the method described above one obtains an estimate $r_3(d) + r_3(-3d) > 0$. Since it appears hard to find curves with very high rank, one may look for curves with rank 2 or 3, such that the invariants s and t appearing in the statement of Theorem 2 are 0.

An example is provided by the curves given as $E_a : y^2 = x^3 + 7a(x+a)^2$ in which a is a non-zero square free integer satisfying $a \equiv 5 \bmod 12$. This curve is the quadratic twist over $Q(\sqrt{a})$ of the one given by $y^2 = x^3 + 7(x + 1)^2$. For any such a the estimate

$$\text{rank } E_a(Q) \leq 1 + r_3(7a) + r_3(-21a)$$

holds. Under the assumption of a parity conjecture which claims that this rank equals modulo 2 the order of vanishing of a certain L-series, one can show following Mazur and Gouvêa that indeed we obtain the 'density exponent' 1/2 already given above. Note however that this is not unconditional! An account on the problem of finding many twists with high rank is given in [14].

A second example is given by the family $E_d : y^2 = x^3 + 16d^3$. From Satgé's paper [10] or from Quer's thesis [8] or using a computation analogous to the one given in the preceding section one shows that for every square free d one has

$$\text{rank } E_d(Q) \leq 1 + r_3(d) + r_3(-3d).$$

These are quadratic twists of the curve given by $y^2 + y = x^3$.

One could hope to use the following idea to get better density exponents. Suppose given a curve $E : y^2 = x^3 + a(t)(x - b)^2$ over the function field $Q(t)$, in which b is a constant and $a(t)$ a polynomial of degree 2. Suppose moreover that the rank of $E(Q(t))$ is 'high'. Then for almost all specializations of t to a rational number c/d, the rank remains at least as high. The specialization can be given by an equation

$$y^2 = x^3 + d^2 a(c/d)(x - bd^2)^2.$$

Writing $A(c, d) = d^2 a(c/d)$ (which is a binary quadratic form), we deal with the family of fields $Q(\sqrt{A(c, d)})$. Although this defines indeed a set of discriminants which has density exponent 1, it seems hard to control the 'error terms' s and t appearing in Theorem 2 for all curves in such a family. In the remaining two sections of this paper a construction of such pairs $a(t), b$ will be given.

One can try to describe constructions appearing in the literature on 3-rank using the language of elliptic curves as above. To illustrate this, consider Shanks's polynomial $D_3(t) = 27t^4 - 74t^3 + 84t^2 - 48t + 12$ (or an other polynomial dealt with by him as, e.g., $D_6(t)$ defined by $4D_6(t) = D_3(2t)$). Shanks has proven in [11] that many of the fields $Q(\sqrt{-D_3(t_0)})$ have 3-rank at least 2. In order to relate this result to elliptic curves, consider $E_{D_3}/Q(t)$ given by $y^2 = x^3 + 108D_3(t)$. Using the results from § 10 in Shioda's paper [12] it follows that this curve has $\overline{Q}(t)$-rank 6. Since the Mordell-Weil group is a module over the endomorphism ring of the curve one deduces from this that the $Q(t)$-rank is at most 3. Moreover, again using [12], one finds a set of generators among the 18 points

$(\rho x_1, \pm y_1)$, $(\rho x_2, \pm y_2)$, $(\rho x_3, \pm y_3)$ in which $\rho^3 = 1$ and x_i is a polynomial of degree 1. The points with x-coordinates $x = 6t$ and $x = 6t - 8$ yield such generators, hence rank $E_{D_3}(\mathbf{Q}(t)) \geq 2$. Apparently Shanks's result can be interpreted as the statement that under certain congruence conditions on $t_0 \in \mathbf{Z}$, these points map to independent elements in the class group part of $H^1(G_{\mathbf{Q}}, T)$.

A similar example, which is one of many given in a paper of Buell [3], is the curve

$$E \ : \quad y^2 = 4x^3 + t^4 + 10t^3 - 305t^2 - 416946t - 3321607.$$

From the theory of Mordell-Weil lattices it follows that the $\mathbf{Q}(t)$-rational points with x-coordinates $4t + 94$, $2t + 252$ and $-6t + 538$ generate $E(\overline{\mathbf{Q}}(t))$ as a module over $\mathrm{End}(E)$.

6 Families of rational 3-isogenies and Mordell-Weil lattices

In this section we will work over an algebraically closed field K of characteristic 0. Suppose an elliptic curve E over $K(t)$ is given, satisfying the conditions of Theorem 1. Multiplying the x and y function by a scalar and changing t to $\alpha t + \beta$ we may assume E is given by an equation

$$y^2 = x^3 + (t^2 + c)(x - 1)^2$$

in which $c \in K$ satisfies $c \neq 0$ and $4c + 27 \neq 0$.

From the work of Kodaira (compare [12]) it follows that one can regard $E/K(t)$ as the generic fibre of an elliptic surface $f \ : \ S \to \mathbf{P}^1$, which is called the Kodaira-Néron model of $E/K(t)$. Moreover, in our case this surface is rational, as is e.g. explained in [12], 10.13-10.14. A result of Oguiso and Shioda [7] implies that the structure of the Mordell-Weil lattice $E(K(t))$ in this situation is usually completely determined by the singular fibres of $f \ : \ S \to \mathbf{P}^1$.

An algorithm for determining these bad fibres and their respective types is given by Tate in [1]. One finds fibres of type II over the zeroes of $t^2 + c = 0$ and fibres of type I_1 over the t's satisfying $4t^2 + 4c + 27 = 0$. The only remaining bad fibre is over $t = \infty$. To compute it, change coordinates

$$\eta = \frac{y}{t^3}, \quad \xi = \frac{x}{t^2}, \quad s = \frac{1}{t}.$$

The equation becomes $\eta^2 = \xi^3 + (cs^2 + 1)(\xi - s^2)^2$ and one concludes easily that over $t = \infty$, which corresponds to $s = 0$ one has a fibre of type I_6.

The Main Theorem in [7] immediately implies that $E(K(t)) \cong A_1^* \oplus A_2^*$ which proves part of Theorem 1. What remains to be proven is that the points with constant x-coordinate yield the desired generators. We first compute these points. If a point with $x = \alpha$ is on the curve, then $(t^2 + c)(\alpha - 1)^2 + \alpha^3$ has to be a square in $K[t]$. In case $\alpha = 1$ this defines a constant, hence a square. In case $\alpha \neq 1$ we deal with a polynomial of degree 2. If it is a square then a square root must be of the form $(\alpha - 1)t + \beta$. One checks $\beta = 0$ and α is a zero of $X^3 + c(X - 1)^2$.

Next we compute the height $h(P)$ of these points. An explicit formula for it appears in [12], 8.12. Using the notations from that paper, one has

$$h(P) = 2 + 2(PO) - \mathrm{contr}_\infty(P).$$

Here (PO) is the intersection number of the sections in S defined by the point P and the point at infinity O. Since $P = (\alpha, *)$ it is clear that these sections do not intersect over any finite value of t. Over $t = \infty$ one uses the ξ, η, s-coordinates given above to conclude that also here is no intersection. Hence $(PO) = 0$.

The remaining term $\mathrm{contr}_\infty(P)$ appearing in the formula for the height is a bit more subtle. Recall that at $t = \infty$ we have a special fibre of type I_6, which is a 6-gon. The term we have to compute depends on which of the components of this 6-gon the section defined by P meets. If it meets the same one as the zero-section, then the contribution is 0. If it meets a component next to this one we have contribution 5/6. The components 'two steps away' from the identity component yield value 4/3, and the one opposite to it 3/2.

Locally the section defined by P is given by $x = \alpha, y = \beta t + \gamma$. Hence in the ξ, η, s-coordinates by $\xi = \alpha s^2, \eta = \beta s^2 + \gamma s^3$. The Kodaira-Néron model is obtained by repeatedly blowing up the singularities of the surface we have, and normalizing. Hence we have to do this process and inspect what happens with the given section. Most of the geometry we need is explained in [5], pp. 28–29.

In $\mathbf{A}^3$ with coordinates ξ, η, s the only singularity of the surface defined locally by $y^2 = \xi^3 + (cs^2 + 1)(\xi - s^2)^2$ becomes visible as the point $\xi = \eta = s = 0$. The blow up of $\mathbf{A}^3$ in this point is locally described in $\mathbf{A}^3 \times \mathbf{A}^2$ by the equations $ux = vy, x = vs, y = us$. Hence the strict transform of our surface is given by $u^2 = v^3 s + (cs^2 + 1)(v - s)^2$. The fibre over $s = 0$ becomes a triangle, of which one sees only two sides using our coordinates. The section we study becomes $v = \alpha s, u = \beta s + \gamma s^2$ and it is clear that in the special fibre we obtain precisely the singular point of the new surface. Since this singularity $u = v = s = 0$ is also the intersection point of the two new components meeting the identity component, it follows that $0 \neq \mathrm{contr}_\infty(P) \neq 5/6$.

We blow up once more by introducing new coordinates w, z satisfying $u = vw, uz = ws, s = zv$. The new equation is $w^2 = v^2 z + (cz^2 v^2 + 1)(1 - z^2)^2$. Our section is now given by $1 = \alpha z, w = \beta z + \gamma z^2 v$. The point

$w = v = 0, z = 1$ is the only singularity of this surface. The part of the special fibre not meeting this singularity is in the $z = 0$-plane. The 'new' part, contained in the plane given by $v = 0$, contains the section we study. One checks that if $\alpha \neq 1$ then the point we obtain in the special fibre is not the singular point. Hence in this case our section meets a component 'two steps away from the identity component', so $\text{contr}_\infty(P) = 4/3$. In case $\alpha = 1$ we do get the singular point, hence our section meets the component which shows up in the next blow up, and which is opposite to the identity component.

It follows that the points with $x = 1$ have height $1/2$ and the remaining ones with constant x-coordinate height $2/3$. Hence these are precisely the points with minimal norm in the Mordell-Weil lattice. This proves Theorem 1.

7 Examples with all sections defined over Q

It is obvious how to deduce the Corollary from Theorem 1. Hence it seems to be more interesting to show how one obtains such examples. We will now give a construction of elliptic curves E over $\mathbf{Q}(t)$ of the kind studied in the previous section, such that all the points in $E(\overline{\mathbf{Q}}(t))$ are already $\mathbf{Q}(t)$-rational.

Let $a(t) \in \mathbf{Q}[t]$ be a polynomial of degree 2 and $b \in \mathbf{Q}^*$. Assume neither $a(t)$ nor $4a(t) + 27b$ has a double zero. Write E ; $y^2 = x^3 + a(t)(x - b)^2$ as before. The Galois group $G_{\mathbf{Q}} = \text{Gal}(\overline{\mathbf{Q}}/\mathbf{Q})$ acts on the Mordell-Weil lattice $E(\overline{\mathbf{Q}}(t)) \cong A_1^* \oplus A_2^*$. The $G_{\mathbf{Q}}$-invariants $E(\mathbf{Q}(t))$ can therefore be decomposed as the ones coming from A_1^* and the ones coming from A_2^*. The first group is infinite cyclic, hence the invariants are either (0) or all of the group. A generator is given by $(x = b, y = b\sqrt{b})$. hence we have invariants here precisely when b is a square.

The generators of A_2^* are the points whose x-coordinate satisfy a certain cubic relation. Hence the $G_{\mathbf{Q}}$-action here is also easily determined. Without loss of generality we may assume $a(t)$ is of the form $a(t) = ct^2 + d$. In order to have invariants among the generators one certainly needs c to be a square. If this is the case, we may as well assume $c = 1$. What remains is that zeroes of $X^3 + d(X - b)^2$ have to be rational.

The number of rational zeroes of the polynomial above is the same as the number of rational points of order 2 on the elliptic curve given by $y^2 = x^3 + d(x - b)^2$. This is exactly a curve as is studied in this paper, namely one which admits a rational isogeny of degree 3. It turns out to be possible to write down a family of elliptic curves having both properties (in fact, the moduli space of elliptic curves having these properties is rational). An example of such a family shows up in a classical problem investigated by Fermat (compare [18]). Namely, for a parameter β, consider the curve defined by the equations

$$1 + \beta^2 x = u^2, \ 1 + (\beta + 1)^2 x = v^2, \ 1 + x = w^2.$$

This defines an elliptic curve; choosing a zero, the group of rational points over $\mathbf{Q}(\beta)$ is in fact $\mathbf{Z}/2\mathbf{Z} \times \mathbf{Z}/6\mathbf{Z}$. Using the classical methods of writing such a curve in Weierstrass form (as explained e.g. in [17], pp. 135–139) one obtains the equation

$$Y^2 = X^3 + (\beta^2 + \beta + 1)^2 \left(X + \frac{\beta^2(\beta+1)^2}{\beta^2 + \beta + 1} \right)^2.$$

Since we prefer a term $(X - b)^2$ in which b is a square we now twist this curve over the field $\mathbf{Q}(\beta, \sqrt{-\beta^2 - \beta - 1})$, which yields precisely the constants listed in the statement of Corollary 2.

As a last remark, note how subtle it seems to obtain fields with positive 3-rank using the family of curves just constructed. Namely, for every specialization we have a curve $y^2 = x^3 + a(x - b)^2$ in which b is a square. Hence the condition b is a square $\mathrm{mod}\, p$ is satisfied for all primes p. This means that the 'error term' $s + t$ appearing in Theorem 2 seems very hard to avoid.

References

1. Birch, B.J. and Kuyk, W. (1975). *Modular Functions of One Variable IV*. LNM 476. Springer-Verlag, New York-Berlin-Heidelberg-Tokyo.

2. Brinkhuis, J. (1990). Normal integral bases and the Spiegelungssatz of Scholz. *Preprint*, Erasmus University Rotterdam.

3. Buell, D.A. (1976). Class groups of quadratic fields. *Math. of Comp.*, **30**, 610–623.

4. Craig, M. (1977). A construction for irregular discriminants. *Osaka J. Math.*, **14**, 365–402.

5. Hartshorne, R. (1977). *Algebraic Geometry*. Springer Verlag, New York-Berlin-Heidelberg-Tokyo.

6. Nekovář, J. (1990). Class numbers of quadratic fields and Shimura's correspondence. *Math. Ann.*, **287**, 577–594.

7. Oguiso, K. and Shioda, T. (1991). The Mordell-Weil lattice of a rational elliptic surface. *Preprint*.

8. Quer, J. (1987). *Sobre el 3-rang dels Cossos Quadràtics i la Corba El·líptica $Y^2 = X^3 + M$*. Ph.D. thesis, Barcelona.

9. Quer, J. (1987). Corps quadratiques de 3-rang 6 et courbes elliptiques de rang 12. *C.R. Acad. Sci. Paris*, **305**, 215–218.

10. Satgé, Ph. (1986). Groupes de Selmer et corps cubiques. *J. Number Theory*, **23**, 294–317.

11. Shanks, D. (1972). New types of quadratic fields having three invariants divisible by three. *J. Number Theory*, **4**, 537–556.

12. Shioda, T. (1990). On the Mordell-Weil Lattices. *Comm. Math. Univ. Sancti Pauli*, **39**, 211–240.

13. Silverman, J.H. (1986). *The Arithmetic of Elliptic Curves.* Springer Verlag, New York-Berlin-Heidelberg-Tokyo.

14. Stewart, C.L. and Top, J. (1991). On twists of elliptic curves having rank at least two. *in preparation.*

15. Uchida, K. (1970). Unramified extensions of quadratic number fields II. *Tôhoku Math. J.*, **22**, 220–224.

16. Vélu, J. (1971). Isogénies entre courbes elliptiques. *C.R. Acad. Sc. Paris*, **273**, 238–241.

17. Weil, A. (1983). *Number Theory.* Birkhäuser, Boston-Basel-Stuttgart.

18. Zagier, D. (1990). Elliptische Kurven: Fortschritte und Anwendungen. *Jber. d. Dt. Math. Verein.*, **92**, 58–76.

Jaap Top
Erasmus University Rotterdam
Vakgroep Wiskunde
P.O. Box 1738, 3000 DR Rotterdam
The Netherlands
e-mail top@cvx.eur.nl

Finite heights and rational points on surfaces

Yuri Tschinkel

Massachusetts Institute of Technology

1   Introduction

In this note we study the distribution of rational points on algebraic surfaces. Geometry and combinatorics are used to establish upper bounds for the number of rational points of bounded height.

Acknowledgments. I would like to thank Yu. I. Manin for many years of generous support and encouragement and the referee for very careful reading and many helpful remarks.

Let V be a smooth projective algebraic variety defined over a number field k, $[k : \mathbb{Q}] < \infty$. Let M_k be the set of all places, M_k^f (resp. M_k^∞) - the set of all finite (resp. infinite) places of k. Let $V(k)$ be the set of k-rational points of V. Denote by K_V the canonical bundle of V. A Fano-variety is a variety with an ample anticanonical bundle. Examples are projective spaces, complete intersections of small degree with a large number of variables etc. One finds many rational points on such varieties.

It was a common perception [2], partially due to the extremely fruitful concept of counting points over finite fields, that one should introduce appropriate counting functions on $V(k)$ and give an interpretation for the number of points of bounded height in simple geometrical terms. Counting functions for points on varieties were known for many years, these are the height functions. V. Batyrev and Yu. Manin [3] proposed a picture of what is to be expected if the variety contains many points. Their conjectures concern mostly the case when $-K_V$ is ample.

Let L be a very ample line bundle on V and $U \subset V$ some subset. Denote by N_{eff}^1 the cone of effective divisors on V. Define the invariants $\alpha(L)$ and $\beta_U(L)$ as

$$\alpha(L) := \inf\{\gamma \in \mathbb{R} \mid \gamma[L] + K_V \in N_{eff}^1\}$$

$$\beta_U(L) := \inf\{\sigma \in \mathbb{R} \mid \sum_{P \in U(k)} H_L(P)^{-s} \text{ conv. for } Re\ s > \sigma\}$$

We see that $\alpha(L)$ is geometrical in nature, whereas $\beta_U(L)$ reflects the arithmetic of V. Denoting by

$$N(U, L, B) = \sharp\{P \in U(k) \mid H_L(P) \le B\}$$

one has for any $\epsilon > 0$

$$N(U, L, B) = O(B^{\beta_U(L)+\epsilon}).$$

Thus $\beta_U(L)$ is a rough measure for the density of rational points on V.

Conjecture. [3] For all $\epsilon > 0$ and any number field k there exists a Zariski open $U = U(L, \epsilon) \subset V$ such that

$$\beta_U(L) \le \alpha(L) + \epsilon.$$

If $V(k)$ is infinite there exists a unique minimal Zariski closed subset Z in V such that

$$\beta_V(L) = \beta_Z(L) > \beta_{V \setminus Z}(L).$$

Of course, Z may coincide with V. If this is not the case we speak about the *accumulating property* and call Z the (minimal) *accumulating subset* (in V with respect to L). For Fano-varieties one expects the *linear growth property* with respect to $-K_V$ ($\beta_U(-K_V) = 1$) for a sufficiently large number field k and any sufficiently small open dense U. The existence of accumulating subvarieties causes serious difficulties in obtaining upper bounds for the number of rational points of bounded height since standard analytical methods do not apply.

Here is the plan of this note: In section 2 we introduce heights and markings, in section 3 we present the method, in section 4 we prove upper bounds for the number of rational points on Del Pezzo surfaces.

2   Heights and markings

In this section we introduce heights and explain the method. We emphasize a new aspect in the theory of heights which is the main ingredient of the proofs.

There are two parallel approaches to the theory of heights: Weil's and Arakelov's versions. The more precise Arakelov setup is not necessary for the proofs and we prefer to work with Weil's heights. Standard references are [4,5,6].

Weil's heights. Let V be a projective variety over k and L a very ample invertible sheaf. Fix a basis of global sections $s = (s_0, ..., s_n)$. For every point $P \in V(k)$ there exists some j such that $s_j(P) \neq 0$. An L-height is a function $V(k) \to \mathbb{R}^+$, defined as

$$H_{L,s}(P) = \prod_{\nu \in M_k} \max_i \{|s_i(P)/s_j(P)|_\nu\}.$$

The definition does not depend on the choice of j by the product formula. It depends on the choice of a basis of global sections $s = (s_0, ..., s_n)$.

Properties: ([4],Ch.10)

1) Let s' be another basis of global sections of L and $H_{L,s'}$ the corresponding height. There exist constants a, b such that

$$aH_{L,s}(P) \leq H_{L,s'}(P) \leq bH_{L,s}(P)$$

for all $P \in V(k)$.

Therefore studying properties which are invariant with respect to this indeterminacy (for example $\beta_U(L)$), we shall denote by H_L an arbitrary L-height.

2) There exists a unique homomorphism

$$
\begin{array}{ccc}
Pic(V) & \to & \mathcal{H}(V)^* \\
L & \to & H_L
\end{array}
$$

such that for very ample L the height H_L coincides with the definition above. $\mathcal{H}(V)^*$ is the (multiplicative) set of equivalence classes of functions $H : V(k) \to \mathbb{R}^+$. Functions H and H' are equivalent if there exists a bounded function $b : V(k) \to \mathbb{R}^+$ such that $H(P) = b(P) \cdot H'(P)$ for all $P \in V(k)$.

3) Functoriality: for $f : V \to W$ we have

$$H_{f \cdot L} \sim H_L \circ f$$

where $\sim$ means that the corresponding functions are equivalent in the sense defined above.

Usually one defines the height as

$$H'_L(P) = H_L(P)^{1/[k:\mathbb{Q}]}$$

Our height is not invariant with respect to field extensions. With the usual normalisation the degree of $[k : \mathbb{Q}]$ appears in the asymptotic. We changed the normalisation to have a uniform formulation of the conjectures. Our choice of normalisation conjecturally insures the linear growth property with respect to $-K_V$.

Finite heights. Let V be a projective variety over k and L a very ample line bundle. Fix a basis of of global sections $s = (s_0, ..., s_n)$ where

$s_i \in \Gamma(L)$. Let $t \in \Gamma(L)$. Let D be the divisor given by $t = 0$. For all rational points $P \in (V \backslash |D|)(k)$ and for all places $\nu \in M_k$ define the local Weil heights of P with respect to the effective divisor D as

$$H_{D,\nu}(P) = \max_i (|s_i(P)/t(P)|_\nu).$$

Fact: For effective divisors D and for all places ν there exist constants $c(D, \nu) > 0$ such that

$$H_{D,\nu}(P) > c(D, \nu)$$

for all points $P \in (V \backslash |D|)(k)$ ([4] Ch.10).

Define Weil's finite height of P with respect to D as

$$H_{D,f}(P) = \prod_{\nu \in M_k^f} H_{D,\nu}(P).$$

It follows that

$$H_{D,f}(P) = N_{k/\mathbb{Q}}(t(P)) / N_{k/\mathbb{Q}}(gcd(s_i(P))).$$

where $gcd(s_i(P))$ is the ideal generated by $s_i(P)$. Define

$$H_{D,\infty}(P) = \prod_{\nu \in M_k^\infty} H_{D,\nu}(P).$$

Finite heights depend on the choice of the basis s and on the choice of the equation for D.

Properties: ([4],Ch.10)

1) Let s' be another basis of global sections of L, t' another equation for D and $H'_{D,f}$ the corresponding finite height. Then there exists a finite valued function $d : V(k) \to \mathbb{Q}^+$ such that

$$H_{D,f}(P) = d(P) H'_{D,f}(P).$$

Therefore, studying properties which are invariant with respect to this indeterminacy we shall denote by $H_{L,f}$ an arbitrary finite D-height.

2) There exists a unique homomorphism, which on effective divisors given by global equations coincides with the definition above

$$\begin{array}{ccc} Pic(V) & \to & \mathcal{H}_f(V)^* \\ D & \to & H_{D,f} \end{array}$$

where $\mathcal{H}_f(V)^*$ is the (multiplicative) set of equivalence classes of functions $H : V(k) \to \mathbb{Q}^+$. Functions H and H' are equivalent, if there exists a finite valued function $d : V(k) \to \mathbb{Q}^+$ such that $H(P) = d(P) \cdot H'(P)$ for all $P \in V(k)$.

3) Functoriality: for $g : V \to W$ we have

$$H_{g \cdot D, f} \sim H_{D, f} \circ g$$

where $\sim$ means that the corresponding functions are equivalent in the sense defined above.

Proposition 1. *Let D be an effective divisor on V. Then*

1. $H_{D, \infty}(P)$ is bounded from below by some constant (not depending on P). It follows that there exists a constant c such that $H_D(P) \geq c \cdot H_{D, f}(P)$.

2. $H_{D, f}(P) \neq 0$ for $P \notin |D|$.

3. For effective divisors D_1, D_2 such that their supports don't intersect, there exists an integer q independent of P such that we have that $\gcd(H_{D_1, f}(P), H_{D_2, f}(P))$ divides q. We say that such finite heights are "almost coprime".

4. Let η be the map $\eta : (\mathbb{P}^n \setminus \cup_{i=0}^n |D_i|)(k) \to \mathbb{Q}_{>0}^{n+1}$ sending P to $(H_{D_0, f}(P), ..., H_{D_n, f}(P))$, where D_i are the hyperplanes defined by $s_i = 0$, $(i = 0, \ldots, n)$. Then the number of points having the same image under η of $\mathcal{O}(1)$ height $\leq B$ is bounded by a constant (independent of B) for $k = \mathbb{Q}$ and $O(B^\epsilon)$ otherwise.

Markings. Let $\{D_i\}_{i \in I}$ be a set of effective divisors on V. We associate to a point $P \in V(k)$ a Weil marking:

$$\mathcal{A}_W := \{H_{D_i, f}(P)\}_{i \in I}$$

A *complete marking* $\{H_{D_i, f}(P)\}$ with respect to a very ample L is a marking $\mathcal{A}$ such that there exists a basis of sections of L with zeros of the form

$$\sum_{y_i \in \mathbb{Z}_{\geq 0}} y_i D_i.$$

We will use the terminology: "one can reconstruct a set of finite heights (J) from another set of finite heights $(I \subset J)$ with finite (resp. $O(B^\epsilon)$) indeterminacy". By definition, this means, that in the sequence

$$V(k) \to \{H_{D_i, f}(P)\}_{i \in J} \to \{H_{D_i, f}(P)\}_{i \in I}$$

the second map has finite (resp. $O(B^\epsilon)$) fibers.

3 The method

Among various possible choices of coordinates (bases for the space of sections of a very ample divisor L) we are trying to find very special ones, preserving the most amount of information about the variety. The suggestion is that one should look for sections with zero divisors consisting of many irreducible components. In this way the inequality $H_L(P) \leq B$ translates into knowledge about finite heights of the points $P \in V(k)$ with respect to many divisors and one can hope to recover the complete marking of P using considerations from geometry and to obtain upper bounds for the number of admissible markings. Additional information comes from the integrality of finite heights.

From now on let V denote a surface. To a rational point on a surface we associate a finite (or infinite) complete marking. This is a vector of integers with components indexed by rational curves on V (in different degrees with respect to L)

Data: $[\,NS, (.), N^1_{eff}, N^1_+\,]$ where NS denotes the Néron-Severi group of V with the intersection pairing $(.)$, N^1_{eff} - the cone of effective divisors and N^1_+ - the ample cone. Choose a family of rational curves $\{D_i\}$. We don't know an optimal choice. One could look for curves of minimal degree with respect to some fixed L (for example $L = -K_V$), or for the union (or parts) of orbits of the symmetry (or Galois) group acting on V (or $V \otimes_k \bar{k}$). One could also include accumulating divisors.

Step 1. Find $R \in N^1_{eff}$ such that $H^0(V, R) = d$ and the equation

$$\sum_{x_i \in \mathbb{Z}_{\geq 0}} x_i D_i = R$$

has $\geq d + 1$ solutions. Linear dependence between sections in $H^0(V, R)$ provides polynomial relations between $H_{D_{i,f}}(*)$ with controlled coefficients.

Step 2. Find a very ample L such that there exists a complete marking with respect to L, consisting of $D_{i,f}$ from Step 1. Now we are able to express $H_{L,f}$ in terms of $H_{D_i,f}$.

Step 3. $\beta_U(L)$ is estimated through the number of all possible complete markings with respect to L. If

$$\beta_U(L) < \max_i \frac{2}{(L, D_i)}$$

we have proved the accumulating property. Outside some of the rational curves D_i we find asymptotically strictly less rational points. The property of being outside D_i is expressed through $H_{D_i,f}(P) \neq 0$. Let us remark that in some cases lower bounds for $\beta_U(L)$ obtained by different methods coincide with upper bounds one establishes using the sketched approach.

Here are distinguished candidates to be used for the reconstruction of a complete marking. Suppose a surface V admits a fibration over a curve C. Let $\pi : V \to C$ be a proper, surjective map. The most interesting case is $C = \mathbb{P}^1$, otherwise the number of rational points on the base is already small. Any three sections s_i ($i = 0, 1, 2$) of $\mathcal{O}(1)$ on $\mathbb{P}^1$ are linearly dependent. The relation $\sum_{i=0}^{2} a_i s_i = 0$, $a_i \in k$ will be lifted to V under π^*.

Application 1. Del Pezzo surfaces of degree ≤ 5 possess a conic bundle structure over $\mathbb{P}^1$ with at least 3 degenerate fibers. The components of the degenerate fibers correspond to exceptional curves on the surface. Choose s_0, s_1, s_2 vanishing at the singular points of the conic fibration on $\mathbb{P}^1$. The zero-divisor of $\pi^*(s_i)$ consists of 2 components: D_i and D_i'. Denote $D_i(P) := H_{D_i, f}(P)$. Suppose the heights $H_{D_i}(P), H_{D_i'}(P)$ are $\leq B$.

Lemma 2. *Knowing* $D_0(P)$, $D_0'(P)$, $D_1(P)$, $D_1'(P)$, *one can reconstruct* $D_2(P)$ *and* $D_2'(P)$ *with an ambiguity of* $O(B^\epsilon)$.

These relations will be called *quadratic relations*.

Application 2. Enriques surfaces admit elliptic fibrations $\pi : V \to \mathbb{P}^1$ with 2 degenerate double fibers and with possibly several ordinary degenerate fibers. Consider sections s_0, s_1 of $\mathcal{O}(1)$ on $\mathbb{P}^1$ vanishing at the singular points of the fibration corresponding to double fibers. Assume all heights with respect to the irreducible components of the degenerate fibers to be $\leq B$.

Lemma 3. *If we know the finite height with respect to the zero divisor of any other section* $s_2(P)$ *of* $\mathcal{O}(1)$ *on* $\mathbb{P}^1$ *we can reconstruct the finite heights with respect to the divisors in the degenerate double fibers up to* $O(B^\epsilon)$.

In [9] we prove Lemma 2 and 3 and use Lemma 3 to establish upper bounds for the number of rational points on some K3 and Enriques surfaces.

4   Del Pezzo surfaces

Del Pezzo surfaces of degree d are smooth projective rational surfaces with an ample anticanonical bundle and $(K_V, K_V) = d$. Over $\overline{k}$ they are $\mathbb{P}^2$, $\mathbb{P}^1 \times \mathbb{P}^1$ or blowing ups of the projective plane $\pi : V_a \to \mathbb{P}^2$, in $a = 9 - d$ (≤ 8) points in general position. We will study Del Pezzo surfaces of degree 3, 4 and 5. Here are some more notations for the exceptional curves (lines in the anticanonical embedding):

D_i — the exceptional curves which are pre-images of the a points,

D_{ij} — exceptional curves which are pre-images of lines on $\mathbb{P}^2$ joining these points,

Q (resp. Q_i) — pre-images of conics on $\mathbb{P}^2$ passing through 5 points on the Del Pezzo surface of degree 4 (and missing the i-th point on $\mathbb{P}^2$ for a Del Pezzo surface of degree 3).

These notations are convenient for the intersection theory on V:

$$
\begin{array}{llll}
(D_i, D_j) & = 0 & (Q_i, D_j) & = 1 \\
(D_i, D_{ij}) & = 1 & (D_{kl}, D_{ij}) & = 1 \\
(Q_i, D_{ij}) & = 1 & (D_{ki}, D_{ij}) & = 0 \\
(Q_i, D_i) & = 0 & C^2 & = -1
\end{array}
$$

for $i \neq j \neq k$ and for all exceptional curves C. Denote by E_a the intersection graph of exceptional curves on V_a. For E_5 the 16 exceptional lines are organised in the following figure

$$
\begin{array}{ll}
D_3 - - - - D_{03} & D_{34} - - - - D_{12} \\
D_2 - - - - D_{02} & D_{24} - - - - D_{13} \\
D_1 - - - - D_{01} & D_{14} - - - - D_{23} \\
D_{04} - - - - D_4 & D_0 - - - - Q
\end{array}
$$

Fig. 1

A Del Pezzo surface will be called *split* if the exceptional curves are defined already over k. Define $U_a = V_a \backslash \{lines\}$.

The following theorem is from [8]. Here we prove one of the statements differently emphasizing the use of finite heights.

Theorem 4. *For a split V_4 and any $\epsilon > 0$, we have*

$$
N(U_4, -K_{V_4}, B) = \begin{cases} O(B(\log B)^6) & for \quad k = \mathbb{Q}, \\ O(B^{1+\epsilon}) & otherwise \end{cases}
$$

For a split V_5, any $\epsilon > 0$ and any k we have

$$
N(U_5, -K_{V_5}, B) = O(B^{5/4+\epsilon}).
$$

For a split V_6, any $\epsilon > 0$ and any k we have

$$
N(U_6, -K_{V_6}, B) = O(B^{5/3+\epsilon}).
$$

Proof. We give a proof for the second statement and only over $\mathbb{Q}$. As a system of effective divisors on V_5 we take the 16 exceptional lines. To each rational point we assign the marking $\{H_{D,f}(P)\}_{lines}$. Denote for brevity $H_{D_i,f}(P) = D_i(P)$. The anticanonical line bundle has a basis of sections with zeros consisting of four lines. They are represented in subgraphs of the intersection graph E_5 of V_5 of the form:

Fig. 2

Note that a marking by exceptional curves is complete with respect to the anticanonical bundle for all Del Pezzo of degree ≤ 5. Since $H_{-K_V,f}(P) \leq H_{-K_V}(P)$ the inequality $H_{-K_V}(P) \leq B$ can be weakened to the following statement: $\max \prod H_{D,f}(P) \leq B$, where the maximum is taken over all subgraphs of the form in Fig. 2 and the product is taken over the finite heights with respect to lines in the corresponding subgraph. The inequality above has $O(B(\log B)^3)$ solutions in integers $H_{D_i,f}(P)$. However, we cannot prove that one can reconstruct the point P even with indeterminacy $O(B^\epsilon)$ from the finite heights with respect to lines in *one* of the subgraphs of the form in Fig. 2. We shall apply the following trick.

Represent $U_5(\mathbb{Q})$ as a finite union of subsets $U^{(D,D')}$. The subsets are indexed by non-intersecting exceptional curves $D, D' \in E_5$. These subsets are defined as follows.

Denote $l(P) = H_{l,f}(P)$. For every rational point P we can order the set of finite heights $\{l(P)\}$:

$$l_1(P) \le l_2(P) \le l_3(P) \le l(P)$$

for all $l \in E_5 \backslash \{l_1, l_2, l_3\}$.

There are 3 possible cases.

1) $l_1 \cap l_2 = \emptyset$

Then, by definition, P belongs to $U^{(l_1,l_2)}$.

2) $l_1 \cap l_2 \ne \emptyset$ but $l_1 \cap l_3 = \emptyset$.

Then P belongs to $U^{(l_1,l_3)}$.

3) $l_1 \cap l_2 \ne \emptyset$ and $l_1 \cap l_3 \ne \emptyset$.

In this case it follows that $l_3 \cap l_2 = \emptyset$ and P belongs to $U^{(l_2,l_3)}$.

By symmetry, it is sufficient to estimate the number of points in the set $U^{(D_3,D_2)}$ which are not in the subset of points $P \in U_5(\mathbb{Q})$ satisfying the height inequality $H_{-K_V}(P) \le B$.

Claim: For all $P \in U^{(D_3,D_2)}$ we have $D_2(P) \le B^{1/4}$.

We will see that there is a hyperplane section of V_5 (in the anticanonical embedding) passing through D_2 and degenerating in 4 lines, such that $D_2(P)$ is the smallest finite height among these 4 corresponding finite heights (for points $P \in U^{(D_3,D_2)}$). Comparing to the height inequality we get the estimate.

In Case 1) the only finite height smaller than $D_2(P)$ is $D_3(P)$. We have

$$D_2(P)D_{02}(P)D_{14}(P)D_{23}(P) \le B.$$

In Case 2) the only finite heights smaller than $D_2(P)$ are $D_3(P)$, $D_{3j}(P)$ for some j or $Q(P)$. If $D_{3j}(P) < D_2(P)$ take

$$D_2(P)D_{24}(P)D_4(P)Q(P) \le B.$$

If $Q(P) < D_2(P)$ take

$$D_2(P)D_{02}(P)D_{34}(P)D_{12}(P) \le B.$$

In Case 3) the only finite heights smaller than $D_2(P)$ are $D_3(P)$ and $D_{23}(P)$. Take

$$D_2(P)D_{24}(P)D_4(P)Q(P) \le B.$$

Now consider (Fig. 1). By our definition of $U^{(D_3,D_2)}$ the only finite height smaller than $D_3(P)$ can be $D_{3j}(P)$ for some j or $Q(P)$. It follows

$D_0(P) > D_3(P)$. Put $D_0'(P) = [D_0(P)/D_3(P)]$. Weaken the height to $D_{03}(P)D_{14}(P)D_{02}(P)D_0'(P)D_3(P) \leq B$. We get $B(\log B)^4$ such 5-tuples. Adding $D_2(P)$ we have $B^{5/4}(\log B)^4$ 6-tuples

$$(D_{02}(P), D_{03}(P), D_0'(P), D_3(P), D_{14}(P), D_2(P)).$$

From this set we recover the rest with an ambiguity of $O(B^\epsilon)$. Quadratic relations give $D_1(P), D_{01}(P), D_{04}(P), D_4(P)$ with an ambiguity of $O(B^\epsilon)$. From

$$
\begin{array}{ccc}
D_0 & ----- & D_{04} \\
D_1 & ----- & D_{14} \\
D_2 & ----- & D_{24} \\
D_3 & ----- & D_{34}
\end{array}
$$

we get $D_0(P) \bmod D_3(P)$ and we can reconstruct $D_0(P)$ with finite ambiguity ($D_{04}(P), D_3(P)$ are almost coprime) and then $D_{34}(P), D_{04}(P)$. From

$$
\begin{array}{ccc}
D_{02} & ----- & D_{14} \\
D_{01} & ----- & D_{24} \\
D_{04} & ----- & D_{12} \\
D_3 & ----- & Q
\end{array}
$$

we get $Q(P)$ and $D_{12}(P)$ and finally from the right column in (Fig. 1) the rest.

Remark. From the algorithmic point of view, we described a procedure to obtain rational points of bounded height on a split intersection of two quadrics in $\mathbb{P}^4$ in $O(B^{5/4+\epsilon})$ time. Previous algorithms required at least $O(B^{2+\epsilon})$. In addition, all integers appearing in our procedure are of the size $O(B)$ and not $O(B^2)$ as in standard techniques. There is a similar method for cubic surfaces.

References

[1] P. Vojta, *Diophantine approximation and value distribution theory*, (1987), SLN 1239, Springer.

[2] J. Silverman, Integral points on curves and surfaces, in *Number Theory, Ulm 1987, Proceedings*,(1989), SLN 1380, Springer, p.202-241.

[3] V. Batyrev, Yu. Manin, (1990), Sur les nombre des points rationels de hauteur bornée des variétés algébriques, Math. Ann **286** p.27-43.

[4] S. Lang, *Fundamentals of Diophantine Geometry*, (1983), Springer.

[5] S. Lang, *Introduction to Arakelov Theory*, (1988), Springer.

[6] A. Weil, L'arithmétique sur les courbes algébriques, (1928), Acta Math. **52**, p. 281-315.

[7] A. Weil, Arithmétique and Géometrie sur les variétés algébriques, (1935), Act. Sc. et Ind. **206**, Hermann, Paris p.3-16.

[8] Yu. Manin, Yu. Tschinkel, Rational points on Del Pezzo surfaces, (1991).

[9] Yu. Tschinkel, Thesis, M.I.T., (1992)

Yuri Tschinkel
Department of Mathematics
Massachusetts Institute of Technology (MIT)
Cambridge, MA 02139
USA
e-mail tschink@math.mit.edu

Chapter IV

Diophantine Approximation

Invited Addresses

and

Contributed Papers

Linear recurrence relations for some generalized Pisot sequences

David W. Boyd

University of British Columbia

Abstract

The sequences of integers which we consider are defined for arbitrary integers $0 < a_0 < a_1$ by $a_{n+2} = \left[\frac{a_{n+1}^2}{a_n} + r \right]$, where r is a fixed real number. We consider mainly the three special cases $r = 0, \frac{1}{2}$, and 1. For $r = \frac{1}{2}$, these are the Pisot sequences; when $r = 1$, they are the sequences recently suggested by Shallit. We show that, for $r = 0$ and 1, the possible recurrence relations for the sequences are more restricted than in the case $0 < r < 1$ and give some examples and experimental evidence for differences between the three types of sequences.

1 Introduction:

Let r be a given real number. For arbitrary integers $0 < a_0 < a_1$, define the sequence of integers $E_r(a_0, a_1)$ by the nonlinear recurrence relation

$$a_{n+2} = \left[\frac{a_{n+1}^2}{a_n} + r \right]. \tag{1}$$

For $r = \frac{1}{2}$, the right member of (1) is the "nearest" integer to $\frac{a_{n+1}^2}{a_n}$. These sequences were introduced by Pisot [Pi] in his study of the distribution of sequences of the form $\lambda \theta^n$ modulo one. Flor [Fl] showed that, if such a sequence satisfies a linear recurrence relation then the defining polynomial $Q(x)$ of that relation must either be $(x - 1)^2$ or else have a single root $\theta > 1$ outside the unit circle. In the latter case, $Q(x)$ factors as $P(x)K(x)$ where P is the minimal polynomial of θ and K is a cyclotomic polynomial with simple roots (not necessarily irreducible) or else $K(x) \equiv 1$. If all the other conjugates of θ are strictly inside the unit circle, then θ is a Pisot number, otherwise θ is a Salem number. Throughout this paper,

This research was supported in part by an operating grant from NSERC we will maintain the convention that P denotes the minimal polynomial of a Pisot or Salem number θ and that K denotes a cyclotomic polynomial with simple roots.

Not all polynomials of the form PK can occur as defining polynomials of a Pisot sequence. Flor showed that $Q(x) = P(x)$ can always occur but that $P(x)(x-1)$ can occur only if $|P(1)| > 2(\theta-1)^2$. Some generalizations of this were obtained in [B3] where criteria were given for deciding whether any particular combination PK can occur. It was shown that, if $\theta > 1 + 2^{d-2}$ then no nontrivial cyclotomic factor can be combined with P to give the defining polynomial of a recurrence for a Pisot sequence.

Recently, Shallit [Sh] asked about linear recurrence relations for the sequences (1) when $r = 1$. These sequences have the pleasant property that the ratios a_{n+1}/a_n form an increasing sequence and indeed, a_{n+2} is the minimal integer for which $a_{n+2}/a_{n+1} > a_{n+1}/a_n$. Although, at first glance, it would seem likely that the restrictions on recurrence relations in this case should be much the same as for Pisot sequences, it turns out that this is not the case as we will indicate below.

Another natural choice of r is $r = 0$, which was mentioned at the end of a paper of Cantor [Ca]. As Cantor pointed out in a conversation with Shallit, if one allows the a_n to take on negative values, then changing a_n to $(-1)^n a_n$ almost interchanges sequences with $r = 0$ and $r = 1$ (except that "floor" should now be changed to "ceiling", which is generally insignificant.) Thus, we would expect certain properties of Shallit sequences to be shared by sequences with $r = 0$, which we call "Truncate" sequences.

In contrast to the situation for Pisot sequences, if $r = 0$ or 1, if d is fixed and if $\theta > 1 + 2^{d-1}$ is a Salem number of degree d, then we show that there is no E_r-sequence for which $\lim \dfrac{a_{n+1}}{a_n} \to \theta$. If $r < 0$ or $r > 1$ and if $\theta > 1 + \max(r, 1-r)2^{d-1}$, and if θ is a Pisot number *or* a Salem number then the same conclusion holds. Thus, for any given degree there are only a finite number of Pisot or Salem numbers occurring as limiting ratios of E_r sequences in this case.

We showed in [B1],[B4] that there are Pisot sequences which satisfy no linear recurrence relation. Specific examples with a_0 small are given in [B2]. The method used there applies equally well to the sequences (1) for any value of r. We do not pursue this here.

For convenience, we shall write $E_r(a_0, a_1)$ as $T(a_0, a_1), E(a_0, a_1)$, and $S(a_0, a_1)$ in the three cases $r = 0, \frac{1}{2}$, and 1.

2   Restrictions on possible linear recurrences.

It is not hard to show, as in [Fl], that for sequences defined by (1), $\lim \dfrac{a_{n+1}}{a_n} = \theta \geq 1$ always exists, and if $a_1 > a_0 + c\sqrt{a_0}$ (where c depends on r), then

$\theta > 1$. We will always assume that $\theta > 1$ here, since otherwise the sequence has defining polynomial $(x-1)^2$. When $\theta > 1$, we also have $\lim \frac{a_n}{\theta^n} = \lambda > 0$. We may rewrite the definition (1) as

$$a_{n+2} - \frac{a_{n+1}^2}{a_n} \in I_r = (r-1, r]. \tag{2}$$

If we write $a_n = \lambda\theta^n + \epsilon_n$ and substitute this into (2), we find that

$$(E - \theta)^2 a_n - \frac{((E-\theta)a_n)^2}{a_n} \in I_r, \tag{3}$$

where E denotes Boole's shift operator for which $Ea_n = a_{n+1}$. From (3), it follows that $(E - \theta)^2 a_n = (E - \theta)^2 \epsilon_n$ is bounded, and hence that ϵ_n is bounded. In fact, if $\bar{I}_r = [r - 1, r]$, then the following are easily proved:

$$\text{LIM}(E - \theta)^2 \epsilon_n \subseteq \bar{I}_r, \tag{3}$$

$$\text{LIM}(E - \theta)\epsilon_n \subseteq \tfrac{1}{\theta-1}\bar{I}_r, \tag{4}$$

and

$$\text{LIM } \epsilon_n \subseteq \tfrac{1}{(\theta-1)^2}\bar{I}_r, \tag{5}$$

where **LIM** denotes the set of limit points of the sequence in question.

From (4) it immediately follows that if a_n satisfies a linear recurrence relation with defining polynomial $Q(x)$, then Q has the single root θ outside the unit circle, the other roots being in $|x| \leq 1$, with those on $|x| = 1$ being simple roots. These are the restrictions discovered by Flor for the case $r = \frac{1}{2}$. As in [Fl], one can show that, if $0 < r < 1$ and if $P(x)$ is the minimal polynomial of any Pisot or Salem number, then $Q(x) = P(x)$ occurs as the defining polynomial for some E_r–sequence.

Following [B3], we have the following simple result:

Proposition 1. *Suppose that a_n is an E_r–sequence with $\lim \frac{a_{n+1}}{a_n} = \theta > 1$, where θ is a Pisot or Salem number. Let d be the degree of θ. If*

$$\theta > 1 + \max(r, 1 - r)2^{d-1} \tag{6}$$

then the sequence a_n satisfies a linear recurrence relation whose defining polynomial is the minimal polynomial $P(x)$ of θ (i.e. no cyclotomic factor $K(x)$ occurs.)

Proof. From (4),

$$\limsup |(E - \theta)a_n| \leq \frac{\max(r, 1 - r)}{\theta - 1}. \tag{7}$$

Write $P(x) = (x - \theta)P_\theta(x)$. Then all roots of P_θ lie in the unit circle, so $P_\theta(x)$ is dominated by $(x + 1)^{d-1}$ and hence the length of P_θ (the sum of the absolute values of its coefficients) satisfies $L(P_\theta) \le 2^{d-1}$. Combining this with (7) shows that, if $b_n = P(E)a_n$, then

$$\limsup |b_n| \le \limsup L(P_\theta)|(E - \theta)a_n| \le \frac{\max(r, 1 - r)}{\theta - 1} 2^{d-1} < 1. \quad (8)$$

But b_n is a sequence of integers so (7) implies $b_n = 0$ eventually, and hence $P(E)a_n = 0$ eventually, showing that a_n satisfies a linear recurrence relation with $P(x)$ as defining polynomial. **QED**

Now let us consider the cases $r = 0$ and $r = 1$. If $r = 0$, then (3) implies that

$$\limsup(E - \theta)^2 \epsilon_n \le 0, \quad (9)$$

while if $r = 1$ then

$$\liminf(E - \theta)^2 \epsilon_n \ge 0. \quad (10)$$

If $Q(x)$ is the defining polynomial of a linear recurrence for a_n, and we let the roots of Q be θ_k, $k = 1, \ldots, m$, with $|\theta_k| \ge |\theta_{k+1}|$ for all k, then we can write $\epsilon_n = \lambda_2 \theta_2^n + \cdots + \lambda_m \theta_m^n$. Then (9) or (10) shows that θ_2 must be real and positive and that $\lambda_2 < 0$ if $r = 0$, while $\lambda_2 > 0$ if $r = 1$.

Thus, if θ is a Pisot number, then one possibility is that θ_2 is a conjugate of θ, in which case $\theta_2 > |\theta_k|$, for $k \ge 3$, which defines a proper subset of the Pisot numbers, which we will denote by PVD. The other possibility is that $\theta_2 = 1$. In case θ is a Salem number, only this latter possibility can occur.

However, by Proposition 1, if (6) holds, then the defining relation for a_n can have no cyclotomic factor, in particular no factor $x - 1$. Thus, if θ is a Pisot or Salem number satisfying (6), then θ must be in PVD.

If $r > 1$ or $r < 0$, then we must have $\liminf |(E - \theta)^2 \epsilon_n| > 0$ and hence $\theta_2 = 1$. Thus, if (6) holds, then there is no E_r–sequence with $r > 1$ or $r < 0$ having limiting ratio θ. So, for any fixed degree, there is only a finite number of such E_r sequences satisfying a recurrence relation of that degree.

Examples.

Given a Pisot sequence, one can check for linear recurrence relations of small degree (say ≤ 200), by using continued fractions in $Q\{\frac{1}{x}\}$, as in [C1], or by the Berlekamp-Massey algorithm, both of which are efficient methods of finding some of the Padé approximants to the generating function $\sum a_n x^n$. Using such methods, Cantor and his student Galyean [Ga] found many apparently non-recurrent Pisot sequences, among them $E(4, 13)$, $E(6, 16)$ and $E(8, 10)$. In some cases these sequences follow "false" recurrences for many terms, e.g $E(8, 10)$ follows $x^6 - x^5 - 1$ for 37 terms. This polynomial has three roots outside the unit circle, but two of them are very close to the unit circle and it is easy to see how this can permit the sequence to satisfy the recurrence relation for a large number of terms.

For $r = 0$ or 1, a similar but different phenomenon can occur. A false recurrence can be due to the recurrence having a second root $\theta_2 < 1$ for which there are other roots θ_k with $|\theta_k| > \theta_2$, but close to θ_2. The most extreme example of this which has been found in our computations is $S(8, 55)$ which seems to have the rational generating function

$$\frac{8 + 7x - 7x^2 - 7x^3}{1 - 6x - 7x^2 + 5x^3 + 6x^4}. \tag{11}$$

However, the reciprocals of the roots of the denominator are:

$$\theta = 6.892070726276\ldots$$

$$\alpha = .95484560059400\ldots$$

$$\beta, \overline{\beta} = .95484787673885 \exp \pm i(165.268\ldots)^\circ.$$

Here θ is certainly a Pisot number, but is not in PVD since $\alpha < |\beta|$. Note that the ratio $|\beta|/\alpha = 1.00000238378\ldots$ is very close to 1. If one writes c_n for the sequence given by (11) and a_n for $S(8, 55)$, then $a_n = c_n$ for $n \le 11055$, but $a_n = c_n + 1$ for $n = 11056$, as Jeff Shallit calculated in 27 hours on a Sparc SLC. This can also be verified by writing

$$c_n = \lambda\theta^n + \mu\alpha^n + \gamma\beta^n + \overline{\gamma}\overline{\beta}^n,$$

where the coefficients are calculated from (11) to, say, 20 decimal places. Then one needs to find the first n for which

$$(E - \theta)^2 \epsilon_n = \mu(\theta - \alpha)^2 \alpha^n + 2\Re(\gamma(\theta - \beta)^2 \beta^n) < 0.$$

This occurs first for $n = 11054$ which agrees with the integer calculation.

3 Families of E_r-sequences.

Experiment shows that E_r-sequences seem to fall into families whose properties depend on $a_1 \bmod a_0^2$. That is, the sequences $E_r(a_0, ka_0^2 + c)$, $k = 0, 1, \ldots$ tend to satisfy similar linear recurrences or else to be (apparently) non-recurrent. Cantor [C2] gave an explanation of this by treating k as a formal parameter and treating a_n as a polynomial in k. Although the emphasis there was on Pisot sequences, his considerations apply equally well to E_r-sequences.

According to that theory, the linear recurrence for $E(a_0, ka_0^2 + c)$, if it exists, should be of the form $F(x)/(A(x) - kxF(x))$, for $k \ge k_0$.

Following this lead, we did some experiments with sequences of the form $S(a_0, ka_0^2 + c)$ and $T(a_0, ka_0^2 + c)$, mainly for $a_0 \le 20$. Remarkably, it seems that $S(a_0, ka_0^2 + c)$ is recurrent for $k \geq 1$ if $0 \leq c \leq a_0$, but not for $c = -1$ or $c = a_0 + 1$. Formally, if $T(a_0, a_1) = \{a_n\}$, then $S(a_0, -a_1) = \{(-1)^n a_n\}$, at least if no ratio a_{n+1}^2/a_n is an integer. This would suggest that the behaviour of the family $T(a_0, ka_0^2 - c)$ should be similar to that of $S(a_0, ka_0^2 + c)$. However, although we find that $T(a_0, ka_0^2 - c)$ does seem to be recurrent for a large interval of the form $0 \leq c \leq \kappa a_0$, for some $\kappa < 1$, it does not seem to be recurrent for all values of $c < a_0$ (we consider the extreme case $c = a_0 - 1$ below.) However, in contrast to S-sequences, $T(a_0, ka_0^2 - c)$ does seem to be recurrent for $c = -1$ or $c = a_0 + 1$.

This suggests looking at some different families than those considered by Cantor, where, rather than fixing a_0, we take $a_0 = a$ to be a formal parameter (to be thought of as an infinitely large integer.) For example, looking at the sequences $S(a, a^2 - 1)$, it is natural to define the formal S-sequences with initial terms $a_0 = a$ and $a_1 = a^2 - 1$ so that a_n is a polynomial of degree $n + 1$ in a which satisfies: $a_{n+1}^2 = a_{n+2}a_n + b_n$, where $\deg b_n \leq \deg a_n$, and where the leading coefficient of b_n is negative. So, a_{n+2} is either the usual quotient in the standard division of the polynomial a_{n+1}^2 by a_n or else this quotient minus a_n, if this is necessary to make the leading coefficient of b_n negative.

We find then that a_n satisfies a degree 4 linear recurrence with generating function:

$$\frac{a - x + x^2(1 - a) + x^3}{1 - xa - x^3(1 - a) - x^4}. \tag{12}$$

However, except for $S(2, 3)$ which satisfies a linear recurrence of degree 2, all of $S(3, 8)$, $S(4, 15)$, etc, seem to be non-recurrent. In fact, for finite a, $S(a, a^2 - 1)$ seems to satisfy (12) for exactly $2a$ terms. The roots of the reciprocal of the denominator of (12) are $a - \frac{1}{a} + O(\frac{1}{a^2})$, $1 - \frac{1}{2a} - \frac{1}{a^2} + O(\frac{1}{a^3})$, $\frac{1}{a} + \frac{1}{a^2} + O(\frac{1}{a^3})$, and $-1 + \frac{1}{2a} - \frac{1}{a^2} + O(\frac{1}{a^3})$. Thus, the negative root dominates the second positive root so this is not PVD for finite a.

On the other hand, $T(a, a^2 + 1)$ satisfies a degree 7 linear recurrence:

$$\frac{a + x - x^3 - x^4 - x^5 - x^6}{1 - ax - x^2 + x^4 + x^5 + x^6 + x^7} = \frac{F(x)}{1 - xF(x)}, \tag{13}$$

which is PVD for all large a. (Here we define a_n as above except so that the leading coefficient of b_n is required to be positive.) This suggests that $T(a, a^2 + 1)$ does satisfy (13) for all large a, which experimentally is the case. However, it seems that $T(a, ka^2 + 1)$ does *not* satisfy the expected recurrence $F(x)/(1 - kxF(x))$ for large k but rather another recurrence.

Another interesting family is $T(a, a^2 - a + 1)$. Here the linear recurrence is of the form $F(x)/(1 + x - xF(x))$, where $F(x) = a + x - x^2 - x^4 - x^{22}$. The denominator is PVD for $a \geq 2849$ and this seems to be the correct recurrence for $T(a, a^2 - a + 1)$ for such a. For smaller a, we find that

$T(4, 13)$ satisfies the recurrence $F(x)/(1 + x - xF(x))$, with $F(x) = 4 + x - x^2 - x^4 - x^{36}$.

The next member of the family, $T(5, 21)$ satisfies a remarkable recurrence of this form, with $F(x) = 5 + x - x^2 - x^4 - x^{26} - x^{2048}$. (Note that the rational function $F(x)/(1 + x - xF(x))$ is not in lowest terms since both numerator and denominator are divisible by $1 + x$. However, it is best left in this form!) The computations to verify that this is the correct recurrence are quite lengthy since they involve power sums of 2047 of the roots of the reciprocal of $G(x) = 1 + x - xF(x)$ to check (3) for n up to 200,000. It was necessary to calculate the roots and other constants to 60 decimal places. Fortunately, it is easy to solve $G(x) = 0$ by Newton's method, since the roots are very close to the 2048th roots of 4.

It seems that the next few members of this family, for $6 \leq a \leq 20$ are probably not recurrent. We note, however, that the example of $T(5, 21)$ shows the danger of concluding that a sequence is non-recurrent by finding possible recurrences up to a given degree. If one were to calculate $T(5, 21)$ up to, say, 1000 terms, and compute Padé approximants, the naïve conclusion would be that $T(5, 21)$ satisfies the linear recurrence given by $F(x)/(1 + x - xF(x))$, where $F(x) = 5 + x - x^2 - x^4 - x^{26}$. Less naïvely, since this is not PVD and hence not a legitimate recurrence for a T-sequence, one might then conclude that the sequence is non-recurrent. However, this sequences *is* recurrent, it is just that the degree of the recurrence is rather larger than expected. Going back to the sequence $S(8, 55)$ discussed earlier, we can only conclude from the calculations done for that sequence that it satisfies no linear recurrence relation of degree ≤ 11055.

Acknowledgment. I would like to thank Jeff Shallit for asking me about his sequences and for encouraging me to investigate them.

References

[B1] D. W. Boyd, Pisot sequences which satisfy no linear recurrence, Acta Arith.,**24** (1977), 89-98.

[B2] D. W. Boyd, Pisot and Salem numbers in intervals of the real line, Math. Comp., **32** (1978), 1244-1260.

[B3] D. W. Boyd, On linear recurrence relations satisfied by Pisot sequences, Acta Arith., **47** (1986), 13-27.

[B4] D. W. Boyd, Pisot sequences which satisfy no linear recurrence II, Acta Arith., **48** (1987), 191-195.

[C1] D. G. Cantor, Investigation of T-Numbers and E-sequences, "Computers in Number Theory", ed. A.O.L. Atkin and B.J. Birch, Academic Press, 1971, 137-140.

[C2] D. G. Cantor, On Families of Pisot E-sequences, Ann. Scient. Éc. Norm. Sup.,(4) **9**(1976), 283-308.

[Fl] P. Flor, Über eine Klasse von Folgen natürlicher Zahlen, Math. Annalen, **140** (1960), 299-307.

[Ga] P. Galyean, On Linear Recurrence Relations for E-Sequences, Ph.D. thesis, University of California at Los Angeles, 1971. ' [Pi] C. Pisot, La répartition modulo 1 et les nombres algébriques, Ann. Scuola Norm. Sup. Pisa, **7**(1938), 205-248.

[Sh] J. Shallit, Problem, Fibonacci Quart., **29** (1991), p.85.

David W. Boyd
Department of Mathematics
University of British Columbia
Vancouver, Canada V6T 1Z2
e-mail David_Boyd@mtsg.ubc.ca

Algebraic Independence of Drinfeld

Exponential and Quasi-Periodic Functions

W. Dale Brownawell

Penn State University

DEDICATED TO PAULO RIBENBOIM

Abstract: We extend previous work of J. Yu to determine natural conditions for the algebraic independence of quasi-periodic Drinfeld functions with respect to A-lattices (this includes the class of Drinfeld exponential functions).

We show that all connected algebraic A-subgroups of power products of $\mathbb{G}_a$ and additive groups with A-action given by non-isogenous A-Drinfeld modules are products of algebraic A-subgroups of the individual powers in which each subgroup H satisfies codim H independent linear relations, modulo its periods, in its tangent space.

We also show that all algebraic relations on quasi-periodic Drinfeld functions arise naturally from two sources:

(1) isogenies (including endomorphisms) and

(2) linear dependence of the de Rham cohomology classes of the biderivations corresponding to the quasi-periodic functions.

Notation

$\mathbb{F}_q$	a finite field of $q = p^s$ elements
$\mathcal{C}$	a smooth projective geometrically irreducible curve over $\mathbb{F}_q$
∞	a closed point on $\mathcal{C}$
k	the function field of $\mathcal{C}$ over $\mathbb{F}_q$
A	the ring of functions in k regular on $\mathcal{C} \setminus \{\infty\}$
k_∞	the completion of k at ∞
$\bar{k}_\infty$	the algebraic closure of k_∞

1. INTRODUCTION

The investigation of the algebraic relations among analytic functions is an interesting pursuit in itself. It is also a necessary first step in the usual investigations of the algebraic independence of the values of such functions. In [K], algebraic subgroups of products of simple algebraic groups were shown to be subproducts whose factors have coordinates satisfying linear relations over the respective endomorphism rings.

Research supported in part by an NSF grant.

In particular, it was shown in [K] and in [BK] (see also [C]) that, for Weierstrass elliptic functions $\wp_i$ corresponding to non-isogenous elliptic curves $\mathcal{E}_i$, the functions $\wp_i(u_{ij}z)$ are algebraically independent exactly when all the numbers u_{ij} are linearly independent over the multiplication ring of each curve $\mathcal{E}_i$. In fact, the main independence result for functions in [BK] included various $\zeta_i(u_{ij}z)$ as well, where the $\zeta_i(z)$ are Weierstrass quasi-periodic functions.

In [Dr], V.G. Drinfeld introduced the very fruitful notion of elliptic modules, now usually called Drinfeld modules, which are rich analogues over global fields of elliptic curves. In a remarkable series of papers, including [Y1],[Y2],[Y3],[Y4], J. Yu has developed a powerful transcendence theory in the setting of Drinfeld modules. For an overview of this and related work, see [W] and the survey [Y5] by J. Yu. In this article, we propose to establish the analogues for Drinfeld functions of the above mentioned results on elliptic and quasi-periodic functions. Since we hope to make our results accessible to as broad an audience as possible, we will not hesitate to dwell on points which are more or less obvious to the Drinfeld journeyman, when we feel that the novice will profit from an explicit discussion.

I wish to thank the referee for extremely helpful comments.

1. Background: Drinfeld Modules, Exponential Functions, and Lattices.

A *Drinfeld A-module over K* is formally defined as a non-trivial representation of A as a ring of endomorphisms on the additive group scheme $\mathbb{G}_{a/K}$ such that the induced representation on the tangent space at zero is simply multiplication. This can all be described rather concretely in terms of an exponential map, its lattice of periods, and the effect on premultiplication by elements of A, as follows:

Let $F = F^1$ be the qth power function (the sth iterate of the Frobenius map $X \mapsto X^p$). Let $\bar{k}_\infty\{F\}$ denote the *twisted* (noncommutative) *polynomial ring* of operators generated by F over $\bar{k}_\infty$. The ring of twisted polynomials $T = \sum a_h F^h$ is evidently isomorphic to the ring of $\mathbb{F}_q$-linear polynomials $P(X) = \sum a_h X^{q^h}$, whose multiplication is composition of polynomials. For clarity, we opt to maintain the rather pedantic distinction between the $\mathbb{F}_q$-linear polynomials and the operations (*twisted polynomials*) given by substituting into them. The ring $\bar{k}_\infty\{F\}$ of twisted polynomials has no zero divisors and in fact is a right (and a left) division ring. In particular every left ideal in $\bar{k}_\infty\{F\}$ is principal.

A (not immediately apparently) equivalent definition of a Drinfeld A-module, with the additional specification of a *rank* $r > 0$, is as an $\mathbb{F}_q$-linear homomorphism $\phi \colon A \to \bar{k}_\infty\{F\}$ with the following property: for each $a \in A, a \neq 0$,

$$(1) \qquad\qquad \phi(a) = aF^0 + a_1 F^1 + \cdots + a_m F^m.$$

for $a_1, \ldots, a_m$ in $\bar{k}_\infty, m = rd(a)$, where d is the valuation associated with the point ∞, and $a_m \neq 0$.

We say that ϕ is *defined over* the field K if $\phi \colon A \to K\{F\}$. We denote by k_ϕ the smallest field over which ϕ is defined. There is also a more analytic way of looking at Drinfeld modules.

A *lattice L of rank r* is a finitely generated discrete A-module of $\bar{k}_\infty$ of projective rank r. Given such an L, Drinfeld defines the following "exponential" function:

$$e_L(z) = z \prod_{\omega \in L}{}' (1 - \frac{z}{\omega}),$$

where $\prod'$ means that no term corresponding to $0 = \omega \in L$ appears in the product. This is an $\mathbb{F}_q$-linear power series, since its partial sums are $\mathbb{F}_q$-linear (cf. Remark A3 of Appendix). Drinfeld showed that $e_L(z)$ is an $\mathbb{F}_q$-linear surjective homomorphism onto $\bar{k}_\infty$. It is not too difficult to show (this is a special case of Lemma 2.1 below) that, for each $a \in A$, there is a unique twisted polynomial $\phi_L(a)$ such that

$$(2) \qquad\qquad e_L(az) = \phi_L(a)e_L(z).$$

The map $a \mapsto \phi_L(a) \in \bar{k}_\infty\{F\}$ gives a Drinfeld module of rank r, and Drinfeld showed [Dr] that all Drinfeld modules can be accounted for in this way. A comparison of coefficients of powers of z in equation (2) using equation (1) shows that, if we write $e_\phi(z) = \sum e_h z^{q^h}$, then $k_\phi = k(\ldots, e_h, \ldots)$, so k_ϕ is also a "field of definition" of $e_\phi(z)$.

For more details on the basic material of this section and surveys of Drinfeld modules, see [DH], [H1], or [H2].

2. Results.

In this note we establish necessary and sufficient conditions for the algebraic independence of the Drinfeld analogues of the Weierstrass functions, i.e. for the exponential (or periodic) and quasi-periodic functions of Drinfeld modules. Independence results for Drinfeld functions had of course already been established by Yu, and our techniques are heavily based upon his.

COMMON HYPOTHESES. *Let* $\Phi = \{\phi\}$ *denote a set of non-isogenous Drinfeld A-modules, whose exponential functions* $e_\phi(z)$ *have period lattices* L_ϕ. *For each* $\phi \in \Phi$, *let* $R_\phi := \{\rho \in \bar{k}_\infty \colon \rho L_\phi \subset L_\phi\}$, *the ring of multiplications of* ϕ, *and let* G_ϕ *denote the additive group with A-module structure coming from* ϕ: $a \cdot g = \phi(a)g$.

We shall see below that ϕ extends naturally to R_ϕ in such a way that R_ϕ becomes naturally isomorphic to the endomorphism ring E_ϕ of ϕ via $\rho \mapsto \phi(\rho)$.

The first main independence result, Theorem 3.1 below, sharpens a previous result of J. Yu [Y4] to show that codim H linear relations modulo periods in the tangent space account for the algebraic A-subgroups H of power products of non-isogenous Drinfeld A-modules.

THEOREM A. *Any connected algebraic subgroup*

$$H < \mathbb{G}_a^{s_0} \times \prod G_\phi^{s(\phi)},$$

where the product $\prod$ here and below is taken over all $\phi \in \Phi$, in which H is closed under the action of A, is of the form

$$H = H_a \times \prod H_\phi,$$

where each subgroup $H_\phi < G_\phi^{s(\phi)}$ is the connected component of the identity of the subgroup whose coordinates $(h_1, \ldots, h_{s(\phi)})$ satisfy codim H_ϕ linearly independent linear relations over the endomorphism ring of ϕ:

$$f_{i1} h_1 + \cdots + f_{is(\phi)} h_{s(\phi)} = 0,$$

$i = 1, \ldots, \text{codim } H_\phi$.

Our second fundamental independence result, which appears below as Theorem 5.1, is the following:

THEOREM B.

(1) *For each $\phi \in \Phi$, let $U_\phi \subset \bar{k}_\infty$ be an R_ϕ-linearly independent set.*
(2) *For each $\phi \in \Phi$, let Δ_ϕ be a set of ϕ-biderivations whose images in $HD_{DR}^*(\phi)$ are $\bar{k}_\infty$-linearly independent.*

Then the $1 + \sum_\phi |\Delta_\phi||U_\phi|$ functions

$$z, F_\delta(uz), \quad \phi \in \Phi, u \in U_\phi, \delta \in \Delta_\phi$$

are algebraically independent over $\bar{k}_\infty$.

We shall see that both these results are best possible in senses that we shall make explicit.

2. ISOGENIES

1. Isogenies, Homotheties, and Twisted Polynomials.

If $\phi: A \to K\{F\}$ and $\psi: A \to K\{F\}$ are Drinfeld modules defined over K, then a *morphism* from ϕ to ψ defined over K is an element T of $K\{F\}$ for which

$$T\phi(a) = \psi(a)T, \quad \forall a \in A.$$

If $T = uF^i +$ higher order terms, $u \neq 0$, then, recalling equation (1), we see that the lowest order terms in the preceding displayed equation are

$ua^{q^i} F^i = auF^i$. Thus, on taking $a \in A$ transcendental over $\mathbb{F}_q$, we see that $i = 0$ and $T = uF^0+$ higher terms, with $u \neq 0$. A non-zero morphism is called an *isogeny*. Much of the material of this section can be found in the literature listed in the bibliography, although not necessarily quite in the form given here.

Let L_ϕ and L_ψ be the A-lattices of periods of the exponential functions $e_\phi(z)$ and $e_\psi(z)$ related to the A-Drinfeld modules ϕ and ψ, respectively. In the next result, we show how to obtain an isogeny from ϕ to ψ, given an element of $\bar{k}_\infty$ carrying L_ϕ to L_ψ, if these lattices have the same rank.

LEMMA 2.1. *Assume that* rank $L_\phi =$ rank L_ψ. *Let* $u \in \bar{k}_\infty^*$ *with* $uL_\phi \subset L_\psi$. *Define the operator* T_u *by*

$$T_u : X \mapsto P(X) = l \prod (X - e_\phi(\lambda)) = uX + \text{higher terms},$$

where λ *runs through a full set of coset representatives for* $\frac{1}{u}L_\psi$ mod L_ϕ *and* $l \in \bar{k}_\infty$ *is determined by the condition that* $P'(0) = u$. *Then*

(1) $P(e_\phi(z)) = e_\psi(uz)$,
(2) $T_u \in u \cdot \bar{k}_\phi\{F\}$, *where* $\bar{k}_\phi$ *denotes the algebraic closure of* k_ϕ *in* $\bar{k}_\infty$, *and*
(3) T_u *is an isogeny from* ϕ *to* ψ.

PROOF: Since L_ϕ and L_ψ have the same rank, L_ϕ is an $A-$sublattice of $\frac{1}{u}L_\psi$ of finite index. Thus the product involved in the definition of $P(X)$ is finite. The resulting monic polynomial $M(X)$ is well-defined, since the value of $e_\phi(\lambda)$ depends only on the residue class of λ modulo L_ϕ. Moreover since distinct residue classes give rise to distinct values $e_\phi(\lambda)$, zero is a simple root of $M(X)$, and therefore the coefficient of X in $M(X)$ is non-zero.

Since the index of L_ϕ in $\frac{1}{u}L_\psi$ is finite, there is a non-zero $a \in A$ such that $a \cdot \frac{1}{u}L_\psi \subset L_\phi$. In particular, for each $\lambda \in \frac{1}{u}L_\psi$, $\phi(a)e_\phi(\lambda) = 0$, and $e_\phi(\lambda)$ is algebraic over k_ϕ. Therefore $M(X) \in \bar{k}_\phi[X]$ and we can choose $l \in u\bar{k}_\phi$, so that $P(X) = uX+$ higher terms $\in u\bar{k}_\phi[X]$.

Now the quotient A-module $\frac{1}{u}L_\psi/L_\phi$ is a finite $\mathbb{F}_q$-vector space, and the same is true for the set of values $e_\phi(\lambda)$, since $e_\phi(cz) = ce_\phi(z)$ for all c in $\mathbb{F}_q$, as $e_\phi(z)$ is $\mathbb{F}_q$-linear. Therefore by Remark A2 of the Appendix, $P(X)$ is $\mathbb{F}_q$-linear and so $T_u \in u\bar{k}_\phi\{F\}$. This establishes the second claim.

By definition, $P(e_\phi(z))$ vanishes exactly at the points of $\frac{1}{u}L_\psi$, to first order since $P(X)$ has no repeated roots. Now, according to Schnirelman's theorem (cf. Remark 4.13 of [Lz]), an entire function on $\bar{k}_\infty$ is determined up to a constant by its zeros, counting multiplicities. So we see that $P(e_\phi(z)) = ce_\psi(uz)$, where the constant c is determined to be equal to 1 by comparing the coefficients of z on both sides of the equation while keeping in mind that the lowest non-zero term of $P(X)$ is uX. In other words,

$$P(e_\phi(z)) = e_\psi(uz).$$

This establishes the first claim.

Now for arbitrary $a \in A$, we apply this identity at the equalities marked by colons to conclude that $T_u \cdot \phi(a) e_\phi(z) = T_u e_\phi(az) = P(e_\phi(az)) := e_\psi(uaz) = \psi(a) e_\psi(uz) =: \psi(a) P(e_\phi(z)) = \psi(a) \cdot T_u e_\phi(z)$. Since $e_\phi(z)$ has infinitely many zeros, it is a transcendental function, and we can conclude that

$$T_u \cdot \phi(a) = \psi(a) \cdot T_u, \quad \forall a \in A,$$

as needed to establish the third claim. $\square$

Thus we have seen that, just as in the case of elliptic curves in $\mathbb{C}$, homotheties taking one lattice into another lattice of the same dimension (automatic in the classical case) induce isogenies of the corresponding A-Drinfeld modules. We now turn to the objective of characterizing isogenies in terms of their effects on the corresponding exponential functions and on their period lattices.

THEOREM 2.2. *Let ϕ and ψ be Drinfeld modules. Let $P(X) = uX + $ higher terms be an $\mathbb{F}_q$-linear polynomial over $\bar{k}_\infty$ with $u \neq 0$, and let $T \in \bar{k}_\infty\{F\}$ be the corresponding twisted polynomial. The following conditions are equivalent:*

(1) *T is an isogeny from ϕ to ψ, i.e.*

$$T\phi(a) = \psi(a)T, \quad \forall a \in A.$$

(2) *$P(e_\phi(z)) = e_\psi(uz)$.*
(3) *$u\mathrm{Ker}(P(e_\phi(z))) = L_\psi$.*

PROOF: Set $e(z) = e_\phi(z)$. Then each of our three conditions implies that $L^* = \mathrm{Ker}\,P(e(z))$ is an A-lattice. Let $e^*(z)$ be the exponential function corresponding to L^* and ψ^* the corresponding Drinfeld module, i.e.

$$\psi^*(a)e^*(z) = e^*(az), \quad \forall a \in A.$$

Note that since $u \neq 0$, all the zeros of $P(e(z)) = uz + $ higher $\mathbb{F}_q$-linear terms are simple. Since both $ue^*(z)$ and $P(e(z))$ have L^* as their set of zeros, which are all simple, and u as their coefficient of z, we see that

$$ue^*(z) = P(e(z)).$$

For non-zero a in A, consider the following diagram (cf. proof of Proposition (2.4) [G1]), where dots indicate multiplication. Note that surjectivity at the back upper right vertex and commutativity of the sides implies commutativity of the right end.

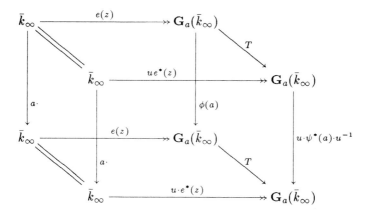

From the commutativity of the right end of this box, we find the first equivalence in the following chain of equivalences: $T\phi(a) = \psi(a)T, \forall a \in A \Leftrightarrow u\psi^*(a)u^{-1}T = \psi(a)T, \forall a \in A \Leftrightarrow u\psi^*(a)u^{-1} = \psi(a), \forall a \in A \Leftrightarrow u\psi^*(a) = \psi(a)u, \forall a \in A$.

Let $L' = uL^*$ and ϕ' be the Drinfeld module corresponding to L' with exponential function $e'(z)$. Then a comparison of zeros and leading coefficients shows that

$$e'(uz) = ue^*(z).$$

Thus $\phi'(a)ue^*(z) = \phi'(a)e'(uz) = e'(uaz) = ue^*(az) = u\psi^*(a)e^*(z)$ for all a in A, and hence for all a in A,

$$\phi'(a)u = u\psi^*(a).$$

Therefore we can continue our chain of equivalent equalities with $\psi(a) = \phi'(a), \forall a \in A \Leftrightarrow L' = L_\psi \Leftrightarrow P(e(z)) = e_\psi(uz)$, with the last equivalence obtained by comparing zeros and coefficients of z for the functions. The last equality is the second item to be shown equivalent, while the third item is the penultimate equality. $\square$

Thus we are able to close the circle begun with Lemma 2.1.

COROLLARY 2.3. (a) If $P(X) = uX+$ higher terms is the the $\mathbb{F}_q$-linear polynomial corresponding to an isogeny from the A-Drinfeld module ϕ to the A-Drinfeld module ψ, then

(1) rank $L_\phi =$ rank L_ψ and
(2) $uL_\phi \subset L_\psi$.

(b) Isogeneity is an equivalence relation.

PROOF: (a): By condition (3) of the theorem, $uL_\phi \subset u\mathrm{Ker}\, P(e_\phi(z)) = L_\psi$, and we have the inclusion claimed. Moreover since $P(X)$ is a polynomial of finite degree, the index of uL_ϕ is finite in L_ψ. Thus rank $L_\phi =$ rank L_ψ.

(b): Reflexivity and transitivity are clear. For symmetry, recall from the proof of Lemma 2.1 that if $uL_\phi \subset L_\psi$ with non-zero u, then there is a non-zero $a \in A$ such that $\frac{a}{u} L_\psi \subset L_\phi$. $\square$

Putting these remarks together, and observing that composition of isogenies corresponds to multiplication of the coefficients of X in the corresponding $\mathbb{F}_q$-linear polynomials, we obtain the following result, which has been extended to t-motives by G. Anderson (cf. Theorem 2.1 of [G2]):

COROLLARY 2.4 (DRINFELD). *There is a category equivalence between A-Drinfeld modules of rank r and A-lattices of rank r in which morphisms correspond to homotheties carrying one lattice into another.*

Theorem 2.2 shows that, when investigating the algebraic independence of Drinfeld exponential functions, one can at will replace functions by the exponential function of any isogenous module, whose argument has been scaled by the coefficient of F^0 in the isogeny. This justifies considering only the algebraic independence of various $e_i(u_{ij}z)$, with the $e_i(z)$ corresponding to non-isogenous Drinfeld modules.

Our goal is to obtain precise conditions for the algebraic independence of various functions $z, e_\phi(uz)$, as well as quasi-periodic functions. Therefore we want to investigate restrictions on the u necessary for algebraic independence of the exponential functions.

2. Endomorphisms and Dependence of Exponential Functions.

When trying to obtain algebraically independent values of $e(u_i z)$ as the u_i run through some set U_ϕ, it is enough to restrict one's attention to the case that the set U_ϕ is A-linearly independent. For if $\sum \alpha_i u_i = 0$ is a linear dependence relation, then $\sum \phi(\alpha_i) e(u_i z) = 0$. The same sort of considerations apply to coefficients α_i from a possibly larger ring than A.

We call $R_\phi := \{c \in \bar{k}_\infty : cL_\phi \subset L_\phi\}$ the *ring of multiplications* of ϕ or L_ϕ. Thus $A \subset R_\phi$, and the non-zero elements of R_ϕ correspond to the isogenies which are endomorphisms of ϕ. Let us first show that ϕ extends uniquely to R_ϕ.

PROPOSITION 2.5. *The Drinfeld A-module ϕ extends uniquely to a ring homomorphism $\phi\colon R_\phi \longrightarrow \bar{k}_\phi\{F\}$ so that $\rho \mapsto \rho F^0 +$ higher terms and $\phi(\rho)e_\phi(z) = e_\phi(\rho z)$, for all $\rho \in R_\phi$.*

PROOF: By the proof of Lemma 2.1, for each $u \in \bar{k}_\infty$ with $uL_\phi \subset L_\phi$, there is an $\mathbb{F}_q$-linear polynomial $P_u(X)$ of the form $P_u(X) = uX +$ higher terms such that $P_u(e_\phi(z)) = e_\phi(uz)$. Since $e_\phi(z)$ is a transcendental function, $P_u(X)$ is unique. In the notation of Lemma 2.1, we simply define $\phi(u) := T_u$, i.e. $\phi(u)X = P_u(X)$. Additivity and multiplicativity of this extension are immediate, $T_u = uF^0 +$ higher terms, and by definition $\phi(r)e_\phi(z) = e_\phi(rz)$ for all $r \in R_\phi$. $\square$

Now we can extend our previous remark that A-linearly dependent u give rise to algebraically dependent functions $e_\phi(uz)$.

PROPOSITION 2.6. *If $u_1, \ldots, u_n$ from $\bar{k}_\infty$ are R_ϕ-linearly dependent, then the functions $e_\phi(u_1 z), \ldots, e_\phi(u_n z)$ are $\bar{k}_\phi$-algebraically dependent.*

PROOF: If $\sum r_i u_i = 0$ with r_i in R_ϕ, not all zero, then

$$0 = e_\phi\left(\sum r_i u_i z\right) = \sum e_\phi(r_i u_i z) = \sum \phi(r_i) e_\phi(u_i z). \square$$

3. SUBGROUPS OF PRODUCTS OF DRINFELD MODULES

Before proceeding to quasi-periodic functions, we examine the exponential case in a more general setting. For that purpose, recall the formal definition of a Drinfeld A-module as a non-trivial representation of A as a ring of endomorphisms on $\mathbb{G}_a$ such that the induced representation on the tangent space at zero is ordinary multiplication. In other words, each Drinfeld A-module ϕ gives the additive group an A-module structure (different from ordinary multiplication by elements of A) via

$$a \cdot g := \phi(a)g,$$

and when we consider $e_\phi(z)$ as a Lie group exponential function from the tangent space $(= \bar{k}_\infty)$ at the identity to the additive group $\bar{k}_\infty$, ordinary multiplication by elements from A on the tangent space is compatible with the A-action on $\bar{k}_\infty$.

By an *additive A-group G*, we mean either a copy of the additive group with the A-action coming from a Drinfeld A-module as above or else the additive group with A acting by ordinary multiplication in $\bar{k}_\infty$: $a \cdot g = ag$. Let $G_1, \ldots, G_m$ be various additive A-groups. Then the product

$$G = G_1 \times \cdots \times G_m$$

is an A-module under the component-wise A-action.

By an *algebraic A-subgroup of G*, we mean an algebraic subgroup of G which is stable under the action of A on G.

Let Λ_i be the lattice of periods of $e_i(z)$ (and $\Lambda_i = 0$ if $G_i = \mathbb{G}_a$ with the ordinary A-action by multiplication) and let R_i be the ring of multiplications of Λ_i: $R_i = \{c \in \bar{k}_\infty : c\Lambda_i \subset \Lambda_i\}$. As we saw above, R_i may be naturally identified with the endomorphism ring of G_i.

THEOREM 3.1. *Let $G_0 = \mathbb{G}_a$ with the ordinary A-action and let $G_1, \ldots, G_n$ be the additive A-groups coming from non-isogenous A-Drinfeld modules $\phi_1, \ldots, \phi_n$, with associated exponential functions $e_1(z), \ldots, e_n(z)$. Set*

$$G = G_0^{s_0} \times G_1^{s_1} \times \cdots \times G_n^{s_n}.$$

(1) *Then any connected algebraic A-subgroup $H < G$ is of the form*

$$H = H_0 \times H_1 \times \cdots \times H_n,$$

where H_i is a connected algebraic A-subgroup of $G_i^{s_i}$, $i = 0, 1, \ldots, n$.

(2) *Let H_i be an algebraic A-subgroup of $G_i^{s_i}$ of codimension κ_i, $i = 1, \ldots, n$. Then there is a rank κ_i $\kappa_i \times s_i$ matrix $(E_{i\sigma}^j)$ of endomorphisms $E_{i\sigma}^j$ of ϕ_i such that for all $\mathbf{h} = (h_{i1}, \ldots, h_{is_i}) \in H_i$,*

$$E_{i1}^j h_1 + \cdots + E_{is_i}^j h_{s_i} = 0,$$

for all $1 \leq j \leq \kappa_i$.

Equivalently in the tangent space, there is a $\kappa_i \times s_i$ matrix $(\rho_{i\sigma}^j)$ over R_i of rank κ_i such that, whenever $e_i(z_{i\sigma}) = h_{i\sigma}$ with $\mathbf{h} = (h_{i1}, \ldots, h_{is_i}) \in H_i$, then

$$\rho_{i1}^j z_{i1} + \cdots + \rho_{is_i}^j z_{is_i} \in \Lambda_i,$$

for all $1 \leq j \leq \kappa_i$.

(3) *Let H_0 be an algebraic A-subgroup of $G_0 = \mathbb{G}_a^{s_0}$ of codimension κ_0. Then there is a $\kappa_0 \times s_0$ matrix $(\rho_{0\sigma}^j)$, of rank κ_0, such that for all $(h_{01}, \ldots, h_{0s_0}) \in H_0$ and all $1 \leq j \leq \kappa_0$,*

$$\rho_{01}^j h_{01} + \cdots + \rho_{0s_0}^j h_{0s_0} = 0.$$

REMARK 3.1: Relations of the above form always define algebraic subgroups of G which are closed under the operation of A.

REMARK 3.2: Of course, part (3) is well-known. It is included here solely for completeness.

REMARK 3.3: Yu has recently [Y4] proved the existence of some relations as in (2), but did not make their rank explicit.

PROOF OF THEOREM 3.1: The algebraic group G has dimension $\nu = \sum s_i$. Let $\kappa = \operatorname{codim}_G H$ and select a set B consisting of $\nu - \kappa$ variables from among

$$X_{01}, \ldots, X_{0s_0}; \cdots; X_{n0}, \ldots, X_{ns_n}$$

which are algebraically independent modulo the ideal $\mathcal{H}$ of H. For each variable $X_{i\sigma} \notin B$, let $B_{i\sigma} = B \cup \{X_{i\sigma}\}$. Then there is a non-zero polynomial $P^{i\sigma}(X)$, involving only the variables of $B_{i\sigma}$, which vanishes on the coordinates of points $\mathbf{h} = (h_{j\tau}) \in H$ and has minimal (positive) degree with respect to $X_{i\sigma}$. (It is not hard to see that this polynomial is unique up to a factor from $\bar{k}_\infty$, but we do not need that fact here.)

For each index $j\tau$ with $X_{j\tau}$ appearing in $P^{i\sigma}$, let $\pi_{j\tau} : H \to \bar{k}_\infty$ denote the projection $\mathbf{h} \mapsto h_{j\tau}$. Then the equation $P^{i\sigma}(\mathbf{h}) = 0$ may be interpreted as an algebraic dependence relation of minimal degree on the additive functions $\pi_{j\tau}$. The proof of a classical theorem of E. Artin's (Theorem 18 of VIII, §11 of [L]) on algebraically dependent additive functions shows that the polynomial $P^{i\sigma}(X)$ underlying such a minimal relation has the form $P^{i\sigma}(X) = \sum P^{i\sigma}_{j\tau}(X_{j\tau})$, where each $P^{i\sigma}_{j\tau}(X_{j\tau})$ is an $\mathbb{F}_q$-linear polynomial involving only the variable $X_{j\tau}$. Thus

$$(3) \qquad P^{i\sigma}(\mathbf{h}) = \sum_{X_{j\tau} \in B_{i\sigma}} P^{i\sigma}_{j\tau}(h_{j\tau}) = 0,$$

with $\mathbb{F}_q$-linear polynomials $P^{i\sigma}_{j\tau}(X_{j\tau})$. Now we will use the fact that H is an A-module to study the polynomial $P^{i\sigma}(X)$.

First Case: If no index j occurring in our relation is positive, then $i = 0$, and the A-action is simply multiplication in each coordinate. Thus relation (3) still holds if we replace $h_{j\tau}$ by $ah_{j\tau}$ for transcendental $a \in A$. Since the degree of the relation (3) is unchanged by this substitution, our minimality assumption on the degree in $X_{0\sigma}$ shows that $P^{0\sigma}(X)$ is unchanged (up to a scalar factor) and is therefore homogeneous of degree, say, p^D:

$$P^{0\sigma}(X) = \sum C^{0\sigma}_{0\tau} X^{p^D}_{0\tau},$$

since our coefficient field is algebraically closed, we can write each $C^{0\sigma}_{0\tau} = (c^{0\sigma}_{0\tau})^{p^D}$, for $c^{0\sigma}_{0\tau} \in \bar{k}_\infty$. But then $P^{0\sigma}(X) = (Q^{0\sigma}(X))^{p^D}$, where the polynomial $Q^{0\sigma}(X) = \sum c^{0\sigma}_{0\tau} X_{0\tau}$ also satisfies $Q^{0\sigma}(\mathbf{h}) = 0$. By the minimality of the degree of our choice of $P^{0\sigma}(X)$ involving only the coordinates of $B_{0\sigma}$, D must equal 0, and $P^{0\sigma}(X)$ gives a linear dependence relation

$$P^{0\sigma}(\mathbf{h}) = \sum C^{0\sigma}_{0\tau} h_{0\tau} = 0$$

on those coordinates of $\mathbf{h}$ with indices lying in $B_{0\sigma}$. As a special case, this establishes Part (3) of our theorem.

Second Case: Otherwise, some index j occurring in $P^{i\sigma}(X)$ is positive. On replacing z by az, for any $a \in A \smallsetminus \mathbb{F}_q$, we obtain a potentially new relation

$$(4) \qquad \sum_{X_{j\tau} \in B_{i\sigma}} Q^{i\sigma}_{j\tau}(h_{j\tau}) = 0,$$

where $Q^{i\sigma}_{j\tau}(X_{j\tau}) = P^{i\sigma}_{j\tau}(\phi_j(a)X_{j\tau})$. This new $\mathbb{F}_q$-linear relation involves exactly the same coordinates as the original relation (3). Now the $\mathbb{F}_q$-linear polynomials $Q^{i\sigma}_{j\tau}(X_{i\sigma})$ for fixed $j\tau$ occurring in $\mathbb{F}_q$-linear relations on the coordinates of $\mathbf{h} \in H$ corresponding to variables in $B_{i\sigma}$ as in (3) give rise to a left ideal in $\bar{k}_\infty\{F\}$. So, since left ideals in $\bar{k}_\infty\{F\}$ are principal, the minimality of the degrees in our original relation (3) forces any non-zero relation such as (4) on the coordinates of H corresponding to $B_{i\sigma}$ involving the given coordinate $X_{j\tau}$ to be a left multiple of (3) by a non-zero term, say $\psi(a)$, from $\bar{k}_\infty\{F\}$. That is, for all a in A,

$$(5) \qquad \psi(a)P^{i\sigma}_{j\tau}(X_{j\tau}) = P^{i\sigma}_{j\tau}(\phi_j(a)X_{j\tau}).$$

Since each $P^{i\sigma}_{j\tau}(X_{j\tau})$ is an $\mathbb{F}_q$-linear polynomial and each $\phi_j : A \rightarrow \bar{k}_\infty\{F\}$ is an $\mathbb{F}_q$-linear homomorphism, it is straight-forward to check that equation (5) defines an $\mathbb{F}_q$-linear homomorphism $\psi : A \rightarrow \bar{k}_\infty\{F\}$. If $\psi(a) \in \bar{k}_\infty$ for some $a \in A$, then

$$\deg P^{i\sigma}_{j\tau}(X_{j\tau}) = \deg P^{i\sigma}_{j\tau}(\phi_j(a)X_{j\tau})$$

i.e. $\deg \phi_j(a)X_{j\tau} = 1$. Since we are dealing with the case that some index j occurring in this relation is positive, then, according to relation (1), $a \in \mathbb{F}_q$, since rank $\phi_j > 0$. In other words, the function $\psi : a \mapsto \psi(a)$ defines an A-Drinfeld module (of positive rank), and, according to equation (5), the twisted polynomial $T^{i\sigma}_{j\tau}$ for which $T^{i\sigma}_{j\tau} X = P^{i\sigma}_{j\tau}(X)$ is an isogeny from ϕ_j to ψ: For all a in A,

$$\psi(a)T^{i\sigma}_{j\tau} = T^{i\sigma}_{j\tau}\phi_j(a).$$

Therefore the Drinfeld modules corresponding to all the positive indices actually occurring in $P^{i\sigma}$ are isogenous. Since however the A-Drinfeld modules of the various G_l, $l = 1, \ldots, n$, are non-isogenous, exactly one positive index l occurs in our minimal relation, and it equals j.

Any $P^{i\sigma}_{0\tau}(X_{0\tau})$ also occurring must in addition satisfy

$$\psi(a)P^{i\sigma}_{0\tau}(X_{0\tau}) = P^{i\sigma}_{0\tau}(aX_{0\tau})$$

for all $a \in A$. As $P^{i\sigma}_{0\tau}(X_{0\tau})$ is $\mathbb{F}_q$-linear, it is either zero or else has positive degree. Since we have seen the operator $\psi(a)$ to lie in $\bar{k}_\infty\{F\} \setminus \bar{k}_\infty F^0$ for $a \in A \setminus \mathbb{F}_q$, the action of $\psi(a)$ produces a polynomial of larger degree when applied to any non-constant polynomial. Therefore, on taking $a \notin \mathbb{F}_q$, we see that the preceding displayed equality can hold only when $P^{i\sigma}_{0\tau}(X_{0\tau}) \equiv 0$.

Thus, if $i \neq 0$, then the minimal relation (3) is actually of the form

$$\sum_{X_{i\tau} \in B_{i\sigma}} P^{i\sigma}_{i\tau}(h_{i\tau}) = 0,$$

for $\mathbb{F}_q$-linear polynomials $P^{i\sigma}_{i\tau}(X_{i\tau})$.

Now since each $T^{i\sigma}_{i\tau}$ obtained from (5) is an endomorphism of ϕ_i (isogeny from ϕ_i to itself), if we write $P^{i\sigma}_{i\tau}(X_{i\tau}) = \rho^{i\sigma}_{i\tau}X_{i\tau}+$ higher terms, we know by the corollary to Theorem 2.2 that $\rho^{i\sigma}_{i\tau}L_{\phi_i} \subset L_{\phi_i}$, so $\rho^{i\sigma}_{i\tau} \in R_i$ and

$$\phi_i(\rho^{i\sigma}_{i\tau})X_{i\tau} = T^{i\sigma}_{i\tau}X_{i\tau} = P^{i\sigma}_{i\tau}(X_{i\tau}).$$

Consequently our minimal relation (3) on the coordinates of H must be a relation solely on the coordinates of $G_i^{s_i}$:

$$\sum_{X_{ir} \in B_{i\sigma}} \phi_i(\rho_{i\tau}^{i\sigma}) h_{i\tau} = 0,$$

as required.

Writing $h_{i\tau} = e_i(z_{i\tau})$, we can restate the preceding equality as

$$\sum_{X_{ir} \in B_{i\sigma}} \phi_i(\rho_{i\tau}^{i\sigma}) e_i(z_{i\tau}) = \sum_{X_{ir} \in B_{i\sigma}} e_i(\rho_{i\tau}^{i\sigma} z_{i\tau})$$

$$= e_i\Big(\sum_{X_{ir} \in B_{i\sigma}} \rho_{i\tau}^{i\sigma} z_{i\tau} \Big)$$

$$= 0,$$

or

$$\sum_{X_{ir} \in B_{i\sigma}} \rho_{i\tau}^{i\sigma} z_{i\tau} \in \Lambda_i.$$

As a special case, this establishes Part (2) of our theorem.

Conclusion of Proof: Now in the situation of Part (1) of our theorem, it is clear that our polynomials $P^{i\sigma}(X)$ lie inside the ideal $\mathcal{H}$ for H and generate an ideal I of dimension at least $\nu - \kappa$, and therefore exactly $\nu - \kappa$, since each $X_{ij} \notin B$ satisfies a non-zero polynomial over $\bar{k}_\infty[X]/I$. It is also clear that I defines an algebraic A-subgroup $K = \prod K_i$ of G having the same dimension as H, where K_i is a subgroup of $G_i^{s_i}$. Therefore, since H is connected and of the same dimension as K, it is the connected component of the identity of K. But that connected component of K is just the direct product of the connected components H_i of the identities of the groups K_i. In other words, $H = \prod H_i$, as claimed. This establishes Part (1) of our theorem. $\square$

4. Quasi-periodic Functions

1. Biderivations.

When the lattice L of periods of the Drinfeld module ϕ has A-rank $r > 1$, then one obtains new functions which are quasi-periodic with respect to L. To describe them, we consider $\bar{k}_\infty\{F\}$ as an A-bimodule, where right multiplication is given by $\phi(a)$ and left multiplication by a itself. Then a $\phi - biderivation$ δ of A into $\bar{k}_\infty\{F\}F$ is an $\mathbb{F}_q$-linear map $\delta \colon A \mapsto \bar{k}_\infty\{F\}F$ such that

$$\delta(ab) = a\delta(b) + \delta(a)\phi(b), \quad \forall a, b \in A.$$

There are three different fundamental types of ϕ-biderivations.

DIFFERENTIALS OF THE FIRST KIND: One ϕ-biderivation is obtained directly from ϕ as

$$\delta_\phi(a) = \phi(a) - aF^0, \ \forall a \in A.$$

The $\bar{k}_\infty$-multiples of δ_ϕ may be thought of as analogues of *differentials of the first kind*.

STRICTLY INNER BIDERIVATIONS: Let $D_{si}(\phi)$ denote the $\bar{k}_\infty$-vector space of *exact* or *strictly inner* ϕ-biderivations $\delta_\phi^{(T)}$, which are those obtained from the Drinfeld module ϕ and any twisted polynomial $T \in \bar{k}_\infty\{F\}F$ via

$$\delta_\phi^{(T)}(a) := T\phi(a) - aT, \ \forall a \in A.$$

Notice that, for non-zero such T,

$$\deg_F \delta_\phi^{(T)} = \deg_F T + \deg_F \delta_\phi(a) > \deg_F \delta_\phi(a).$$

STRICTLY REDUCED BIDERIVATIONS: When $r > 1$, there are biderivations δ, said to be *strictly reduced*, for which $\deg_F \delta(a) < \deg_F \delta_\phi(a)$. Let $D_{sr}(\phi)$ denote the $\bar{k}_\infty$-vector space of strictly reduced biderivations.

Then the Hodge decomposition, equation (2.14) in [G2], is that $D(\phi)$, the full $\bar{k}_\infty$-vector space of ϕ-biderivations, decomposes as

$$D(\phi) = D_{sr}(\phi) \oplus \bar{k}_\infty \delta_\phi \oplus D_{si}(\phi).$$

Thus the de Rham cohomology satisfies

$$H_{DR}^*(\phi) := D(\phi)/D_{si}(\phi) \cong D_{sr}(\phi) \oplus \bar{k}_\infty \delta_\phi.$$

P. Deligne and G. Anderson, (see also §5 [G2]) noticed that

$$\dim_{\bar{k}_\infty} H_{DR}^*(\phi) = \operatorname{rank} \phi = r.$$

E.-U. Gekeler proves the de Rham isomorphism in [G2]

$$H_{DR}^*(\phi) \cong \operatorname{Hom}_A(L_\phi, \bar{k}_\infty),$$

by constructing the quasi-periodic functions inducing given elements of the module $\operatorname{Hom}_A(L_\phi, \bar{k}_\infty)$, while Anderson had a "shtuka" proof.

2. Quasi-Periodic Functions.

Given a biderivation δ, there is a unique (up to a non-zero scalar) entire $\mathbb{F}_q$-linear solution (e.g. [G2]) $Y = F_\delta(z)$ of the functional equation

(6) $$Y(az) - aY(z) = \delta(a)e_\phi(z), \ \forall a \in A,$$

satisfying the "initial condition" $Y(z) \equiv 0 \mod z^q$. The function F_δ is said to be *quasi-periodic* with respect to δ and L, since

(1) $F_\delta(z + \omega) = F_\delta(z) + F_\delta(\omega), \ \forall z \in \bar{k}_\infty, \omega \in L,$
(2) $F_\delta(z)$ is A-linear on L.

We can already account for the quasi-periodic functions arising from the *inner* derivations, i.e. those of the form $\delta_\phi^{(T)}$, with $T \in \bar{k}_\infty\{F\}$, possibly with non-zero coefficient of F^0. The related quasi-periodic functions are easily verified to be of the form

$$F_{\delta_\phi^{(T)}}(z) := Te_\phi(z) - uz,$$

where $T = uF^0 +$ higher terms. This justifies the use of the word "exact" to describe the biderivations $\delta \in D_{si}(\phi)$. Note that when $T \in \bar{k}_\infty F\{F\}$, then the corresponding quasi-periodic function is simply $Te_\phi(z)$, since $u = 0$.

REMARK 4.1: $F_\delta(z) \in \bar{k}_\infty(e_\phi(z), z)$, whenever δ is strictly inner. Consequently, for the purposes of finding algebraically independent functions, we necessarily restrict our attention to $z, e_\phi(z)$, and $F_\delta(z)$, for various δ representing linearly independent de Rham cohomology classes, i.e. linearly independent strictly reduced biderivations. (Although these $F_\delta(z)$ are much less accessible to direct description, as mentioned Gekeler shows in [G2] for his proof of the de Rham isomorphism how to construct the reduced $F_\delta(z)$ corresponding to a given element of $\mathrm{Hom}\,(L_\phi, \bar{k}_\infty)$.)

3. Algebraic Independence and de Rham Cohomology.

THEOREM 4.1. *Let Δ be a set of ϕ-biderivations of A into $\bar{k}_\infty\{F\}F$. Let $\{F_\delta\}_{\delta \in \Delta}$ be the corresponding quasi-periodic functions. Then the following statements are equivalent:*

(1) *$e_\phi(z)$ and the functions $F_\delta(z)$ are algebraically dependent over $\bar{k}_\infty$.*
(2) *Some non-trivial $\bar{k}_\infty$-linear combination of the functions $F_\delta(z)$ is equal to $P(e_\phi(z))$, for some $\mathbb{F}_q$-linear polynomial $P(X)$.*
(3) *Some non-trivial $\bar{k}_\infty$-linear combination of the functions $F_\delta(z)$ is quasi-periodic with respect to a strictly inner biderivation from $D_{si}(\phi)$.*
(4) *The images of the δ in $H_{DR}^*(\phi)$ are $\bar{k}_\infty$-linearly dependent.*

Using the Hodge decomposition, one can rephrase this theorem as follows:

THEOREM 4.2. *Let ∇ be a set of strictly reduced ϕ-biderivations with corresponding quasi-periodic functions $\{F_\delta(z)\}_{\delta \in \nabla}$. Then the following statements are equivalent:*

(1) *$z, e_\phi(z)$, and the functions $F_\delta(z)$ are algebraically dependent over $\bar{k}_\infty$.*
(2) *$\{F_\delta(z)\}_{\delta \in \nabla}$ is $\bar{k}_\infty$-linearly dependent.*
(3) *∇ is $\bar{k}_\infty$-linearly dependent.*

To deduce this result from the preceding theorem, it suffices to set $\Delta = \nabla \cup \{\delta_\phi\}$.

(1) Then z, $e_\phi(z)$, and the $F_\delta(z)$ are algebraically dependent exactly when $e_\phi(z)$, $e_\phi(z) - z$, and the $F_\delta(z)$ are algebraically dependent. But these are $e_\phi(z)$ and the quasi-periodic functions associated to Δ.

(2) Any $\sum_{\delta \in \Delta} c_\delta F_\delta(z)$ is a quasi-periodic function with respect to the strictly reduced biderivation $\sum c_\delta \delta$, which, by the Hodge decomposition, is strictly inner only if it is identically zero, i.e. exactly when $\sum_{\delta \in \nabla} c_\delta F_\delta(z) = 0$.

(3) Again by the Hodge decomposition, Δ will be $\bar{k}_\infty$-linearly dependent exactly when ∇ is $\bar{k}_\infty$-linearly dependent.

Thus the three conditions of Theorem 4.2 reduce to special cases of conditions (1),(3),(4) of Theorem 4.1.

PROOF OF THEOREM 4.1: $(4) \Rightarrow (3)$ If the images in $H^*_{DR}(\phi)$ of the δ are $\bar{k}_\infty$-linearly dependent, then there are elements $c_\delta \in \bar{k}_\infty$, not all zero, and a twisted polynomial $T \in \bar{k}_\infty\{F\}F$ such that

$$\sum c_\delta \delta = \delta_\phi^{(T)}.$$

We claim that $\sum c_\delta F_\delta$ is quasi-periodic with respect to $\delta_\phi^{(T)}$. For, according to (6), when a in A, $\sum c_\delta F_\delta(az) - a \sum c_\delta F_\delta(z) = \sum c_\delta \delta(a) e_\phi(z) = \delta_\phi^{(T)}(a) e_\phi(z)$. This verifies (3) from (4).

$(3) \Rightarrow (2)$ Assume that, say, $\sum c_\delta F_\delta(z)$ is quasi-periodic with respect to $\delta_\phi^{(T)}$, where $T \in \bar{k}_\infty\{F\}F$. As we noted earlier, the quasi-periodic function related to $\delta_\phi^{(T)}$ is $Te_\phi(z)$, Thus, by the unicity of $F_{\delta_\phi^{(T)}}$,

$$\sum c_\delta F_\delta(z) = Te_\phi(z),$$

and we have what we want with $P(X) := TX$. This verifies (2) from (3).

$(2) \Rightarrow (1)$ Obvious.

$(1) \Rightarrow (4)$ When the $\mathbb{F}_q$-linear functions are algebraically dependent, we again use Artin's theorem (Theorem 18 of VIII, §11 of [L]) on the minimal relations for additive functions to write a minimal dependence relation in the form

$$\sum P_\delta(F_\delta(z)) + P_0(e_\phi(z)) = 0,$$

involving the fewest functions and, for those functions, having $\mathbb{F}_q$-linear polynomials P_δ of least degree (provisionally for arbitrary fixed δ'). Since the twisted polynomials corresponding to the position δ' occurring in any such $\mathbb{F}_q$-linear relation on our minimal set of functions form a left ideal in $\bar{k}_\infty\{F\}$, then each such relation is in fact a left multiple from $\bar{k}_\infty\{F\}$ of our minimal one. Now substituting az for z and using the functional equations for the various F_δ shows that for every a in A, we obtain

$$\sum P_\delta(aF_\delta(z)) + \sum P_\delta(\delta(a)e_\phi(z)) + P_0(\phi(a)e_\phi(z)) = 0.$$

Since for an $\mathbb{F}_q$-linear polynomial $P(X) = \sum c_i X^{q^i}$, $P(aX) = \sum c_i a^{q^i} X^{q^i}$ and the degrees in $F_\delta(z)$ remain the same after our substitution and rewriting, the preceding displayed equation must be a multiple from $\bar{k}_\infty$ of the minimal equation. However since the multiplying factor a^{q^i} for each monomial in the various P_δ depends on the degree i of the monomial, these degrees must *all* be the same, say q^h, as is seen by taking $a \in A$ transcendental over $\mathbb{F}_q$: $P_\delta(X) = p_\delta X^{q^h}$, with $p_\delta \in \bar{k}_\infty$. (Some $p_\delta \neq 0$, since $e_\phi(z)$, as a transcendental function, cannot be a root of a non-zero $P_0(X)$.) Then, on comparing the terms involving $e_\phi(z)$ in the two relations, we see that

$$\sum p_\delta(\delta(a)e_\phi(z))^{q^h} + P_0(\phi(a)e_\phi(z)) = a^{q^h} P_0(e_\phi(z)),$$

or, since $e_\phi(z)$ is a transcendental function,

$$\sum p_\delta\left(\delta(a)X\right)^{q^h} + P_0(\phi(a)X) = a^{q^h} P_0(X).$$

If $P_0(X) = c_l X^{q^l} +$ higher terms, then comparing orders of vanishing at $z = 0$ in the original dependence relation shows that $l > h$ for any $c_l \neq 0$. Since $P_0(X)$ is $\mathbb{F}_q$-linear, with no terms of degree lower than q^{h+1}, we may take the qth roots of coefficients as in Remark A4 to find the $\mathbb{F}_q$-linear polynomial $Q(X) \in \bar{k}_\infty[X]$ (with no terms of degree less than q) such that $(-1)^{q^h} P_0(X) = Q(X)^{q^h}$. Finally on taking q^hth roots c_δ of p_δ, we can rewrite our relation as

$$\left[\sum c_\delta\delta(a)X - Q(\phi(a)X) + aQ(X)\right]^{q^h} = 0, \ \forall a \in A$$

or finally

$$\sum c_\delta\delta(a) = T\phi(a) - aT = \delta_\phi^{(T)}(a), \ \forall a \in A,$$

where $T \in \bar{k}_\infty\{F\}F$ is the twisted polynomial associated to $Q(X)$. In other words, algebraic dependence of $e_\phi(z)$ and the $F_\delta(z)$ leads to a $\bar{k}_\infty$-linear dependence relation on the δ modulo $D_{si}(\phi)$, which is what we wanted to show. $\square$

5. ALGEBRAIC INDEPENDENCE OF DRINFELD EXPONENTIAL AND QUASI-PERIODIC FUNCTIONS

1. Independence of Quasi-Periodic Functions.

Finally we are in a position to state our second main result.

THEOREM 5.1. *Let $\Phi = \{\phi\}$ denote a set of non-isogenous Drinfeld A-modules, whose exponential functions $e_\phi(z)$ have period lattices L_ϕ. For each $\phi \in \Phi$, let $R_\phi := \{\rho \in \bar{k}_\infty : \rho L_\phi \subset L_\phi\}$ be the ring of multiplications of ϕ.*

(1) *For each $\phi \in \Phi$, let $U_\phi \subset \bar{k}_\infty$ be an R_ϕ-linearly independent set.*
(2) *For each $\phi \in \Phi$, let Δ_ϕ be a set of ϕ-biderivations whose images in $HD^*_{DR}(\phi)$ are $\bar{k}_\infty$-linearly independent.*

Then the $1 + \sum_\phi |\Delta_\phi||U_\phi|$ functions

$$z, F_\delta(uz), \quad \phi \in \Phi, u \in U_\phi, \delta \in \Delta_\phi$$

are all algebraically independent over $\bar{k}_\infty$.

We first remark that, since $e_\phi(z) - z$ is quasi-periodic with respect to the differential of the first kind δ_ϕ, this statement includes algebraic independence of exponential functions. In order to make this point explicitly, we restate the theorem in the following form:

THEOREM 5.1'. *Let $\Phi = \{\phi\}$ denote a set of non-isogenous Drinfeld A-modules, whose exponential functions $e_\phi(z)$ have period lattices L_ϕ. For each $\phi \in \Phi$, let R_ϕ denote $\{\rho \in \bar{k}_\infty : \rho L_\phi \subset L_\phi\}$, the ring of multiplications of ϕ.*

(1) *For each $\phi \in \Phi$, let $U_\phi \subset \bar{k}_\infty$ be an R_ϕ-linearly independent set.*
(2) *For each $\phi \in \Phi$, let Δ_ϕ be a set of $\bar{k}_\infty$-linearly independent strictly reduced ϕ-biderivations.*

Then the $1 + \sum_\phi (1 + |\Delta_\phi|)|U_\phi|$ functions

$$z, e_\phi(uz), F_\delta(uz), \quad \phi \in \Phi, u \in U_\phi, \delta \in \Delta_\phi$$

are all algebraically independent over $\bar{k}_\infty$.

REMARKS 5.1:

i   We have seen through Theorem 2.2 that the isogeny assumption is actually a normalization for the independence of the $e_\phi(uz)$ rather than a restriction. By Remark 5.3 below it is also a normalization for the functions $F_\delta(uz)$ as well.

ii We have seen that the conclusion of Theorem 5.1' fails to hold, with coefficients in the field $\bar{k}_\phi$, whenever either indented hypothesis fails.

iii We shall see below that any finite set of $z, e_\phi(uz), F_\delta(uz)$ is contained in a field generated by algebraically independent functions of the type given in Theorem 5.1'.

iv Finally we observe that, say in the notation of Theorem 5.1, $|\Delta_\phi| \leq r_\phi = \mathrm{rank}\ L_\phi$, and that any set Δ_ϕ of ϕ-biderivations, whose images in $HD^*_{DR}(\phi)$ are $\bar{k}_\infty$-linearly independent may be augmented to a set of biderivations of cardinality equal to r_ϕ and whose images in $HD^*_{DR}(\phi)$ are linearly independent.

As a special case, we can take $|\Phi| = 1$ and $|\Delta_\phi| = 0$ to obtain Theorem 3.1 of [Y1]:

COROLLARY 5.2 (YU). *Let $u_1, \ldots, u_n$ lie in $\bar{k}_\infty$. Then z and*

$$e_\phi(u_1 z), \ldots, e_\phi(u_n z)$$

are algebraically independent over $\bar{k}_\infty$ if and only if $u_1, \ldots, u_n$ are R_ϕ-linearly independent.

Similarly we obtain the following result, which is implicit in Yu's proof of Theorem 3.1 of [Y2]:

COROLLARY 5.3 (YU). *If $\delta \in D(\phi) \smallsetminus D_{si}(\phi)$, then the functions $F_\delta(z)$ and $e_\phi(z)$ are algebraically independent over $\bar{k}_\infty$.*

PROOF OF THEOREM 5.1′: We prove the contrapositive. The proof is essentially an extension of the argument for the implication (1) $\Rightarrow$ (4) in Theorem 4.1. Since all the functions involved are $\mathbb{F}_q$-linear, by Artin's theorem, any minimal dependence relation will be of the form

$$P_0(z) + \sum_{\phi \in \Phi} \sum_{u \in U_\phi} P_{\phi,u}(e_\phi(uz)) + \sum_{\phi \in \Phi} \sum_{\delta \in \Delta_\phi, u \in U_\phi} P_{\delta,u}(F_\delta(uz)) = 0,$$

with $\mathbb{F}_q$-linear polynomials $P_{\phi,u}, P_{\delta,u}$. Assume that this relation involves only the $e_\phi(z)$ if that is possible. In any case, assume that the relation involves a minimal set (with respect to containment) of algebraically dependent functions and that, for these functions, this relation is of minimal degree with respect to one, and therefore all, non-zero polynomials actually occurring. On replacing z by az, we obtain the new relation

$$P_0(az) + \sum_{\phi \in \Phi} \sum_{u \in U_\phi} P_{\phi,u}(\phi(a)e_\phi(uz)) + \sum_{\phi \in \Phi} \sum_{\delta \in \Delta_\phi, u \in U_\phi} P_{\delta,u}(F_\delta(auz)) = 0.$$

Now the course of our argument must depend on which sorts of functions are involved in our dependence relation.

CASE 1: All $P_{\delta,u}(X) = 0$ in some minimal dependence relation. This case has been dealt with in Theorem 3.1, whence it follows that our first hypothesis fails to hold.

CASE 2: Some $P_{\delta,u}(X) \neq 0$ in every minimal dependence relation. In this case, the exponential functions $e_\phi(uz)$ are all algebraically independent. Moreover we know that the degrees in the $F_\delta(uz)$ are not raised by our substitution of az for z (and rewriting), and that the polynomial $P_{\delta,u}(X)$ is multiplied by a to the degree of $P_{\delta,u}(X)$, say by a^{q^h}. This means that

$P_0(X) = p_0 X^{q^h}$ and, by the functional equation for F_δ with argument uz, that each $P_{\delta,u}(X) = p_{\delta,u} X^{q^h}$, for the same fixed h, and finally that

(7)
$$\sum_{\phi,u} P_{\phi,u}(\phi(a)e_\phi(uz)) + \sum_\phi \sum_{\delta,u} p_{\delta,u}\{\delta(a)(e_\phi(uz))\}^{q^h}$$
$$= a^{q^h} \sum_{\phi,u} P_{\phi,u}(e_\phi(uz)).$$

Write $P_{\phi,u}(X) = p_{\phi,u} X^{q^{l(\phi,u)}}$ + higher terms, $p_{\phi,u} \neq 0$. Recall that $\phi(a)e_\phi(uz) = ae_\phi(uz)$ + higher terms in $e_\phi(uz)$ and that the various $e_\phi(uz)$ are algebraically independent. Note that, for fixed u and ϕ, the lowest terms on the left in the preceding displayed equation are $p_{\phi,u}(ae_\phi(uz))^{q^{l(\phi,u)}}$ and $p_{\delta,u}(ae_\phi(uz))^{q^h}$, while the lowest terms on the right are $a^{q^h} p_{\phi,u}e_\phi(uz)^{q^{l(\phi,u)}}$. Taking $a \in A$ transcendental over $\mathbb{F}_q$, we realize that each $l(\phi,u) \geq h$.

As in Remark A4, this fact allows us to write each

$$P_{\phi,u}(X) = Q_{\phi,u}(X)^{q^h},$$

for $\mathbb{F}_q$-linear polynomials $Q_{\phi,u}(X) \in \bar{k}_\infty[X]$. Thus for the q^hth roots $q_{\delta,u}$ of the $p_{\delta,u}$, we rewrite (7) as

$$0 = \left\{ \sum_{\phi,u} Q_{\phi,u}(\phi(a)e_\phi(uz)) + \sum_\phi \sum_{\delta,u} q_{\delta,u}\delta(a)e_\phi(uz) - a \sum_{\phi,u} Q_{\phi,u}(e_\phi(uz)) \right\}^{q^h}$$

and thus

$$0 = \sum_{\phi,u} Q_{\phi,u}(\phi(a)e_\phi(uz)) + \sum_\phi \sum_{\delta,u} q_{\delta,u}\delta(a)e_\phi(uz) - a \sum_{\phi,u} Q_{\phi,u}(e_\phi(uz)).$$

Therefore, on splitting off any linear terms of the $Q_{\phi,u}$ to the left side, we find that

$$-\sum_\phi \sum_{\delta,u} q_{\delta,u}\delta(a)e_\phi(uz)$$
$$= \sum_{\phi,u} \{[T_{\phi,u}\phi(a) - aT_{\phi,u}] + [q_{\phi,u}\phi(a) - aq_{\phi,u}]\} e_\phi(uz),$$

where $q_{\phi,u}F^0 + T_{\phi,u}$ is the twisted polynomial whose value at X is $Q_{\phi,u}(X)$ and $T_{\phi,u} \in \bar{k}_\infty\{F\}F$. Since the distinct $e_\phi(uz)$ occurring are algebraically independent, this means that for each such ϕ and u,

$$-\sum_{\delta \in \Delta_\phi} q_{\delta,u}\delta(a) = \delta_\phi^{(T_{\phi,u})}(a) + q_{\phi,u}\delta_\phi(a),$$

for all a in A, which shows that our second hypothesis fails to hold. This establishes Theorem 5.1'. $\square$

2. Reduction of General Case to Normal Form.

The preceding discussion encourages us to make the following definition: Let Φ be a set of Drinfeld modules, and for each $\phi \in \Phi$, let $U_\phi \subset \bar{k}_\infty$ and $\nabla_\phi \subset D_{sr}(\phi)$. We say the set of pairs $(U_\phi, \nabla_\phi)_{\phi \in \Phi}$ is in *normal form (for independence)* if

(1) the Drinfeld modules $\phi \in \Phi$ are pair-wise non-isogenous,
(2) each U_ϕ is R_ϕ-linearly independent, and
(3) each set ∇_ϕ of strictly reduced biderivations is $\bar{k}_\infty$-linearly independent.

We saw in Theorem 4.1' that condition (3) is necessary for the algebraic independence of $z, e_\phi(uz)$, and the various $F_\delta(uz), \delta \in \nabla_\phi, u \in U_\phi$, for fixed $\phi \in \Phi$. We saw in Proposition 2.6 that condition (2) is necessary for the algebraic independence of the $e_\phi(uz)$. We have seen (in condition (2) of Theorem 2.2) that condition (1) here is merely a normalization as far as the independence of the $e_\phi(uz)$ is concerned. We now want to show that condition (1) is also a normalization as far as the independence of the $F_\delta(uz)$ are concerned and to determine the role of condition (2) with regard to the $F_\delta(uz)$.

Now let T be an isogeny from the Drinfeld module ϕ to the Drinfeld module ψ, i.e. $T\phi = \psi T$, and let δ be a ψ-biderivation. Then

$$\delta(ab)T = a\delta(b)T + \delta(a)\psi(b)T = a\delta(b)T + \delta(a)T\phi(b).$$

Therefore $\delta \cdot T : a \mapsto \delta(a)T$ is a ϕ-biderivation. Thus we have a contravariant functor D from the category of Drinfeld modules to the category of finite dimensional $\bar{k}_\infty$-vector spaces which associates to any $T \in \mathrm{Hom}(\psi, \phi)$ the element $DT \in \mathrm{Hom}(D(\psi), D(\phi))$ given by $DT : \delta \mapsto \delta \cdot T$. Since each DT maps $D_{si}(\psi)$ into $D_{si}(\phi)$, each DT induces a map on de Rham cohomology:

$$H_{DR}^* T : H_{DR}^*(\psi) \to H_{DR}^*(\phi),$$

and so we obtain the following observation.

REMARK 5.2:

(1) The correspondence $D : \phi \mapsto D(\phi)$ is a contravariant functor on Drinfeld modules.
(2) The correspondence $H_{DR}^* : \phi \mapsto H_{DR}^*(\phi)$ is a contravariant functor on Drinfeld modules.

Moreover as $F_{\delta \cdot T}$ is quasi-periodic with respect to ϕ, by Theorem 2.2

$$F_{\delta \cdot T}(az) - aF_{\delta \cdot T}(z) = \delta(a)Te_\phi(z) = \delta(a)e_\psi(uz),$$

for all a in A, where $T = uF^0 +$ higher terms. This shows by the unicity of the solutions of the functional equations for quasi-periodic functions that, quite analogously to what happens for the exponential function and isogenies T (cf. condition (2) of Theorem 2.2),

$$(8) \qquad\qquad F_{\delta \cdot T}(z) = F_\delta(uz).$$

Since quasi-periodic functions with respect to ψ are thus quasi-periodic (with appropriately scaled arguments) with respect to isogenous Drinfeld modules ϕ, we have the following:

REMARK 5.3: Condition (1) in the definition of normal form is merely a normalization of notation for exponential and quasi-periodic functions.

Now the situation analogous to Proposition 2.6, giving necessary and sufficient conditions for algebraic independence involving quasi-periodic functions, is not nearly so easily stated. The matter would be easily dealt with (as over A), if only we could simply extend every ϕ-biderivation δ to one defined on all of R_ϕ so that (6) holds for $Y = F_\delta$. That extension however is generally possible only for a certain subspace, as

$$\dim_{\bar{k}_\infty} H^*_{DR}(\phi|_{R_\phi}) = \operatorname{rank}_{R_\phi} \Lambda = \frac{1}{[K_\phi : k]} \operatorname{rank}_A \Lambda$$

$$= \frac{1}{[K_\phi : k]} \dim_{\bar{k}_\infty} H^*_{DR}(\phi|_A),$$

where K_ϕ is the field of quotients of R_ϕ.

However the identities

$$(9) \quad F_{\delta'+\delta''}(z) = F_{\delta'}(z) + F_{\delta''}(z) \quad \text{and} \quad F_{cT}(z) = cF_T(z), \forall c \in \bar{k}_\infty$$

will allow us, in any given case, to reduce to a related normal form.

REMARK 5.4: If ϕ and ψ are isogenous A-Drinfeld modules, then R_ϕ and R_ψ have the same field of quotients.

For if $u\Lambda_\psi \subset \Lambda_\phi$, with rank Λ_ψ = rank Λ_ϕ, then there is a non-zero $a \in A$ with $a\Lambda_\phi \subset u\Lambda_\psi$. Thus

$$aR_\psi \cdot \Lambda_\phi \subset R_\psi u\Lambda_\psi \subset u\Lambda_\psi \subset \Lambda_\phi,$$

and $aR_\psi \subset R_\phi$. Similarly there is a non-zero $a' \in A$ with $a'R_\phi \subset R_\psi$, and both $a, a' \in R_\phi \cap R_\psi$. $\square$

Now assume that we are given a finite set Ψ of Drinfeld A-modules and, for each $\psi \in \Psi$, finite sets $V_\psi \subset \bar{k}_\infty$ and $\nabla_\psi \subset D(\psi)$. Choose a set Φ of representatives ϕ for the isogeny classes $<\psi>$ in Ψ, and, for each $\psi \in <\phi>$, fix an isogeny $T_\psi = u_\psi F^0+$ higher terms from ϕ to ψ. For each representative ϕ, choose a maximal R_ϕ-linearly independent subset U'_ϕ of $\bigcup_{\psi \in <\phi>} u_\psi^{-1} V_\psi$, and, for each $\psi \in <\phi>$, express each of the finitely many given $v \in V_\psi$ via $au_\psi^{-1}v = \sum_u \rho_u(v)u$, $u \in U'_\phi$, with $\rho_u(v) \in R_\psi$ and with fixed non-zero $a \in A$, independent of ψ. (This is possible because all R_ψ with ψ from a fixed $<\phi>$ have the same field of quotients, which is a finite dimensional extension of the field of quotients of A.) Then, for every $v \in V_\psi$ and for every $\partial \in \bigcup_{\psi \in <\phi>} \nabla_\psi$, first use the extension of ψ to R_ψ given in Proposition 2.5 with (8) and then T_ψ with (8) to rewrite

$$F_\partial(avz) = F_\partial(\sum_u \rho_u(v)u_\psi uz) = \sum_u F_{\partial \cdot \psi(\rho_u(v))}(u_\psi uz)$$

$$= \sum_u F_{\partial \cdot \psi(\rho_u(v))T_\psi}(uz).$$

Since $\psi(\rho_u(v))T_\psi$ is an isogeny from ϕ to ψ, we see, by our remarks on the functoriality of D, that $\partial \cdot \psi(\rho_u(v))T_\psi \in D(\phi)$.

Next choose a $\bar{k}_\infty$-basis $\Delta_\phi = \{\delta\}$ for each $D_{sr}(\phi)$ and rewrite each $\partial \in \nabla_\psi, \psi \in <\phi>$ as

$$\partial \cdot \psi(\rho_u(v))T_\psi = \sum_\delta c_{\partial,v}(\delta, u)\delta \, v + c_{\partial,v}(u)\delta_\phi + \delta_{\partial,v}(u)$$

with $c_{\partial,v}(\delta, u), c_{\partial,v}(u) \in \bar{k}_\infty$ and $\delta_{\partial,v}(u) \in D_{si}(\phi)$. Therefore by (9), each

$$F_\partial(avz) = \sum_\delta \sum_u c_{\partial,v}(\delta, u)F_\delta(uz) + \sum_u c_{\partial,v}(u)\{e_\phi(uz) - uz\} +$$

$$\sum_u P_{\partial,v}(e_\phi(uz)),$$

for explicit $\mathbb{F}_q$-linear polynomials $P_{\partial,v}(X)$. Note that in addition,

$$e_\psi(avz) = e_\psi(\sum_u \rho_u(v)u_\psi uz)$$

$$= \sum_u \psi(\rho_u(v))e_\psi(u_\psi uz) = \sum_u \psi(\rho_u(v))e_\phi(uz).$$

Thus we have shown (on replacing z by z/a in each of these formulas and letting $W_\phi := \{u/a : u \in U'_\phi\}$):

PROPOSITION 5.4. *Given any finite set Ψ of Drinfeld A-modules and, for each $\psi \in \Psi$, finite sets $V_\psi \subset \bar{k}_\infty$ and $\nabla_\psi \subset D(\psi)$, there is a set $(W_\phi, \Delta_\phi)_{\phi \in \Phi}$ in normal form for independence such that*

$$z, e_\psi(vz), F_\partial(vz), \quad \forall \psi \in \Psi, v \in V_\psi, \partial \in \nabla_\psi,$$

are elements of the field $\bar{k}_\infty(z, e_\phi(wz), F_\delta(wz))_{\phi \in \Phi, w \in W_\phi, \delta \in \Delta_\phi}$.

We call $(W_\phi, \Delta_\phi)_{\phi \in \Phi}$ as constructed above a *normal form for* $(V_\psi, \nabla_\psi)_{\psi \in \Psi}$.

Using this result, we are able to determine all dependence relations on the given functions, since we have proved the lack of algebraic relations on $z, e_\phi(wz)$ and the functions $F_\delta(wz)$ with $(W_\phi, \Delta_\phi)_{\phi \in \Phi}$ in normal form.

6. Appendix: $\mathbb{F}_q$-linear Functions

An $\mathbb{F}_q - linear\, power\, series\, P(X)$ is a power series of the following form: $P(X) = \sum_{h=0}^{\infty} a_h X^{q^h}$. The terminology stems from the fact that for $c \in \mathbb{F}_q$, $P(X+cY) = P(X)+cP(Y)$, since such an identity holds term-by-term.

REMARK A1. *If P is $\mathbb{F}_q$-linear in X, then for any t so is*

$$\prod_{c \in \mathbb{F}_q} P(X + ct) = \prod_{c \in \mathbb{F}_q} \{P(X) + cP(t)\} = P(X)^q - P(X)P(t)^{q-1}.$$

REMARK A2. *The product*

$$L_d(X) = \prod_{c_i \in \mathbb{F}_q} (X + c_1 t_1 + \cdots + c_d t_d)$$

is an $\mathbb{F}_q$-linear polynomial in X, as we can write inductively

$$L_d(X) = \prod_{c_d \in \mathbb{F}_q} L_{d-1}(X + c_d t_d).$$

REMARK A3. *If L is a discrete A-module, then it is a countable $\mathbb{F}_q$-linear vector space, and*

$$e_L(z) = z \prod_{w \in L}{}' (1 - \frac{z}{\omega})$$

is $\mathbb{F}_q$-linear, where the dash means that we omit the term involving zero in the denominator, since the partial products are $\mathbb{F}_q$-linear.

REMARK A4. *Taking q^lth roots term-by-term, we see that if $P(X) = \sum_{h=l}^{\infty} c_h X^{q^h}$, then there is a unique $\mathbb{F}_q$-linear power series $Q(X)$ satisfying*

$$P(X) = Q(X)^{q^l}.$$

Bibliography

[A] G. W. Anderson, *t-Motives*, Duke Math. J. **53** (1986), 457-502.

[BK] W.D. Brownawell and K.K. Kubota, *The algebraic independence of Weierstrass functions and some related numbers*, Acta Arith. **33** (1977), 111-149.

[C] R.F. Coleman, *On a stronger version of the Schanuel-Ax theorem*, Am. J. Math. **102** no. 4 (1980), 595-624.

[DH] P. Deligne and D. Husemöller, *Survey of Drinfeld modules*, Contemporary Math. **67** (1987), 25-91.

[Dr] V. G. Drinfeld, *Elliptic modules*, Mat. Sbornik **94** (1974), 594-627. Engl. transl. Math. USSR Sbornik **23** (1974), 561-592.

[G1] E.-U. Gekeler, *Arithmetik von Drinfeld-Moduln*, Math. Ann. **262** (1983), 167-182.

[G2] E.-U. Gekeler, *De Rham isomorphism for Drinfeld modules*, J. für die reine und angew. Math. **401** (1989), 188-208.

[H1] D.R. Hayes, *Explicit class field theory in global function fields*, in Studies in Algebra and Number Theory (G.-C. Rota, ed.), Academic Press, New York, 1979.

[H2] D.R. Hayes, *A brief introduction to Drinfeld modules*, in The Arithmetic of Function Fields (D. Goss, D.R. Hayes, M. Rosen, eds.), Walter De Gruyter, Berlin, New York, 1992.

[K] E. Kolchin, *Algebraic groups and algebraic independence*, Amer. J. Math. **90** (1968), 1151-1164.

[L] S. Lang, *Algebra*, Addison-Wesley, Reading, MA, 1965.

[Lz] M. Lazard, *Les zéros d'une fonction analytique d'une variable sur un corps valué complet*, Inst. Hautes Etudes Sci. Publ. Math. **14** (1964), 47-75.

[W] M. Waldschmidt, *Transcendence problems connected with Drinfeld modules.* in Arf Symposium, Siliviri.

[Y1] J. Yu, *Transcendence and Drinfeld modules: several variables*, Duke Math. J. **58** (1989), 559-575.

[Y2] J. Yu, *On periods and quasi-periods of Drinfeld modules*, Compositio Math. **74** (1990), 235-245.

[Y3] J. Yu, *Transcendence and special zeta values in characteristic p*, Annals of Math. **134** (1991), 1-23.

[Y4] J. Yu, *Analytic homomorphisms into Drinfeld modules*, preprint (1991).

[Y5] J. Yu, *Transcendence in finite characteristic*, in The Arithmetic of Function Fields (D. Goss, D.R. Hayes, M. Rosen, eds.), Walter De Gruyter, Berlin, New York, 1992.

Dale W. Brownawell
Mathematics Department
Penn State University
University Park, PA 16802
e-mail wdb@math.psu.edu

Estimates for Discriminants and Resultants of Binary Forms

Jan-Hendrik Evertse

University of Leiden

The *resultant* $R(F, G)$ of two binary forms

$$F(X, Y) = a_0 X^r + a_1 X^{r-1} Y + \ldots + a_r Y^r$$

and

$$G(X, Y) = b_0 X^s + b_1 X^{s-1} Y + \ldots + b_s Y^s$$

is defined by the $(r + s) \times (r + s)$ determinant

$$R(F, G) = \begin{vmatrix} a_0 & a_1 & \cdots & a_r & & & \\ & a_0 & a_1 & \cdots & a_r & & \\ & & \ddots & & & \ddots & \\ & & & a_0 & a_1 & \cdots & a_r \\ b_0 & b_1 & \cdots & \cdots & b_s & & \\ & \ddots & & & & \ddots & \\ & & b_0 & b_1 & \cdots & \cdots & b_s \end{vmatrix}$$

where the first s rows contain coefficients of F and the last r rows coefficients of G. The *discriminant* $D(F)$ of the binary form F satisfies

$$D(F) = (-1)^{\frac{r(r-1)}{2}} r^{-(r-2)} R\left(\frac{\partial F}{\partial X}, \frac{\partial F}{\partial Y}\right) \quad \text{(cf. [20], p.335)}.$$

The *height* $H(P)$ of a polynomial P with complex coefficients is defined as the maximum of the absolute values of these coefficients. From Hadamard's inequality it follows easily that

(1.1)
$$\begin{cases} |D(F)| \leq r^{2r-1} H(F)^{2r-2}, \\ |R(F, G)| \leq (r + 1)^{s/2}(s + 1)^{r/2} H(F)^s H(G)^r. \end{cases}$$

Binary forms can be factored into linear forms over an algebraically closed field. If $F(X, Y) = \prod_{i=1}^{r}(\alpha_i X + \beta_i Y), G(X, Y) = \prod_{j=1}^{s}(\gamma_j X + \delta_j Y)$, then

(1.2) $\quad D(F) = \prod_{1 \leq i < j \leq r} (\alpha_i \beta_j - \alpha_j \beta_i)^2, \quad R(F, G) = \prod_{i=1}^{r} \prod_{j=1}^{s} (\alpha_i \delta_j - \beta_i \gamma_j).$

This research was made possible by a fellowship of the Royal Netherlands Academy of Arts and Sciences (K.N.A.W.)

From now on, we consider only binary forms with coefficients in $\mathbf{Z}$. Two such binary forms F, F' are said to be *equivalent*, notation $F \sim F'$, if $F'(X, Y) = F(aX + bY, cX + dY)$ for some matrix $\left(\begin{smallmatrix} a & b \\ c & d \end{smallmatrix}\right) \in SL_2(\mathbf{Z})$. Two pairs of binary forms $(F, G), (F', G')$ with coefficients in $\mathbf{Z}$ are said to be equivalent, notation $(F, G) \sim (F', G')$, if $F'(X, Y) = F(aX + bY, cX + dY)$ and $G'(X, Y) = G(aX + bY, cX + dY)$ for some matrix $\left(\begin{smallmatrix} a & b \\ c & d \end{smallmatrix}\right) \in SL_2(\mathbf{Z})$. From (1.2) it follows that

$$(1.3) \qquad \begin{cases} F \sim F' \Longrightarrow D(F) = D(F'); \\ (F, G) \sim (F', G') \Longrightarrow R(F, G) = R(F', G'). \end{cases}$$

This implies that there are binary forms with given discriminant and arbitrarily large height, and pairs of binary forms with given resultant and arbitrarily large heights. It will be convenient to call a binary form $F \in \mathbf{Z}[X, Y]$ *reduced* if $H(F) \leq H(F')$ for every $F' \sim F$, and to call a pair of binary forms (F, G) in $\mathbf{Z}[X, Y]$ with $\deg F = r$, $\deg G = s$ reduced if $H(F)^s H(G)^r \leq H(F')^s H(G')^r$ for every pair $(F', G') \sim (F, G)$. We are interested in estimates of the type

$$(1.4) \qquad H(F) \leq C(r, D(F)) \text{ if } F \text{ is a reduced binary form in } \mathbf{Z}[X, Y]$$

and

$$(1.5) \qquad H(F)^s H(G)^r \leq C(r, s, R(F, G))$$

if (F, G) is a reduced pair of binary forms in $\mathbf{Z}[X, Y]$.

In §2 of this paper we give a survey of results in this direction several of which are new. Further, as an application we give some estimates for the size of small values of binary forms, and for the difference of algebraic numbers. The complete proofs will be published in [8], [9], [12]. In the remainder of this paper, we denote by $C_i^{\mathrm{eff}}(...), C_i^{\mathrm{ineff}}(...)$ positive numbers depending only on the parameters between the parentheses; if there are no parameters, these numbers are absolute constants. The superscripts 'eff' and 'ineff' indicate whether or not the number can be computed effectively by using the method of proof involved.

§2. Results.

From classical results of Lagrange [18] on binary quadratic forms and Hermite [17] on binary cubic forms, it follows that for reduced binary forms $F \in \mathbf{Z}[X, Y]$ of degree 2 or 3 and non-zero discriminant one has

$$(2.1) \qquad H(F) \leq C_1^{\mathrm{eff}} |D(F)|.$$

It is a much more difficult problem to derive estimates like (1.4) for binary forms of degree $r \geq 4$. It was not until 1972 that Birch and Merriman [2] were able to prove that for every $r \geq 4$ and $D \neq 0$, there are only finitely many equivalence classes of binary forms $F \in \mathbf{Z}[X, Y]$ of degree r and discriminant D. This implies that every reduced binary form F satisfies (1.4) for some number $C(r, D(F))$. In fact, the dependence on r is not necessary, as Győry [15] proved that

$$(2.2) \qquad r \le 3 + \frac{2 \log |D(F)|}{\log 3}.$$

In their proof, Birch and Merriman used the following result, implicitly proved by Siegel [27]: for every algebraic number field K and every non-zero $\alpha_1, \alpha_2 \in K$, there are only finitely many units $\varepsilon_1, \varepsilon_2$ in K with

$$(2.3) \qquad \alpha_1 \varepsilon_1 + \alpha_2 \varepsilon_2 = 1.$$

It follows implicitly from the work of Baker [1] that all units $\varepsilon_1, \varepsilon_2$ in K satisfying (2.3) can be determined effectively. Győry [15] derived an explicit bound for the heights of $\varepsilon_1, \varepsilon_2$. However, this is not enough to make the proof of Birch and Merriman effective, i.e. from their proof it is only possible to show (1.4) with an ineffective number $C(r, D(F))$.

Independently of Birch and Merriman, Győry [14] showed that every *monic* reduced binary form $F \in \mathbf{Z}[X, Y]$ (i.e. with $F(1,0) = 1$) of degree r and with splitting field L (the smallest number field over which F factors into linear forms) satisfies

$$(2.4) \qquad H(F) \le C_2^{\mathrm{eff}}(r, D_L)|D(F)|^{C_3^{\mathrm{eff}}(r, D_L)},$$

where D_L is the discriminant of L. In Győry's method of proof, the restriction that F be monic was essential, and it required a new idea to get rid of this. Recently, Győry and the author [10] were able to show that (2.4) holds for *every* reduced binary form $F \in \mathbf{Z}[X, Y]$ of degree $r \ge 2$ with non-zero discriminant. Using simple algebraic number theory, one can estimate $|D_L|$ from above effectively in terms of $|D(F)|$. Hence (1.4) holds for every reduced binary form $F \in \mathbf{Z}[X, Y]$, with an effective number $C(r, D(F))$, but the dependence of this number on $|D(F)|$ is no longer polynomial.

We would like to have more precise information about how the height of a reduced form F compares with its discriminant. Obviously, the exponent $C_3^{\mathrm{eff}}(r, D_L)$ is far too large and in view of (1.1) one would rather expect something like $c/(r-1)$. It is indeed possible to derive an upper bound with such an exponent, but it is partially ineffective.

Theorem 1. *Let $F \in \mathbf{Z}[X, Y]$ be a reduced binary form of degree $r \ge 2$ with splitting field L and non-zero discriminant $D(F)$. Then*

$$(2.5) \qquad H(F) \le C_4^{\mathrm{ineff}}(r, L)|D(F)|^{\frac{21}{r-1}}.$$

One of the features of this bound is that up to a constant, the exponent on $|D(F)|$ is best possible, but one of its disadvantages, besides being ineffective, is that the constant involved depends on the splitting field L of F. It would be of great interest to find out whether or not this dependence on L is necessary. The proof of Theorem 1 will be given in [8].

For every algebraic number α, there is up to sign a unique binary form $F \in \mathbf{Z}[X, Y]$ of minimal degree and with coefficients having gcd 1 such that $F(\alpha, 1) = 0$. Put $H(\alpha) := H(F), D(\alpha) := D(F)$. Two algebraic numbers α, β are called equivalent, notation $\alpha \sim \beta$, if their corresponding binary forms are equivalent, which is the same as $\beta = (a\alpha + b)/(c\alpha + d)$ for some $\left(\begin{smallmatrix} a & b \\ c & d \end{smallmatrix}\right) \in SL_2(\mathbf{Z})$. Put

$$H^*(\alpha) = \min\{H(\beta) : \beta \sim \alpha\}.$$

For a given algebraic number field K, denote by $\psi(K)$ the set of numbers

$$\frac{\log |D(\alpha)|}{\log H^*(\alpha)}$$

for all $\alpha \in K$ with $\mathbf{Q}(\alpha) = K$, $H^*(\alpha) > 1$ (note that there are only finitely many equivalence classes of algebraic numbers α of given degree with $H^*(\alpha) = 1$). As usual, denote by $\limsup A, \liminf A$ the largest, smallest limit point, respectively, of a set A of real numbers. Then we have:

Corollary 1. *Let $r = [K : \mathbf{Q}]$. Then:*
(i). $\limsup \psi(K) \leq 2r - 2$; (ii). $\liminf \psi(K) \geq \frac{r-1}{21}$.

(i) follows at once from (1.1) and (ii) from (2.5). Instead of $\psi(K)$, one may consider the set $\psi(r)$ of numbers $\log |D(\alpha)|/\log H^*(\alpha)$ for algebraic numbers α of degree $r \geq 2$ with $H^*(\alpha) > 1$. From (1.1) it follows that $\limsup \psi(r) \leq 2r - 2$ but it is as yet unknown whether $\liminf \psi(r) > 0$.

We now consider pairs of binary forms with given resultant. Győry [16] proved that for every algebraic number field L and for every $m \in \mathbf{Z}, m \neq 0$, there are only finitely many equivalence classes of pairs of binary forms (F, G) such that FG has splitting field L, FG has no multiple factors in $L[X, Y]$, F and G are monic, $\deg F \geq 2$, $\deg G \geq 3$, and $R(F, G) = m$. Győry's proof was ineffective. Győry's result can be made more precise in several respects. First, it is possible to drop the restriction that F, G be monic and second, one can bound the discriminants of F, G in terms of their resultant.

Theorem 2. *Let $F, G \in \mathbf{Z}[X, Y]$ be binary forms of degrees $r \geq 3, s \geq 3$ respectively, such that FG has splitting field L and FG has no multiple factors in $L[X, Y]$. Then*

$$(2.6) \qquad |R(F, G)| \geq C_5^{\text{ineff}}(r, s, L)\left(|D(F)|^{\frac{s}{r-1}}|D(G)|^{\frac{r}{s-1}}\right)^{\frac{1}{18}}.$$

The dependence on L is necessary. Namely, assume that $s \leq r$. Then for every binary form $H(X,Y) \in \mathbf{Z}[X,Y]$ of degree $r - s$ one has $R(F + GH, G) = R(F, G)$, whereas $|D(F+GH)|$ can be arbitrarily large. Further, it is necessary to assume that $r \geq 3, s \geq 3$. For instance, let ε be an algebraic unit of degree r with conjugates $\varepsilon^{(1)}, ..., \varepsilon^{(r)}$, say, over $\mathbf{Q}$ and put

$$F_m(X,Y) = (X - \varepsilon^{(1)^m}Y)...(X - \varepsilon^{(r)^m}Y), \quad G(X,Y) = XY.$$

Note that $F_m \in \mathbf{Z}[X,Y]$ and that all F_m have the same splitting field. From a result of Győry [14] it follows that $\lim_{m \to \infty} |D_{F_m}| = \infty$. On the other hand,

$$|R(F_m, G)| = |F(1,0)F(0,1)| = 1 \quad \text{for} \quad m \in \mathbf{Z}.$$

Using Theorems 1 and 2 one can show that, roughly speaking, a binary form has at most two small values. For a given binary form $F \in \mathbf{Z}[X,Y]$, define the set

$$S_F = \{(x,y) \in \mathbf{Z}^2 : \gcd(x,y) = 1, y > 0 \text{ or } (x,y) = (1,0), \ F(x,y) \neq 0\}.$$

We may express S_F as a sequence $\{(x_1, y_1), (x_2, y_2), (x_3, y_3), ...\}$ with

$$0 < |F(x_1, y_1)| \leq |F(x_2, x_2)| \leq |F(x_3, y_3)| \leq \cdots .$$

Corollary 2. *Let $F \in \mathbf{Z}[X,Y]$ be a binary form of degree $r \geq 3$ with splitting field L and non-zero discriminant $D(F)$.*
(i). $|F(x_2, y_2)| \leq C_6^{\text{ineff}}(r, L)|D(F)|^{21/(r-1)}$;
(ii). $|F(x_3, y_3)| \geq C_7^{\text{ineff}}(r, L)|D(F)|^{1/18(r-1)}$.

Proof. The constants implied by $\ll, \gg$ depend on r and L.
(i). We may and shall assume that F is reduced. There is an $m_1 \in \{0, 1, ..., r\}$ with $F(1, m_1) \neq 0$. Now Theorem 1 implies that

$$0 < |F(1, m_1)| \ll H(F) \ll |D(F)|^{21/(r-1)}.$$

Similarly, there is an $m_2 \in \{0, -1, ..., -r\}$ with $F(m_2, 1) \neq 0$ and $0 < |F(m_2, 1)| \ll |D(F)|^{21/(r-1)}$. Clearly $(1, m_1), (m_2, 1)$ are different pairs in S_F. This implies (i).
(ii). Put $G(X,Y) = (y_1 X - x_1 Y)(y_2 X - x_2 Y)(y_3 X - x_3 Y)$. Then $|D(G)| \geq 1$ and by Theorem 2,

$$|F(x_3, y_3)|^3 \geq |F(x_1, y_1)F(x_2, y_2)F(x_3, y_3)| = |R(F, G)| \gg |D(F)|^{3/18(r-1)}$$

This implies (ii). $\qquad\square$

It is not known, whether in (i) the dependence on L is necessary. However, in (ii) the dependence on L is needed as is shown by the example $F(X,Y) = (X - a_1 Y)...(X - a_r Y) + Y^r$, where $a_1, ..., a_r$ are distinct integers: here $F(a_i, 1) = 1$ for $i = 1, ..., r$ and hence $|F(x_r, y_r)| = 1$, whereas the discriminant of F can be arbitrarily large.

By slightly modifying the arguments of Bombieri and Schmidt [6], one can show that if $F(X,Y) \in \mathbf{Z}[X,Y]$ is a binary form of degree $r \geq 400$ with

$$|D(F)| \geq \exp\{80r(r-1)\}m^{120(r-1)},$$

then the inequality $|F(x,y)| \leq m$ has at most $6r$ solutions $(x,y) \in S_F$ (cf. [11]). This implies that for every binary form $F \in \mathbf{Z}[X,Y]$ of degree $r \geq 400$ with non-zero discriminant,

$$(2.7) \qquad |F(x_{6r+1}, y_{6r+1})| \geq \exp\{-\frac{2}{3}r\}|D(F)|^{\frac{1}{120(r-1)}}.$$

We now consider the relation between $|R(F,G)|$ and the heights of F and G. From Theorems 1 and 2 it is possible to derive the following:

Theorem 3. *Let $F,G \in \mathbf{Z}[X,Y]$ be binary forms as in Theorem 2 and suppose that the pair (F,G) is reduced (cf. §1). Then*

$$(2.8) \qquad |R(F,G)| \geq C_8^{\text{ineff}}(r,s,L)\left(H(F)^s H(G)^r\right)^{\frac{1}{760}}.$$

Again the conditions $r \geq 3, s \geq 3$ and the dependence on L are necessary. Hence a bound as in (1.5) (independent of L) does not exist. Theorem 2 will be proved in [12] and Theorem 3 in [9].

If one fixes one of the binary forms F,G then it is possible to derive a lower bound for $|R(F,G)|$ independent of L. Wirsing [28] proved that if $F,G \in \mathbf{Z}[X,Y]$ are binary forms of degrees r, s respectively such that FG has no multiple factors in $\mathbf{C}[X,Y]$ then for all $\varepsilon > 0$,

$$(2.9) \qquad |R(F,G)| \geq C_9^{\text{ineff}}(F,s,\varepsilon)H(G)^{r-\varphi(s)-\varepsilon}$$

where

$$\varphi(s) = 2s(1 + \frac{1}{3} + ... + \frac{1}{2s-1}).$$

Schmidt [23] proved a stronger result with $2s$ instead of $\varphi(s)$, but under the additional hypothesis that F is not divisible by any binary form in $\mathbf{Z}[X,Y]$ of degree $\leq s$ (e.g. $2s < r$ and F is irreducible).

It is possible to derive some results from Theorems 1 and 3 about the difference of two algebraic numbers. For $\xi, \eta \in \mathbf{C}$, put

$$\|\xi - \eta\| = |\xi - \eta|$$

if both $\xi, \eta \in \mathbf{R}$ or if $\eta = \bar{\xi}$;

$$\|\xi - \eta\| = |\xi - \eta|^2$$

otherwise.

Let $\xi \in \mathbf{C}$ denote an algebraic number of degree r and $\eta \in \mathbf{C}$ an algebraic number of degree s. If η is not a conjugate of ξ, then

$$(2.10) \qquad \|\xi - \eta\| \gg H(\xi)^{-s} H(\eta)^{-r}$$

where, for the moment, the constants implied by $\gg$ are effectively computable and depend only on r, s. Further, if $\xi^{(p)}, \xi^{(q)}$ are distinct conjugates of ξ, then

$$(2.11) \qquad \|\xi^{(p)} - \xi^{(q)}\| \gg H(\xi)^{-(r-1)}.$$

Inequalities of this type with $|.|$ instead of $\|\cdot\|$ have been derived in [3] and [21] (but see also [22]). (2.10) and (2.11) can be proved as follows. Choose irreducible binary forms $F, G \in \mathbf{Z}[X, Y]$ with coefficients having gcd 1, such that $F(\xi, 1) = G(\eta, 1) = 0$. We have

$$F(X, Y) = \prod_{i=1}^{r} (\alpha_i X + \beta_i Y),$$

$$G(X, Y) = \prod_{j=1}^{s} (\gamma_j X + \delta_j Y),$$

where $-\beta_1/\alpha_1 = \xi, -\delta_1/\gamma_1 = \eta$. If ξ, η are not both real, we may assume that $\bar{\xi} = -\beta_p/\alpha_p, \bar{\eta} = -\delta_q/\gamma_q$ with $(p, q) = (1, 2)$ or $(2, 2)$. Hence with $I = \{(1, 1)\}, \{(1, 1), (1, 2)\}$ or $\{(1, 1), (2, 2)\}$ we have

$$(2.12) \qquad \|\xi - \eta\| = \prod_{(i,j) \in I} \left| \frac{\beta_i}{\alpha_i} - \frac{\delta_j}{\gamma_j} \right|.$$

Put $\theta_{ij} = |\alpha_i \delta_j - \beta_i \gamma_j|, f_i = \max(|\alpha_i|, |\beta_i|), g_j = \max(|\gamma_j|, |\delta_j|)$ for $i = 1, ..., r, j = 1, ..., s$. Let $S = \{1, ..., r\} \times \{1, ..., s\}$. Then

(2.13)

$$\prod_{(i,j)\in S} \theta_{ij} = |R(F,G)|, \quad \prod_{i=1}^{r} f_i \ll H(F) = H(\xi), \quad \prod_{j=1}^{s} g_j \ll H(G) = H(\eta)$$

(cf [19], Chap 3, §2). We have

(2.14) $$\theta_{ij} \leq 2f_i g_j.$$

By combining (2.12), (2.13), (2.14) and $|R(F,G)| \geq 1$ we get

(2.15)
$$\|\xi - \eta\| \geq \prod_{(i,j)\in I} \frac{\theta_{ij}}{f_i g_j} \gg \prod_{(i,j)\in S} \frac{\theta_{ij}}{f_i g_j}$$

$$\gg \frac{|R(F,G)|}{(f_1\ldots f_r)^s (g_1\ldots g_s)^r} \gg H(\xi)^{-s} H(\eta)^{-r}.$$

(2.11) can be proved in a similar way. We may assume that

$$\|\xi^{(p)} - \xi^{(q)}\| = \prod_{(i,j)\in J} |\xi^{(i)} - \xi^{(j)}|$$

with $J = \{(1,2)\}, \{(1,2),(1,3)\}$ or $\{(1,2),(3,4)\}$. Put $\Delta_{ij} = |\alpha_i \beta_j - \alpha_j \beta_i|$ for $1 \leq i < j \leq r$, and let $T = \{(i,j) : 1 \leq i < j \leq r\}$. Then

$$\|\xi^{(p)} - \xi^{(q)}\| \geq \prod_{(i,j)\in J} \frac{\Delta_{ij}}{f_i f_j} \gg \prod_{(i,j)\in T} \frac{\Delta_{ij}}{f_i f_j}$$

$$= \frac{|D(F)|^{\frac{1}{2}}}{(f_1\ldots f_r)^{r-1}} \gg H(\xi)^{-(r-1)}.$$

By using Theorems 3 and 1 it is possible to improve (2.10 and (2.11).

Corollary 3. *Let K be an algebraic number field.*
(i). Let ξ, η be elements of K such that $\deg \xi = r \geq 3, \deg \eta = s \geq 3$ and η is not a conjugate of ξ. Then

$$\|\xi - \eta\| \geq C_{10}^{\text{ineff}}(K)\left(H(\xi)^{-s} H(\eta)^{-r}\right)^{1-\frac{1}{760}}.$$

(ii). Let ξ be an element of K of degree $r \geq 4$. Then for each pair of distinct conjugates $\xi^{(p)}, \xi^{(q)}$ of ξ one has

$$\|\xi^{(p)} - \xi^{(q)}\| \geq C_{11}^{\text{ineff}}(K)\left(H(\xi)^{-(r-1)}\right)^{1-\frac{1}{42}}.$$

Proof. (i). Let (F^*, G^*) be a reduced pair of binary forms equivalent to (F, G). Then
$$U(F^*, G^*) \leq U(F, G),$$
where $U(F, G) = H(F)^s H(G)^r$ and $U(F^*, G^*) = H(F^*)^s H(G^*)^r$. Choose $\begin{pmatrix} a & b \\ c & d \end{pmatrix} \in SL(\mathbf{Z})$ such that $F^*(X, Y) = F(aX + bY, cX + dY)$ and $G^*(X, Y) = G(aX + bY, cX + dY)$. Then

$$F^*(X, Y) = \prod_{i=1}^{r} (\alpha_i^* X + \beta_i^* Y), \quad G^*(X, Y) = \prod_{j=1}^{s} (\gamma_j^* X + \delta_j^* Y),$$

where $(\alpha_i^*, \beta_i^*) = (\alpha_i, \beta_i)\begin{pmatrix} a & b \\ c & d \end{pmatrix}$, $(\alpha_j^*, \delta_j^*) = (\gamma_j, \delta_j)\begin{pmatrix} a & b \\ c & d \end{pmatrix}$. Put

$$f_i^* = \max(|\alpha_i^*|, |\beta_i^*|), \qquad g_j^* = \max(|\gamma_j^*|, |\delta_j^*|).$$

Note that
$$\theta_{ij} = |\alpha_i \delta_j - \beta_i \gamma_j| = |\alpha_i^* \delta_j^* - \beta_i^* \gamma_j^*|.$$

Hence (2.14) can be sharpened to $\theta_{ij} \leq 2\min(f_i g_j, f_i^* g_j^*)$. By inserting this instead of (2.14) into (2.15) we get

(2.16)
$$\|\xi - \eta\| \gg \frac{|R(F, G)|}{\Lambda},$$

with

$$\Lambda = \prod_{(i,j) \in I} (f_i g_j) \prod_{(i,j) \in S \setminus I} \min(f_i g_j, f_i^* g_j^*)$$

where now the constants involved in $\gg$ depend on K and may be ineffective. From $I \subseteq \{(1,1), (1,2), (2,2)\}$ and (2.13) it follows that

$$\Lambda \leq f_1 g_1 f_1 g_2 f_2 g_2 f_2 g_3 f_3 g_1 f_3 g_3 \cdot f_1^* g_3^* f_2^* g_1^* f_3^* g_2^* \cdot \prod_{\max(i,j) \geq 4} (f_i g_j)^{2/3} (f_i^* g_j^*)^{1/3}$$

$$= \left\{ (f_1 \cdots f_r)^s (g_1 \cdots g_s)^r \right\}^{2/3} \left\{ (f_1^* \cdots f_r^*)^s (g_1^* \cdots g_s^*)^r \right\}^{1/3}$$

$$\ll U(F, G)^{2/3} U(F^*, G^*)^{1/3}.$$

By inserting this in (2.16) and using Theorem 3, we get

$$\|\xi - \eta\| \gg \frac{|R(F^*, G^*)|}{U(F, G)^{2/3} U(F^*, G^*)^{1/3}} \gg U(F, G)^{-2/3} U(F^*, G^*)^{\frac{1}{760} - \frac{1}{3}}$$

$$\geq U(F, G)^{\frac{1}{760} - 1} = \left(H(\xi)^{-s} H(\eta)^{-r} \right)^{1 - \frac{1}{760}}.$$

(ii). Let F be a reduced binary form equivalent to F, and let $f_1^*, \ldots, f_r^*$ have the same meaning as above. Similarly as above, one obtains

$$\|\xi^{(p)} - \xi^{(q)}\| \gg \frac{|D(F)|^{1/2}}{M}$$

with

$$M = \prod_{(i,j)\in J} (f_i f_j) \prod_{(i,j)\in T\setminus J} \min(f_i f_j, f_i^* f_j^*)$$

and

$$M \le (H(F)^{2/3} H(F^*)^{1/3})^{r-1}.$$

Together with Theorem 1 this gives

$$\|\xi^{(p)} - \xi^{(q)}\| \gg \frac{|D(F^*)|^{1/2}}{(H(F)^{2/3}H(F^*)^{1/3})^{r-1}} \gg$$

$$(H(F)^{-\frac{2}{3}} H(F^*)^{\frac{1}{42}-\frac{1}{3}})^{r-1} \ge \left(H(F)^{-(r-1)}\right)^{1-\frac{1}{42}}.$$

This completes the proof of Corollary 3. $\qquad\square$

Liouville's inequality (2.10) can be improved to

$$\|\xi - \eta\| \gg H(\xi)^{-s'} H(\eta)^{-r'},$$

where $r' = [\mathbf{Q}(\alpha,\beta) : \mathbf{Q}(\beta)]$, $s' = [\mathbf{Q}(\alpha,\beta) : \mathbf{Q}(\alpha)]$. This is better than Corollary 3, (i) if $[\mathbf{Q}(\alpha,\beta) : \mathbf{Q}] < rs$. So Corollary 3 (i) is of interest only if $[\mathbf{Q}(\alpha,\beta) : \mathbf{Q}] = rs$, where $r = [\mathbf{Q}(\alpha) : \mathbf{Q}] \ge 3$, $s = [\mathbf{Q}(\beta) : \mathbf{Q}] \ge 3$. Under these hypotheses, Corollary 3, (i) seems to be the only inequality in the literature of the type

$$(2.17) \qquad \|\xi - \eta\| \ge C_{13}^{\text{ineff}}(K, \lambda, \mu) H(\xi)^{-\lambda} H(\eta)^{-\mu} \quad (K = \mathbf{Q}(\alpha,\beta))$$

where at the same time $\lambda < s$ and $\mu < r$. It is possible to derive inequalities of type (2.17) where one of the exponents λ, μ is smaller but the other (much) larger. From Theorem 4 of [4] (taking $t = \sqrt{(2-\theta)/r}$ with $\theta = c_1\varepsilon^2/r(\sqrt{2r+\varepsilon})^2, \tau = t - 2/(\sqrt{2r}+\varepsilon)$, where c_1 is a sufficiently small absolute constant) it follows that (2.17) holds with

$$\mu = \sqrt{2r} + \varepsilon, \quad \lambda = c_2(\sqrt{2r}+\varepsilon)^3 \cdot \varepsilon^{-2} \cdot s$$

for a sufficiently large absolute constant c_2 and for all $\varepsilon > 0$. From Theorem 1 of [5] Corrigenda (taking $n = c_3(1+\varepsilon^{-2})\log r, \eta = n^{-n}$ where c_3 is a sufficiently large absolute constant) it follows that (2.17) holds with

$$\mu = 2 + \varepsilon, \quad \lambda = s\exp\{c_4(1+\varepsilon^{-2})\log r[\log(1+\varepsilon^{-2}) + \log\log r]\}$$

for a sufficiently large absolute constant c_4 and for all $\varepsilon > 0$. Schmidt [26] conjectured that for every ξ, η in a number field K with $\xi \ne \eta$,

$$|\xi - \eta| \ge C(K,\varepsilon)\max(H(\xi), H(\eta))^{-2-\varepsilon} \quad \text{for all } \varepsilon > 0.$$

Perhaps, in this conjecture one may replace $|\xi - \eta|$ by $\|\xi - \eta\|$.

The main tool in the proofs in the Theorems 1, 2 and 3 is the following. Let K be an algebraic number field of degree d, and denote by $x \mapsto x^{(i)}(i = 1, ..., d)$ the $\mathbf{Q}$-isomorphisms of K. Define the K-height of $\mathbf{x} = (x_1, ..., x_n) \in K^n$ by

$$H_K(\mathbf{x}) = H_K(x_1, ..., x_n) = \prod_{i=1}^{r} \max(|x_1^{(i)}|, ..., |x_n^{(i)}|).$$

Let $\mathcal{O}_K$ denote the ring of integers of K. Then we have:

Proposition. *Let* $x_0, ..., x_n$ *be elements of* $\mathcal{O}_K$ *with*

$$x_0 + ... + x_n = 0, \sum_{i \in I} x_i \neq 0$$

for each proper, nonempty subset I *of* $\{0, ..., n\}$. *Then for all* $\varepsilon > 0$,

$$(2.18) \qquad H_K(x_0, ..., x_n) \leq C_{12}^{\text{ineff}}(K, \varepsilon)|N_{K/\mathbf{Q}}(x_0...x_n)|^{1+\varepsilon}.$$

This is a special case of Theorem 1 of [7]. Thus, this Proposition is a consequence of Schmidt's Subspace Theorem (cf. [24, 25]). We mention that for $n = 2$, this Proposition follows from Roth's theorem over number fields (cf. [19], Chap 7).

The proof of Theorem 1 goes roughly as follows. Let $F(X, Y)$ be the reduced binary form in Theorem 1. We can express F as

$$F(X, Y) = \prod_{i=1}^{r} (\alpha_i X + \beta_i Y).$$

Put $\Delta_{ij} = \alpha_i \beta_j - \alpha_j \beta_i$ for $1 \leq i, j \leq r$. Using for instance geometry of numbers one can show that $H(F)$ is bounded above by some expression depending on the numbers $|\Delta_{ij}|$. We have some freedom in choosing the linear forms $\alpha_i X + \beta_i Y$. If we were able to choose these linear forms such that the numbers $|\Delta_{ij}|$ are approximately equal, then these numbers would be approximately equal to $(\prod_{i<j} |\Delta_{ij}|)^{2/r(r-1)} = |D(F)|^{1/r(r-1)}$ and an upper bound for $H(F)$ in terms of $|D(F)|$ would follow.

For $r = 2, 3$ this is easily done. If $r = 2$ then we have only one number $|\Delta_{12}|$. If $r = 3$ and $F(X, Y) = \prod_{i=1}^{3} (\alpha_i' X + \beta_i' Y)$, then take $\alpha_i X + \beta_i Y = |\Delta_{jk}'||D(F)|^{-1/6}(\alpha_i' X + \beta_i' Y)$ for $i = 1, 2, 3$ where $\{i, j, k\} = \{1, 2, 3\}$ and $\Delta_{jk}' = \alpha_j' \beta_k' - \alpha_k' \beta_j'$. Then indeed $|\Delta_{ij}| = |D(F)|^{1/6}$ for $1 \leq i < j \leq 3$. For $r \geq 4$ it is not obvious that the linear forms $\alpha_i X + \beta_i Y$ can be chosen such that the numbers $|\Delta_{ij}|$ are close to each other. Namely, if the $|\Delta_{ij}|$ are close to each other, then the cross ratios $|\Delta_{ij} \Delta_{pq}/\Delta_{ip} \Delta_{jq}|$ $(i, j, p, q$ distinct) are close to 1, but these cross ratios are independent of the choices of the linear forms. However, by applying the Proposition to the identities

$$(2.19) \qquad \Delta_{ij} \Delta_{pq} + \Delta_{pi} \Delta_{jq} + \Delta_{iq} \Delta_{jp} = 0$$

it is possible to derive upper bounds for the above mentioned cross ratios in terms of $|D(F)|$. This eventually leads to an upper bound for $H(F)$ in terms of $|D(F)|$. Note that the expressions in (2.19) belong to the splitting field L of F. So the ineffective constants coming from (2.18) depend on L. This explains why the ineffective constant in Theorem 1 depends on L.

In the proof of Theorem 2, we express F and G as

$$F(X,Y) = \prod_{i=1}^{r}(\alpha_i X + \beta_i Y), \quad G(X,Y) = \prod_{j=1}^{s}(\gamma_j X + \delta_j Y).$$

Put $\Delta_{ij} = \alpha_i\beta_j - \alpha_j\beta_i$, $E_{ij} = \gamma_i\delta_j - \gamma_j\delta_i$, $\theta_{ij} = \alpha_i\delta_j - \beta_i\gamma_j$. From the product rule of determinants it follows that

$$\Delta_{ij}E_{pq} = \theta_{ip}\theta_{jq} - \theta_{iq}\theta_{jp}.$$

Hence $|\Delta_{ij}| \cdot |E_{pq}| \leq 2\max(|\theta_{ip}\theta_{jq}|, |\theta_{iq}\theta_{jp}|)$. By taking the product over all quadruples (i,j,p,q) with $1 \leq i < j \leq r, 1 \leq p < q \leq s$, we obtain

$$|D(F)|^{\frac{s(s-1)}{2}}|D(G)|^{\frac{r(r-1)}{2}} \ll B, \quad B = \prod_{\substack{i \leq i < j \leq r \\ 1 \leq p < q \leq s}} \max(|\theta_{ip}\theta_{jq}|, |\theta_{iq}\theta_{jp}|).$$

We estimate B from above by using an idea of Győry, Stewart and Tijdeman [13]. Assume that $\{i,j\}, \{p,q\} \subset \{1,2,3\}$. The 3×3 determinant $\det (\theta_{ij})_{1 \leq i,j \leq 3}$ equals 0. By expanding the determinant we get a six-term expression

(2.20) $u_1 + u_2 + u_3 + u_4 + u_5 + u_6 = 0,$

where each u_i is a product of three numbers θ_{ij}. The left-hand side of (2.20) might have vanishing subsums. By considering all possible partitions of the left-hand side of (2.20) into minimal vanishing subsums and applying the Proposition to these subsums, one obtains in all cases an upper bound for $\max(|\theta_{ip}\theta_{jq}|, |\theta_{iq}\theta_{jp}|)$. This yields an upper bound for B in terms of $|R(F,G)|$. In this way one proves Theorem 2.

Theorem 3 is derived from Theorems 1 and 2 in the following way. We prove inequality (2.8) for a pair (F, G) in which one of the binary forms F and G, F, say, is reduced, which clearly suffices. The other binary form G is not necessarily reduced. However, the quantities Δ_{ij}, θ_{ij} defined above satisfy the relations

(2.21) $\begin{cases} \Delta_{jk}\theta_{ip} + \Delta_{ki}\theta_{jp} + \Delta_{ij}\theta_{kp} = 0 \\ \text{for } 1 \leq i, j, \; k \leq r, \; 1 \leq p \leq s. \end{cases}$

Each expression $\Delta_{jk}\theta_{ip}$ is a linear form in γ_p, δ_p, whose coefficients are expressible in terms of F. This implies that γ_p, δ_p can be expressed as linear combinations of the terms in (2.21), with coefficients depending on F. By estimating these coefficients from above, and applying the Proposition to all identities (2.21), we obtain upper bounds for $|\gamma_p|, |\delta_p|, (p = 1, ..., s)$ and hence of $H(G)$, in terms of $H(F), |R(F, G)|$. But by Theorem 1, $H(F)$ can be estimated from above in terms of $|D(F)|$ and by Theorem 2, $|D(F)|$ in terms of $|R(F, G)|$. Hence both $H(F), H(G)$ can be bounded above in terms of $|R(F, G)|$. This gives Theorem 3. □

References

[1] A. Baker, *Bounds for the solutions of the hyperelliptic equations*, Proc. Cambridge Philos. Soc. 65 (1969), 439–444.

[2] B. J. Birch, J. R. Merriman, *Finiteness theorems for binary forms with given discriminant*, Proc. London Math. Soc. 25 (1972), 385–394.

[3] E. Bombieri, *Sull' approssimazione di numeri algebrici mediante numeri algebrici*, Boll. Un. Mat. Ital. (3), 13 (1958), 351–354.

[4] E. Bombieri, *On the Thue-Siegel-Dyson theorem*, Acta. Math. 148 (1982), 255–296.

[5] E. Bombieri, A. J. V. D. Poorten, *Some quantitative results related to Roth's theorem*, J. Austral. Math. Soc. (Series A) 45 (1988), 233–248, Corrigenda, ibid, 48 (1990), 154–155.

[6] E. Bombieri, W. M. Schmidt, *On Thue's equation*, Invent. Math. 88 (1987), 69–81.

[7] J. H. Evertse, *On sums of S-units and linear recurrences*, Compositio Math. 53 (1984), 225–244.

[8] J. H. Evertse, *Estimates for reduced binary forms*, J. reine angew. Math., to appear.

[9] J. H. Evertse, *Lower bounds for resultants II*, in preparation.

[10] J. H. Evertse, K. Győry, *Effective finiteness results for binary forms with given discriminant*, Compositio Math. 79 (1991), 169–204.

[11] J. H. Evertse, K. Győry, *Thue inequalities with a small number of solutions*, in : The Mathematical Heritage of C. F. Gauss, G. R. Rassias, ed., 204–224. World Scientific Publ., Singapore, 1991.

[12] J. H. Evertse, K. Győry, *Lower bounds for resultants I*, submitted for publication.

[13] J. H. Evertse, K. Győry, C. L. Stewart, R. Tijdeman, *On S-unit equations in two unknowns*, Invent. Math. 92 (1988), 461–477.

[14] K. Győry, *Sur les polynômes à coefficients entiers et de discriminant donné*, Acta Arith. 23 (1973), 419–426.

[15] K. Győry, *Sur les polynômes à coefficients entiers et de discriminant donné II*, Publ. Math. Debrecen 21 (1974), 125–144.

[16] K. Győry, *On arithmetic graphs associated with integral domains*, in: A Tribute to Paul Erdős, A. Baker, B. Bollobás, A. Hajnal, eds. 207–222, Cambridge Un. Press, 1990.

[17] C. Hermite, *Sur l'introduction des variables continues dans la théorie des nombres*, J. reine angew Math. 41 (1851), 191-216.

[18] J. L. Lagrange, *Recherches d'arithmétique*, Nouv. Mém. Acad. Berlin, 1773, 265–312, Oeuvres, III, 693–758.

[19] S. Lang, *Fundamentals of Diophantine Geometry*, Springer-Verlag, New York, 1983.

[20] D. J. Lewis, K. Mahler, *Representation of integers by binary forms*, Acta. Arith. 6 (1961), 333–363.

[21] K. Mahler, *An inequality for the discriminant of a polynomial*, Michigan Math. J. 11 (1964), 257–262.

[22] M. Mignotte, M. Payafar, *Distance entre les racines d'un polynôme*, RAIRO, Analyse numérique 13 (1979), 183–192.

[23] W. M. Schmidt, *Inequalities for Resultants and for Decomposable Forms*, in: Proc. Conf. Diophantine Approximations and its Applications, ed. by C. F. Osgood, 235–254, Academic Press, New York etc., 1973.

[24] W. M. Schmidt, *Simultaneous Approximation to Algebraic Numbers by Elements of a Number Field*, Monatsh. Math. 79, 55–66.

[25] W. M. Schmidt, *Diophantine Approximation*, Lecture Notes in Mathematics 785, Springer Verlag, Berlin etc., 1980.

[26] W. M. Schmidt, *Open problems in Diophantine approximation*, in: Approximations Diophantiennes et Nombres Transcendents (Luminy, 1982), D. Bertrand, M. Waldschmidt, eds. Progr. Math. 31, 271 - 287. Birkhäuser, Boston etc., 1983.

[27] C. L. Siegel (Under the pseudonym X) *The integer solutions of the equation $y^2 = ax^n + bx^{n-1} + \ldots + k$*, J. London. Math. Soc. 1 (1926), 66–68.

[28] E. Wirsing, *On Approximations of Algebraic Numbers by Algebraic Numbers of Bounded Degree*, in : Proc. Symp. Pure Math. 20, 1969 Number Theory Institute, D. J. Lewis, ed., 213–247, AMS, Providence, RI, 1971.

J.H. Evertse
University of Leiden
Department of Mathematics and Computer Science
P.O. Box 9512
2300 RA Leiden
The Netherlands
e-mail evertse@rulcri.LeidenUniv.nl

A note on Thue's inequality

with few coefficients

Julia Mueller

Fordham University, New York

§1. Introduction

Let F be a binary form of degree $r \geq 3$, irreducible over the rationals, with $s + 1$ nonzero coefficients. Recently, C.L. Stewart [5] conjectured that if the discriminant D of F is nonzero, then there exists an absolute constant c_0, and a constant c_1 which depends on F, such that if $h > c_1$ then the number of primitive solutions of the Thue equation

$$(1.1) \qquad F(x, y) = h$$

is at most c_0. Here a solution (x, y) is called primitive if x and y are coprime.

A proof of this conjecture or even a weaker version such as the number of primitive solutions of (1.1) depends on r or s is out of reach at the present time. What we are able to prove is the following.

Theorem 1. Let F be as above. Suppose $r > 3s$, then the number of primitive solutions (x, y) of the Thue inequality

$$(1.2) \qquad |F(x, y)| \leq h,$$

with

$$\min(|x|, |y|) \geq Y_S,$$

where

$$(1.3) \qquad Y_S = ((e^6 s)^r R^{2s} h)^{\frac{1}{r-3s}}, \quad \text{and} \quad R = e^{800 \log^3 r},$$

is

$$\ll s \left(1 + \frac{\log s}{\log\left(\frac{r}{s} - 1\right)}\right).$$

Here and below the constant implicit in $\ll$ will be absolute and effectively computable.

Research supported by NSF grants HRD 910-3318 and DMS 900-2876

When the discriminant D of F is nonzero and when h is small relative to D, then the number of primitive solutions of both (1.1) and (1.2) can be bounded independently of h. Stewart [5] has shown that for any $\varepsilon > 0$, the number of primitive solutions of (1.1) is at most $2800\,(r+(8\varepsilon)^{-1})$ provided

$$h^{\frac{2}{r+\varepsilon}} \le |D|^{\frac{1}{r(r-1)}}.$$

As for Thue's inequality, we quote from [5] the following result due to Evertse and Győry [1] and [2]: let

$$(N(r), \delta(r)) = \left(6r7^{2\binom{r}{3}}, \frac{5}{6}r(r-1)\right)$$

for $3 \le r < 400$ and

$$(N(r), \delta(r)) = (6r, 120(r-1))$$

for $r > 400$. Then the number of primitive solutions of (1.2) is at most $N(r)$ provided $h^{\delta(r)} \exp(80r(r-1)) \le |D|$.

Let

$$(1.4) \qquad F(x,y) = a_r\, x^r + a_{r-1}x^{r-1}y + \cdots + a_0 y^r$$

with $a_0 a_r \ne 0$ and with $s + 1$ nonzero coefficients. Let $H = \max|a_i| = |a_q|$, and $t = \max(q, r - q)$. In [3] the authors conjectured that when $h \le H^{1-\frac{s}{r}-\varepsilon}$, then the number of primitive solutions of (1.2) is $\ll c(s, \varepsilon)$. This conjecture appears to be true when F is given by (1.4) with few coefficients. More precisely, we have the following.

Theorem 2. Let F be given by (1.4). Suppose

$$(1.5) \qquad r \ge \max(3s, s\log^3 s),$$

and

$$(1.6) \quad r^2 h^{\frac{2}{r}} \le \min\left(Y_S h^{\frac{1}{r}}, \begin{cases} H^{\frac{1}{t}(1-(\frac{s}{r})-\varepsilon)} & \text{if } t \ne \frac{r}{2}, \\[2mm] (H^{\frac{2}{r}(1-(\frac{s}{r})-\varepsilon)})/(1 + \frac{4}{r}\log H) & \text{if } t = \frac{r}{2} \end{cases}\right)$$

where $0 < \varepsilon < 1 - \frac{s}{r}$ and where

$$(1.7) \qquad Y_S = \left((e^6 s)^r R^{2s}\right)^{\frac{1}{r-2s}}, \qquad R = e^{800\log^3 r}.$$

Then the number of primitive solutions of (1.2) is

$$(1.8) \qquad \ll s^{\frac{2r-2s}{r-2s}} \left(1 + \frac{\log \varepsilon^{-1}}{\log\left(\frac{r}{s} - 1\right)}\right).$$

I would like to thank the University of Geneva for its generous support and hospitality while this manuscript was prepared.

§2. Preliminaries

Relative to two constants Y_L and Y_S which will be specified below we call a solution (x, y) of (1.2)

large if $\max(|x|, |y|) > Y_L$,

medium if $\max(|x|, |y|) \leq Y_L$ and $\min(|x|, |y|) \geq Y_S$,

small if $\min(|x|, |y|) < Y_S$.

In Proposition 1 of [3] it has been shown that when Y_L is given by (2.9) of [3], then the number of primitive large solutions of (1.2) is $\ll s$. To obtain Theorem 1 we have to show that the number of primitive medium solutions is $\ll s^2$. Such a bound will be given in Lemma 1. Theorem 2 follows from Proposition 1 of [3] and Lemmas 2 to 5. We remark that the diophantine approximation method, which was crucial in bounding the number of large and medium solutions, fails to be useful when the solutions are small. In Lemmas 3 and 4 we show that when F has few coefficients, and when h is small relative to the height H of F, then both the constant Y_S and the measure of the set (x, y) in R^2 with (1.2) can be bounded in terms of s. This allows us to bound the number of small solutions of (1.2) in terms of s.

Lemma 1. *Let Y_L and Y_S be given by (2.9) of [3] and (1.3). Suppose $r > 3s$, then the number of primitive medium solutions of (1.2) is*

$$(2.1) \qquad \ll s\left(1 + \frac{\log s}{\log\left(\frac{r}{s} - 1\right)}\right).$$

Proof: We remark first that the proof of this lemma follows closely that of Proposition 2 for medium solutions on p. 235–6 of [3]. Let α be given as in Lemma 17 of [3] and let

$$\left\{\frac{x_0}{y_0}, \cdots, \frac{x_\nu}{y_\nu}\right\}$$

be the set of primitive medium solutions of (1.2) that satisfies (9.4) of [3] and ordered such that $Y_S \leq y_0 \leq \cdots \leq y_\nu \leq Y_L$. Let K be given by (9.5) of [3]. That is,

$$\begin{aligned} K &= 2R\,(rs)^2\,(4e^3 s)^{\frac{r}{s}}\,h^{\frac{1}{s}}\,H^{\frac{1}{r}-\frac{1}{s}} \\ &< R^2 (e^5 s)^{\frac{r}{s}}\,h^{\frac{1}{s}}\,H^{\frac{1}{r}-\frac{1}{s}} \\ (2.2) \qquad &< R^2 (e^5 s)^{\frac{r}{s}}\,h^{\frac{1}{s}}\,H^{-\frac{1}{2s}}, \end{aligned}$$

where R is given by (1.3). Set $\ell = \frac{r}{s}$, then it has been shown (p. 236 of [3]) that the y_i's satisfy the gap principle

$$y_{i+1} \geq K^{-1} y_i^{\ell-1} \geq \left(K^{-1} Y_S^{\ell-2}\right) y_i, \quad i = 0, \cdots, \nu,$$

here Y_S is given by (1.3) and

(2.3)
$$K^{-1} Y_S^{\ell-2} \geq e^{\frac{r}{s}} H^{\frac{1}{2s}} h^{\frac{1}{r-3s}} > 1.$$

Applying this gap principle to $i = 0, \cdots, \nu$ successively and set $\mu = 1 + (\ell - 1) + \cdots + (\ell - s)^{\nu-1}$, then since $\mu < (\ell - 1)^{\nu}(\ell - 2)^{-1}$ and $y_0 \geq Y_s$ we get

(2.4)
$$y_\nu \geq K^{-\mu} y_0^{(\ell-1)^\nu} > (K^{-1} Y_S^{\ell-2})^{(\ell-1)^\nu(\ell-2)^{-1}}$$

Since $(\ell - 2)^{-1} = s(r - 2s)^{-1}$, (2.4) in conjunction with (2.3) yields

$$\log y_\nu > (\ell - 1)^\nu (r - 2s)^{-1} (r + \frac{1}{2} \log H + s(r - 3s)^{-1} \log h).$$

As for the upper bound, one can find on p. 236 of [3] that

(2.5)
$$\log y_\nu \ll (r^{\frac{1}{2}} + \log H + r^{-1} \log h).$$

Since

$$\frac{r^{\frac{1}{2}} + \log H + r^{-1} \log h}{r + \frac{1}{2} \log H + s(r - 3s)^{-1} \log h} \ll 1,$$

we get from the above bounds that

$$(\ell - 1)^\nu \ll (r - 2s).$$

Hence

(2.6)
$$\nu \ll \frac{\log(r - 2s)}{\log(\ell - 1)} = \frac{\log\left(\frac{r}{s} - 2\right) s}{\log\left(\frac{r}{s} - 1\right)} \ll 1 + \frac{\log s}{\log\left(\frac{r}{s} - 1\right)}.$$

Now (2.1) follows from (2.6) and Lemma 17 of [3]. This proves Lemma 1.

The following lemma is crucial in the proof of Theorem 2. What it says is that under the hypotheses (1.5) and (1.6), we may take Y_S to be independent of h and still obtain the result of Lemma 1.

Lemma 2. *Let Y_L be given by (2.9) of [3], and let Y_S be given by (1.7). Suppose (1.5) and (1.6) hold, then the number of primitive medium solutions of (1.2) is*

$$(2.7) \qquad \ll s \left(1 + \frac{\log s \varepsilon^{-1}}{\log \left(\frac{r}{s} - 1 \right)} \right)$$

Proof: We remark first that (1.6) implies

$$(2.8) \qquad h \leq H^{1 - \frac{s}{r} - \varepsilon}.$$

Next, we note that since $r \geq 3s$ by (1.5), we have $2(rs)^{\frac{2s}{r}} < r^2$. Now suppose (1.6) holds, then by assuming $|y| \geq Y_S$, where Y_S is given by (1.7), we get $|y| \geq 2(rs)^{\frac{2s}{r}} h^{\frac{1}{r}}$. But this last inequality is (8.6) of [3]. We may now apply Lemma 17 of [3]. In fact, the proof of this lemma follows closely that of Lemma 1. First from (2.2) and (2.8) we get

$$K^{-1} \geq R^{-2} (e^5 s)^{-\frac{r}{s}} H^{\frac{\varepsilon}{s}}$$

and

$$(2.9) \qquad K^{-1} Y_S^{\ell - 2} \geq e^{\frac{r}{s}} H^{\frac{\varepsilon}{s}} > 1.$$

Next, (2.4) in conjunction with (2.9) yields

$$(2.10) \qquad \log y_\nu > (\ell - 1)^\nu (r - 2s)^{-1} (r + \varepsilon \log H).$$

On the other hand, (2.5) in conjunction with (2.8) yields

$$(2.11) \quad \log y_\nu \ll r^{\frac{1}{2}} + \left(1 + r^{-1}(1 - \left(\frac{s}{r} \right) - \varepsilon) \right) \log H \ll r^{\frac{1}{2}} + 2 \log H.$$

Since

$$\frac{r^{\frac{1}{2}} + 2 \log H}{r + \varepsilon \log H} \ll \varepsilon^{-1},$$

we get from (2.10) and (2.11) that

$$(\ell - 1)^\nu \ll (r - 2s) \varepsilon^{-1}.$$

Hence,

$$(2.12) \quad \nu \ll \frac{\log(r - 2s)\varepsilon^{-1}}{\log(\ell - 1)} = \frac{\log \left(\frac{r}{s} - 2 \right) s \varepsilon^{-1}}{\log \left(\frac{r}{s} - 1 \right)} \ll 1 + \frac{\log s \varepsilon^{-1}}{\log \left(\frac{r}{s} - 1 \right)}.$$

Now (2.7) follows from (2.12) and Lemma 17 of [3]. This proves Lemma 2.

Lemma 3.   *Let Y_S be given by (1.7). Suppose (1.5) holds, then*

$$(2.13) \qquad\qquad Y_S \ll s^{\frac{r}{r-2s}},$$

Proof: From (1.7) we have

$$(2.14) \qquad\qquad Y_S = (e^6 s)^{\frac{r}{r-2s}} \left(e^{800 \log^3 r} \right)^{\frac{2s}{r-2s}}.$$

We remark that from (1.5) and the fact that $\log^3 s < 3$ for $s \leq 4$ and $\log^3 s > 3$ for $s > 4$, we get $r \geq 3s$ for $s \geq 1$, and

$$(2.15) \qquad\qquad r \geq s \log^3 s \qquad \text{for } s > 4.$$

From $r \geq 3s$ one can show easily that

$$\left(1 - 2\frac{s}{r}\right)^{-1} \leq 3 \qquad \text{and} \qquad \frac{2s}{r - 2s} \leq 6\frac{s}{r}.$$

Hence we have

$$(2.16) \qquad \left(e^{800 \log^3 r} \right)^{\frac{2s}{r-2s}} \leq e^d \text{ with } d = \frac{4800\, s \log^3 r}{r}.$$

In view of (2.14) and (2.16) one sees easily that to show (2.13) it suffices to show $d \ll 1$. When $s \leq 4$, we clearly have $d \ll 1$. When $s > 4$, we see that since $\frac{\log^3 r}{r}$ is a decreasing function, (2.15) implies that

$$
\begin{aligned}
\frac{s \log^3 r}{r} &\leq \frac{s \log^3 (s \log^3 s)}{s \log^3 s} \\
&\ll \frac{\log^3 s + (3 \log\log s)^3}{\log^3 s} \\
&\ll 1,
\end{aligned}
$$

and hence $d \ll 1$. This proves (2.13) and Lemma 3.

The following lemma may be regarded as the discrete version of the Corollary of Theorem 4 in [4]. Our proof relies heavily on the proof of Theorem 4 in [4].

Lemma 4.   *Let $F(x, y)$ be given by (1.4). For a given y, let $\mu(y)$ be the measure of the set of real numbers x with (1.2). Then*

$$(2.17) \qquad \sum_{y \neq 0} \mu(y) \leq \begin{cases} 60\, r^2 s^2 h^{\frac{2}{r}} H^{-\frac{1}{t}} & \text{if } t \neq \frac{r}{2} \\[2mm] 60\, r^2 s^2 h^{\frac{2}{r}} H^{-\frac{2}{r}}\left(1 + \frac{4}{r} \log H\right) & \text{if } t = \frac{r}{2}. \end{cases}$$

Proof: Let $\xi = h^{\frac{-1}{r}}x$ and $\eta = h^{\frac{-1}{r}}y$. For a given η, let $\mu(\eta)$ be the measure of the set of real numbers ξ with $|F(\xi, \eta)| \leq 1$. Then it is clear that

$$(2.18) \qquad \mu(y) = h^{\frac{1}{r}}\mu(\eta).$$

We recall that in sections 3 and 6 of [4] we have shown the following: Let a_u and a_v be coefficients of F. Then there exists a transformation $\xi \longrightarrow e^\sigma \xi$ and $\eta \longrightarrow e^\tau \eta$ such that the coefficients a_u^* and a_v^* of $F^*(\xi, \eta) = F(e^\sigma \xi, e^\tau \eta)$ corresponding to a_u and a_v have modulus 1. Furthermore, the quantities φ and ψ (defined in Theorem 4 of [4]) are transformed into

$$(2.19) \qquad \varphi^* = \varphi - \frac{(\sigma + \tau)r}{2}, \qquad \psi^* = \psi - \frac{(\sigma + \tau)r}{2}.$$

From now on we will assume that the integers u and v mentioned above are those given in section 6 of [4]. Let $\mu^*(\eta)$ be the measure of the set of real numbers ξ with $|F^*(\xi, \eta)| \leq 1$. Then we see that $\mu^*(\eta)$ and $\mu(\eta)$ are related by

$$(2.20) \qquad \mu(\eta) = e^\sigma \mu^* (e^{-\tau}\eta).$$

Define, for $\eta \neq 0$,

$$f(\eta) = \min \left(|\eta|^{-\alpha}, |\eta|^{-\gamma}\right),$$

where $0 \leq \alpha < 1 < \gamma$. Then clearly f is a piecewise monotonic function and we have

$$(2.21) \qquad \sum_{\eta \neq 0} f(\eta) \leq \int_{-\infty}^{\infty} f(\eta)\, d\eta.$$

From (2.18), (2.20), (2.21) and the following inequality (see section 8 of [4])

$$(2.22) \qquad \mu^*(\eta) \leq 20rs^2 f(\eta),$$

we get, with $\eta^* = e^{-\tau}\eta$, that
$$(2.23)$$

$$\sum_{y \neq 0} \mu(y) = h^{\frac{1}{r}}e^\sigma \sum_{y \neq 0} \mu^*(\eta^*) \leq h^{\frac{1}{r}}e^\sigma 20rs^2 \int_{-\infty}^{\infty} f(\eta^*)\, dy \leq h^{\frac{2}{r}}e^{\sigma + \tau} A,$$

where $A = 20rs^2 \int_{-\infty}^{\infty} f(\eta^*)\, d(\eta^*)$ has been shown (see section 8 of [4]) to be bounded above by the upper bound in (15) or (16) of [4], but with φ

and ψ replaced by φ^* and ψ^*. From (2.19) we see that replacing φ or ψ by φ^* or ψ^* introduces the factor $e^{-\sigma-\tau}$. But this factor will cancel the factor $e^{\sigma+\tau}$ in (2.23). So the conclusion is that $\sum_{y\neq 0} \mu(y)$ is bounded by the upper bound in (15) or (16) of [4]. Now (2.17) can be derived in the same way that the Corollary was derived from Theorem 4 of [4]. This proves Lemma 4.

In estimating the number of small solutions of (1.2), the following lemma will be needed.

Lemma 5. *Let g be a polynomial with $s + 1$ nonzero coefficients and $g(0) \neq 0$, then g has $\leq 2s$ real zeros.*

Proof: We will prove this lemma by induction on s. When $s = 0$, g is a constant and so g has no zeros. When $s > 0$, we have $g'(x) = x^m h(x)$ where $h(0) \neq 0$ and h has s nonzero coefficients. By induction, h has $\leq 2s - 2$ real zeros, hence g' has $\leq 2s - 1$ and g has $\leq 2s$ real zeros.

§3. Proof of Theorem 1 and Theorem 2.

We remark that Theorem 1 follows from Proposition 1 of [3] and Lemma 1, and Theorem 2 will follow from Proposition 1 of [3], Lemma 2 and the assertion that the number of primitive small solutions of (1.2) is

$$(3.1) \qquad\qquad \ll s^{\frac{2r-2s}{r-2s}}.$$

To estimate the number of small solutions, it will suffice by symmetry (see the definition of a small solution) to estimate the number of solutions with

$$(3.2) \qquad\qquad 0 \leq y < Y_S,$$

where Y_S is given by (1.7).

For a given y, let $Z(y)$ be the number of integers x with (1.2) and with $(x, y) = 1$. When $y = 0$, the only integers x with $(x, 0) = 1$ are ± 1. Therefore, $Z(0) = 2$. When $y \neq 0$, it follows from Lemma 5 that the number of x with $F(x, y) = h$ (or $-h$) is at most $2s$. Therefore $Z(y)$ with (1.2) is

$$\leq \mu(y) + 4s,$$

where $\mu(y)$ is the measure of the set of real numbers x with (1.2). In view of (3.2), this implies that the number of primitive small solutions of (1.2) is

$$(3.3) \qquad\qquad \ll 2 + \sum_{y\neq 0} \mu(y) + 8\,(Y_S + 1)s.$$

To estimate the second and the third terms in (3.3), we remark first that (2.17) in conjunction with (1.6) yields

$$\sum_{y \neq 0} \mu(y) \leq 60s^2.$$

Next, from (2.13) we get

(3.4) $$8(Y_S + 1)s \ll s^{1+\frac{r}{r-2s}} = s^{\frac{2r-2s}{r-2s}}.$$

Since $\frac{2r-2s}{r-2s} > 2$ the bound in (3.3) is given by that of (3.4). This proves (3.1). Now one sees easily that (1.8) follows from (2.7) and (3.1). This completes the proof of Theorem 2.

References

[1]   Evertse, J.H., *"Upper bounds for the numbers of solutions of dio-phantine equations"*. M.C. – Tract **168**, Centre of Mathematics and Computer Science, Amsterdam, (1983).

[2]   Evertse, J.H. and Györy, K., *"Thue inequalities with a small number of solutions"*. The Math. Heritage of C. F. Gauss, World Scientific Publ. Co., Singapore, (1991), 204 – 224.

[3]   Mueller, J. and Schmidt, W.M., *"Thue's equation and a conjecture of Siegel"*. Acta Math. **160** (1988), 207 – 247.

[4]   Mueller, J. and Schmidt, W.M., *"On the Newton Polygon"*. Monatsh. f. Math., **113**, (1992), 33 – 50.

[5]   Stewart, C.L., *"On the number of solutions of polynomial congruences and Thue equations"*. J. Amer. Math. Soc. **4**, no. 4, (1991), 793 – 835.

Julia Mueller
Department of Mathematics
Columbia University
New York, N.Y. 10027

    and

Permanent address:
Department of Mathematics
Fordham University
New York, N.Y. 10458

K-nombres de Pisot et de Salem

M. J. Bertin

Université P. et M. Curie

1 Introduction

A. M. Bergé et J. Martinet [1] ont introduit en 1987, à propos de régulateurs relatifs, la notion de hauteur relative d'un nombre algébrique par rapport à un corps de nombres K.

Cette notion permet la généralisation des notions et problèmes classiques: nombres de Pisot, nombres de Salem et problème de Lehmer [4].

Dans le premier paragraphe, après avoir donné la définition et des exemples nous énonçons une caractérisation des K-nombres de Pisot lorsque le corps K est totalement réel.

Dans le deuxième paragraphe, après avoir expliqué brièvement la méthode utilisée, nous donnons une liste des $\mathbb{Q}(\sqrt{d})$-nombres de Pisot de mesure de Mahler inférieure à 2.6 pour tout $d > 0$. Nous en déduisons que les polynômes minimaux des nombres de Pisot inférieurs à $\sqrt{2.6}$ sont irréductibles sur tout corps quadratique réel.

Dans le dernier paragraphe enfin, nous terminons par une application à la construction de polynômes réciproques de petite mesure.

2 Définitions, exemples, caractérisation

Définition 1. *Soit K un corps de nombres. Un entier algébrique $\theta \in \overline{\mathbb{Q}}^*$ est un K-nombre de Salem (resp. Pisot) si au-dessus de tout plongement τ de K dans $\mathbb{C}$, θ possède un unique conjugué θ_τ de module supérieur à 1 et au moins un (resp. aucun) conjugué de module 1.*

Remarques 2. 1) Les $\mathbb{Q}$-nombres de Pisot (resp. Salem) sont les nombres de Pisot (resp. Salem).

2) Un entier rationnel $n, n > 1$, est un nombre de Pisot. Mais un entier de K n'est pas forcément un K-nombre de Pisot.

Par exemple, si $K = \mathbb{Q}(\sqrt{5})$, le nombre $\theta = (3 + \sqrt{5})/2$ est un entier de K mais n'est pas un K-nombre de Pisot. En effet, son polynôme minimal sur K

$$X - (3 + \sqrt{5})/2$$

a pour polynôme conjugué

$$X - (3 - \sqrt{5})/2$$

dont l'unique racine est de module strictement inférieur à 1.

3) Tout comme dans le cas $K = \mathbb{Q}$, les unités de K ne sont pas des K-nombres de Pisot.

4) Un $\mathbb{Q}$-nombre de Pisot n'est pas forcément un K-nombre de Pisot.

Considérons par exemple le plus petit nombre de Pisot θ_0 racine du polynôme irréductible $X^3 - X - 1$ dont les autres racines θ' et θ'' sont de module strictement inférieur à 1. Prenons $K = \mathbb{Q}(\theta')$. Alors le polynôme minimal de θ_0 sur K est $(X - \theta'')(X - \theta_0)$ qui admet pour polynôme conjugué $(X - \theta'')(X - \theta')$, polynôme n'ayant que des racines de module strictement inférieur à 1.

Par suite, le nombre de Pisot θ_0 n'est pas un $\mathbb{Q}(\theta')$-nombre de Pisot.

Plus généralement un nombre de Pisot θ est un K-nombre de Pisot si et seulement si $\mathbb{Q}(\theta)$ et K sont linéairement disjoints.

Par exemple l'entier algébrique $\theta = 2.31651243\ldots$, unique racine de module supérieur à 1 du polynôme

$$\frac{\sqrt{5}+1}{2} + \frac{\sqrt{5}+1}{2}z - z^2$$

est un nombre de Pisot mais n'est pas un $\mathbb{Q}(\sqrt{5})$-nombre de Pisot car le polynôme

$$\frac{\sqrt{5}-1}{2} + \frac{\sqrt{5}-1}{2}z + z^2$$

a toutes ses racines de module strictement inférieur à 1. Par suite, les corps $\mathbb{Q}(\theta)$ et $\mathbb{Q}(\sqrt{5})$ ne sont pas linéairement disjoints.

5) *De même un $\mathbb{Q}$-nombre de Salem τ est K-nombre de Salem si et seulement si les corps $\mathbb{Q}(\tau)$ et K sont linéairement disjoints.*

Par exemple, le nombre de Salem $\tau = 1.359997117\ldots$ de la table de Boyd [5], de polynôme minimal sur $\mathbb{Q}$

$$z^8 - z^7 + z^6 - 2z^5 + z^4 - 2z^3 + z^2 - z + 1$$

n'est pas un $\mathbb{Q}(\sqrt{5})$-nombre de Salem; en effet son polynôme minimal sur $\mathbb{Q}(\sqrt{5})$ est le polynôme

$$z^4 - \frac{\sqrt{5}+1}{2}z^3 + z^2 - \frac{\sqrt{5}+1}{2}z + 1$$

dont le polynôme conjugué ne possède que des racines de module strictement inférieur à 1.

Par suite, les corps $\mathbb{Q}(\tau)$ et $\mathbb{Q}(\sqrt{5})$ ne sont pas linéairement disjoints.

6) Le polynôme minimal P d'un nombre de Pisot est unitaire à coefficients dans $\mathbb{Z}$; d'où $\mid P(0) \mid \geq 1$.

Le polynôme minimal P_K sur K d'un nombre de Pisot relatif est à coefficients dans $\mathbb{Z}_K$, anneau des entiers de K mais l'on n'a pas forcément $\mid P_K(0) \mid \geq 1$.

Toutefois si l'on suppose K totalement réel galoisien, il existe un plongement réel τ de K dans $\mathbb{R}$ tel que $\mid P_{\tau(K)}(0) \mid \geq 1$; comme $\tau(K) = K$, le polynôme $P_{\tau(K)}$ fournit un K-nombre de Pisot $\tau(\theta)$ dont le polynôme minimal vérifie $\mid P_K(0) \mid \geq 1$.

Dans toute la suite, on identifiera un K-nombre de Pisot θ avec l'ensemble $(\theta_\tau)_\tau$ de ses conjugués dans $A = \mathbb{R}^{r_1} \times \mathbb{C}^{r_2}$, où (r_1, r_2) est la signature du corps K.

On désigne par Λ le réseau image dans A de l'anneau des entiers de $\mathbb{Z}_K$.

Si K est totalement réel, le théorème suivant donne une caractérisation des K-nombres de Pisot analogue à celle des nombres de Pisot [3].

Théorème 3. *Soit K un corps de nombres réel et θ un K-nombre de Pisot, de K-degré différent de 2.*

Alors il existe une suite $(a_n)_{n \geq 0}$ d'entiers de K et un élément λ, $\lambda \neq 0$, $\lambda \in K(\theta)$ tel que pour tout plongement réel τ, on ait pour tout $n \neq 0$

$$\mid \lambda_\tau \theta_\tau^n - a_{n,r} \mid = \mid \varepsilon_{n,\tau} \mid$$

$$\sum_{n \geq 0} \mid \varepsilon_{n,\tau} \mid^2 < +\infty \quad et \quad 0 < \mid \lambda_\tau \mid < \mid \theta_\tau \mid$$

et pour tout plongement imaginaire τ de conjugué $\overline{\tau}$, il existe $\mu_\tau \in \mathbb{C}$ vérifiant

$$\mid \varepsilon_{n,\tau} \mid = \mid \mu_\tau \theta_\tau^n + \overline{\mu}_\tau \overline{\theta}_\tau^n - (a_{0,\tau} \overline{a}_{n,\tau} + a_{1,\tau} \overline{a}_{n-1,\tau} + \cdots + a_{n,\tau} \overline{a}_{0,\tau}) \mid$$

$$\sum_{n \geq 0} \mid \varepsilon_{n,\tau} \mid^2 < +\infty$$

$$\mid \mu_\tau \mid \leq (\mid \theta_\tau \mid + 1).$$

Réciproquement, soit K un corps de nombres totalement réel et $(\theta_\tau)_\tau \in A$ vérifiant $\mid \theta_\tau \mid > 1$ pour tout τ et $\theta_\tau \neq \theta_{\tau'}$ pour $\tau \neq \tau'$.

S'il existe pour tout $n \geq 0$ un élément $(a_{n,\tau})_\tau \in \Lambda$ et $(\lambda_\tau)_\tau \in A$, vérifiant pour tout plongement τ, $\lambda_\tau \neq 0$ et

$$\sum_{n \geq 0} \mid \lambda_\tau \theta_\tau^n - a_{n,\tau} \mid^2 < +\infty$$

alors $\theta = (\theta_\tau)_\tau$ est un K-nombre de Pisot.

3 Nombres de Pisot quadratiques de mesure inférieure à 2.6

Dans ce paragraphe, on désigne par nombre de Pisot quadratique un K-nombre de Pisot où K est un corps quadratique réel quelconque.

La détermination de ces nombres est basée sur une généralisation de l'algorithme de Schur utilisé par Dufresnoy et Pisot [6] dans la détermination des plus petits nombres de Pisot au cas de l'ensemble des fractions rationnelles $f = A/Q$, A et Q à coefficients entiers de K telles que $Q(0) = 1$, f ne possédant dans $\mid z \mid \leq 1$ qu'un seul pôle simple α, $0 < \alpha < 1$, et vérifiant $\mid f(z) \mid \leq 1$ sur $\mid z \mid = 1$.

De même, les familles infinies sont trouvées par un argument généralisant aux corps de nombres totalement réels une méthode due à Dufresnoy et Pisot [6].

On donne ces nombres par mesure croissante et l'on indique pour chacun le nombre, la mesure et le polynôme minimal sur le corps quadratique.

$$(\sqrt{2}, -\sqrt{2}) \qquad\qquad 2$$
$$z - \sqrt{2}$$

$$(1.198669\ldots, -1.791852\ldots) \quad 2.147837\ldots$$
$$z^4 - z^3(\sqrt{5} - 1)/2 - 1$$

$$(1.115082\ldots, -2.016106\ldots) \quad 2.248124\ldots$$
$$z^3 - (z^2 + 1)(\sqrt{5} - 1)/2$$

$$(1.532512\ldots, -1.532512\ldots) \quad 2.348593\ldots$$
$$z^4 - z^2 - \sqrt{2}z - 1$$

$$(1.280601\ldots, -1.859147\ldots) \quad 2.380825\ldots$$
$$z^4 - \sqrt{2}z^3 + z - 1$$

$$(1.188179\ldots, 2.0325969\ldots) \quad 2.4190956\ldots$$
$$z^3 - z^2 - z(1 - \sqrt{5})/2 - 1$$

$$(1.105281\ldots, 2.257298\ldots) \quad 2.494948\ldots$$
$$z^3 - z^2(1 + \sqrt{5})/2 - z - 1$$

$$(1.580094\ldots, -1.580094\ldots) \quad 2.496697\ldots$$
$$z^4 - z^2 - \sqrt{3}z - 1$$

$(1.24063\ldots, -2.021312\ldots)$ 2.5077003
$z^6 - z^5(\sqrt{5}-1)/2 - z^4 + z^2 - 1$

$(1.287488\ldots, 1.952184\ldots)$ 2.513414\ldots
$z^3 - z^2(\sqrt{2}-1)/2 - \sqrt{2}z - 1$

$(1.158851\ldots, -2.209347\ldots)$ 2.56030398\ldots
$z^3 - z^2(1-\sqrt{2}) - 1$

$(1.204478\ldots, -2.125736\ldots)$ 2.56040\ldots
$z^5 - z^4(\sqrt{5}-3)/2 - (z^3 + z^2 + z + 1)(\sqrt{5}-1)/2$

$(1.163262\ldots, -2.21257\ldots)$ 2.57379\ldots
$z^5 - (z^4 - z^3)(\sqrt{5}-1)/2 - z^2 - (\sqrt{5}-1)/2$

Familles infinies

(1) $z^n(z^2 - z(\sqrt{5}-1)/2 - 1) + \epsilon(z^2 - 1)$

(2) $z^n(z^2 - z\sqrt{2} - 1) + \epsilon(z^2 + z - 1)$

(3) $z^n(z^2 - z(3-\sqrt{5})/2 - 1) + \epsilon(z^2 - 1)$

(4) $z^n(z^2 - z(2-\sqrt{2}) - 1) + \epsilon(z^2 - z(\sqrt{2}-1) - 1)$

Le deuxième (resp. neuvième) $\mathbb{Q}(\sqrt{d})$-nombre de Pisot fait partie de la famille infinie (1) pour $\epsilon = 1$ et $n = 2$ (resp. $n = 4$).

Le cinquième (resp. onzième) $\mathbb{Q}(\sqrt{d})$-nombre de Pisot fait partie de la famille infinie (2) pour $\epsilon = 1$ et $n = 2$ (resp. $n = 1$).

Le septième $\mathbb{Q}(\sqrt{d})$-nombre de Pisot fait partie de la famille infinie (3) pour $\epsilon = 1$ et $n = 1$.

Le dixième $\mathbb{Q}(\sqrt{d})$-nombre de Pisot fait partie de la famille infinie (4) pour $\epsilon = 1$ et $n = 1$.

On trouve également tous les nombres de Pisot inférieurs à $\sqrt{2.6}$. On en déduit donc que les corps $\mathbb{Q}(\theta)$ pour θ nombre de Pisot inférieur à $\sqrt{2.6}$ et $\mathbb{Q}(\sqrt{d})$ sont linéairement disjoints.

Remarques

1) On peut également remarquer que le nombre d'or $(1+\sqrt{5})/2$ n'étant pas un $\mathbb{Q}(\sqrt{5})$-nombre de Pisot, l'ensemble des $\mathbb{Q}(\sqrt{5})$-nombres de Pisot n'est pas un ensemble fermé.

2) D'après Schinzel [8], si K est un corps de nombres totalement réel, P un polynôme unitaire non réciproque à coefficients entiers de K et $P(0) \neq 0$, alors

$$\max_{i=1,\cdots,k} \prod_j \max(1, |a_{i,j}|) \geq \theta_0,$$

où k est le degré de K et $a_{i,j}$ les zéros du polynôme P_i polynôme conjugué de P, $\theta_0 = 1.32\ldots$ désignant le plus petit nombre de Pisot.

Par suite, si θ est un nombre de Pisot inférieur à θ_0^2, le couple $(\sqrt{\theta}, -\sqrt{\theta})$ n'est pas un $\mathbb{Q}(\sqrt{d})$-nombre de Pisot, pour tout $d > 0$.

En outre les résultats précédents montrent que les seuls nombres de Pisot θ inférieurs à 2.6 tels que $(\sqrt{\theta}, -\sqrt{\theta})$ soit un $\mathbb{Q}(\sqrt{d})$-nombre de Pisot sont l'entier 2 et les nombres de Pisot $2.348593\ldots$ et $2.496697\ldots$, leurs racines étant respectivement des $\mathbb{Q}(\sqrt{2})$ et un $\mathbb{Q}(\sqrt{3})$-nombre de Pisot.

4   Perspectives

La construction de Salem [7] se généralise et montre que tout K-nombre de Pisot est limite d'une suite de K-nombres de Salem. Mais nous pouvons espérer davantage.

Prenons, par exemple $K = \mathbb{Q}(\sqrt{2})$.

Alors $(\sqrt{2}, -\sqrt{2})$ est un $\mathbb{Q}(\sqrt{2})$-nombre de Pisot et le polynôme minimal de $\sqrt{2}$ sur $\mathbb{Q}(\sqrt{2})$ est $z - \sqrt{2}$.

Le polynôme

$$(1) \qquad z^n(z - \sqrt{2}) + 1 - \sqrt{2}z.$$

possède pour $n \geq 6$ une unique racine τ_n extérieure au cercle unité et effectivement des racines de module 1.

Il en est de même du polynôme conjugué

$$(2) \qquad z^n(z + \sqrt{2}) + 1 + \sqrt{2}z.$$

Or pour $n = 2$ le polynôme (2) n'a pas de racine extérieure au disque unité et le polynôme (1) possède la racine $1.883204\ldots$.

Par suite le nombre $1.883204\ldots$ est un nombre de Salem de degré 4. On le trouve effectivement dans les tables de Boyd [5].

De même, pour $n = 4$, le polynôme (2) est sans racine extérieurer au disque unité tandis que le polynôme (1) possède la racine $1.60545\ldots$ qui est donc un nombre de Salem de degré 8 se trouvant également dans les tables de Boyd [5].

En résumé, la construction de Salem appliquée à des nombres de Pisot relatifs peut fournir des entiers algébriques réciproques de petite mesure.

L'un de mes étudiants T. Zaimi vient de prouver que l'ensemble des 2-Pisot réciproques est identique à l'ensemble des $\mathbb{Q}(\sqrt{d})$-nombres de Pisot réciproques et que leur mesure est supérieure à 3. Il achève la détermination des $\mathbb{Q}(\sqrt{d})$-nombres de Pisot, pour $d < 0$, de mesure inférieure à 2 et étudie les K-nombres de Pisot de petite mesure lorsque K est un corps cubique totalement réel de discriminant inférieur à 10000.

La plus petite mesure m_K d'un K-nombre de Pisot semble d'ailleurs augmenter avec le discriminant du corps de nombres K.

5 Références

1. Bergé, A. M. et Martinet, J. (1987). Notions relatives de régulateurs et de hauteurs. *Séminaire de théorie des nombres de Bordeaux.*

2. Bertin, M. J. (1990). Hauteurs et nombres de Pisot relatifs. *Groupe d'étude en théorie analytique des nombres. Publications de l'Université d'Orsay.*

3. Bertin, M. J., Decomps-Guilloux, A., Grandet-Hugot, M., Pathiaux-Delefosse, M. et Schreiber, J. P. (à paraître). *Pisot and Salem numbers.* Birkhaüser.

4. Bertin, M. J. et Pathiaux-Delefosse, M. (1989). *Conjecture de Lehmer et petits nombres de Salem.* Queen's papers in pure and applied Mathematics, **81**, Kingston.

5. Boyd, D. W. (1980). Reciprocal polynomials having small measure. *Math. Comp.*, **35**, 1361-77.

6. Dufresnoy, J. et Pisot, C. (1953). Etude de certaines fonctions méromorphes bornées sur le cercle unité. Application à un ensemble fermé d'entiers algébriques. *Ann. Sc. Ec. Norm. Sup.*, **70**, 105-33.

7. Salem, R. (1945). Power series with integral coefficients. *Duke Math. J.*, **12**, 153-71.

8. Schinzel, A. (1973). On the product of the conjugates outside the unit circle of an algebraic number. *Acta arithmetica*, **24**, 385-99.

M. J. Bertin
Mathèmatiques
Université P. et M. Curie
4, Place Jussieu
F-75005 Paris
France

Cubic Inequalities of Additive Type

J. Brüdern and R.J. Cook

Universität Göttingen
and
University of Sheffield

1 Introduction

The first result on cubic inequalities is due to Davenport and Roth [7]. They proved that if $\lambda_1, \ldots, \lambda_8$ are real numbers with at least one ratio λ_i/λ_j irrational then for any $\epsilon > 0$ the inequality

$$|\lambda_1 x_1^3 + \ldots + \lambda_8 x_8^3| < \epsilon \tag{1.1}$$

has infinitely many solutions in integers $x_1, \ldots, x_8$. No progress has been made on reducing the number of variables, however the result can be made more precise. In [2] Brüdern proved that the inequality

$$|\lambda_1 x_1^3 + \ldots + \lambda_8 x_8^3| < (\max |x_i|)^{-\sigma} \tag{1.2}$$

has infinitely many solutions in integers $x_1, \ldots, x_8$ for any $\sigma < 1/4$ (the details were only given for $\sigma < 3/14$, the stronger result follows easily on combining the methods of [2] with Vaughan's sixth power estimate [11, Theorem 4.4].

Recently [3] we considered two cubic diophantine inequalities. Suppose that $\lambda_1, \ldots, \lambda_{15}, \mu_1, \ldots, \mu_{15}$ are real coefficients, and let

$$\begin{aligned} F(\mathbf{x}) &= \lambda_1 x_1^3 + \ldots + \lambda_{15} x_{15}^3, \\ G(\mathbf{x}) &= \mu_1 x_1^3 + \ldots + \mu_{15} x_{15}^3. \end{aligned} \tag{1.3}$$

With the forms F, G we associate ternary linear forms

$$L_{ijk}(\boldsymbol{\theta}) = \det \begin{pmatrix} \theta_i & \theta_j & \theta_k \\ \lambda_i & \lambda_j & \lambda_k \\ \mu_i & \mu_j & \mu_k \end{pmatrix}. \tag{1.4}$$

In [3] we proved

Theorem 1. *Suppose that $\lambda_1, \ldots, \lambda_{15}, \mu_1, \ldots, \mu_{15}$ are algebraic numbers, and suppose that not all the ternary linear forms L_{ijk} have coefficients linearly dependent over the rationals. Then, for $\sigma < 3/56$ the simultaneous inequalities*

$$\max(|F(\mathbf{x})|, |G(\mathbf{x})|) < \max(|x_i|)^{-\sigma} \tag{1.5}$$

have infinitely many solutions in integers $x_1, \ldots x_{15}$.

We define the *order* Ω of a linear form

$$L(\mathbf{x}) = \sum \ell_i x_i \tag{1.6}$$

as the supremum of the values α for which the inequalities

$$|L(\mathbf{x})| < U^{-\alpha}, \quad \max|x_i| \leqslant U \tag{1.7}$$

are soluble for all sufficiently large U. Clearly any linear form of finite order has coefficients linearly independent over $\mathbb{Q}$. The proof of Theorem 1 yields some result provided that the ternary linear forms L_{ijk} do not all have order ∞. A result of Schmidt [9] shows that when the coefficients are algebraic, and linearly independent over $\mathbb{Q}$, the ternary linear forms have order at most 2.

In [4] we considered more general systems of additive diophantine inequalities

$$|\lambda_{i1} x_1^k + \ldots + \lambda_{iN} x_N^k| < \epsilon, \quad i = 1, \ldots, R. \tag{1.8}$$

We write Λ for the matrix of coefficients (λ_{ij}) and for $0 \leqslant d \leqslant R - 1$ we denote by $\mu(d) = \mu(d, \Lambda)$ the maximum number of columns from Λ lying in a linear subspace of dimension d. With the matrix Λ we associate linear forms in $R + 1$ variables. Let $\mathcal{J} \subset \{1, 2, \ldots, N\}$ with $|\mathcal{J}| = R + 1$. Let $A_{\mathcal{J}}$ be the $(R + 1) \times (R + 1)$ matrix formed with the columns

$$(\xi_j \lambda_{1j} \ldots \lambda_{Rj})^T \quad j \in \mathcal{J} \tag{1.9}$$

and put

$$L_{\mathcal{J}}(\boldsymbol{\xi}) = \det A_{\mathcal{J}} \tag{1.10}$$

where $\boldsymbol{\xi}_{\mathcal{J}} = (\xi_j)_{j \in \mathcal{J}}$.

In [4] we showed that there is a function $n_0(k)$ such that $n_0(2) = 4$, $n_0(3) = 8$, $n_0(4) = 12, \ldots$ and

$$n_0(k) = 2k(\log k + O(1)) \text{ as } k \to \infty \tag{1.11}$$

with the following property.

Theorem 2. *Suppose that $\lambda_{ij} \in \mathbb{R}$, that $N = n_0(k)R + 1$, and that at least one of the linear forms (1.10) is of finite order. Suppose that $\mu(d) \leqslant n_0 d$*

holds for $0 \leqslant d \leqslant R - 1$. Then the number θ_1 of solutions to the diagonal inequalities (1.8) in integers with $|x_i| \leqslant P$ satisfies, for infinitely many positive integers P,

$$\theta_1 \gg P^{N - Rk}. \tag{1.12}$$

In the particular case $k = 3$ we obtained some improvement on this which shows, roughly speaking, that in Theorem 2 one may take $n_0(3) = 7$.

Theorem 3. *Let $k = 3$ and $N = 7R + 1$. Suppose that at least one of the linear forms (1.10) is of finite order. Suppose that $\mu(d, \Lambda) \leqslant 11\ d/2$ for $0 \leqslant d \leqslant R - 1$. Then, with θ_1 as in Theorem 2, we have $\theta_1 \gg P^{4R+1}$, for infinitely many positive integers.*

The main purpose here is to illustrate the techniques of [3] and [4], combining them to obtain a more precise result for simultaneous cubic inequalities

$$|F_i(x)| = |\lambda_{i1} x_1^3 + \ldots + \lambda_{iN} x_N^3| < (\max |x_i|)^{-\sigma}. \tag{1.13}$$

Theorem 4. *Suppose $N \geqslant 7R + 1$, the coefficients λ_{ij} in (1.13) are real and satisfy $\mu(d, \Lambda) \leqslant 11d/2$ for $1 \leqslant d \leqslant R - 1$. Suppose further that at least one of the associated linear forms L_J is of finite order Ω. Then the inequalities (1.13) have infinitely many solutions in integers $x_1, \ldots, x_N$ for any*

$$\sigma < \frac{3}{\Omega + 3R(R + 1)(\Omega + 1)}. \tag{1.14}$$

In the case when the coefficients are algebraic, a result of Schmidt [9] shows that the order Ω is bounded by R and we obtain the following variant in this case.

Theorem 5. *Suppose $N \geqslant 7R + 1$, the coefficients λ_{ij} in (1.13) are real and satisfy $\mu(d, \Lambda) \leqslant 11d/2$ for $1 \leqslant d \leqslant R - 1$. Suppose further that at least one of the associated linear forms L_J has algebraic coefficients which are linearly independent over $\mathbb{Q}$. Then the inequalities (1.13) have infinitely many solutions in integers $x_1, \ldots, x_N$ for any*

$$\sigma < \frac{3}{R + 3R(R + 1)^2}. \tag{1.15}$$

In proving Theorem 4 we shall suppose that $N = 7R+1$, and take excess variables as zero. Clearly we may also suppose that the other conditions are still satisfied.

2 Preliminary normalization

We may suppose that the first $R+1$ columns of Λ give a linear form $L = L_{\mathcal{J}}$ (where $\mathcal{J} = \{1, 2, \ldots, R+1\}$) of finite order Ω and having coefficients linearly independent over $\mathbb{Q}$. In particular, the coefficients of L are non-zero so these $R+1$ columns of Λ are in general position in $\mathbb{R}^R$. Using elementary row operations we can bring the cubic forms $F_1, \ldots, F_R$ to a system $G_1, \ldots, G_R$ with coefficient matrix

$$G = [I, \boldsymbol{\lambda}, B] \tag{2.1}$$

where I is the $R \times R$ identity, $\boldsymbol{\lambda}$ is a column vector

$$\boldsymbol{\lambda} = (\lambda_1, \ldots, \lambda_R)^T \tag{2.2}$$

and B is $R \times (N - (R+1))$.

The effect of these row operations on the linear forms $L_{\mathcal{J}}$ can only be to multiply them by non-zero constants. This leaves their order unaltered. It is now enough to prove the result for the system G, the result for the system $F_1, \ldots, F_R$ is easily recovered using a different σ for $G_1, \ldots, G_R$ to overwhelm any constants brought in by row operations. Henceforth we shall suppose that $F_1, \ldots, F_R$ are normalized in this way and that their coefficient matrix Λ is of the shape $[I, \boldsymbol{\lambda}, B]$. The linear form L is then

$$L(\boldsymbol{\xi}) = (-1)^{R+1}(\lambda_1 \xi_1 + \ldots + \lambda_R \xi_R - \xi_{R+1}) \tag{2.3}$$

and L has finite order Ω.

We denote the columns of Λ by $\mathbf{c}_1, \ldots \mathbf{c}_N$ (so that $\mathbf{c}_{R+1} = \boldsymbol{\lambda}$) and define linear forms

$$\Lambda_j = \Lambda_j(\boldsymbol{\alpha}) = \mathbf{c}_j . \boldsymbol{\alpha}, \quad j = 1, \ldots, N \tag{2.4}$$

where $\boldsymbol{\alpha} \in \mathbb{R}^R$. Thus

$$\Lambda_j = \alpha_j \text{ for } j = 1, \ldots R \tag{2.5}$$

and

$$L(\boldsymbol{\xi}) = (-1)^{R+1}(\Lambda_{R+1}(\xi_1, \ldots, \xi_R) - \xi_{R+1}). \tag{2.6}$$

Changing some of the variables from x_i to $-x_i$, if necessary, we may suppose that the equations

$$F_1(\mathbf{y}) = \ldots = F_R(\mathbf{y}) = 0 \tag{2.7}$$

have a non-singular real solution $\mathbf{y}$ such that all variables y_i satisfy the condition $0 < y_j < 1$. We choose a constant $C > 0$ so that $C < y_j^3 < 1$ for all coordinates of this non-singular solution. We write $e(\theta)$ for $\exp(2\pi i\theta)$ and define exponential sums

$$f(\alpha) \quad = \quad \sum_{CP < x \leqslant P} e(\alpha x^3), \tag{2.8}$$

$$g(\alpha) \quad = \quad \sum_{x \in \mathcal{A}} e(\alpha x^3) \tag{2.9}$$

where $\mathcal{A}$ is the set of positive integers $x \leqslant P$ and free of prime factors exceeding P^η, where $\eta > 0$ may be chosen later. We take

$$H(\boldsymbol{\alpha}) = f(\Lambda_1) \dots f(\Lambda_{R+1}) \prod_{j=R+2}^{N} g(\Lambda_j), \tag{2.10}$$

and, for any $\sigma > 0$ satisfying (1.14),

$$\tau = P^{-\sigma}. \tag{2.11}$$

To count solutions of the inequalities (1.13) we choose, for a given $\phi > 1$, a kernel $K : \mathbb{R} \to [0, \infty)$ with $K(\alpha) = K(-\alpha)$, $K(\alpha) \ll \min(1, |\alpha|^{-\phi})$ and whose Fourier transform $\hat{K}$ satisfies

$$\left. \begin{array}{ll} \hat{K}(\alpha) = 0 & (|\alpha| > \tau) \\[2mm] \hat{K}(\alpha) = 1 & (|\alpha| \leqslant \tau/3) \end{array} \right\} \tag{2.12}$$

and $0 \leqslant \hat{K}(\alpha) \leqslant 1$ for all $\alpha \in \mathbb{R}$. Davenport [6, Lemma 1] showed that such a function exists. Writing

$$K(\boldsymbol{\alpha}) = K(\tau \alpha_1) \dots K(\tau \alpha_R) \tag{2.13}$$

and

$$\Theta = \int_{\mathbb{R}^R} H(\boldsymbol{\alpha}) K(\boldsymbol{\alpha}) d\boldsymbol{\alpha} \tag{2.14}$$

we see that $\tau^R \Theta$ does not exceed the number of solutions to the simultaneous inequalities

$$\max |F_i(\mathbf{x})| < \tau, \qquad 0 < x_j \leqslant P. \tag{2.15}$$

We shall prove Theorem 4 by showing that $\tau^R \Theta \to \infty$ as $P \to \infty$ through some suitable sequence.

We begin by replacing the integral (2.14) with a finite integral. Let $B(X)$ denote the box defined by $\max|\alpha_i| \leqslant X$ in $\mathbb{R}^R$, let $\psi > 0$ be small and put

$$t = \mathbb{R}^R \backslash B(\tau^{-1} P^\psi). \tag{2.16}$$

Choosing ϕ sufficiently large (in terms of ψ^{-1})

$$\int_t H(\alpha)K(\alpha)d\alpha \quad \ll \quad P^N \int_t \prod_{j=1}^R (1 + \tau|\alpha_i|)^{-\phi} d\alpha$$

$$\ll \quad 1. \tag{2.17}$$

The box $B(\tau^{-1}P^\psi)$ is now partitioned into the major arc $\mathcal{M} = B(P^{-9/4})$ and the minor arc $m = B(\tau^{-1}P^\psi)\backslash\mathcal{M}$.

3   The minor arc

We begin with a Weyl-type estimate for the exponential sums $f_1, \ldots, f_{R+1}$, and recall that in the normalized situation $\Lambda_1, \ldots \Lambda_{R+1}$ satisfy (2.5) and (2.6). Our next lemma is a variant on Lemma 2 of Cook [5], and may be generalized to kth powers.

Lemma 6. *For any*

$$\delta < \frac{3R}{\Omega + 3R(R+1)(\Omega+1)} \tag{3.1}$$

there is an infinite sequence $\mathcal{P}$ of integers such that for all $P \in \mathcal{P}$ and all $\alpha \in m$ we have

$$\min_{j=1,\ldots,R+1} |f(\Lambda_j)| \ll P^{1-\delta}. \tag{3.2}$$

Proof. Suppose that such a sequence P does not exist. Then for all large $\mathcal{P}$ there is an $\alpha \in m$ with

$$|f(\Lambda_j)| > P^{1-\delta} \text{ for } j = 1, \ldots R+1. \tag{3.3}$$

Now $R \geqslant 1$ and $\Omega \geqslant 1$ so $\delta < 1/4$. Therefore, from Lemma 29 of Davenport and Lewis [8], we see that each Λ_j has a rational approximation a_j/q_j satisfying

$$1 \leqslant q_j \ll P^{3\delta}, \ |\Lambda_j - a_j/q_j| < q_j^{-1/3} P^{\delta-3} \tag{3.4}$$

for $j = 1, \ldots, R+1$. Since $\alpha \in m$ we also have

$$|a_j| \ll q_j \tau^{-1} P^\psi. \tag{3.5}$$

We set $Q = q_1 q_2 \ldots q_{R+1}$ and $y_j = a_j Q/q_j$, so that

$$y_j \ll \tau^{-1} P^{3\delta(R+1)+\psi}. \tag{3.6}$$

Now, for $j = 1, \ldots, R$,

$$\Lambda_j = \alpha_j = \frac{a_j}{q_j} + O(q_j^{-1/3} P^{\delta-3}) \tag{3.7}$$

and

$$
\begin{aligned}
\Lambda_{R+1} &= \frac{a_{R+1}}{q_{R+1}} + O(q_{R+1}^{-1/3} P^{\delta-3}) \\
&= \lambda_1 \alpha_1 + \ldots + \lambda_R \alpha_R \\
&= \lambda_1 \frac{a_1}{q_1} + \ldots + \lambda_R \frac{a_R}{q_R} + O(P^{\delta-3} \max q_j^{-1/3}) \tag{3.8}
\end{aligned}
$$

so that

$$
\begin{aligned}
L(y_1, \ldots, y_{R+1}) &= (-1)^{R+1} \left(\lambda_1 \frac{a_1 Q}{q_1} + \ldots + \lambda_R \frac{a_R Q}{q_R} - \frac{a_{R+1} Q}{q_{R+1}} \right) \\
&= O(Q P^{\delta-3} \max q_j^{-1/3}) \\
&= O(P^{3(R+1)\delta-3}). \tag{3.9}
\end{aligned}
$$

Thus we have integers $y_1, \ldots, y_{R+1}$ satisfying

$$\max |y_j| \ll \tau^{-1} P^{3\delta(R+1)+\psi}. \tag{3.10}$$

Recalling (1.14) and (2.11), it is convenient at this point to fix the relationship between τ and δ by taking

$$\sigma = (\delta/R) - \rho \tag{3.11}$$

with $\rho = 2R\psi > 0$. Then

$$\max |y_j| \ll P^{(3R(R+1)+1)\delta/R+\psi-\rho} \tag{3.12}$$

and

$$|L(y_1, \ldots, y_{R+1})| \ll P^{3(R+1)\delta-3}. \tag{3.13}$$

Thus L has order $\omega - \eta$ for some small $\eta > 0$ where

$$\omega \geqslant \frac{R(3 - 3(R+1)\delta)}{(3R(R+1)+1)\delta} > \Omega, \tag{3.14}$$

on rearranging (3.1). For small η, $\omega - \eta > \Omega$ to give a contradiction, and η will be small provided that ψ is. Choosing ψ small requires ϕ to be large for the kernel K but does not affect other parameters, so we obtain the lemma.

We shall use Lemma 6 to establish the estimate

$$\int_m H(\alpha)K(\alpha)d\alpha \ll P^{4R+1-\rho+R\psi+\epsilon}. \tag{3.15}$$

The contribution of the major arcs $\mathcal{M}$ satisfies

$$\int_{\mathcal{M}} H(\alpha)K(\alpha)d\alpha \gg P^{4R+1} \tag{3.16}$$

and, since $\rho = 2R\psi$, we obtain the required solutions.

We denote by U a cube of volume 1 in $\mathbb{R}^R$ with sides parallel to the coordinate hyperplanes.

Lemma 7. *Let $A = (a_{ij})$ be an $R \times 7R$ matrix of real numbers such that $\mu(d, A) \leqslant 11d/2$ for $d = 0, 1, \ldots, R$. Suppose that for $0 \leqslant \ell \leqslant 6$ the matrices (a_{ij}), $1 \leqslant i \leqslant R$, $\ell R < j \leqslant (\ell+1)R$ are non-singular. Let*

$$\lambda_j = \sum_{i=1}^{R} a_{ij}\alpha_i, \quad j = 1, \ldots, 7R, \tag{3.17}$$

and suppose that U is a unit cube as described above. Then,

$$\int_U |f(\lambda_1)\ldots f(\lambda_R)g(\lambda_{R+1})\ldots g(\lambda_{7R})|d\alpha \ll P^{4R+\epsilon}, \tag{3.18}$$

for any $\epsilon > 0$. The implied constant depends on a_{ij} and ϵ only.

This is Lemma 5 of Brüdern and Cook [4].

We require the following combinatorial result. An $r \times rn$ matrix over a field is called *partitionable* if its columns can be grouped into n disjoint blocks, with each block forming a non-singular $r \times r$ matrix.

Lemma 8. *(Aigner's criterion). An $r \times nr$ matrix M is partitionable if and only if for $0 \leqslant d \leqslant r - 1$ one has $\mu(d, M) \leqslant nd$.*

For a proof see Aigner [1].

To complete the treatment of the minor arcs we observe that for $P \in \mathcal{P}$ and $\alpha \in m$ at least one of $f(\Lambda_j)$, $j = 1, \ldots, R+1$ satisfies $|f(\Lambda_j)| \ll P^{1-\delta}$ and the remaining $7R$ columns are partitionable and therefore satisfy the conditions of Lemma 7. Thus for each unit cube U

$$\int_{U \cap m} H(\alpha)K(\alpha)d\alpha \ll P^{4R+1-\delta+\epsilon}. \tag{3.19}$$

Since $B(P^\psi)$ may be covered with $O(P^{R\psi})$ unit cubes U we obtain the estimate (3.15) for the contribution of the minor arcs.

4 The major arc

The major arc techniques are standard so we outline the proof of the lower bound (3.16), a more detailed treatment is given in [3].

On $\mathcal{M}$ we approximate $f(\alpha)$ by

$$I(\alpha) = \int_{CP}^{P} e(\alpha\beta^3)d\beta \tag{4.1}$$

which satisfies

$$f(\alpha) - I(\alpha) \ll 1 \tag{4.2}$$

and

$$I(\alpha) \ll P(1 + P^3|\alpha|)^{-1}. \tag{4.3}$$

Thus

$$\begin{aligned}
f(\Lambda_1)\ldots f(\Lambda_R) &= f(\alpha_1)\ldots f(\alpha_R) \\
&\ll P^R \prod_{j=1}^{R}(1 + P^3|\alpha_j|)^{-1}.
\end{aligned} \tag{4.4}$$

Let $N = B(P^{-3}(\log P)^A)$, for some $A > 0$. Then

$$\int_{\mathcal{M}\setminus N} \prod_{j=1}^{R}|f(\Lambda_j)|^4 d\alpha \ll P^R(\log P)^{-3A}. \tag{4.5}$$

We take

$$H^*(\alpha) = \prod_{j=R+2}^{7R+1} g(\Lambda_j), \tag{4.6}$$

the corresponding coefficient matrix partitions into six non-singular matrices. Using Vaughan's stronger form of Hua's Lemma for eighth powers, ([10], Theorem 2), and factorizing the integral, we see that

$$\int_{\mathcal{M}} |H^*(\alpha)|^{4/3}d\alpha \ll P^{5R}. \tag{4.7}$$

Then Hölder's inequality, and the trivial bound for one factor f, shows that

$$\int_{\mathcal{M}\setminus N} |H(\alpha)|K(\alpha)d\alpha \ll P^{4R+1}(\log P)^{-A/2} \tag{4.8}$$

so we now have to obtain the lower bound for the integral over N.

From (4.2) we have

$$\int_N H(\alpha)K(\alpha)d\alpha = \int_N \prod_{j=1}^{R+1} I(\Lambda_j)H^*(\alpha)K(\alpha)d\alpha + O(P^{4R+1/2}) \quad (4.9)$$

and the next step is to approximate to the sums $g(\Lambda_j)$. Let

$$J(\alpha) = \frac{1}{3} \int_{(CP)^3}^{P^3} \rho\left(\frac{\log\beta}{3\delta\log P}\right) \beta^{-2/3} e(\beta\alpha) d\beta, \quad (4.10)$$

where $\rho(u)$ denotes Dickman's function. Then, see [3],

$$H^*(\alpha) - \prod_{j=R+2}^{7R+1} J(\Lambda_j) \ll P^{4R}(\log P)^{-1/4}. \quad (4.11)$$

Replacing $H^*(\alpha)$ in (4.9) by the approximating product $\Pi J(\Lambda_j)$ in (4.11), and noting that the volume of N is $P^{-3R}(\log P)^{AR}$, we obtain

$$\int_N H(\alpha)K(\alpha)d\alpha = V(N) + O(P^{4R+1}(\log P)^{-A/2}) \quad (4.12)$$

where

$$V(B) = \int_B \prod_{j=1}^{R+1} I(\Lambda_j) \cdot \prod_{j=R+2}^{7R+1} J(\Lambda_j)K(\alpha)d\alpha. \quad (4.13)$$

Standard methods show that

$$V(N) = V(\mathbb{R}^R) + O(P^{4R+1}(\log P)^{-A/2}). \quad (4.14)$$

Fourier's inversion formula interprets $V(\mathbb{R}^R)$ as a singular integral, representing the weighted volume of the manifold of real solutions of the corresponding real equations in the box

$$U(P) = \{\beta \in \mathbb{R}^{7R+1} : (CP)^3 < \beta_i < P^3\}. \quad (4.15)$$

Our original choice of C, involving a non-singular solution of cubic equations, leads to the lower bound

$$V(\mathbb{R}^R) \gg P^{4R+1} \quad (4.16)$$

and this completes the proof of Theorem 4.

References

1. M. Aigner, Combinatorial Theory, Springer, Berlin (1979).

2. J. Brüdern, Cubic diophantine inequalities, Mathematika 35 (1988), 51-58.

3. J. Brüdern and R.J. Cook, On pairs of cubic diophantine inequalities, Mathematika 38 (1991), 250-263.

4. J. Brüdern and R.J. Cook, On Simultaneous diagonal equations and inequalities, Acta Arith. (to appear).

5. R.J. Cook, Simultaneous quadratic inequalities, Acta Arith. 25 (1974), 337-346.

6. H. Davenport, Indefinite quadratic forms in many variables, Mathematica 3 (1956), 81-107.

7. H. Davenport and K.F. Roth, The solubility of certain diophantine inequalities, Mathematika 2 (1955), 81-96.

8. H. Davenport and D.J. Lewis, Cubic equations of additive type, Phil. Trans. Roy. Soc. London 264A (1969), 557-595.

9. W.M. Schmidt, Linear forms with algebraic coefficients, I, J. Number Theory 3 (1971), 253-277.

10. R.C. Vaughan, On Waring's problem for cubes, Crelle 365 (1986), 122-170.

11. R.C. Vaughan, A new iterative method in Waring's problem, Acta Math. 162 (1989), 1-71.

J. Brüdern
Mathematisches Institut
Universität Göttingen
3400 Göttingen
Germany

and

Roger J. Cook
Department of Pure Mathematics
University of Sheffield
Sheffield S3 7RH
England
e-mail pm1rc@primea.sheffield.ac.uk

Discrepance et Diaphonie en Dimension Un

Henri Faure

Université de Provence, Marseilles

Abstract

Dans cette communication, nous présentons les derniers développements concernant la discrépance quadratique T et la diaphonie F des deux familles de suites connues pour leurs faibles irrégularités de distribution: suites $(\{n\alpha\})$ et suites de van der Corput.

1 Mesures d'irrégularités de distribution

Rappelons les principales définitions concernant la **discrépance**: Soit $X = (x_n)$ une suite infinie de l'intervalle $[0,1]$, N un entier au moins égal à un et α un élément de $[0,1]$; notons $A(\alpha, N, X)$ le nombre d'indices n tels que $(1 \leq n \leq N$ et $x_n \in [0, \alpha[$ et posons $E(\alpha, N, X) = A(\alpha, N, X) - N\alpha$. La fonction E s'appelle le reste (ou écart) à l'équirépartition de la suite X.

Alors, pour $1 \leq p \leq \infty$, la **discrépance** L^p, est définie par

$$D^{(p)}(N, X) = \left(\int_0^1 |E(\alpha, N, X)|^p \, d\alpha \right)^{1/p} \text{, si } p < \infty$$

et par $D^{(\infty)}(N, X) = \sup_\alpha |E(\alpha, N, X)|$.

La discrépance quadratique $D^{(2)}$ est souvent désignée par la lettre T, et la discrépance extrême $D^{(\infty)}$ par D^*.

Une autre notion est la **diaphonie** F, définie par

$$F(N, X) = \left(2 \sum_{m=1}^{\infty} \left| \sum_{n=1}^{N} e^{2i\pi m x_n} \right|^2 / m^2 \right)^{1/2}.$$

et reliée à T par la formule (Koksma, [6] p. 110):

$$T^2(N, X) = \left(\sum_{n=1}^{N} (1/2 - x_n) \right)^2 + F^2(N, X)/(4\pi^2);$$

Noter que pour toute suite X, on a $F(N, X) \in \Omega(\sqrt{Log\,N})$ d'après [7] et $T(N, X) \in \Omega(\sqrt{Log\,N})$ d'après K.F. Roth ([6],p.105).

411

2 Suites $(n\alpha)$ et leurs symétriques

Résultats de Y.J. Xiao [12]:

2.1 Enoncé des résultats

Soit α un nombre irrationnel à quotients partiels bornés. L'objet de ce travail est d'étudier la discrépance quadratique et la diaphonie des suites $\theta_\alpha = (\{n\alpha\})$, où $\{\cdot\}$ désigne la partie fractionnaire, et des suites symétriques qui s'en déduisent, à savoir

$$\tilde{\theta}_\alpha = (\{(-1)^{n+1}[(n+1)/2]\alpha\})\,, \quad \text{où } [.] \text{ désigne la partie entière.}$$

Proinov a obtenu des estimations de ces quantités en fonction de la borne supérieure M des quotients partiels de α. Nous nous proposons ici d'une part d'affiner les estimations de Proinov et d'autre part d'étudier plus précisément le cas où les quotients partiels sont constants, en particulier le cas du nombre d'or.

Pour alléger les énoncés, nous nous limiterons à l'étude du comportement asymptotique et nous noterons pour $x \geq 0$:

$$S(x) = \Sigma_{1 \leq n < \infty}(n+x)^{-2}.$$

Théorème 2.1. *Soit α un nombre irrationnel à quotients partiels bornés par $M > 0$; posons*

$$A(M) = M^2 + 4M + \frac{1}{4}(\sqrt{M^2+4M}+M)^2 S\left(\frac{4}{M(1+\sqrt{1+4/M})^2}\right)\ ;$$

on a alors:

$$f(\theta_\alpha) = \limsup_{N\to\infty}\frac{F^2(N,\theta_\alpha)}{Log\,N} \leq \frac{A(M)}{Log((1+\sqrt{5})/2)} < 5.5(M^2+M)$$

et

$$t(\tilde{\theta}_\alpha) = \limsup_{N\to\infty}\frac{T^2(N,\tilde{\theta}_\alpha)}{Log\,N} \leq \frac{A(M)}{\pi^2\,Log((1+\sqrt{5})/2)} < 0.557(M^2+M).$$

Remarque: Proinov [7] obtient les estimations suivantes:

$$f(\theta_\alpha) \leq (M+1)^2(16+2\pi^2/3) < 22.58(M+1)^2 \text{ et}$$

$$t(\tilde{\theta}_\alpha) \leq (M+1)^2(16/\pi^2+2/3) < 2.29(M+1)^2.$$

Dans le cas où les quotients partiels de α sont constants, les résultats peuvent être précisés sous la forme suivante:

Théorème 2.2. *Soit* $\alpha = [0 \; ; a, a, \ldots a, \ldots]$ *un nombre irrationnel à quotients partiels constants; posons*

$$A'(\alpha) = \frac{(1 + \alpha^2)^2 + S(1 - \alpha)}{-\alpha^2 \operatorname{Log} \alpha};$$

on a alors:
Si $a \geq 2$:

$$f(\theta_\alpha) \leq A'(\alpha) \qquad et \qquad t(\tilde{\theta}_\alpha) \leq A'(\alpha)/\pi^2 \; ;$$

si $a = 1$ *(α est alors le nombre d'or $(\sqrt{5} - 1)/2$):*

$$f(\theta_\alpha) \leq A'(\alpha)/2 < 7.5 \qquad et \qquad t(\tilde{\theta}_\alpha) \leq A'(\alpha)/(2\pi^2) < 0.76.$$

Remarque: l'estimation de Proinov donne dans le cas du nombre d'or:

$$f(\theta_\alpha) \leq 90.32 \quad et \quad t(\tilde{\theta}_\alpha) \leq 9.16.$$

2.2 Indications sur les démonstrations

Le lemme suivant, dû à Proinov [10], permet de passer des estimations sur $F(N, \theta_\alpha)$ aux estimations sur $T(N, \theta_\alpha)$:

Lemme 2.1. *Soit* $\tilde{X}$ *la suite symétrisée d'une suite* X; *alors pour tout* $N \geq 1$, *on a la majoration:*

$$T^2(N, \tilde{X}) \leq (F(n, X)/\pi + c)^2 + d^2$$

où $n = [N/2], c = d = 0$ *si* N *est pair et* $c = 1/(2\sqrt{3}), d = 1/2$ *si* N *est impair.*

Pour estimer $F(N, \theta_\alpha)$, on part de l'inégalité

$$F^2(N, \theta_\alpha) \leq 2G(N, \theta_\alpha)N^2$$

avec

$$G(N, \theta_\alpha) = \sum_{h=1}^{\infty} \frac{1}{h^2} \min(1, \frac{1}{4N^2 \langle h\alpha \rangle^2}),$$

où $\langle \cdot \rangle$ désigne la distance à l'entier le plus proche.

On coupe la somme en deux parties, de 1 à N puis de $N + 1$ à l'infini. Pour la somme infinie, on majore avec le lemme suivant obtenu par les méthodes de Proinov:

Lemme 2.2. *Pour tout réel* $X \geq 1$:

$$\sum_{h \geq X} \frac{1}{h^2} \min(1, \frac{1}{4N^2 \langle h\alpha \rangle^2}) \leq \frac{12}{XNC(\alpha)}.$$

où $C(\alpha) = \inf_{h \geq 1} h\langle h\alpha \rangle > 0$.

Pour la somme finie de 1 à N, on introduit les suites $(p_n)_{n \geq -1}$ et $(q_n)_{n \geq -1}$ définies par:

$p_{-1} = 1, p_0 = a_0$ et pour $n \geq 0$: $p_{n+1} = a_{n+1}p_n + p_{n-1}$, puis

$q_{-1} = 0, q_0 = 1$ et pour $n \geq 0$: $q_{n+1} = a_{n+1}q_n + q_{n-1}$,

p_n et q_n sont respectivement le numérateur et le dénominateur du rationnel $[a_0 ; a_1, \ldots, a_n]$ écrit sous forme irréductible et on pose, pour $n \geq 1$:

$$\delta_n = [0 ; a_n, \ldots, a_1], \text{ et } \theta_n = q_n|a_n\alpha - p_n|.$$

On démontre alors le lemme:

Lemme 2.3. *Pour tout entier* $n \geq 1$:

$$\sum_{q_{n-1} \leq h < q_n} \frac{1}{\langle h\alpha \rangle^2} \leq 2q_{n-1}^2 \left(\frac{1}{\theta_{n-1}^2} + \sum_{k=2}^{\infty} \frac{1}{(k\delta_n - \delta_n\delta_{n+1})^2} \right).$$

De plus, si α *est le nombre d'or, on peut supprimer le facteur 2 dans le membre de droite.*

Des propriétés classiques des fractions continues on déduit que:

$$\delta_n = \frac{q_{n-1}}{q_n}$$

et

$$\theta_n = \frac{1}{\frac{1}{\alpha_n'} + \frac{q_{n-1}}{q_n}},$$

où $\alpha_n' = [0 ; a_{n+1}, a_{n+2}, \ldots]$. A partir de ces formules, on étudie facilement le comportement asymptotique des suites (δ_n) et (θ_n) sous les hypothèses des théorèmes 2.1 et 2.2. On en déduit, grâce au lemme 2.3, une majoration de $f(\theta_\alpha)$.

3   Suites de van der Corput et leurs symétriques

Résultats de H. Chaix et H. Faure [2],[3],[4]:

3.1   Définitions

— *Suites en base b, pour b entier au moins égal à 2:* Soit $\sum = (\sigma_j)_{j \geq 0}$ une suite de permutations de l'ensemble $\{0, 1, \ldots, b-1\}$; la suite de van der Corput en base b associée à Σ, notée S_b^Σ, est définie [5] par:

$$S_b^\Sigma(n) = \sum_{j=0}^{\infty} \sigma_j(a_j(n))b^{-j-1}$$

où $a_j(n)$ est la j-ième décimale de $(n-1)$ développé en base b:

$$n - 1 = \sum_{j=0}^{\infty} a_j(n) b^j \text{ avec } n \geq 1 \text{ et } 0 \leq a_j(n) < b.$$

La suite originelle de van der Corput s'obtient pour $b = 2$ et $\Sigma = I$, suite constante égale à la permutation identique; certains auteurs appellent suites de v.d. Corput-Halton les suites S_b^I.

— *Suites en base variable:* $B = (b_j)_{j \geq 0}$, avec $b_0 = 1$ et $b_j \geq 2$ pour $j \geq 1$: De façon analogue, on définit ([5], 3.4) la suite de van der Corput en base B associé à Σ:

$$S_B^{\Sigma}(n) = \sum_{j=0}^{\infty} \sigma_j(a_j(n)) / B_{j+1},$$

où $\Sigma = (\sigma_j)_{j \geq 0}$ avec σ_j permutation de l'ensemble $\{0, 1, \ldots, b_{j+1} - 1\}$, $B_j = \prod_{i=0}^{j} b_i$ et $n - 1 = \sum_{j=0}^{\infty} a_j(n) B_j$ (développement de $(n-1)$ en base variable B).

3.2 Cas des bases fixes

Théorème 3.1. *Dans le cas particulier des suites de van der Corput-Halton ($\sigma = I$), on obtient, pour p réel, $1 \leq p \leq 2$:*

$$d^{(p)}(S_b^I) = \begin{cases} (b^2 + b - 2)/(8(b+1) \operatorname{Log} b), & \text{si } b \text{ est pair} \\ (b^2 - 1)/(8b \operatorname{Log} b), & \text{si } b \text{ est impair}, \end{cases}$$

avec $d^{(p)}(X) = \lim\sup_{N \to \infty} (D^{(p)}(N, X) / \operatorname{Log} N)$.

Théorème 3.2. *Dans le cas particulier des suites S_b^I, on a:*

$$f(S_b^I) = \pi^2 (b^3 + b^2 + 4)/(48(b+1) \operatorname{Log} b), \text{ si } b \text{ est pair},$$

$$f(S_b^I) = \pi^2 (b^4 + 2b^2 - 3)/(48 b^2 \operatorname{Log} b), \text{ si } b \text{ est impair},$$

avec $f(X) = \lim\sup_{N \to +\infty} (F^2(N, X) / \operatorname{Log} N)$.

Théorème 3.3. *Pour $2 \leq b \leq 11$, la plus faible diaphonie est obtenue pour les permutations données dans le tableau suivant:*

b	permutation σ	minorant de $f(S_b^{\sigma})$	majorant de $f(S_b^{\sigma})$
2	(0 1)	$1.58209 \cdots$	$1.58209 \cdots$
3	(0 1 2)	$1.99637 \cdots$	$1.99637 \cdots$
4	(0 2 1 3)	$1.58209 \cdots$	$1.58209 \cdots$
5	(0 3 1 2 4)	$1.59440 \cdots$	1.667
6	(0 4 2 1 3 5)	$1.74867 \cdots$	1.824
7	(0 3 5 1 4 2 6)	$1.42613 \cdots$	1.479
8	(0 3 5 1 6 2 4 7)	$1.51146 \cdots$	1.636
9	(0 5 3 7 1 6 2 4 8)	$1.37528 \cdots$	1.451
10	(0 4 7 2 8 5 1 6 3 9)	$1.33525 \cdots$	1.433
11	(0 7 3 5 9 1 8 4 2 6 $\overline{10}$)	$1.43887 \cdots$	1.592

Théorème 3.4. *Avec les couples (b, σ) du théorème 3.3 on obtient les majorations suivantes pour la discrépance quadratique des suites symétriques $\tilde{S}_b^\sigma$:*

b	2	3	4	5	6	7	8	9	10	11
majorant de $t(\tilde{S}_b^\sigma)$	.161	.203	.161	.169	.185	.150	.166	.148	.146	.162

où $t(X) = \lim\limits_{N \to \infty} \sup \, (T^2(N, X)/\operatorname{Log} N)$.

3.3 Cas des bases variables

Théorème 3.5. *Pour toute base B et toute suite de permutations Σ, on a:*

$$T(N, S_B^\Sigma) \in 0(\operatorname{Log} N) \ \text{ et } \ D^*(N, S_B^\Sigma) \in 0(\operatorname{Log} N) \ \text{ si } \ \sum_{j=1}^{n} b_j \in 0(n).$$

Théorème 3.6. *Pour toute base B et pour tout réel p vérifiant $1 \le p \le \infty$, on a:*

$$D^{(p)}(N, S_B^I) \in 0(\operatorname{Log} N) \ \text{ si et seulement si } \ \sum_{j=1}^{n} b_j \in 0(n).$$

Théorème 3.7. *Pour toute base B et toute suite Σ, on a:*

$$F^2(N, S_B^\Sigma) \in 0(\operatorname{Log} N) \ \text{ si } \ \sum_{j=1}^{n} b_j^2 \in 0(n).$$

Théorème 3.8. *Pour toute base B, on a:*

$$F^2(N, S_B^I) \in 0(\operatorname{Log} N) \ \text{ si et seulement si } \ \sum_{j=1}^{n} b_j^2 \in 0(n).$$

Remarque:

Les théorèmes 3.6 et 3.8 ont été montrés par Proinov et Atanassov [1] et [9], le théorème 3.7 par Proinov [8]; nous les retrouvons ici avec de meilleures constantes explicites.

Théorème 3.9. *Soit $B = (b_j)$ une base variable vérifiant les deux hypothèses*

- *"$\sum_{j=1}^{n} b_j \notin 0(n)$" et*

- *"il existe $K > 0$ tel que, pour tout entier $n \geq 2, b_n \leq \left(\prod_{i=1}^{n-1} b_i \right)^{K}$ "*;

alors il existe des suites Σ_0 de permutations telles que, pour $N \geq 1$,

$$D^*(N, S_B^{\Sigma_0}) \leq (K+1) \operatorname{Log} N / \operatorname{Log} 2 + 1.$$

Théorème 3.10. *Soit $B = (b_j)$ une base variable vérifiant les deux hypothèses*

- *"$\sum_{j=1}^n b_j^2 \notin 0(n)$ " et*
- *"il existe $K > 0$ tel que, pour tout entier $n \geq 1$, $\sum_{j=1}^n (\operatorname{Log} b_j)^2 \leq Kn$ ";*

alors il existe des suites Σ_0 de permutations telles que, pour $N \geq 1$,

$$F^2(N, S_B^{\Sigma_0}) \leq 2\pi^2 K(\operatorname{Log} N / \operatorname{Log} b + 1)/(\operatorname{Log} 2)^2 + 4\pi^2/3.$$

3.4 Indications et conclusion

Tous les résultats asymptotiques des théorèmes précédents découlent de formules exactes sous forme de séries pour la discrépance et la diaphonie; les fonctions intervenant dans ces séries sont attachées aux couples (b, σ) formés d'une base b et d'une permutation des chiffres de la base; il serait trop long de détailler ici la construction et les propriétés de ces fonctions; les lecteurs intéressés pourront se reporter aux Notes aux Comptes Rendus de l'Académie des Sciences ([2], [3], [4]) ou à l'article détaillé à paraître annoncé dans ces Notes; à titre d'exemple nous donnons ci-dessous le cas particulièrement simple de la base 2 où la fonction φ est la distance à l'entier le plus proche (c'est à dire $\langle \cdot \rangle$):

$$F^2(N, S_2^I) = 4\pi^2 \left(\sum_{j=1}^{\infty} \varphi^2(N/b^j)/b - \sum_{j=1}^{\infty} \left(\varphi(N/b^j) \right)^2 /b^2 \right).$$

$$T^2(N, S_2^I) = \left(\sum_{j=1}^{\infty} \varphi(N/b^j) \right)^2 /b^2 + 0(\operatorname{Log} N).$$

Dans le cas d'une base fixe et d'une permutation, l'écriture est formellement la même pour T^2 et elle est légèrement différente pour F^2.

Les valeurs numériques des théorèmes 3.3 et 3.4 représentent les meilleures majorations actuellement connues pour la diaphonie et la discrépance quadratique; nos résultats améliorent et couvrent tous les travaux précédemment effectués sur ce sujet, excepté celui d'un des deux auteurs sur la suite de van der Corput symétrique en base 2.

En conclusion de ces études, on constate, comme c'était déjà le cas pour la discrépance extrême, que les suites de van der Corput se laissent mieux approcher que les suites $(n\alpha)$ et qu'elles ont de plus faibles irrégularités de distribution, particulièrement quand on fait agir des permutations sur les chiffres de la base; il est à noter que jusqu'à présent, la considération de suites à bases variables n'a pas apporté d'amélioration en ce qui concerne les plus faibles discrépances.

Bibliographie

[1] E. Y. Atanassov, Note on the discrepancy of the van der Corput generalized sequences, C.R. Acad. Bulgare Sci., 42, n.3, 1989, p. 41-44.

[2] H. Chaix et H. Faure, Discrépance et diaphonie des suites de van der Corput généralisées, C.R. Acad. Sci. Paris, 310, série I, 1990, p. 315-320.

[3] H. Chaix et H. Faure, Discrépance et diaphonie des suites de van der Corput généralisées (II), C.R. Acad. Sci. Paris, 311, série I, 1990, p. 65-68.

[4] H. Chaix et H. Faure, Discrépance et diaphonie des suites de van der Corput généralisées (III), C.R. Acad. Sci. Paris, 312, Série I, 1991, p. 755-758..

[5] H. Faure, Discrépance de suites associées à un système de numération (en dimension un), Bull. Soc. Math. France, 109, 1981, p. 143-182.

[6] L. Kuipers et H. Niederreiter, Uniform distribution of sequences, Wiley, New York, 1974.

[7] P. D. Proinov, On irregularities of distribution, C.R. Acad. Bulgare Sci., 39, n.9, 1986, p. 31-34.

[8] P .D. Proinov, Symmetrization of the van der Corput generalized sequences, Proc. Japon Acad. Ser. A, 1988, p. 159-162.

[9] P .D. Proinov and E. Y. Atanassov, On the distribution of the van der Corput generalized sequences, C.R. Acad. Sci. Paris, t.307, série I, p. 895-900, 1988.

[10] P. D. Proinov and V. S. Grozdanov, Symmetrization of the van der Corput-Halton sequence, C.R. Acad. Bulgare Sci., 40, n.8, 1987, p. 5-8.

[11] P .D. Proinov and V. S. Grozdanov, On the diaphony of the van der Corput-Halton sequence, J. Number, Theory, 30, 1988, p. 94-104.

[12] Y. J. Xiao, Suites équiréparties associées aux automorphismes du tore, C.R. Acad. Sci. Paris, 311 série I, 1990, p. 579-582.

UFR - MIM et URA 225
Université de Provence
3 Place Victor-Hugo
13331 Marseille cedex 3
France

Une Relation Entre les Points Entiers sur une Courbe Algébrique et les Points Rationnels de la Jacobienne

Noriko Hirata-Kohno

Tokyo Institute of Technology

1 Introduction

Nous donnons ici des minorations de la distance entre des points rationnels distincts d'une variété abélienne. Ces minorations impliquent une majoration de la hauteur des points rationnels de dénominateurs bornés sur une courbe algébrique de genre ≥ 1 (cf. le théorème 3.1). Nos résultats donnent alors une version quantitative du théorème de Siegel sur la finitude des points entiers de telle courbe.

Soient K un corps de nombres de degré fini sur $\mathbb{Q}$ et $\mathcal{C}$ une courbe de genre ≥ 1, définie sur K, complète et non singulière. Considérons les points rationnels de la courbe $\mathcal{C}$ de dénominateurs bornés. Alors on montre que la hauteur de ces points est majorée par une fonction qui dépend de K, $\mathcal{C}$, J en notant J la jacobienne de $\mathcal{C}$, des dénominateurs des points, des plongements de $\mathcal{C}$ et J dans des espaces projectifs, d'une base de l'espace tangent de J à l'origine, d'une base du réseau de périodes de J, du rang et d'un système générateur de groupe de Mordell-Weil de J.

On ne disposait pas, jusqu'à présent, d'une telle estimation de la hauteur des points entiers sur une courbe algébrique de genre ≥ 1, hormis dans certains cas particuliers: courbes d'équations de formes spéciales, courbes de genres 1 [Ba] [Ba-C] [Sch], lorsque J a des multiplications complexes [Mas 1], lorsque l'on fixe une isogénie de J sur un produit de facteurs simples [Be]. De plus, notre majoration est meilleure que celles de [Mas 1] [Be]. Nos estimations sont conséquences d'une minoration de formes linéaires de logarithmes sur les groupes algébriques commutatifs de [H], et l'idée essentielle de la déduction de formes linéaires est originalement montrée par A. Baker, J. Coates, S. Lang et D. W. Masser dans [Ba] [Ba-C] [L 1] [L 2] [Mas 1]. La minoration de [H] que nous utilisons est la meilleure connue pour notre situation, et nous remarquons qu'il est aussi possible d'appliquer l'estimation de [P-W] pour obtenir une majoration de la hauteur des points entiers, ce qui est moins bonne que notre résultat ici. On remarque encore que la finitude des points entiers de la variété abélienne a été démontrée dans [F], sans estimation pour le nombre ni pour la hauteur de tels points. Une majoration pour le nombre des points rationnels d'une courbe algébriqbue de genre > 1 est donnée dans [Bo] [V].

Nous montrons dans le paragraphe 2 des minorations de la distance entre des points rationnels d'une variété abélienne. Le paragraphe 3 donne le théorème principal pour les points rationnels de dénominateurs bornés sur une courbe de genre ≥ 1.

L'auteur remercie infiniment Professeur D. Bertrand et le referee pour les remarques sur ce papier.

2   Points rationnels sur une variété abélienne

(1) Distance

Notons $\overline{\mathbb{Q}}$ la clôture algébrique de $\mathbb{Q}$ dans $\mathbb{C}$. Soit A une variété abélienne de dimension $g \geq 1$, définie sur $\overline{\mathbb{Q}}$, supposée plongée sur $\overline{\mathbb{Q}}$ dans un espace projectif $\mathbb{P}^N$ par un diviseur très ample et pair. On fixe un tel plongement X.

Fixons une base $(t_1, \cdots, t_g)$ définie sur $\overline{\mathbb{Q}}$, de l'espace tangent de A à l'origine de A. Notons $j : \mathbb{C}^g \longrightarrow T_A(\mathbb{C})$ l'isomorphisme donné par $j(z_1, \cdots, z_g) = z_1 t_1 + \cdots + z_g t_g$. On note $\exp_A$ l'application exponentielle de A; alors l'application $X \circ \exp_A \circ j : \mathbb{C}^g \longrightarrow \mathbb{P}^N$ est un homomorphisme analytique dont le noyau Ω est un réseau de $\mathbb{C}^g$. Notons encore cette application $\exp_A$; on voit que $\exp_A$ induit un isomorphisme entre $\mathbb{C}^g/\Omega$ et $A(\mathbb{C})$. Dans le reste, nous identifions $\mathbb{C}^g$ et $T_A(\mathbb{C})$ par j.

Pour $z = (z_1, \cdots, z_g) \in \mathbb{C}^g$, on note $\| \cdot \|$ la norme définie par $\| z \| = (|z_1|^2 + \cdots + |z_g|^2)^{1/2}$ et celle qu'elle induit sur $T_A(\mathbb{C})$ via l'identification par j. On définit $d(P)$, la distance entre $P \in A$ et l'origine O de A, par la formule

$$d(P) := \min\{\| u_P \| \, ; \, \exp_A(u_P) = P\}.$$

Enfin, choisissons $(\omega_1, \cdots, \omega_{2g})$ un système fondamental de périodes dans $\Omega = \mathrm{Ker}\, \exp_A$. On fixe ces éléments $\omega_1, \cdots, \omega_{2g}$ qui forment une $\mathbb{R}$-base de Ω.

(2) Points rationnels

Soit K_0 un corps de nombres sur lequel la variété A, le plongement de A dans $\mathbb{P}^N$ et la base fixée de $T_A(\mathbb{C})$ soient tous définis. Soit K une extension finie de K_0. Notons $A(K)$ l'ensemble des points K-rationnels de A. On sait que $A(K)$, le groupe de Mordell-Weil de A, est un groupe commutatif de type fini (voir par exemple [L 4] p. 138). Soit r le rang de la partie libre de $A(K)$.

(3) Hauteurs

Soit M_K l'ensemble des valeurs absolues inéquivalentes de K, normalisées de sorte que l'on ait pour $x \in \mathbb{Q}, v \in M_K$ et p nombre premier dans $\mathbb{Z}$: $\mid x \mid_v = \max(x, -x)$ si v est archimédienne et $\mid p \mid_v = 1/p$ si v prolonge la valeur absolue p-adique.

On pose encore M_K^∞ l'ensemble des valeurs absolues archimédiennes dans M_K.

Pour $P \in A(K) \subset \mathbb{P}^N(K)$, on note

$$(X_0(P), \cdots, X_N(P)) \in K^{N+1} - \{(0, \cdots, 0)\}$$

un système de coordonnées projectives et on définit $H_K(P)$ par

$$H_K(P) = \prod_{v \in M_K} \max\{\mid X_0(P) \mid_v, \cdots, \mid X_N(P) \mid_v\}^{N_v}$$

où N_v désigne le degré local en v, c'est-à-dire $N_v = [K_v : \mathbb{Q}_v]$. On sait que cette définition est indépendante du choix des coordonnées projectives de $P \in \mathbb{P}^N(K)$ (cf. [Si 2] Chap. 8, §5).

Soit $h(P)$ la hauteur (de Weil) logarithmique absolue; nous avons

$$h(P) = \frac{1}{[K : \mathbb{Q}]} \log H_K(P) \text{ (voir [W] Chap. 1 ou [Si 2] Chap. 8, §6).}$$

Soit q la hauteur de Néron-Tate. On sait que q est une forme quadratique définie positive. De plus, par hypothèse, le plongement fixé de A dans $\mathbb{P}^N$ est associé à un diviseur très ample et pair, on a donc $h(P) = q(P) + f(P)$ pour $P \in A(K)$ où f est une fonction bornée à valeur réelle (voir [L 4] Chap. 5). Cette fonction f satisfait

$$(2.1) \qquad \mid f(P) \mid \leq c_1 \text{ pour tout } P \in A(K)$$

où c_1 est une constante effectivement calculable ne dépendant que de g, du plongement de A et des équations définissant A (voir [Man-Zar] et aussi [Si 1]). Pour tout point de torsion $P \in A(K)_{\text{tors}}$, on a $q(P) = 0$, donc d'après (2.1), on obtient

$$(2.2) \qquad h(P) \leq c_2 \quad (P \in A(K)_{\text{tors}})$$

avec c_2 une constante effective ne dépendant que de g, du plongement de A et des équations définissant A. Ceci entraîne que le choix de générateurs pour la partie de torsion n'offre qu'un nombre fini de possibilité. On déduit facilement de (2.2) une majoration de $\#A(K)_{\text{tors}}$, mais nous utiliserons la majoration plus fine de [Mas 2] qui s'énonce

$$(2.3) \qquad \#A(K)_{\text{tors}} \leq c_3 D^{c_4}$$

où $D = [K : \mathbb{Q}]$ et les constantes c_3, c_4 sont effectivement calculables ne dépendant que de g, du plongement de A et des équations définissant A, mais indépendantes de K. (Il existe des estimations de $\#A(K)_{\text{tors}}$ encore plus raffinées; voir par exemple [Mas 3] p.111, [Se] p.17.)

Remarquons ainsi que l'on peut supposer

$$(2.4) \qquad\qquad m - r \leq 2g,$$

car pour un nombre naturel n, on a $A(K)_{\text{tor}} \subset A_n$ en notant

$$A_n = \{P \in A(\overline{K}) : nP = O\}$$

qui possède $2g$ générateurs $\exp_A(\frac{\omega_i}{n})$ $(1 \leq i \leq 2g)$. On note encore

$$< P, Q > = q(P + Q) - q(P) - q(Q,$$

et, enfin, on pose

$$R(P_1, \cdots, P_r) = \det(< P_i, P_j >, 1 \leq i, j \leq r)$$

pour $(P_1, \cdots, P_r)$ un système générateur de la partie libre de $A(K)$.

(4) Énoncé d'une minoration de $d(P)$

Théorème 2.5 *Soit K_0 un corps de nombres de degré fini sur $\mathbb{Q}$. Soit A une variété abélienne de dimension $g \geq 1$ définie sur K_0, supposée plongée sur K_0 dans un espace projectif $\mathbb{P}^N$ par un diviseur très ample et pair. On fixe ce plongement et encore une base définie sur K_0 de l'espace tangent de A à l'origine. Fixons $(\omega_1, \cdots, \omega_{2g})$ un système fondamental de périodes de Ω. Il existe une constante $C_1 > 0$ effectivement calculable ayant la propriété suivante. Soit K une extension finie de K_0. Soit $P \in A(K)$ un point rationnel. Soit r le rang de la partie libre de $A(K)$. Soit $(P_1, \cdots, P_r, P_{r+1}, \cdots, P_m)$ un système générateur de $A(K)$ où $P_1, \cdots, P_r$ sont indépendants et $P_{r+1}, \cdots, P_m$ sont de torsion. Soient $n_1, \cdots, n_m$ des entiers rationnels tels que le point P s'écrive par rapport au générateur $(P_1, \cdots, P_m)$:*

$$P = n_1 P_1 + \cdots + n_m P_m.$$

Posons $N = \max(\mid n_1 \mid, \cdots, \mid n_r \mid, e^e)$. Soit $u_i \in \mathbb{C}^g$ tel que $\exp_A(u_i) = P_i$ pour chaque $1 \leq i \leq m$. Soient $q_{i,j} (1 \leq i \leq m, 1 \leq j \leq 2g)$ des nombres réels tels que $u_i = q_{i,1}\omega_1 + \cdots + q_{i,2g}\omega_{2g}$. On pose

$$Q = \max_{1 \leq i \leq m, 1 \leq j \leq 2g} \{1, \mid q_{i,j} \mid\}.$$

Posons encore $D = [K : K_0]$. Soient $V_1, \cdots, V_m, V$ des nombres réels vérifiant

$$\log V_i \geq \max(h(P_i), \| \, u_i \, \|^2 / D, \, 1/D) \quad (1 \leq i \leq r)$$

$$\log V_i \geq \max(\| \, u_i \, \|^2 / D, \, 1/D) \quad (r < i \leq m)$$

$$V = \max V_i \quad (1 \leq i \leq m)$$

Alors ou bien $P = O$, *ou bien on a*

$$\log d(P) \; > \; -C_1 D^{2g(m+g)+2}(\log(NQ) + \log(D \log V))$$
$$\times (\log \log(NQ) + \log(D \log V))^{g(m+2g)+1} \left(\prod_{i=1}^{m} (\log V_i)^g \right).$$

Remarque 2.6: Sans expliciter la dépendance en u_i $(1 \leq i \leq m)$, si $P \neq O$, on a avec (2.4):

$$\log d(P) > -C_2 D^{2g(3g+r)+2}(\log N + \log D)(\log \log N + \log D)^{g(4g+r)+1}.$$

Remarque 2.7: La constante C_1 ne dépend que de g, m, r, K_0, de la base de l'espace tangent de A (donc des coefficients des équations différentielles que vérifient les fonctions thêta), du plongement de A dans l'espace projectif, des éléments $\omega_1, \cdots, \omega_{2g}$ et des équations définissant A. En fait, m est majoré dans (2.4). On peut théoriquement expliciter C_1 grâce aux travaux [D].

(5) Démonstration du théorème 2.5

Pour $u \in \mathbb{C}^g$ vérifiant $\exp_A(u) = P$, on a $u \equiv n_1 u_1 + \cdots + n_m u_m$ (mod Ω) car l'application $\exp_A$ est un homomorphisme. Posons

$$U \; = \; D^{2g(m+g)+2}(\log(NQ) + \log(D \log V))$$
$$\times (\log \log(NQ) + \log(D \log V))^{g(m+2g)+1} \left(\prod_{i=1}^{m} (\log V_i)^g \right).$$

On note $c_5, c_6, \cdots$ des nombres réels positifs qui ne dépendent que des données, mais indépendants des paramètres D, Q, N, V_i $(1 \leq i \leq m)$.

Nous supposons pour $P \neq O$ la négation de l'énoncé, c'est-à-dire $d(P) \leq \exp(-C_1 U)$, ainsi on a $d(P) < 1$.

Considérons la période $\omega_0 \in \Omega$ telle que

$$d(P) = \min_{\omega \in \Omega} \| \, n_1 u_1 + \cdots + n_m u_m - \omega \, \| = \| \, n_1 u_1 + \cdots + n_m u_m - \omega_0 \, \|$$

et notons

$$\omega_0 = s_1 \omega_1 + \cdots + s_{2g} \omega_{2g}$$

pour $s_1, \cdots, s_{2g} \in \mathbb{Z}$. Posons

$$S = \max_{1 \leq j \leq 2g} | \, s_j \, |.$$

Supposons par exemple $S = |\ s_1\ |$. Comme on a $\boldsymbol{u}_i = q_{i,1}\boldsymbol{\omega}_1 + \cdots + q_{i,2g}\boldsymbol{\omega}_{2g}$, on obtient

$$n_1\boldsymbol{u}_1 + \cdots + n_m\boldsymbol{u}_m = \sum_{1 \le j \le 2g}\ \sum_{1 \le i \le m} n_i q_{i,j}\boldsymbol{\omega}_j,$$

donc on en déduit

$$|\sum_{1 \le i \le m} n_i q_{i,1} - s_1\ |\ \|\ \boldsymbol{\omega}_1\ \| \le c_5\ \|\ n_1\boldsymbol{u}_1 + \cdots + n_m\boldsymbol{u}_m - \boldsymbol{\omega}_0\ \| \le c_5 d(P) \le c_5$$

avec c_5 dépendant des $\boldsymbol{\omega}_1 \cdots, \boldsymbol{\omega}_{2g}$.

En conséquence, on a $|\ s_1\ |\ \|\ \boldsymbol{\omega}_1\ \| \le c_5 + |\ \sum_{1 \le i \le m} n_i q_{i,1}\ |\ \|\ \boldsymbol{\omega}_1\ \|$, d'où $S \le c_6(1 + mQ\max\{N, c_3 D^{c_4}\})$ où c_6 dépend des $\boldsymbol{\omega}_1, \cdots, \boldsymbol{\omega}_{2g}$. Comme $P \ne O$, $\boldsymbol{u}_i \ne 0$ $(1 \le i \le r)$ et $\boldsymbol{\omega}_j \ne 0$ $(1 \le j \le 2g)$, on va utiliser le théorème 2.1 et la remarque 2.9 de [H] pour minorer $\|\ n_1\boldsymbol{u}_1 + \cdots + n_m\boldsymbol{u}_m - s_1\boldsymbol{\omega}_1 - \cdots - s_{2g}\boldsymbol{\omega}_{2g}\ \|$ en remarquant dans ce théorème que l'on peut remplacer l'hypothèse (2.6) par $L(\boldsymbol{v}) \ne 0$ car $L(\boldsymbol{v}) \ne 0$ implique (2.6). Selon les notations du théorème 2.1 de [H], considérons

$$k = m + 2g,\ G = \mathbb{G}_a \times A^{m+2g},\ d = d_2 = g(m + 2g), E = e,$$

$$\boldsymbol{v} = (1, \boldsymbol{u}_1, \cdots, \boldsymbol{u}_m, \boldsymbol{\omega}_1, \cdots, \boldsymbol{\omega}_{2g}).$$

Pour chaque k, $1 \le k \le g$, on note la k-ième coordonnée du nombre $n_1\boldsymbol{u}_1 + \cdots + n_m\boldsymbol{u}_m - s_1\boldsymbol{\omega}_1 - \cdots - s_{2g}\boldsymbol{\omega}_{2g}$ par $L_k(\boldsymbol{v})$ où L_k est la forme linéaire sur T_G à $g(m+2g)+1$ coefficients choisis dans $0, n_1, \cdots, n_m, -s_1, \cdots, -s_{2g}$.

Maintenant, on modifie le théorème 2.1 de [H] de la manière suivante. D'abord, on distingue K et K_0 comme suit. Dans §2(1) et §3(1) de [H], on introduit $G_{-1}, G_0, \cdots, G_k$ définis sur K_0, en supposant, pour chaque $-1 \le i \le k$, que le plongement de G_i que l'on fixe dans un espace projectif et la base que l'on fixe de T_{G_i} sont définis sur K_0. Dans §4(8) de [H], on prend $\beta_0, \cdots, \beta_d, \beta_{d+1}, \cdots, \beta_M$ des générateurs de K sur K_0 où $\beta_0, \cdots, \beta_d$ provenant des coefficients de la forme linéaire, et on considère une base de K sur K_0 $\{\eta_1 = 1, \eta_2, \cdots, \eta_D\}$ formée d'éléments de l'ensemble

$$\left\{\prod_{i=0}^{M} \beta_i^{a_i}; 0 \le a_i < [K_0(\beta_i) : K_0], 0 \le i \le M, a_0 + \cdots + a_M < D\right\}.$$

Notons c_7 le degré de K_0 sur $\mathbb{Q}$. Soit $\{\zeta_1, \cdots, \zeta_{c_7}\}$ une base de K_0 sur $\mathbb{Q}$ et soit $\{\xi_1, \cdots, \xi_{c_7 D}\}$ une base de K sur $\mathbb{Q}$ formée d'éléments $\eta_i\zeta_j$; $1 \le i \le D$, $1 \le j \le c_7\}$. Alors on a la modification de la propriété 4.19 de [H] qui nous montre $h(1, \xi_1, \cdots, \xi_{c_7 D}) \le C_0^{-3}U_0/c_7 D$ et $\operatorname{Max}\{\log|\ \xi_i\ |; 1 \le i \le c_7 D\} \le C_0^{-3}U_0$ par les propriétés de la hauteur logarithmique et un choix de C_0. On peut remplacer encore D par $c_7 D$ dans le lemme 4.20, dans le lemme 4.31, et dans la proposition 4.24. Alors ces modifications dissus nous permettent de remplacer les premières 8 lignes du théorème 2.1 de [H] par les suivantes. Il existe une constante $C_1' > 0$ effectivement calculable ne dépendant que des données et des choix précédents, ayant la propriété suivante. Soit $\mathcal{L}(z) = \beta_0 z_0 + \cdots + \beta_d z_d$ une forme linéaire non nulle à coefficients dans $\overline{\mathbb{Q}}$, sur l'espace tangent $T_{\mathfrak{g}}(\mathbb{C})$. Notons $\mathcal{W}$ son noyau. Pour chaque $1 \leq i \leq k$, soit $\boldsymbol{u}_i \in T_{G_i}(\mathbb{C})$, $\boldsymbol{u}_i \neq 0$ tel que $\gamma_i = \exp_{G_i}(\boldsymbol{u}_i)$ appartient à $G_i(\overline{\mathbb{Q}})$. Soit K le corps engendré sur K_0 par $\beta_0, \cdots, \beta_d$ et toutes les coordonnées projectives des $\gamma_1, \cdots, \gamma_k$. Notons $\boldsymbol{v} = (1, \boldsymbol{u}_1, \cdots, \boldsymbol{u}_k) \in T_{\mathfrak{g}}(\mathbb{C})$ et $D = [K : K_0]$. Soient $B, E, V_1, \cdots, V_k$ des nombres réels vérifiant ....

Ensuite, on justifie la remarque 2.9 de [H]. Dans la démonstration du théorème 2.1 de [H], les endroits où on utilise l'hypothèse (2.2) de ce théorème sont seulement les deux suivants; la démonstration de la propriété 4.9 et celle de la proposition 4.18. Si l'on modifie les paramètres L_{-1}, L_0 et T dans la démonstration du théorème de [H] selon la remarque 2.9 de [H] avec $E = e$, on peut alors supprimer l'hypothèse (2.2) en énonçant l'estimation de la remarque (2.9), car cette modification établit encore la propriété 4.9 et la proposition 4.18.

Ainsi on peut adapter ici le théorème 2.1 de [H] avec les deux modifications dessus. Grâce à (2.3), on a $\log \max(|n_{r+1}|, \cdots, |n_m|) < c_8 \max(1, \log D)$. En remarquant l'estimation (2.2) de $h(P_j)$ pour $r < j \leq m$, on en déduit pour un i $(1 \leq i \leq g)$; $\log |L_i(\boldsymbol{v})| > -c_9^{(i)}U$, ce qui entraîne

$$\| n_1 \boldsymbol{u}_1 + \cdots + n_m \boldsymbol{u}_m - s_1 \boldsymbol{\omega}_1 - \cdots - s_{2g} \boldsymbol{\omega}_{2g} \| > \exp(-c_{10}U)$$

d'où une contradiction pour $C_1 > c_{10}$.

(6) Minoration en termes de hauteurs

Maintenant, on présente une version du théorème 2.5 en exprimant la minoration par la hauteur du point rationnel.

Théorème 2.8 *Soit K_0 un corps de nombres de degré fini sur $\mathbb{Q}$. Soit A une variété abélienne de dimension $g \geq 1$ définie sur K_0, supposée plongée sur K_0 dans un espace projectif $\mathbb{P}^N$ par un diviseur très ample et pair. On fixe ce plongement et encore une base définie sur K_0 de l'espace tangent de A à l'origine. Fixons $(\boldsymbol{\omega}_1, \cdots, \boldsymbol{\omega}_{2g})$ un système fondamental de périodes de Ω. Il existe une constante $C_3 > 0$ ayant la propriété suivante. Soit K une extension finie de K_0. Soit r le rang de la partie libre de $A(K)$. Soit $(P_1, \cdots, P_r, P_{r+1}, \cdots, P_m)$ un système générateur de $A(K)$ où $P_1, \cdots, P_r$ sont indépendants et $P_{r+1}, \cdots, P_m$ sont de torsion. Soit $P \in A(K)$ un point rationnel de hauteur logarithmique absolue $h(P) \leq \log H$ avec $\log \log \log H \geq 1$. Soit $\boldsymbol{u}_i \in \mathbb{C}^g$ tel que $\exp_A(\boldsymbol{u}_i) = P_i$ pour chaque $1 \leq i \leq m$. Soient $q_{i,j}$ $(1 \leq i \leq m, 1 \leq j \leq 2g)$ des nombres réels tels que $\boldsymbol{u}_i = q_{i,1} \boldsymbol{\omega}_1 + \cdots + q_{i,2g} \boldsymbol{\omega}_{2g}$. On pose*

$$Q = \max_{1 \leq i \leq m, 1 \leq j \leq 2g} \{1, |q_{i,j}|\}.$$

Posons encore $D = [K : K_0]$. *Soient* $V_1, \cdots, V_m, V$ *des nombres réels vérifiant*

$$log\, V_i \geq \max(h(P_i),\ \|u_i\|^2 / D,\ 1/D) \quad (1 \leq i \leq r)$$

$$log\, V_i \geq \max(\|u_i\|^2 / D,\ 1/D) \quad (r < i \leq m)$$

$$V = \max V_i \quad (1 \leq i \leq m).$$

On pose encore

$$Y = \max\left(e, \log\left(\frac{\prod_{i=1}^r q(P_i)}{R(P_1, \cdots, P_r) \cdot \min_{1 \leq i \leq r} q(P_i)}\right)\right).$$

Alors ou bien $P = O$, *ou bien on a*

$$\log d(P) \;>\; -C_3 D^{2g(m+g)+2} (\log\log H + \log Q + \log(D \log V) + Y)$$
$$\times (\log\log\log H + \log\log Q + \log(D \log V) + \log Y)^{g(m+2g)+1}$$
$$\times \prod_{i=1}^m (\log V_i)^g.$$

Remarque 2.9: La constante C_3 est effectivement calculable en fonction de g, m, r, K_0, de la base de $T_A(\mathbb{C})$, du plongement que nous prenons de A dans $\mathbb{P}^N$, du choix de $\omega_1, \cdots, \omega_{2g}$ et des équations définissant A. Notons que $m \leq 2g + r$ dans (2.4). On a une majoration pour le rang r dans [O-T].

(7) Démonstration du théorème 2.8

On note c_{11}, c_{12} des nombres réels positifs qui ne dépendent que des données, mais sont indépendants des paramètres $D, H, Q, Y, V_1, \cdots, V_m$. Soient $n_1, \cdots, n_m \in \mathbb{Z}$ tels que $P = n_1 P_1 + \cdots + n_m P_m$ par rapport au générateur $(P_1, \cdots, P_m)$. D'après le §2(3), il existe une forme quadratique définie positive q et une fonction f vérifiant $h(P) = q(P) + f(P)$. Posons $N_1 = \max(|n_1|, \cdots, |n_r|)$. Sans perte de généralité, on suppose $N_1 = |n_1|$. On note $R = R(P_1, P_2, \cdots, P_r)$ et $\rho = R(P, P_2, \cdots, P_r)$. Comme N_1 est l'indice du réseau engendré par $P, P_2, \cdots, P_r$ dans celui engendré par $P_1, P_2, \cdots, P_r$, on a $\rho/R = N_1^2$. D'autre part, d'après l'inégalité de Hadamard, on obtient

$$\rho \leq 2^r q(P) q(P_2) \cdots q(P_r).$$

Ainsi on en déduit

$$q(n_1 P_1 + \cdots + n_r P_r) \geq \frac{R N_1^2}{2^r q(P_2) \cdots q(P_r)} \geq \frac{R N_1^2 \min_{1 \leq i \leq r} q(P_i)}{2^r \prod_{1 \leq i \leq r} q(P_i)}.$$

(voir [L 4] Chap. 5, §7 et aussi [Si 2] p.274).

Comme $h(n_1 P_1 + \cdots + n_r P_r) \le h(P) + h(n_{r+1}P_{r+1} + \cdots + n_m P_m) + \log 2$, grâce à (2.1) et (2.2), on a $q(n_1 P_1 + \cdots + n_r P_r) \le \log H + c_{11}$. En conséquence, on obtient $\log N_1 \le c_{12}(\log\log H + Y)$ où c_{12} ne dépend que de g, r, du plongement de A et des équations définissant A. Le théorème 2.5 fournit l'énoncé pour $P \ne O$.

3 Points rationnels de dénominateurs bornés sur une courbe de genre ≥ 1

(1) Points rationnels d'une courbe

Soit K un corps de nombres. Soit $\mathcal{C}$ une courbe de genre $g \ge 1$, définie sur K, complète, non singulière, supposée plongée dans un espace projectif $\mathbb{P}^n$ par un plongement défini sur K. Posons $\mathcal{C}(K) = \mathcal{C} \cap \mathbb{P}^n(K)$ l'ensemble des points K-rationnels de $\mathcal{C}$. Pour $(X_0, \cdots, X_n) \in \mathbb{P}^n$, quitte à changer des coordonnées, on peut supposer que la courbe $\mathcal{C}$ n'est pas contenue dans l'hyperplan $X_n = 0$. Considérons un point rationnel $P = (X_0(P), \cdots, X_{n-1}(P), 1) \in \mathcal{C}(K)$. Notons $H_K(P), h(P)$ les hauteurs définies comme dans le §2(3), associées au plongement $\mathcal{C} \subset \mathbb{P}^n$ que l'on prend. Nous désignons une mesure de P par

$$\mu_K(P) = \prod_{v \in M_K^\infty} \max\{|X_0(P)|_v, \cdots, |X_{n-1}(P)|_v, 1\}^{N_v}$$

et nous définissons encore le dénominateur de P par

$$\delta_K(P) = \prod_{v \in M_K \setminus M_K^\infty} \max\{|X_0(P)|_v, \cdots, |X_{n-1}(P)|_v, 1\}^{N_v}.$$

Posons $\Delta_K(P) = \max(\delta_K(P), e^{e^e})$ et $H_K^*(P) = \max(H_K(P), e^{e^e})$. On a évidemment $H_K(P) = \mu_K(P) \cdot \delta_K(P)$, donc

$$h(P) = \frac{1}{[K:\mathbb{Q}]}(\log \mu_K(P) + \log \delta_K(P)).$$

(2) Jacobiennes

Soit J la jacobienne de la courbe $\mathcal{C}$, qui est une variété abélienne, et soit ψ une application rationnelle $\mathcal{C} \longrightarrow J$. Si $\mathcal{C}(K) \ne \emptyset$ alors on sait que ψ et J sont définis sur K (voir le théorème 8 dans Chap. II, §2 de [L 3]). Supposons $\mathcal{C}(K) \ne \emptyset$ (sinon, le théorème est vide). On suppose que J est plongé dans un espace projectif $\mathbb{P}^N$ par un plongement défini sur K, correspondant à un diviseur très ample et pair. Posons $J(K)$ l'ensemble des points K-rationnels de J, associés à ce plongement fixé. Pour un point P_∞ à l'infini de $\mathcal{C}$, on a $\psi(P_\infty) \in J(\overline{\mathbb{Q}})$ (cf. [Mas 1] p.562). Soient ψ_i ($0 \leq i \leq N$) des fonctions rationnelles de $\mathcal{C}$, définies sur $\overline{\mathbb{Q}}$ telles que

$$
\begin{aligned}
\psi(P) &= (\psi_0(P), \cdots, \psi_N(P)) \\
&= (\psi_0(X_0(P), \cdots, X_{n-1}(P), 1), \cdots, \psi_N(X_0(P), \cdots, X_{n-1}(P), 1)).
\end{aligned}
$$

Soit t un paramètre local sur $\mathcal{C}$ au P_∞, tel que ψ_i ($0 \leq i \leq N$) soient holomorphes en t au point P_∞ et pour tout $0 \leq i \leq n-1$, X_i se développe en une série de Laurent en t avec au moins un indice j, $0 \leq j \leq n-1$, X_j étant celle de Laurent qui commence par la partie d'exposant négatif.

(3) Énoncé du théorème pour les points rationnels

Voici le résultat principal dans ce papier.

Théorème 3.1 *Soit K un corps de nombres. Soit $\mathcal{C}$ une courbe de genre $g \geq 1$, définie sur K, complète, non singulière, supposée plongée dans un espace projectif $\mathbb{P}^n$ par un plongement défini sur K. On fixe ce plongement. Soit J la jacobienne de la courbe, supposée plongée dans un espace projectif $\mathbb{P}^N$ par un plongement défini sur K, correspondant à un diviseur très ample et pair. On fixe ce plongement. Soit ψ une application rationnelle $\mathcal{C} \to J$. On suppose $\mathcal{C}(K) \neq \emptyset$.*

Notons $(P_1, \cdots, P_r, P_{r+1}, P_m)$ un système générateur du groupe de Mordell-Weil $J(K)$, où $P_1, \cdots, P_r$ sont indépendants et $P_{r+1}, \cdots, P_m$ sont de torsion, en désignant par r le rang de $J(K)$. Supposons $\psi(P_\infty) \in J(K)$ pour tout point P_∞ à l'infini de $\mathcal{C}$. Enfin, on fixe aussi une base de l'espace tangent de J en l'origine, qui est supposée définie sur K, et on fixe un système fondamental du réseau de périodes de J.

Il existe une constante $C_4 > 0$ dépendant des données et des choix, ayant la propriété suivante. Pour tout point rationnel $P \in \mathcal{C}(K)$ tel que $X_n(P) \neq 0$, on a

$$
h(P) < C_4\{\log \delta_K(P) + \log\log \Delta_K(P) \cdot (\log\log\log \Delta_K(P))^{g(m+2g)+1}\}.
$$

Remarque 3.2: La constante C_4 dépend de K, de l'application ψ, du plongement que l'on prend de $\mathcal{C}$ dans $\mathbb{P}^n$, du plongement que nous fixons de J dans $\mathbb{P}^N$, des équations définissant $\mathcal{C}$ et J, du genre g, du choix de la base de l'espace tangent de J, du système fondamental que nous prenons du réseau de périodes de J et du rang et d'un système générateur du groupe de Mordell-Weil $J(K)$. On a une majoration de m dans (2.4).

Corollaire 3.3 *Si nous prenons $\delta_K(P) = 1$ dans le théorème 3.1, alors on a une majoration pour la hauteur des points entiers sur la courbe $\mathcal{C}$. D'où on déduit la finitude de tels points dans le théorème de Siegel.*

(4) Démonstration du théorème 3.1

On emploie l'argument dans [Mas 1]. Soit $P \in \mathcal{C}(K)$ avec $X_n(P) \neq 0$. Pour tout $v \in M_K^\infty$, si $\max\{|X_0(P)|_v, \cdots, |X_{n-1}(P)|_v, 1\}$ est borné, il n'y a rien à démontrer. Ainsi on suppose que le nombre

$$\max_{v \in M_K^\infty} \{|X_0(P)|_v, \cdots, |X_{n-1}(P)|_v, 1\}$$

est suffisamment grand. D'abord considérons le cas où $\max\{|X_0(P)|, \cdots, |X_{n-1}(P)|, 1\}$ est suffisamment grand. Ceci montre que le point P est proche d'un point P_∞ à l'infini de $\mathcal{C}$. Comme on a $\psi(P_\infty) \in J(K)$ par hypothèse, quitte à composer à ψ une translation par un point de $J(K)$, on suppose $\psi(P_\infty) = O$ en notant O l'origine de la jacobienne J. On a donc $\psi(P) \in J(K)$. On considère maintenant la distance $d(\psi(P))$ sur J concernant la norme $\|\cdot\|$ sur $T_J(\mathbb{C})$ comme dans le §2(1). Selon la notation du §3(2), si $P \neq P_\infty$ est suffisamment proche du point P_∞ tel que les séries de Laurent X_i convergent à P, alors $\psi(P)$ est très proche de l'origine O et on obtient en notant t un paramètre local sur $\mathcal{C}$ au P_∞ choisi comme dans le §3(2),

$$(3.4) \qquad \log d(\psi(P)) \leq c_{14} \log |t|.$$

Posons

$$U^* = (\log \log H_K^*(P))(\log \log \log H_K^*(P))^{g(m+2g)+1}.$$

Nous utilisons ici le théorème 2.8 pour déduire

$$(3.5) \qquad \log d(\psi(P)) > -c_{15} U^*.$$

En combinant les deux inégalités (3.4) (3.5), on a

$$(3.6) \qquad \log |t| > -c_{16} U^*.$$

Par définition de t, en mettant la minoration (3.6) dans la série de Laurent $X_j(P)$ que l'on a prise, nous obtenons

$$(3.7) \qquad \max\{|X_0(P)|, \cdots, |X_{n-1}(P)|, 1\} < \exp(c_{17} U^*).$$

On peut montrer la même estimation que (3.7) pour tout $v \in M_K^\infty$ en considérant les conjugués de K, par conséquent,

$$\mu_K(P) < \exp(c_{18} U^*)$$

ce qui implique:

$$\mu_K(P) < \exp\{c_{19}(\log \log \Delta_K(P))(\log \log \log \Delta_K(P))^{g(m+2g)+1}\}$$

d'où l'énoncé.

Remarque 3.8: On peut donc théoriquement obtenir tous les points entiers de la courbe elliptique dès que l'on connaît une majoration du rang et encore un système générateur du groupe de Mordell-Weil (cf. [Si 2] p. 262 et [Zag]).

Références

[Ba] Baker, A.: Transcendental Number Theorey (Cambridge Math. Library series), Cambridge Univ. Press, Cambridge New York, 1990

[Ba-C] Baker, A., Coates, J.: Integer points on curves of genus 1. Proc. Camb. Philos. Soc. **67**, 595-602 (1970)

[Be] Bertrand, D.: La théorie de Baker revisitée. Dans "Problèmes Diophantiens 1984/85", Publ. Univ. Paris VI **73**, exposé II, 1-25.

[Bo] Bombieri, E.: The Mordell conjecture revisited. Preprint.

[D] David, S.: Théorie de Baker pour des familles de groupes algébriques commutatifs. Thèse, Univ. Paris VI, 1989, pp. 32-63

[F] Faltings, G.: Diophantine approximation on abelian varieties. Annals of Math. **133**, 549-576 (1991)

[H] Hirata-Kohno, N.: Formes linéaires de logarithmes de points algébriques sur les groupes algébriques. Invent. Math. **104**, 401-433 (1991)

[L 1] Lang, S.: Diophantine approximation on toruses. Amer. J. Math. **86**, 521-533 (1964)

[L 2] Lang, S.: Higher dimensional Diophantine problems. Bulletin Amer. Math. Soc., **80** No. 5, 779-787 (1974)

[L 3] Lang, S.: Abelian varieties, Springer, Berlin Heidelberg New York, 1983

[L 4] Lang, S.: Fundamentals of Diophantine Geometry, Springer, Berlin Heidelberg New York, 1983

[Man-Zar] Manin, Ju. I., Zarhin, Ju. G.: Height on families of abelian varieties. (In Russian) Math. Sbornik **89**, 171-181 (1972), (English translation) Math. USSR Sbornik **18**, No. 2, 169-179 (1972)

[Mas 1] Masser, D. W.: Linear forms in algebraic points of abelian functions III. Proc. London Math. Soc. 33, 549-564 (1976)

[Mas 2] Masser, D. W.: Small values of the quadratic part of the Néron-Tate height. Sém. de Théorie des Nombres, Paris 1979/80, Progr. Math. **12**, Birkhäuser, Boston, 1981, pp. 213-222

[Mas 3] Masser, D. W.: Small values of heights on families of abelian varieties. In: Wüstholz G. (ed.) Diophantine approximation and transcendence theory (Lect. Notes Math. **1290**), Springer, Berlin Heidelberg New York, 1986, pp. 109-148

[O-T] Ooe, T., Top, J.: On the Mordell-Weil rank of an abelian variety over a number field. J. Pure and Applied Algebra **58**, 261-265 (1989)

[P-W] Philippon, P., Waldschmidt, M.: Formes linéaires de logarithmes sur les groupes algébriques commutatifs. Illinois J. Math., **32**, 281-314 (1988)

[Se] Serre, J.-P.: Lectures on the Mordell-Weil theorem, Vieweg, Braunschweig Wiesbaden, 1989

[Sch] Schmidt, W.M.: Integer points on curves of genus 1. Compositio Math. **81**, 33-59 (1992)

[Si 1] Silverman, J. H.: Heights and the specialization map for families of abelian varieties. J. Reine Angew. Math. **342**, 197-211 (1983)

[Si 2] Silverman, J. H.: The arithmetic of elliptic curves (GTM **106**), Springer, Berlin Heidelberg New York, 1986

[V] Vojta, P.: Siegel's theorem in the compact case. Annals of Math. **133**, 509-548 (1991)

[W] Waldschmidt, M.: Nombres transcendants et groupes algébriques, Astérisque **69/70** (1979)

[Zag] Zagier, D.: Large integral points on elliptic curves. Math. Computation **48**, No. 177. 425-436 (1987)

Noriko Hirata-Kohno
Department of Mathematics
Tokyo Institute of Technology
Oh-Okayama, Meguro, Tokyo 152
Japan
e-mail hirata@math.titech.ac.jp

Problèmes de Favard et de Lehmer Generalisés

Michel Langevin

Université de Lille

Abstract

Let $(z_1, \cdots, z_d)$ be a complete set of conjugates on a number field K, where z_1 is an algebraic integer. When $K = \mathbf{Q}$, the inequalities

$$\sup |z_i - z_j| \geq \sqrt{3} \quad (d > 1) \quad \text{and} \quad \sup |z_i - z_j| > 2 - \varepsilon \quad (d > D(\varepsilon))$$

are known. In this work, we describe how to extend these results in the case of any number field and how to introduce the problem in the context of the generalized Lehmer's problem ; moreover, we give some improvements in the case of self-inversive polynomials.

1 Notations et introduction

Soit K un corps de nombres. On note $Z(K)$ l'anneau des entiers de K, $MI(K)$ l'ensemble des polynômes unitaires et irréductibles à coefficients dans $Z(K)$, $S(K)$ l'ensemble de cardinal $(K : \mathbf{Q})$ des plongements de K dans le corps des complexes $\mathbf{C}$. Pour tout élément s de $S(K)$ et tout polynôme P à coefficients dans K, on note P_s le polynôme à coefficients complexes images par s de ceux de P et $e(P_s) = \dim(P_s^{-1}(0))$ l'écart maximal entre les zéros dans $\mathbf{C}$ de P_s. Enfin, on pose

$$e(P) = \sup_s e(P_s), \qquad e'(P) = (\prod_s e(P_s))^{1/(K:\mathbf{Q})}$$

Il est clair que les quantités $e(P)$ et $e'(P)$ définies ci-dessus ne dépendent que du corps engendré par les coefficients de P.

Avec ces notations, on associe à tout corps de nombres K, la fonction $F_1(K)$ (resp. $F_2(K)$) égale à la borne inférieure (resp. la limite inférieure quand d_0 tend vers l'infini) des $e(P)$ quand P décrit l'ensemble des éléments de $MI(K)$ de degré > 1 (resp. $> d_0$). On définit de même à partir de $e'(P)$ des fonctions F_1' et F_2'. En considérant les discriminants des divers polynômes P_s d'une part, et des polynômes n'ayant pour zéros que des racines de l'unité d'autre part, on vérifie aisément que

$$1 \leq F_1'(K) \leq \inf(F_1(K), F_2'(K)) \leq \sup(F_1(K), F_2'(K)) \leq F_2(K) \leq 2 .$$

De plus, si K est un sous-corps d'un corps de nombres L, il est clair que

$$F_1(L) \leq F_1(K), \ F_1'(L) \leq F_1'(K), \ F_2(L) \leq F_2(K), \ F_2'(L) \leq F_2'(K) \ .$$

Le but de l'exposé est de décrire les propriétés des fonctions F_1, F_2, F_1', F_2' introduites ci-dessus qui correspondent aux problèmes de Favard généralisés et de montrer, dans le cas des fonctions F_1, F_2, comment cette étude s'insère dans la généralisation naturelle du problème de Lehmer. On renvoie à [3] pour le détail des démonstrations.

2 Rappels sur les problèmes de Favard et problèmes de Favard généralisés

Si $K = \mathbf{Q}$ ou si K est quadratique imaginaire, les fonctions e et e', F_1 et F_1', F_2 et F_2' coïncident. Dans ces cas, on sait qu'on a (cf. Comptes-Rendus des Conférences de Québec (1987,[1]) et Banff (1988, [2]) où l'on trouvera de plus toutes les références bibliographiques) : $F_1(\mathbf{Q}) = \sqrt{3}$ et, si K est quadratique imaginaire, $F_2(K) = F_2(\mathbf{Q}) = 2$.

De plus, cette dernière égalité, dans le cas quadratique imaginaire, est équivalente, via la théorie de Fekete et Szegö, à l'inégalité

$$(T) \qquad\qquad\qquad 2t(X) \leq \operatorname{diam}(X)$$

pour toute partie compacte X du plan complexe, $t(X)$ désignant le diamètre transfini de X et $\operatorname{diam}(X)$ le diamètre usuel.

Les égalités précédentes sur $F_1(\mathbf{Q})$ et $F_2(\mathbf{Q})$ résolvent respectivement les premier et second problèmes de Favard. On va montrer dans cet exposé que l'introduction des fonctions $e(P)$ et $e'(P)$ (suivant une technique souvent utilisée par Jacques Martinet et à sa suggestion) permet d'étendre la notion d'écart maximal entre conjugués d'un entier algébrique et donc de généraliser les problèmes de Favard. Le théorème suivant résout le second :

Théorème 1. *Pour tout corps de nombres K, $F_2'(K) = F_2(K) = 2$. Plus précisément, on montre, pour tout réel $\varepsilon > 0$, l'existence d'un rang $D(\varepsilon)$ (fonction de ε seulement) tel que, pour tout polynôme P élément d'un $MI(K)$ et de degré au moins $D(\varepsilon)$, l'on ait $e(P) \geq e'(P) > 2 - \varepsilon$.*

Le théorème 1 se déduit d'un lemme technique valable pour tout polynôme à coefficients complexes permettant de majorer le discriminant de ce polynôme en fonction de son degré, du module de son coefficient dominant et du diamètre de l'ensemble de ses zéros dans $\mathbf{C}$. Ce lemme technique non trivial est lui-même conséquence de (T) de sorte que le théorème 1 s'insère dans la suite d'implications ci-dessous (où l'expression "Th. 1 faible" signifie qu'il existe un corps quadratique imaginaire K vérifiant $F_2'(K) = F_2(K) = 2$) :

$$(T) \Rightarrow (\text{Théorème 1}) \Rightarrow (\text{Th. 1 faible}) \Rightarrow (T)$$

Le théorème 2 suivant résout partiellement le premier problème de Favard généralisé :

Théorème 2. *(i)* $F_1(\mathbf{Q}(\sqrt{d})) = 1$ *(resp.* $\leq \sqrt{2}$, $\leq \sqrt{3}$) *si* $d \equiv -1$ *(resp.* $-2, -3$) *modulo* 4.

(ii) Tout corps quadratique admet une extension K *de degré au plus 2 vérifiant* $F_1(K) = 1$.

Le problème est donc en fait de caractériser les corps K pour lesquels $F_1(K) = 1$; autrement dit, d'après le théorème de Kronecker, de rechercher les éléments P de $MI(K)$ pour lesquels $\mathrm{disc}(P)$ est une racine de 1 (autre que 1).

3 Apport des théorèmes 1 et 2

Les théorèmes 1 et 2 ne sont pas seulement des généralisations des résultats résolvant les problèmes de Favard. Ils apportent en effet des renseignements sur les diverses factorisations des polynômes de $MI(\mathbf{Q})$. On a en effet la proposition suivante :

Proposition 1. *Soit* P *(resp.* U*) le polynôme minimal d'un entier algébrique* x *sur un corps de nombres* K *(resp.* $\mathbf{Q}$*). On a alors :*

$$\prod_{s \in S(K)} P_s = U^{(K(x):\mathbf{Q}(x))}$$

Exemple : x racine de 1 d'ordre 12, $U(X) = X^4 - X^2 + 1$ peut se factoriser en

$$\begin{aligned}(X^2 + \sqrt{3}X + 1)(X^2 - \sqrt{3}X + 1) &= (X^2 + iX - 1)(X^2 - iX - 1) \\ &= (X^2 + j)(X^2 + j^2).\end{aligned}$$

avec $j = \exp(2i\pi/3)$. On obtient $e(U) = 2$ et respectivement $1, \sqrt{3}, 2$ pour $e(P)$ suivant la factorisation retenue.

4 Problème de Lehmer généralisé

On rappelle qu'une "hauteur" sur les polynômes éléments de $MI(K)$ est une fonction de cet ensemble à valeurs réelles pour laquelle tout ensemble d'éléments de $MI(K)$ de hauteur et de degré bornés est fini (exemple : $K = \mathbf{Q}$, hauteur d'un polynôme = borne supérieure des modules de ses coefficients). Le problème de Lehmer généralisé est le problème de l'étude de la répartition des hauteurs des éléments de $MI(K)$ (le cas classique étant celui où la hauteur est la mesure de Mahler et où $K = \mathbf{Q}$).

En travaillant non sur $MI(K)$ mais sur son quotient où l'on ne distingue pas $P(X)$ et $P(X + h)$, h désignant un quelconque élément de $Z(K)$, on peut montrer :

Proposition 2. *La fonction associant $e(P)$ à la classe de P est une hauteur (contrairement à $e'(P)$).*

Remarque : Cette distinction n'avait pas lieu d'être dans les cas étudiés antérieurement ($K = \mathbf{Q}$ et K quadratique imaginaire) où la démonstration était évidente puisque $Z(K)$ était alors discret.

Le problème de Favard généralisé s'insère donc dans le cadre général du problème de Lehmer dont on sait que le cas classique est résolu pour les polynômes non réciproques (cf. [5]). Curieusement, quand $K = \mathbf{Q}$, le premier problème de Favard a été résolu d'abord pour les polynômes réciproques (et sans faire usage de tables de discriminants (cf. [4])). Comme pour le théorème 1, on peut dégager le lemme technique sous-jacent où n'intervient aucune hypothèse arithmétique ; c'est le théorème 3 suivant où le deuxième terme de l'alternative ne peut être vérifié quand le polynôme réciproque P appartient à $MI(\mathbf{Q})$:

Théorème 3. *Soit P un polynôme de degré $d > 1$ à coefficients réels vérifiant $P(X) = \pm X^d P(1/X)$. On a alors :*

$$\operatorname{diam} P^{-1}(0) \geq \sqrt{3} \quad ou \quad \inf(|(P(1)|, |P(-1)|) < |P(0)| \,.$$

Références

1. M. Langevin, E. Reyssat, G. Rhin : Transfinite diameter and Favard's problems on diameters of algebraic integers, *dans Number Theory, ed. by J.M. de Koninck and C. Levesque, 1989, de Gruyter, p. 574-578.*

2. M. Langevin : Systèmes complets de conjugués sur les corps quadratiques imaginaires et ensembles de largeur constante, *dans Number Theory and Applications, ed. by R.A. Mollin, NATO-ASI Series, 1989, Kluwer, p. 445-457.*

3. M. Langevin : Problèmes de Favard pour les corps de nombres, à *paraître.*

4. C.W. Lloyd-Smith, Problems on the distribution of conjugates of algebraic numbers, Ph. D., Univ. of Adelaide, 1980 - voir aussi P.E. Blanksby, C.W. Lloyd-Smith, M.J.McAuley : On diameters of algebraic integers, *Acta Arith., LII, 1989, p.1-9.*

5. C. Smyth : On the product of the conjugates outside the unit circle of an algebraic integer, *Bull. London Math. Soc., 3, 1971, p. 169-175.*

Université de Lille
UFR de Mathématiques F. 59655
Villeneuve d'Ascq – Cedex
France

On the v-adic independence of algebraic numbers

Damien Roy

Université Laval,Quebéc

1   Introduction

Let $\bar{\mathbf{Q}}$ be an algebraic closure of $\mathbf{Q}$, let $\bar{\mathbf{Q}}^\times$ be its multiplicative group, and let v be a place of $\mathbf{Q}$. If v is infinite, we denote by $\mathcal{U}_v$ the group $\bar{\mathbf{Q}}^\times$. Otherwise, we denote by $\mathcal{U}_v$ the subgroup of $\bar{\mathbf{Q}}^\times$ consisting of the elements of $\bar{\mathbf{Q}}^\times$ which are prime to v. A conjecture of J.-F. Jaulent, based on Schanuel's conjecture, allows an arithmetic description of the v-adic rank of the finitely generated subgroups of $\mathcal{U}_v$ which are stable under the Galois group of $\bar{\mathbf{Q}}$ over $\mathbf{Q}$ (Theorem 2 of [2]). This conjecture implies Leopoldt's conjecture. It has been proved for those subgroups contained in an abelian extension of $\mathbf{Q}$ (Theorem 3 of [2]), and in some other situations. In fact, the transcendence methods apply just as well to Jaulent's conjecture as they apply to Leopoldt's conjecture (see [3]).

We present here a conjecture which gives an arithmetic description of the v-adic rank of the finitely generated subgroups of $\mathcal{U}_v$, without Galois hypothesis. Our conjecture implies Jaulent's conjecture and, as we show in §3, it is a consequence of Schanuel's conjecture. We deduce from it a (conjecturally) necessary and sufficient condition for the v-adic independence of a finite family of elements of $\mathcal{U}_v$. The study of this condition brings us to formulate a general result concerning modules over algebras which are direct products of skew fields. We show in §4 that, if for some integer $m \geq 1$ our conjecture is true for the finitely generated subgroups of $\mathcal{U}_v$ of rank $\leq m$ which are contained in a Galois extension of $\mathbf{Q}$ of degree $\leq 3^{m^4}$, then it is true for all finitely generated subgroups of $\mathcal{U}_v$ of rank $\leq m$. In §5, we consider the subgroups of $\mathcal{U}_v$ which are stable under the Galois group of $\bar{\mathbf{Q}}$ over $\mathbf{Q}$. We show that Leopoldt's conjecture for dihedral extensions of $\mathbf{Q}$ of degree 6, 8 and 12 is equivalent to the conjectural statement: any set of two independent units is v-adically independent.

The author thanks professor M. Waldschmidt for having suggested him the problem of extending Jaulent's conjecture. He also thanks professor C. Levesque for having been invited to continue his researches at Université

Laval, and for financial support. Some of the results which are collected here can be found in the thesis of the author [4].

2 Preliminaries

Given a field K, we denote by $K^\times$ its multiplicative group. If V is a vector space over a field K, and S a subset of V, we denote by KS the subspace of V generated by S over K.

Let v be a place of $\mathbf{Q}$. We denote by $\mathbf{C}_v$ the completion of $\bar{\mathbf{Q}}$ with respect to an absolute value extending a v-adic absolute value of $\mathbf{Q}$. We write D_v for the disk of convergence of the exponential series in $\mathbf{C}_v$, and $\exp_v \colon D_v \to \mathbf{C}_v^\times$ for the group homomorphism defined by this series, D_v being viewed as a subgroup of $\mathbf{C}_v$. Finally, we define L_v as the $\mathbf{Q}$-subspace of $\mathbf{C}_v$ generated by $\exp_v^{-1}(\bar{\mathbf{Q}}^\times)$.

Schanuel's conjecture for logarithms says that elements of L_v which are linearly independent over $\mathbf{Q}$ are algebraically independent over $\mathbf{Q}$. A consequence of this conjecture, known as the four exponentials conjecture, can be stated as follows: if V is a subspace of $\mathbf{C}_v^2$ such that $V \cap \mathbf{Q}^2 = 0$, then $\dim_{\mathbf{Q}}(V \cap L_v^2) \leq 1$. Beside this last conjecture, we have the following result (Theorem 2 of [1]) obtained by M. Emsalem on the basis of a transcendence result of M. Waldschmidt; it will be needed in §4:

Theorem 1. *Let m be a positive integer and let V be a subspace of $\mathbf{C}_v^m$. If $V \cap \mathbf{Q}^m = 0$, then we have $\dim_{\mathbf{Q}}(V \cap L_v^m) \leq m(m-1)$.*

We introduce the following definition:

Definition 2. *Let N be a finitely generated subgroup of $\mathcal{U}_v$. A section of $\exp_v$ on N is a group–homomorphism $\lambda \colon N \to \mathbf{C}_v$ such that $\exp_v \circ \lambda$ is defined and coincide with the identity on a subgroup of N of finite index.*

It is easy to show that, if N is a finitely generated subgroup of $\mathcal{U}_v$, there exists a least one section of $\exp_v$ on N and that the kernel of such a section is the torsion-subgroup of N. If v is a finite place corresponding to a prime p, such a section is unique and coincides with the restriction to N of the p-adic logarithm $\log_p \colon \mathbf{C}_p^\times \to \mathbf{C}_p$ as defined in Proposition 5.4 of [10]. If v is the infinite place of $\mathbf{Q}$, the various sections of $\exp_v$ over N differ by an element of $\mathrm{Hom}_{\mathbf{Z}}(N, \mathbf{Q}\pi i)$.

Let M be a finitely generated subgroup of $\mathcal{U}_v$, and let N be the subgroup of $\mathcal{U}_v$ generated by the elements of M and their conjugates. Let K be a Galois extension of $\mathbf{Q}$ of finite degree which contains N, and let G be its Galois group. Since N is stable under G, it is a $\mathbf{Z}[G]$-submodule of $K^\times$.

Let $\lambda: N \to \mathbf{C}_v$ be a section of $\exp_v$ on N. We choose a generating system $\alpha_1, \ldots, \alpha_m$ of the group M and we form the matrix

$$\left(\lambda(\sigma \alpha_i) \right)_{\sigma \in G; \ 1 \le i \le m} \tag{2.1}$$

Its rank does not depend on the choice of K nor on the choice of the α_i's. We define the *v-adic rank* of M as the minimum of the ranks of the matrices (2.1) associated with the various sections λ of $\exp_v$ on N; we denote it by $\mathrm{rk}_v(M)$. The following properties are clear:

Proposition 3. *If M is a finitely generated subgroup of $\mathcal{U}_v$, it satisfies $\mathrm{rk}_v(M) \le \mathrm{rk}(M)$. Moreover, we have $\mathrm{rk}_v(M) \ge 1$ if $\mathrm{rk}(M) \ge 1$. If $M_1 \subset M$ are finitely generated subgroups of $\mathcal{U}_v$, they satisfy $\mathrm{rk}_v(M_1) \le \mathrm{rk}_v(M)$ with equality if $\mathrm{rk}(M_1) = \mathrm{rk}(M)$. Moreover, we have $\mathrm{rk}_v(M) < \mathrm{rk}(M)$ if $\mathrm{rk}_v(M_1) < \mathrm{rk}(M_1)$.*

Remark. M. Waldschmidt has defined the notion of v-adic rank in algebraic groups [9]; in our case the group which arises is $\mathbf{G}_m$. For the infinite place v of $\mathbf{Q}$, our definition of the v-adic rank differs somewhat from his. Nevertheless, it is easy to check that, if Schanuel's conjecture is true, both definitions are equivalent.

We say that elements $\alpha_1, \ldots, \alpha_m$ of $\mathcal{U}_v$ are *v-adically independent* if the v-adic rank of the subgroup of $\mathcal{U}_v$ which they generate is m.

We recall the statement of Leopoldt's and Jaulent's conjectures (§5.5 of [10] and §2 of [2]):

Leopoldt's conjecture. *If E_k is the unit group of a number field K, then we have $\mathrm{rk}_v(E_k) = \mathrm{rk}(E_k)$.*

Jaulent's conjecture. *Let M be a finitely generated subgroup of $\mathcal{U}_v$ which is stable under the Galois group of $\bar{\mathbf{Q}}$ over $\mathbf{Q}$. Then, we have $\mathrm{rk}_v(M) = \mathrm{rk}(M)$ if and only if M contains an element which, together with its conjugates, generates a subgroup of finite index of M.*

It would suffice to prove Leopold's conjecture for Galois number fields. Since such a field contains a Minkowski's unit, we see that Jaulent's conjecture implies Leopoldt's conjecture. We emphasize that the latter is true for v infinite, while this is not known for the former.

3 The v-adic rank under Schanuel's conjecture

Let v be a place of $\mathbf{Q}$ and let M be a finitely generated subgroup of $\mathcal{U}_v$. Our aim is to give an arithmetical description of the v-adic rank of M, assuming Schanuel's conjecture. To this end, we consider the subgroup N of $\mathcal{U}_v$ generated by the elements of M and their conjugates. We denote by K the extension of $\mathbf{Q}$ generated by N, and by G its Galois group.

Conjecture. The v-adic rank of M is the largest integer r for which there exists a $\mathbf{Q}[G]$-homomorphism ϕ from $\mathbf{Q} \otimes_{\mathbf{Z}} N$ to $\mathbf{Q}[G]$ such that $\dim_{\mathbf{Q}}(\phi(\mathbf{Q} \otimes_{\mathbf{Z}} M)) = r$.

Theorem 4. *Let r be the integer attached to M by the conjecture. We have $\mathrm{rk}_v(M) \leq r$ with an equality if Schanuel's conjecture is true.*

Proof. Let $\lambda \colon N \to \mathbf{C}_v$ be a section of $\exp_v$ on N, as in Definition 2. We denote by $\lambda_{\mathbf{Q}} \colon \mathbf{Q} \otimes_{\mathbf{Z}} N \to \mathbf{C}_v$ the $\mathbf{Q}$-linear map for which $\lambda_{\mathbf{Q}}(1 \otimes \alpha) = \lambda(\alpha)$ $(\alpha \in N)$, and we consider the $\mathbf{Q}[G]$-homomorphism $\Phi \colon \mathbf{Q} \otimes_{\mathbf{Z}} N \to \mathbf{C}_v[G]$ given by

$$\Phi(x) = \sum_{\sigma \in G} \lambda_{\mathbf{Q}}(\sigma x) \sigma^{-1} \qquad (x \in \mathbf{Q} \otimes_{\mathbf{Z}} N).$$

The v-adic rank of M satisfies

$$\mathrm{rk}_v(M) \leq \dim_{\mathbf{C}_v}(\mathbf{C}_v \Phi(\mathbf{Q} \otimes_{\mathbf{Z}} M)). \tag{3.1}$$

By the definition of $\mathrm{rk}_v(M)$, both sides of (3.1) are equal for at least one choice of λ.

Let $n = \dim_{\mathbf{Q}}(\mathbf{Q} \otimes_{\mathbf{Z}} N)$. Since $\lambda_{\mathbf{Q}}$ is injective, it can be written as a sum

$$\lambda_{\mathbf{Q}} = t_1 \lambda_1 + \cdots + t_n \lambda_n$$

for a basis $t_1, \ldots, t_n$ of $\lambda_{\mathbf{Q}}(\mathbf{Q} \otimes_{\mathbf{Z}} N)$ over $\mathbf{Q}$, and a basis $\lambda_1, \ldots, \lambda_n$ of $\mathrm{Hom}_{\mathbf{Q}}(\mathbf{Q} \otimes_{\mathbf{Z}} N, \mathbf{Q})$ over $\mathbf{Q}$. For $i = 1, \ldots, n$, we let $\phi_i \colon \mathbf{Q} \otimes_{\mathbf{Z}} N \to \mathbf{Q}[G]$ be the $\mathbf{Q}[G]$-homomorphism given by

$$\phi_i(x) = \sum_{\sigma \in G} \lambda_i(\sigma x) \sigma^{-1} \qquad (x \in \mathbf{Q} \otimes_{\mathbf{Z}} N).$$

Then, $\phi_1, \ldots, \phi_n$ constitute a basis of $\mathrm{Hom}_{\mathbf{Q}[G]}(\mathbf{Q} \otimes_{\mathbf{Z}} N, \mathbf{Q}[G])$ over $\mathbf{Q}$, and we get the decomposition

$$\Phi = t_1 \phi_1 + \cdots + t_n \phi_n.$$

Let $A_1, \ldots, A_n$ be the matrices of the restrictions of $\phi_1, \ldots, \phi_n$ to $\mathbf{Q} \otimes_{\mathbf{Z}} M$, with respect to a choice of basis of $\mathbf{Q} \otimes_{\mathbf{Z}} M$ and $\mathbf{Q}[G]$ over $\mathbf{Q}$. The integer r attached to M by the conjecture is

$$r = \max\{\mathrm{rk}(u_1 A_1 + \cdots + u_n A_n) \; ; \; (u_1, \ldots, u_n) \in \mathbf{Q}^n\},$$

while

$$\dim_{\mathbf{C}_v}(\mathbf{C}_v \Phi(\mathbf{Q} \otimes_{\mathbf{Z}} M)) = \mathrm{rk}(t_1 A_1 + \cdots + t_n A_n).$$

From these equalities, we deduce

$$\dim_{\mathbf{C}_v}(\mathbf{C}_v \Phi(\mathbf{Q} \otimes_{\mathbf{Z}} M)) \leq r. \qquad (3.2)$$

Combining (3.1) and (3.2), we get $\mathrm{rk}_v(M) \leq r$.

If we assume Schanuel's conjecture, the elements $t_1, \ldots, t_n$ which belong to L_v are algebraically independent over $\mathbf{Q}$ since they are linearly independent over $\mathbf{Q}$. Then, we get an equality in (3.2) and, choosing λ to have an equality in (3.1), we obtain $\mathrm{rk}_v(M) = r$.

Remark 1. If $X_1, \ldots, X_n$ are indeterminates over $\mathbf{Q}$, a result of M. Waldschmidt gives $\mathrm{rk}(t_1 A_1 + \cdots + t_n A_n) \geq \frac{1}{2} \mathrm{rk}(X_1 A_1 + \cdots + X_n A_n)$ (see the proof of Corollary 2.2.p of [8]). The right member of this inequality is simply $\frac{1}{2} r$. Thus, choosing λ to have an equality in (3.1), we get $\mathrm{rk}_v(M) \geq \frac{1}{2} r$.

Remark 2. The proof shows that, under Schanuel's conjecture, the left member of (3.2) does not depend on the choice of λ : any section $\lambda: N \to \mathbf{C}_v$ of $\exp_v$ can be used to compute $\mathrm{rk}_v(M)$.

Corollary. If Schanuel's conjecture is true, the following conditions are equivalent:
(i) $\mathrm{rk}_v(M) = \mathrm{rk}(M)$;
(ii) there exists a $\mathbf{Q}[G]$-homomorphism $\phi: \mathbf{Q} \otimes_{\mathbf{Z}} N \to \mathbf{Q}[G]$ which is injective on $\mathbf{Q} \otimes_{\mathbf{Z}} M$.

If M is stable under G, the condition (ii) of Corollary amounts to the existence of an injective $\mathbf{Q}[G]$-homomorphism from $\mathbf{Q} \otimes_{\mathbf{Z}} M$ to $\mathbf{Q}[G]$. By Theorem 1 of [2], such a homomorphism exists if and only if $\mathbf{Q} \otimes_{\mathbf{Z}} M$ is a principal $\mathbf{Q}[G]$-module. We recover in this way Jaulent's conjecture.

The following theorem gives a necessary and sufficient condition for the condition (ii) of Corollary to hold, when the semi-simple $\mathbf{Q}$-algebra $\mathbf{Q}[G]$ is a direct product of skew fields; it applies for example when G is abelian.

Theorem 5. *Let k be a field of infinite cardinality, A a semi-simple k-algebra of finite dimension, N a finitely generated A-module, and M a k-subspace of N. Assume that A is a direct product of skew fields. Then, the following conditions are equivalent:*
(i) there exists an A-homomorphism $\phi: N \to A$ which is injective on M;
(ii) for each two-sided ideal I of A, we have $\dim_k(M \cap IN) \leq \dim_k(I)$.

Proof of the implication $(i) \Rightarrow (ii)$. Let $\phi \in \mathrm{Hom}_A(N, A)$. For each two-sided ideal I of A, we have $\phi(M \cap IN) \subset IA = I$. If ϕ is injective on M, this implies $\dim_k(M \cap IN) \leq \dim_k(I)$.

To prove the reverse implication, we use the following results, the first one being an easy consequence of Proposition 3 of [5] :

Lemma 6. *Let k be a field of infinite cardinality, let U, V be vector spaces of finite dimension over k, and let T be a k-subspace of $\operatorname{Hom}_k(U, V)$. Endow T with the Zariski topology of vector spaces of finite dimension over k. Let U_1 be a k-subspace of U, and let $\mathcal{O}_1$ be the set of all $\phi \in T$ for which $\dim_k(\phi(U_1))$ is maximum. Then, $\mathcal{O}_1$ is a Zariski open set of T. Moreover, for each $\phi \in \mathcal{O}_1$ and each $\psi \in T$, we have $\psi(\ker(\phi) \cap U_1) \subset \phi(U_1)$.*

Lemma 7. *Let A be a semi-simple algebra of finite dimension over a field k, let N be a finitely generated A-module, and let E be a k-subspace of N. Then, there is a smallest right ideal I of A such that $E \subset IN$: it is the sum of the k-subspaces $\psi(E)$ of A for all $\psi \in \operatorname{Hom}_A(N, A)$.*

Proof of Lemma 7. Let I be the sum of the k-subspaces $\psi(E)$ of A for all $\psi \in \operatorname{Hom}_A(N, A)$. It is clear that this is a right ideal of A. Since A is semi-simple, there exists an integer $n \geq 1$ and A-homomorphisms $j: N \to A^n$, $\pi: A^n \to N$ such that $\pi \circ j$ is the identity on N. By looking at the projections of $j(E)$ on each factor of A^n, we get $j(E) \subset I^n$. Since I is a right ideal of A, we have $I^n = IA^n$, and so we get $E \subset \pi(IA^n) = IN$. Finally, if a right ideal J of A satisfies $E \subset JN$, we have $\psi(E) \subset J$ for each $\psi \in \operatorname{Hom}_A(N, A)$, and so I is contained in J. This shows that I has the required minimal property.

Proof of the implication $(ii) \Rightarrow (i)$. Assume that (ii) is satisfied. The set $\operatorname{Hom}_A(N, A)$ is a k-subspace of $\operatorname{Hom}_k(N, A)$. We endow it with the Zariski topology of vector spaces of finite dimension over k. Let S be the set of right ideals of A. Since A is a direct product of skew fields, these are two-sided ideals of A, and the set S is finite. For each $I \in S$, we denote by $\mathcal{O}_I$ the set of all $\phi \in \operatorname{Hom}_A(N, A)$ for which $\dim_k(\phi(M \cap IN))$ is maximal. By Lemma 6, the sets $\mathcal{O}_I$ ($I \in S$) are Zariski open sets of $\operatorname{Hom}_A(N, A)$. Since S is finite, their intersection $\mathcal{O}$ is not empty (it is even dense).

Let $\phi \in \mathcal{O}$, and let $E = \ker(\phi) \cap M$. By Lemma 7, there exists a smallest ideal I in S such that $E \subset IN$. Since $\phi \in \mathcal{O}_I$, Lemma 6 gives

$$\psi(E) \subset \phi(M \cap IN)$$

for all $\psi \in \operatorname{Hom}_A(N, A)$. By Lemma 7 and the choice of I, this implies

$$I \subset \phi(M \cap IN).$$

Since by hypothesis we have $\dim_k(M \cap IN) \leq \dim_k(I)$, this in turn implies $\ker(\phi) \cap (M \cap IN) = 0$. We thus get $E = 0$, showing that condition (i) is satisfied for our choice of ϕ.

4 Case of rank m

Let v be a place of $\mathbf{Q}$, and let M be a finitely generated subgroup of $\mathcal{U}_v$. We have the following results :

Theorem 8. *Suppose* $\mathrm{rk}(M) = 2$. *If there exists a subgroup of* M *of finite index which is contained in* $\mathbf{Q}^\times$ *or in the group of elements of norm 1 of a quadratic number field, then* $\mathrm{rk}_v(M) = 1$. *Otherwise, if we assume the four exponentials conjecture,* $\mathrm{rk}_v(M)$ *is 2.*

Theorem 9. *Suppose* $\mathrm{rk}(M) = 2$ *and* $\mathrm{rk}_v(M) = 1$. *Then, there exists a subgroup of* M *of finite index which is contained in a Galois extension of* $\mathbf{Q}$ *whose group is cyclic of order* n *or dihedral of order* $2n$ *with* $n = 1, 2, 3, 4$ *or 6.*

Theorem 10. *Let* m *be the rank of* M. *Suppose* $\mathrm{rk}_v(M) < m$. *Suppose also* $\mathrm{rk}_v(M') = \mathrm{rk}(M')$ *for all subgroups* M' *of* M *with* $\mathrm{rk}(M') < m$. *Then, there exists a subgroup of* M *of finite index which is contained in a Galois extension of* $\mathbf{Q}$ *whose group is isomorphic with a subgroup of* $\mathrm{GL}_{m(m-1)}(\mathbf{Z}/3\mathbf{Z})$.

Before proving these theorems we give a corollary :

Corollary. If the conjecture of §3 is true for any finitely generated subgroup of $\mathcal{U}_v$ of rank $\leq m$ which is contained in a Galois extension of $\mathbf{Q}$ of degree $\leq 3^{m^4}$, then it is true for any finitely generated subgroup of $\mathcal{U}_v$ of rank $\leq m$.

Proof. Suppose that the conjecture of §3 is false for some finitely generated subgroup M of $\mathcal{U}_v$ of rank $\leq m$. Let N be the subgroup of $\mathcal{U}_v$ generated by the elements of M and their conjugates, let K be the extension of $\mathbf{Q}$ generated by N, and let G be the Galois group of K over $\mathbf{Q}$. Then, there exists a $\mathbf{Q}[G]$-homomorphism $\phi \colon \mathbf{Q} \otimes_{\mathbf{Z}} N \longrightarrow \mathbf{Q}[G]$ and a subgroup M_1 of M with $\mathrm{rk}(M_1) > \mathrm{rk}_v(M)$ such that ϕ is injective on $\mathbf{Q} \otimes_{\mathbf{Z}} M_1$. Since $\mathrm{rk}_v(M) \geq \mathrm{rk}_v(M_1)$, we get $\mathrm{rk}(M_1) > \mathrm{rk}_v(M_1)$. Let M_2 be a subgroup of M_1 of smallest rank satisfying $\mathrm{rk}(M_2) > \mathrm{rk}_v(M_2)$. By Theorem 10, there exists a subgroup M_3 of M_2 of finite index which is contained in a Galois extension of $\mathbf{Q}$ of degree $\leq 3^{m^4}$. Since $\mathrm{rk}(M_3) > \mathrm{rk}_v(M_3)$ and since ϕ is injective on $\mathbf{Q} \otimes_{\mathbf{Z}} M_3$, M_3 does not satisfy the conjecture of §3.

Proof of Theorems 8 to 10. Let m be the rank of M, and let M_0 be a subgroup of M of finite index such that the subgroup N_0 of $\mathcal{U}_v$ generated by the elements of M_0 and their conjugates is torsion-free. We denote by K_0 the extension of $\mathbf{Q}$ generated by the elements of N_0, and by G_0 its Galois group. Let $\alpha_1, \ldots, \alpha_m$ be a basis of M_0, and let A be the subgroup of N_0^m generated by the m-tuples $(\sigma\alpha_1, \ldots, \sigma\alpha_m)$ with $\sigma \in G_0$. The group

G_0 acts on A componentwise. This action is faithful since an element of G_0 which fixes all elements of A fixes also all elements of N_0 and so fixes all elements of K_0. More generally, we see that, as $\mathbf{Z}[G_0]$-modules, A and N_0 have the same annihilator.

Let $\lambda \colon N_0 \to \mathbf{C}_v$ be a section of $\exp_v$ on N_0 for which the subgroup Y of $\mathbf{C}_v{}^m$ generated by the m-tuples

$$\big(\lambda(\sigma\alpha_1), \ldots, \lambda(\sigma\alpha_m)\big) \qquad \text{with } \sigma \in G_0$$

satisfies

$$\mathrm{rk}_v(M) = \mathrm{rk}_v(M_0) = \dim_{\mathbf{C}_v}(\mathbf{C}_v Y). \tag{4.1}$$

We have $Y \subset L_v^m$. Moreover, since N_0 is torsion-free, λ is injective on N_0 and the product map $\lambda^m \colon N_0^m \to \mathbf{C}_v{}^m$ induces, by restriction, a group isomorphism from A to Y. In particular, A and Y have the same rank.

If M_0 is contained in $\mathbf{Q}^\times$ or in the group of elements of norm 1 of a quadratic field, then the rank of A is ≤ 1. We then have $\mathrm{rk}(Y) \leq 1$, and the relation (4.1) implies $\mathrm{rk}_v(M) \leq 1$, with an equality if $\mathrm{rk}(M) \neq 0$. This proves the first half of Theorem 8.

Assume, from now on, $m \geq 2$ and $\mathrm{rk}_v(M) < m$. Assume also $\mathrm{rk}_v(M') = \mathrm{rk}(M')$ for all subgroups M' of M with $\mathrm{rk}(M') < m$. This last assumption is automatically satisfied if $m = 2$. It implies

$$(\mathbf{C}_v Y) \cap \mathbf{Q}^m = 0. \tag{4.2}$$

Indeed, suppose on the contrary that (4.2) is false. Then, $\mathbf{C}_v Y$ contains the kernel of a surjective linear map $s \colon \mathbf{C}_v{}^m \to \mathbf{C}_v{}^{m-1}$ satisfying $s(\mathbf{Z}^m) \subset \mathbf{Z}^{m-1}$. Letting $Y' = s(Y)$, we thus get $\dim_{\mathbf{C}_v}(\mathbf{C}_v Y') < m - 1$. But, as a group, Y' is generated by the $(m-1)$-tuples $(\lambda(\sigma\alpha_1'), \ldots, \lambda(\sigma\alpha_{m-1}'))$ with $\sigma \in G_0$, for a basis $\alpha_1', \ldots, \alpha_{m-1}'$ of a subgroup M' of M of rank $m - 1$. Since $\mathrm{rk}_v(M') \leq \dim_{\mathbf{C}_v}(\mathbf{C}_v Y')$, we get $\mathrm{rk}_v(M') < \mathrm{rk}(M')$, in contradiction with the last assumption on M.

If $m = 2$ and if we assume the four exponentials conjecture, the relation (4.2) implies $\mathrm{rk}(Y) = 1$; so $\mathrm{rk}(A)$ is 1. Since A is torsion-free, $\mathrm{Aut}_{\mathbf{Z}}(A)$ is a cyclic group of order 2 generated by the action of -1. Since G_0 acts faithfully on A, we deduce that G_0 is a cyclic group of order ≤ 2, and that, if G_0 is generated by an element σ of order 2, then $\sigma + 1$ annihilates A and so, annihilates N_0. Thus, K_0 is an extension of $\mathbf{Q}$ of degree ≤ 2 and, if $K_0 \neq \mathbf{Q}$, M_0 is contained in the group of elements of K_0 of norm 1. This proves the second part of Theorem 8.

By Theorem 1, the relation (4.2) implies $\mathrm{rk}(Y) \leq m(m-1)$. We thus get $\mathrm{rk}(A) \leq m(m-1)$. Since A is a free $\mathbf{Z}$-module on which G_0 acts faithfully, we deduce that G_0 is isomorphic with a subgroup of $\mathrm{GL}_{m(m-1)}(\mathbf{Z})$.

If $m = 2$, G_0 is isomorphic with a finite subgroup of $\mathrm{GL}_2(\mathbf{Z})$. We deduce that G_0 is cyclic of order n or dihedral of order $2n$ with $n = 1, 2,$

3, 4 or 6. Indeed, let H be a subgroup of $\mathrm{GL}_2(\mathbf{Z})$ which is isomorphic to G_0. In $\mathrm{GL}_2(\mathbf{R})$, H is conjugate to a finite group H' consisting of orthogonal transformations of $\mathbf{R}^2$. This group H' is composed of rotations and symmetries. The first ones constitute a cyclic subgroup H'_1 of H' of order n for some n. The symmetries, if there are any, form a coset of H'_1. Thus, H is cyclic of order n or dihedral of order $2n$. Let σ be a generator of H'_1. Since σ is conjugate to an element of H, its minimal polynomial has integer coefficients. Since this polynomial divides $X^n - 1$ but does not divide $X^m - 1$ for any proper divisor m of n, it should be divisible by the nth cyclotomic polynomial; and since its degree is ≤ 2, this implies $n = 1$, 2, 3, 4 or 6. Theorem 9 is proved.

In general, a theorem of Minkowski (Theorem 3, §12 of [7]) shows that, for any integer $n \geq 1$, the homomorphism of reduction modulo 3 from $\mathrm{GL}_n(\mathbf{Z})$ to $\mathrm{GL}_n(\mathbf{Z}/3\mathbf{Z})$ is injective on the torsion-subgroup of $\mathrm{GL}_n(\mathbf{Z})$. Thus, G_0 is isomorphic to a subgroup of $\mathrm{GL}_{m(m-1)}(\mathbf{Z}/3\mathbf{Z})$, which proves Theorem 10.

5 Galois case

In this section, v denotes a fixed place of $\mathbf{Q}$.

Theorem 11. *Let M be the subgroup of $\mathcal{U}_v$ generated by an element α and its conjugates. Let K be the extension of $\mathbf{Q}$ generated by M, and let G be its Galois group. Assume that each irreducible linear representation of G over $\mathbf{C}$ is realizable over $\mathbf{Q}$ and of degree $\leq d$. Then, if $\mathrm{rk}_v(M) < \mathrm{rk}(M)$, there exists a subgroup M' of M with $\mathrm{rk}_v(M') < \mathrm{rk}(M') \leq d$.*

Proof. Let $\lambda \colon M \to \mathbf{C}_v$ be a section of $\exp_v$ on M for which the $\mathbf{C}_v[G]$-homomorphism $\Phi \colon \mathbf{C}_v \otimes_{\mathbf{Z}} M \to \mathbf{C}_v[G]$ given by

$$\Phi(1 \otimes x) = \sum_{\sigma \in G} \lambda(\sigma x)\sigma^{-1} \qquad (x \in M)$$

satisfies

$$\mathrm{rk}_v(M) = \dim_{\mathbf{C}_v}(\Phi(\mathbf{C}_v \otimes_{\mathbf{Z}} M)).$$

For each subgroup M' of M, we have $\mathrm{rk}_v(M') \leq \dim_{\mathbf{C}_v}(\Phi(\mathbf{C}_v \otimes_{\mathbf{Z}} M'))$. In particular, we have $\mathrm{rk}_v(M') < \mathrm{rk}(M')$ if Φ is not injective on $\mathbf{C}_v \otimes_{\mathbf{Z}} M'$.

Suppose $\mathrm{rk}_v(M) < \mathrm{rk}(M)$. We then have $\ker(\Phi) \neq 0$. Let W be a simple submodule of $\ker(\Phi)$. By hypothesis, W is isomorphic to $\mathbf{C}_v \otimes_{\mathbf{Q}} I$ for a simple left ideal I of $\mathbf{Q}[G]$ of dimension $\leq d$. Let V be the sum of the submodules of $\mathbf{Q} \otimes_{\mathbf{Z}} M$ which are isomorphic to I, and let s be the multiplicity of I in V. Since $\mathbf{Q} \otimes_{\mathbf{Z}} M$ is a principal $\mathbf{Q}[G]$-module, generated by $1 \otimes \alpha$, we have $s \leq d$. Write

$$V = I x_1 \oplus \ldots \oplus I x_s$$

for elements $x_1, \ldots, x_s$ of $\mathbf{Q} \otimes_{\mathbf{Z}} M$, and let us identify $\mathbf{C}_v \otimes_{\mathbf{Z}} M$ with $\mathbf{C}_v \otimes_{\mathbf{Q}} (\mathbf{Q} \otimes_{\mathbf{Z}} M)$. Since W is a submodule of $\mathbf{C}_v \otimes_{\mathbf{Z}} M$ which is isomorphic to $\mathbf{C}_v \otimes_{\mathbf{Q}} I$, we have $W \subset \mathbf{C}_v \otimes_{\mathbf{Q}} V$, and so

$$W = (\mathbf{C}_v \otimes_{\mathbf{Q}} I). \sum_{i=1}^{s} a_i \otimes x_i$$

for elements $a_1, \ldots, a_s$ of $\mathbf{C}_v$. Let x_0 be a non-zero element of I, and let M' be a subgroup of M such that

$$\mathbf{Q} \otimes_{\mathbf{Z}} M' = \mathbf{Q} x_0 x_1 \oplus \ldots \oplus \mathbf{Q} x_0 x_s.$$

Since $\mathbf{C}_v \otimes_{\mathbf{Z}} M'$ contains $\sum_{i=1}^{s} a_i \otimes (x_0 x_i)$, we have $W \cap (\mathbf{C}_v \otimes_{\mathbf{Z}} M') \neq 0$. This implies $\mathrm{rk}_v(M') < \mathrm{rk}(M')$. The rank of M' being $s \leq d$, M' has the required properties.

Corollary. If one of the following statements is true, the other is true:
(i) two multiplicatively independent units are v-adically independent;
(ii) for each dihedral extension K of $\mathbf{Q}$ of degree 6, 8 or 12, the group E_K of units of K satisfies $\mathrm{rk}_v(E_K) = \mathrm{rk}(E_K)$.

Proof. Assume (i) is false. Let M be a subgroup of rank 2 of $\mathcal{U}_v$ generated by two units of $\bar{\mathbf{Q}}$, for which $\mathrm{rk}_v(M) = 1$. By Theorem 9, there exists a subgroup of M of finite index which is contained in the group of units E_K of a Galois extension K of $\mathbf{Q}$, whose Galois group G is cyclic of order n or dihedral of order $2n$ with $n = 1, 2, 3, 4$ or 6. This implies $\mathrm{rk}_v(E_K) < \mathrm{rk}(E_K)$. Since Leopoldt's conjecture is true for abelian extensions of $\mathbf{Q}$, the group G must be dihedral of order 6, 8 or 12. Thus, (ii) is false.

Assume (ii) is false. Let K be a dihedral extension of $\mathbf{Q}$ of degree $2n$ with $n = 3, 4$ or 6, whose group of units E_K satisfies $\mathrm{rk}_v(E_K) < \mathrm{rk}(E_K)$. We claim that E_K contains a subgroup M' of rank 2 with $\mathrm{rk}_v(M') = 1$, and thus (i) is false. To show this, we first observe that E_K contains a subgroup M of finite index which is generated by a Minkowski's unit and its conjugates. Then, in virtue of Theorem 11, it suffices to verify that each irreducible $\mathbf{C}$-linear representation of the Galois group G of K has degree ≤ 2 and is realizable over $\mathbf{Q}$. Indeed, the character table given in §5.3 of [6] shows that the irreducible $\mathbf{C}$-linear representations of a dihedral group have degree ≤ 2, and that those of degree 1 are realizable over $\mathbf{Q}$ since their characters take values in $\mathbf{Q}$. Now, let ζ be a primitive nth root of unity in $\mathbf{C}$, and let r, s be generators of G with $r^n = 1$, $s^2 = 1$ and $rs = sr^{-1}$.

The fact that each irreducible **C**-linear representation of degree 2 of G is realizable over **Q** follows from the fact that its character coincides with the character of one of the **Q**-linear representations ρ^k of G in **Q**[ζ] defined, for an integer $k \geq 1$, by letting r act on **Q**[ζ] by multiplication by ζ^k and by letting s act on **Q**[ζ] as complex conjugation.

6 References

1. Emsalem, M. (1987). Sur les idéaux dont l'image par l'application d'Artin dans une $\mathbf{Z}_p$-extension est triviale. *J. reine angew. Math.*, *382*, 181–198.

2. Jaulent, J.-F. (1985). Sur l'indépendance *l*-adique de nombres algébriques. *J. Number Theory, 20*, 149–158.

3. Laurent, M. (1989). Rang p-adique d'unités et action de groupes. *J. reine angew. Math., 399*, 81–108.

4. Roy, D. (1988). *Sous-groupes minimaux de* $\mathbf{R}^n$ *et applications arithmétiques*. Ph. D. thesis, Université Laval, Québec.

5. Roy, D. (1990). Matrices dont les coefficients sont des formes linéaires. *Sém. Théorie des Nombres Paris 1987–88, Progress in Math., 81*, 273–281.

6. Serre, J. P. (1977). *Linear Representations of Finite Groups*. Springer-Verlag, New-York.

7. Suprunenko, D. A. (1976). *Matrix Groups*. Translations of Mathematical Monographs, **45**, Amer. Math. Soc., Providence.

8. Waldschmidt, M. (1981). Transcendance et exponentielles en plusieurs variables. *Invent. math., 63*, 97–127.

9. Waldschmidt, M. (1983). Dépendance de logarithmes dans les groupes algébriques. *Approximations diophantiennes et nombres transcendants, Progress in Math.*, **31**, 289–328.

10. Washington, L. C. (1982). *Introduction to Cyclotomic Fields*. Springer-Verlag, New-York.

Départment de Mathématiques
Université Laval
Ste–Foy
P. Québec G1K 7P4
Canada
e-mail droy@mat.ulaval.ca

Continued fractions of formal power series

A. J. van der Poorten

ceℵTℛe, Macquarie University

> *Though my travels took a long time,*
> *I hope Paulo will think it is fine*
> *For my remarks to be short;*
> *'Cause the point is the thought*
> *That I write this for P. Ribenboim*

1 Introduction

I will discuss continued fractions of formal power series, not for their own sake, but in terms of their use in obtaining explicit continued fraction expansions of classes of numbers. As we will see, the approach I outline accounts for essentially all the interesting examples of the past dozen years.

2 First principles

My viewpoint is formal. A continued fraction is an expression of the shape

$$a_0 + \cfrac{1}{a_1 + \cfrac{1}{a_2 + \cfrac{1}{a_3 + \ddots}}}$$

which one denotes in a space-saving flat notation by

$$[a_0 , a_1 , a_2 , a_3 , \ldots \ldots].$$

Everything follows from the correspondence whereby we have for $h = 0$, 1, 2, ...

$$\begin{pmatrix} a_0 & 1 \\ 1 & 0 \end{pmatrix} \begin{pmatrix} a_1 & 1 \\ 1 & 0 \end{pmatrix} \cdots \cdots \begin{pmatrix} a_h & 1 \\ 1 & 0 \end{pmatrix} = \begin{pmatrix} p_h & p_{h-1} \\ q_h & q_{h-1} \end{pmatrix}$$

if and only if

$$\frac{p_h}{q_h} = [a_0 , a_1 , \ldots \ldots , a_h] \quad \text{for } h = 0, 1, 2, \ldots.$$

453

I was first motivated to observe this useful relationship in [10] by remarks of Stark [15]. It goes back at least to [5].

Of course, a quotient p/q defines p and q only up to common factors; our correspondence can only refer to some appropriate choice of p and q.

Taking the transpose in the correspondence we see that

$$[a_h, a_{h-1}, \ldots, a_1] = \frac{q_h}{q_{h-1}}$$

and, taking determinants, that

$$p_h q_{h-1} - p_{h-1} q_h = (-1)^{h+1} \quad \text{so} \quad \frac{p_h}{q_h} = \frac{p_{h-1}}{q_{h-1}} + (-1)^{h-1} \frac{1}{q_{h-1} q_h},$$

whence

$$\frac{p_h}{q_h} = a_0 + \frac{1}{q_0 q_1} - \frac{1}{q_1 q_2} + \cdots + (-1)^{h-1} \frac{1}{q_{h-1} q_h}.$$

The regular continued fraction expansion of a real number has *partial quotients* a_h that are positive integers (other than perhaps for a_0 which may take any integer value); zero, negative and fractional partial quotients are termed *inadmissible*. Similarly, the admissible partial quotients a_h of a formal series in X^{-1} are polynomials of degree at least 1 (except perhaps for a_0 which may be constant).

If $a_1, a_2, \ldots$ are positive integers then $q_{h+1} = a_{h+1} q_h + q_{h-1}$ (and $q_{-1} = 0$, $q_{-2} = 1$) entails that the sequence (q_h) is increasing, so it follows that for a regular continued fraction expansion $p_h/q_h \longrightarrow \alpha \in \mathbb{R}$ with

$$\alpha - \frac{p_h}{q_h} = (-1)^h \left(\frac{1}{q_h q_{h+1}} - \frac{1}{q_{h+1} q_{h+2}} + \cdots \cdots \right).$$

Similarly, if the partial quotients a_h are polynomials of degree at least 1 then the *convergents* p_h/q_h converge to a formal series in X^{-1}. To see that momentarily surprising fact, just notice that

$$\frac{p_h}{q_h} = a_0 + \frac{x^{-(\deg q_0 + \deg q_1)}}{x^{-(\deg q_0 + \deg q_1)} q_0 q_1} - \frac{x^{-(\deg q_1 + \deg q_2)}}{x^{-(\deg q_1 + \deg q_2)} q_1 q_2} + \cdots$$

$$+ (-1)^{h-1} \frac{x^{-(\deg q_{h-1} + \deg q_h)}}{x^{-(\deg q_{h-1} + \deg q_h)} q_{h-1} q_h},$$

Let $\mathbb{L} = \mathbb{K}((X^{-1}))$ denote the field of formal Laurent series in X^{-1} over a field $\mathbb{K}$. Then $f \in \mathbb{L}$ has a continued fraction expansion

$$[a_0, a_1, \ldots, a_{h-1}, f_h],$$

with f_h the h-th *complete quotient* in $\mathbb{L}$. The continued fraction algorithm proceeds by taking the next partial quotient a_h to be the 'polynomial part' of f_h, to wit those terms in X (rather than in X^{-1}) — including the constant term — and one defines $f_{h+1} = (f_h - a_h)^{-1}$, observing that it is again an element of $\mathbb{L}$. We have seen above that

$$q_h f - p_h = q_h(-1)^h \left(\frac{1}{q_h q_{h+1}} - \cdots \right),$$

which ensures that

$$\deg(q_h f - p_h) = -\deg q_{h+1} < -\deg q_h .$$

In fact, the convergents p_h/q_h are characterised by the 'locally best approximation property': if $\deg s < \deg q_h$ then $\deg(q_h f - p_h) < \deg(sf - r)$ for all $r \in \mathbb{K}[X]$. To see this, suppose without loss of generality that $\deg q_{h-1} < \deg s < \deg q_h$ and note that, because the matrix

$$\begin{pmatrix} p_h & p_{h-1} \\ q_h & q_{h-1} \end{pmatrix}$$

is unimodular, there are polynomials a and b so that

$$\begin{aligned} s &= aq_h + bq_{h-1} \\ r &= ap_h + bp_{h-1} . \end{aligned}$$

Then

$$sf - r = a(q_h f - p_h) + b(q_{h-1}f - p_{h-1}),$$

and the evident fact that $\deg b > \deg a$ (there is, as usual, the forced convention that the identically zero polynomial has degree $-\infty$) shows that, indeed

$$\deg(sf - r) > \deg(q_{h-1}f - p_{h-1}) > \deg(q_h f - p_h),$$

showing also that only convergents are locally best approximations. We should also note that if $\deg s = \deg q_h$, but s is not a constant multiple of q_h, then necessarily $\deg(sf - r) > \deg(q_h f - p_h)$. For otherwise there is a $\mathbb{K}$-linear combination of s and q_h of lower degree yielding as good an approximation of f as does q_h. Hence we have:

Criterion. *If* $\deg(qf - p) < -\deg q$ *then* p/q *is a convergent of* f.

Remark. It is customary, but evidently pleonastic to add the qualification 'and if p and q are coprime'. I won't add the qualification but, throughout, I do of course suppose when referring to convergents that the quoted numerator and denominator are in fact coprime.

Proof. Suppose $\deg s < \deg q$. Then there is a polynomial r so that

$$0 \neq qr - ps = s(qf - p) - q(sf - r),$$

whence $\deg(sf - r) > 0$, so $\deg(sf - r) \geq -\deg q > \deg(qf - p)$ entails that p/q is indeed a locally best approximation.

Central to my subsequent observations is the following invaluable lemma:

Folding formula.

$$\frac{p_h}{q_h} + \frac{(-1)^h}{xq_h^2} = [a_0, \overrightarrow{w}, x - \frac{q_{h-1}}{q_h}] = [a_0, \overrightarrow{w}, x, -\overleftarrow{w}].$$

Here $\overrightarrow{w}$ is a convenient abbreviation for the word $a_1, a_2, \ldots, a_h$ and, accordingly, $-\overleftarrow{w}$ denotes the word $-a_h, -a_{h-1}, \ldots, -a_1$.

Proof. Let $\longleftrightarrow$ denote the correspondence between matrix products and continued fractions. Then

$$[a_0, \overrightarrow{w}, x - \frac{q_{h-1}}{q_h}] \longleftrightarrow \begin{pmatrix} p_h & p_{h-1} \\ q_h & q_{h-1} \end{pmatrix} \begin{pmatrix} x - q_{h-1}/q_h & 1 \\ 1 & 0 \end{pmatrix}$$

$$= \begin{pmatrix} xp_h - (p_hq_{h-1} - p_{h-1}q_h)/q_h & p_h \\ xq_h & q_h \end{pmatrix} \longleftrightarrow \frac{p_h}{q_h} + \frac{(-1)^h}{xq_h^2}$$

since $(p_hq_{h-1} - p_{h-1}q_h) = (-1)^{h-1}$; and, of course, $x - q_{h-1}/q_h = [x, -\overleftarrow{w}]$.

My nomenclature is based on the observation that iterated application of the formula leads to a pattern of signs corresponding to the creases in a sheet of paper repeatedly folded in half. For details in context see [11]; paperfolding is surveyed in [4].

3   Folded continued fractions

Because $1 + X^{-1} = [1, X]$, it follows from the folding formula that

$$1 + X^{-1} + X^{-3} = [1, X] + \frac{(-1)}{xX^2},$$

with $x = -X$; so

$$1 + X^{-1} + X^{-3} = [1, X, -X, -X].$$

Ultimately,

$$1 + X^{-1} + X^{-3} + X^{-7} + X^{-15} + X^{-31} + \cdots =$$
$$= [1, X, -\underline{X}, -X, -\underline{X}, X, X, -X, -\underline{X}, X, -X, -X, X,$$
$$X, X, -X, -\underline{X}, X, -X, \ldots].$$

The pattern of signs is exactly that of the pattern of creases in a sheet of paper folded in half right half under left an appropriate number of times and finally right half over left.

Regardless of how one folds (that is, under or over) it is a property of paper that the creases in the odd-numbered places alternate in sign. Moreover, changing the fold just changes the sign of the term being added; it changes the continued fraction expansion to the extent of changing the sign of the terms marked $-X$, and those induced from them: but all of those occur in the even-numbered places. Hence, rather more generally, we can specialise X to 2, say, and conclude that the uncountably many numbers $2\sum_0^\infty \pm 2^{-2^n}$ (where we suppose for convenience that the signs for $n = 0, 1$ in the sums both are $+$) all have continued fraction expansions of the shape

$$[1, 2, a, -2, b, 2, c, -2, d, 2, e, -2, f, \ldots],$$

with $a, b, c, d, \ldots = \pm 2$.

These expansions are severely polluted by inadmissible partial quotients. However,

$$
\begin{aligned}
-y &= 0 + -y \\
-1/y &= -1 + (y-1)/y \\
y/(y-1) &= 1 + 1/(y-1) \\
y - 1 &= -1 + y \\
1/y &= 0 + 1/y \\
y &= y \qquad\qquad \text{so} \quad -y = [0, \bar{1}, 1, \bar{1}, 0, y].
\end{aligned}
$$

Here $\bar{1}$ of course means -1. Hence

$$[\ldots, A, -B, C, \ldots] = [\ldots, A, 0, \bar{1}, 1, \bar{1}, 0, B, -C, -\ldots].$$

This doesn't seem an improvement, but easily

$$
\begin{pmatrix} D & 1 \\ 1 & 0 \end{pmatrix}
\begin{pmatrix} 0 & 1 \\ 1 & 0 \end{pmatrix}
\begin{pmatrix} E & 1 \\ 1 & 0 \end{pmatrix}
=
\begin{pmatrix} D+E & 1 \\ 1 & 0 \end{pmatrix},
$$

so

$$[\ldots, D, 0, E, \ldots] = [\ldots, D+E, \ldots].$$

Hence

$$
\begin{aligned}
[\ldots, A, -B, C, \ldots] &= [\ldots, A-1, 1, B-1, -C, -\ldots] \\
&= [\ldots, A-1, 1, B-1, 0, \bar{1}, 1, \bar{1}, 0, C, \ldots] = \\
&= [\ldots, A-1, 1, B-2, 1, C-1, \ldots].
\end{aligned}
$$

Thus

$$[\ldots, y, -2, z, \ldots] = [\ldots, y-1, 1, 0, 1, z-1, \ldots] =$$

$$= [\ldots, y - 1, 2, z - 1, \ldots],$$

and we see that

$$2 \sum_{0}^{\infty} \pm 2^{-2^n} =$$

$$= [1, 2, a - 1, 2, b - 1, 2, c - 1, 2, d - 1, 2, e - 1, 2, f - 1, \ldots].$$

We are not done yet, because a partial quotient $y - 1 = -3$ is inadmissible. However,

$$[\ldots, 2, -3, 2, \ldots] = [\ldots, 2 - 1, 1, 1, 1, 2 - 1, \ldots],$$

whilst a further -3 yields

$$[\ldots, 1, 1, -3, 2, \ldots]$$
$$= [\ldots, 1, 1 - 1, 1, 1, 1, 1, \ldots] = [\ldots, 2, 1, 1, 1, \ldots].$$

Thus, remarkably, all the numbers $2 \sum \pm 2^{-2^n}$ have regular continued fraction expansions requiring the partial quotients 1 and 2 alone. In [11] Shallit and I give a precise description of these expansions as folded sequences of words for all choices of sign in the series.

The first examples of explicit continued fraction expansions with bounded partial quotients were noticed independently by Kmošek [6] and by Shallit [14] some dozen years ago. Indeed, by precisely the argument just sketched, the sums $\sum_{0}^{\infty} \pm x^{-2^n}$ all have expansions of the shape

$$[0, x, a, -x, b, x, c, -x, d, x, e, -x, \ldots],$$

with $a, b, c, d, \ldots = \pm 1$. It is now not difficult to verify that only a small number of different partial quotients appear once we make the partial quotients admissible. Some details, arising from a slightly different viewpoint, appear at §2.3 of the survey [4] whilst at §6 of [11] we explain that it is practicable, using an idea of Raney [13], explicitly to divide the expansions of the numbers $x \sum_{0}^{\infty} \pm x^{-2^n}$ by x to obtain those earlier results. Mendès France and Shallit [9] provide yet a different context in which these continued fractions appear.

4   Specialisation

From my viewpoint, the genesis of the ideas just sketched is an observation of Blanchard and Mendès France [3] to the effect that if E is the set $\{0, 1, 4, 5, 16, 17, \ldots\}$ of nonnegative integers which are sums of distinct powers of 4 then

$$\chi = 3 \sum_{h \in E} 10^{-h} \text{ entails } \chi^{-1} = 3 \sum_{h \in 2E} 10^{-h-1} .$$

This is remarkable since, generally speaking, given an irrational real number represented by its decimal expansion, it is not practicable to explicitly represent its reciprocal.

It is easy to see that

$$\chi = 3 \prod_{n=0}^{\infty} (1 + 10^{-4^n})$$

and one then notices that the partial products yield every second convergent of χ. That readily yields an explicit continued fraction expansion. In [8] Mendès France and I ask, and answer, just which formal products $\prod_{n=0}^{\infty} (1 + X^{-\lambda_n})$ share the property that their truncations yield exactly every second convergent. We find that inter alia the cases $\lambda_n = k^n$ behave that way, *provided that $k > 2$ is even*, and that then the partial quotients are polynomials with integer coefficients. That means that we can reduce the continued fraction expansions modulo p at every prime and obtain the expansion for the formal product defined over the finite field $\mathbb{F}_p$. Alternatively, we may substitute an integer $x \geq 2$ for X and obtain the regular continued fraction expansion of the product; in other words, we can *specialise*. The polynomial partial quotients appearing for k even are of increasing degree.

It is of course exactly the phenomenon of specialisability, equivalently that of good reduction everywhere, that allows the approach described in the previous section.

It was momentarily a surprise to discover that the product $\prod_{n=0}^{\infty}(1+X^{-3^n})$ studied in [7] has a continued fraction expansion that has good reduction almost nowhere, the exception being $p = 3$, when the product in fact reduces to the quadratic irrational $(1 + X^{-1})^{-1/2}$.

Explicit computation in characteristic zero yields

$$\prod_{h=0}^{\infty} (1 + X^{-3^h})$$

$$= [1, X, -X + 1, -\tfrac{1}{2}X - \tfrac{1}{4}, 8X + 4, \tfrac{1}{16}X - \tfrac{1}{16}, -16X + 16,$$
$$-\tfrac{1}{32}X - \tfrac{1}{16}, 32X - 32, \tfrac{1}{64}X + \tfrac{5}{256}, \tfrac{1024}{5}X - \tfrac{256}{5},$$
$$-\tfrac{25}{2048}X + \tfrac{25}{2048}, -\tfrac{2048}{35}X - \tfrac{4096}{245}, \tfrac{343}{4096}X + \tfrac{245}{4096}, \ldots \ldots].$$

The partial quotients all appear to be linear, but their coefficients grow in complexity at a furious rate — the 30th partial quotient is

$$-\frac{1374389534720}{15737111}X - \frac{13743895347200}{456376219}$$

— and seem quite intractable. Nevertheless, Allouche, Mendès France and I [1] prove that these partial quotients are indeed all linear and, implicitly, we give a relatively easy technique for the recursive computation of the coefficients. We also notice that our argument shows that in general for odd k, 2 of every 3 partial quotients are of degree 1, and that those expansions too have good reduction almost nowhere, with exception, to our continuing surprise, again at $p = 3$ and, seemingly, nowhere else.

A few moments thought shows that the continued fraction expansion of almost every formal power series (defined over $\mathbb{Z}$, say) fails to have good reduction anywhere and has almost all its partial quotients of degree 1. It is thus our results for even $k > 2$ that are truly surprising.

5   A specialised continued fraction

A dozen or so years ago, Jeff Shallit, evidently in thrall to Fibonacci, noticed the continued fraction expansion

$$2^{-1} + 2^{-2} + 2^{-3} + 2^{-5} + \cdots + 2^{-F_h} + \cdots$$
$$= [0\,,1\,,10\,,6\,,1\,,6\,,2\,,14\,,4\,,124\,,2\,,1\,,2\,,2039\,,1\,,9\,,1\,,1\,,$$
$$1\,,262111\,,2\,,8\,,1\,,1\,,1\,,3\,,1\,,536870655\,,4\,,16\,,3\,,$$
$$1\,,3\,,7\,,1\,,140737488347135\,,\ldots]\,.$$

The increasing sequence of very large partial quotients demands explanation; the truncations of the sum do not yield convergents and the shape of the very good approximations is not immediately obvious. However, it turns out that a correct context for the cited expansion can be discovered in the remarks of mine and Mendès France [8] summarised above, wherein we consider continued fractions of formal Laurent series and then *specialise* the variable to an appropriate integer. Indeed, at the time we were finding the arguments detailed in [11], we noticed experimentally that

$$X^{-1} + X^{-2} + X^{-3} + X^{-5} + \ldots + X^{-F_h} + \cdots$$
$$= [0\,,X-1\,,X^2+2X+2\,,X^3-X^2+2X-1\,,-X^3+X-1\,,-X\,,$$
$$-X^4+X\,,-X^2\,,-X^7+X^2\,,-X-1\,,X^2-X+1\,,X^{11}-X^3\,,$$
$$-X^3-X\,,-X\,,X\,,X^{18}-X^5\,,-X\,,X^3+1\,,X\,,-X\,,-X-1\,,$$
$$-X+1\,,-X^{29}+X^8\,,X-1\,,\ldots]\,.$$

The limited number of shapes for the partial quotients, the phenomenon of self-similarity whereby bits and pieces from early in the sequence of partial quotients reappear subsequently, and of most importance the fact that all of the partial quotients have rational integer coefficients, all demand explanation and generalisation. Shallit and I provide that in [12] .

Above, and in the sequel, (F_h) denotes the popular sequence of Fibonacci numbers defined by the recurrence relation $F_{h+2} = F_{h+1} + F_h$ and the initial values $F_0 = 0$, $F_1 = 1$.

We did not find it easy to find an explanation for the phenomena just observed, until we accepted the fact that all we knew was the folding formula. Setting $s_h = X^{-1} + X^{-2} + X^{-3} + X^{-5} + \ldots + X^{-F_h}$ and $s_h = [0\,,f_h]$ we have

$$s_{h+1} = s_h + X^{-F_{h+1}} = [0\,,f_h] + \frac{1}{X^{-F_{h-2}}q^2}\,.$$

We use the identity $F_{h+1} = 2F_h - F_{h-2}$, whilst $q = X^{F_h}$ denotes the denominator of the final partial quotient of s_h. Let q' denote the denominator of the next to last partial quotient. Then, supposing that $|f_h|$ is even, the formula states that

$$s_{h+1} = [0\,,f_h\,,X^{-F_{h-2}} - q'/q]\,.$$

The point is that in this example we happen to be able to show fairly readily[1] that

$$q'/q = s_{h-1} - X^{-F_{h-3}} - X^{-F_h}\,,$$

with the critical relationship being $2F_h + F_{h-1} = F_{h+2}$. After repeated application of such facts we ultimately show that, once $h \geq 11$,

$$s_{h+1} = [0\,,f_h\,,0\,,-f_{h-4}\,,-X^{L_{h-4}}\,,\overleftarrow{f_{h-4}}\,,0\,,-f_{h-3}\,,X^{F_{h-4}}\,,\overleftarrow{f_{h-3}}]$$

$$= [0\,,g_h\,,X^{F_{h-5}}\,,\overleftarrow{f_{h-4}}\,,0\,,-f_{h-4}\,,-X^{L_{h-4}}\,,\overleftarrow{f_{h-4}}\,,$$

$$0\,,-f_{h-3}\,,X^{F_{h-4}}\,,\overleftarrow{f_{h-3}}]$$

$$= [0\,,g_h\,,X^{F_{h-5}} - X^{L_{h-4}}\,,\overleftarrow{f_{h-4}}\,,0\,,-f_{h-3}\,,X^{F_{h-4}}\,,\overleftarrow{f_{h-3}}] =$$

$$= [0\,,g_{h+1}\,,X^{F_{h-4}}\,,\overleftarrow{f_{h-3}}]\,,$$

and

$$s_\infty = X^{-1} + X^{-2} + X^{-3} + X^{-5} + \cdots = \lim_{h\to\infty} [0\,,g_h]\,.$$

Above, we have replaced f_h by

$$f_{h-1}\,,0\,,-f_{h-5}\,,-X^{L_{h-5}}\,,\overleftarrow{f_{h-5}}\,,0\,,-f_{h-4}\,,X^{F_{h-5}}\,,\overleftarrow{f_{h-4}}\,,$$

so

$$g_h = f_{h-1}\,,0\,,-f_{h-5}\,,-X^{L_{h-5}}\,,\overleftarrow{f_{h-5}}\,,0\,,-f_{h-4}\,.$$

These results explain the experimental data completely. For example, the large partial quotients in the numerical expansion arise from $X^{F_{h-5}} - X^{L_{h-4}}$ after making the specialised partial quotients admissible.

---

[1] That is, after laboriously discovering the formula from the experimental evidence, it slowly dawned on us that there is a straightforward argument allowing one to find q' 'spontaneously'.

An obstacle to our finding a workable argument was my conviction that the argument would apply generally as follows: Suppose (U_h) is a integer recurrence sequence, that is the solution of a linear homogeneous recurrence relation

$$U_{h+n} = s_1 U_{h+n-1} + \cdots + s_n U_h \qquad h = 0, 1, \ldots,$$

with integer coefficients $s_1, \ldots, s_n$ and integer initial values $U_0, \ldots, U_{n-1}$. Suppose further that the sequence (U_h) is strictly increasing with

$$\lim_{h \to \infty} U_{h+1}/U_h = \rho > 1.$$

I had guessed on the basis of the Fibonacci example that the series

$$X^{-U_0} + X^{-U_1} + X^{-U_2} + \cdots$$

is likely to have a specialisable continued fraction expansion.

With $\rho > 2$ this is trivially true by the folding lemma, perhaps with the qualification that one must omit some initial terms of the series to ensure that always $U_{h+1}/U_h \geq 2$ (and then $\rho = 2$ will do).

However, careful inspection of the arguments of [12] suggests that the properties of the Fibonacci numbers actually used are that the sequence (F_h) is strictly increasing with $F_{h-2} + F_{h-1} \leq F_h$ and $2F_{h-1} = F_{h-3} + F_h$, and of course that the initial partial quotients have integer coefficients. It follows immediately that, subject to that last condition — but it seems to be satisfied as soon as one chooses an appropriate starting point for the sequence (in the example we start with F_2), the arguments apply to strictly increasing Lucas sequences generally.

To my chagrin, because, *pace* Paulo, I shy away from matters Fibonacci, it seems that the technique Shallit and I discovered applies rather rarely. When $1 < \rho < 2$ we have not as yet noticed any examples, except for cases relying on the identity $2U_{h+n} = U_{h+n+1} + U_h$. I am moved to admit:

> *I'm allied to one of the factions,*
> *But I cannot accept its distractions*
> *I'm forced to agree*
> *With Fibonacci,*
> *When it's a matter of continued fractions.*

The phenomenon of specialisability seems only to apply to the Polynacci (*sic*) series:

Conjecture. *Let (T_n) be an increasing sequence of nonnegative integers satisfying a recurrence relation*

$$T_{h+d} = T_{h+d-1} + T_{h+d-2} + \cdots + T_h \text{ with } d > 1,$$

and set

$$s_n = X^{-T_d} + X^{-T_{d+1}} + X^{-T_{d+2}} + \cdots + X^{-T_n} \; ; \quad s_n = [0, t_n] .$$

Then, subject to appropriate initial conditions on the T_h, the words t_h consist of polynomials with integer coefficients, which is to say that s_∞ has a specialisable continued fraction expansion.

Remark. The point is that it is easy to see that one has $2T_h = T_{h+1} + T_{h-d}$ and $T_{h-2} + T_{h-1} \leq T_h$. Moreover, computations Shallit and I carried out show that for small $d = 3, 4, 5, 6, \ldots$ and initial values $0, \ldots, 0, 1$ the commencing partial quotients are specialisable.

We sketch arguments in [12] proving the validity of the conjecture for $d = 3$ and $d = 4$ and suggesting its truth for larger d. And we do not know what weight to give to our negative evidence as regards further examples of specialisability. That evidence is not utterly compelling because one must adjust sequences tested to have them start with some 'appropriate' term. However, in any case our arguments partially vindicating the conjecture suggest that our present techniques are not up to *constructing* further favourable examples, if there are any.

6   Some symmetric continued fractions

In [16] Jun-Ichi Tamura displays the continued fraction expansions of certain series

$$\sum_{h=0}^{\infty} \frac{1}{f_0(x) f_1(x) \cdots f_h(x)} ,$$

with f in $\mathbb{Z}[X]$ a polynomial with positive leading coefficient and of degree at least 2; f_h denotes the h-th iterate of f: so $f_0(X) = X$ and $f_h(X) = f(f_{h-1}(X))$. The genesis of his observations is apparently the fact that the case $f(x) = x^2 - 2$ (and x an integer at least 3) yields a quadratic irrational with a symmetric period. Tamura determines those f for which the specialisations at x of the truncations of the cited series have a symmetric continued fraction expansion.

The relevance of the folding formula is manifest. We set

$$s_n = \sum_{h=0}^{n} \frac{1}{f_0(X) f_1(X) \cdots f_h(X)} \text{ and } s_n = [0 , g_n] ,$$

with the word g_n supposed of odd length and a palindrome[2]. Then

$$s_{n+1} = [0 , g_n] + \frac{1}{f_0(X) f_1(X) \cdots f_{n+1}(X)}$$

---

[2] I have said it before, but once more cannot hurt too much: 'A palindrome is never even; it is a toyota'. The second comment I owe to Rick Mollin.

$$= [0, g_n, -\frac{f_{n+1}(X)}{f_0(X)f_1(X)\cdots f_n(X)} - q'/q],$$

by the folding formula.

However, we know that $p/q = s_n = [0, g_n]$ entails that $q'/q = [0, \overleftarrow{g_n}]$, so by symmetry we see that the continued fraction expansion of q'/q is simply that of s_n; that is, $q' = p$. So in the formula $pq' - p'q = (-1)^{k+1}$, where $k = |g_n|$, we have $p^2 - (-1)^{k+1} = p'q$.

It is now convenient to be more explicit, say by setting

$$s_n(X) = A_n(X)/B_n(X);$$

so $p = q' = A_n$ and $q = B_n$. It is easy to verify that $A_{n+1} = f_{n+1}A_n + 1$; of course, $B_{n+1} = f_{n+1}B_n$. Hence, certainly $k = |g_n|$ is odd as we had supposed, for on specialising X to 0 and noting that $X \mid B_n$ we have $(A_n(0))^2 - (-1)^{k+1} = 0$; and k even would contradict reality. So the final formula of the previous paragraph asserts that $B_n \mid (A_n^2 - 1)$.

Furthermore, $A_{n+1}^2 - 1 = f_{n+1}(f_{n+1}A_n^2 + 2A_n)$. The left hand side is divisible by $B_{n+1} = f_{n+1}B_n$, so B_n divides $f_{n+1}A_n^2 + 2A_n$. Since we know that $B_n \mid (A_n^2 - 1)$ it follows that in fact $q = B_n$ divides $f_{n+1} + 2A_n$.

We may now return to the folded formula to observe that

$$s_{n+1} = [0, g_n, -\frac{f_{n+1}(X)}{f_0(X)f_1(X)\cdots f_n(X)} - q'/q]$$
$$= [0, g_n, -f_{n+1}/B_n - A_n/B_n]$$
$$= [0, g_n, -(f_{n+1} + 2A_n)/B_n + A_n/B_n]$$
$$= [0, g_n, -(f_{n+1} + 2A_n)/B_n, g_n].$$

It is congenial to remove minus signs that will prove inadmissible after specialisation. Accordingly, set $(f_{n+1} + 2A_n)/B_n = d_n$ and note that

$$[0, g, -d, g] = [0, g, 0, \overline{1}, 1, d - 2, 1, \overline{1}, 0, g],$$

therewith recovering Tamura's principal result.

Finally, we ask for conditions on f implied by our assumption of symmetry in the formal power series case. We see that $B_n \mid (f_{n+1} + 2A_n)$ is $X \mid (f(X) + 2)$ for $n = 0$ and entails $f(0) = -2$. For $n = 1$ it is $Xf(X) \mid (f(f(X)) + 2f(X) + 2)$, whence $f(-2) = 2$; and similarly $n = 2$ yields $Xf(X)f(f(X)) \mid (f(f_2(X)) + 2f_2(X)(f(X) + 1) + 2)$, which entails $f(2) = 2$. So, certainly f is of the shape

$$f(X) = X(X - 2)(X + 2)g(X) + (X^2 - 2)$$

for some polynomial $g \in \mathbb{Z}[X]$. Since Tamura shows that this suffices for the continued fraction expansions cited here to be symmetric, this shape for f is equivalent to the symmetry of the continued fraction expansions. Of course, the cited expansions are specialisable.

7 Acknowledgments

This work was partly supported by grants from the Australian Research Council. I am grateful to my colleagues at the University of British Columbia for their assistance towards my attendance at the 3rd Meeting of the Canadian Number Theory Association.

I am indebted to Gerry Myerson as always, but in particular for his attempts to improve the scansion of the poetry appearing here. I did not accept his advice to open with the lines:

> *I travelled from here to Des Moines*
> *To pen a few lines for the doyen*
> *... etc.*

Of course, my 'tĭme' and 'fĭne' are to be pronounced so as to rhyme with Paulo.

8 References

1. J.-P. Allouche, M. Mendès France and A. J. van der Poorten, 'An infinite product with bounded partial quotients', to appear in *Acta Arithmetica*.

2. L. Auteurs, Letter to the editor, *The Mathematical Intelligencer* **5** (1983), p5.

3. André Blanchard et Michel Mendès France, 'Symétrie et transcendance', *Bull. Sc. Math.* 2e série **106** (1982), 325–335.

4. Michel Dekking, Michel Mendès France and Alf van der Poorten, 'FOLDS!', *The Mathematical Intelligencer* **4** (1982), 130-138; II: 'Symmetry disturbed', *ibid.* 173-181; III: 'More morphisms', *ibid.* 190-195.

5. J. S. Frame, 'Continued fractions and matrices', *Amer. Math. Monthly* **56** (1949), 98–103.

6. M. Kmošek, *Master's Thesis*, Warsaw, 1979.

7. M. Mendès France and A. J. van der Poorten, 'From geometry to Euler identities', *Theoretical Computer Science* **65** (1989), 213–220.

8. M. Mendès France and A. J. van der Poorten, 'Some explicit continued fraction expansions', *Mathematika* **38** (1991), 1–9.

9. M. Mendès France and J. Shallit, 'Wire Bending', *J. Combinatorial Theory* **50** (1989), 1–23.

10. A. J. van der Poorten, 'An introduction to continued fractions', in J. H. Loxton and A. J. van der Poorten eds., *Diophantine Analysis* (Cambridge University Press, 1986), 99–138.

11. A. J. van der Poorten and J. Shallit, 'Folded continued fractions', to appear in *J. Number Theory*.

12. A. J. van der Poorten and J. Shallit, 'A specialised continued fraction', submitted to *Canad. J. Math.*.

13. G. N. Raney, 'On continued fractions and finite automata', *Math. Ann.* **206** (1973), 265–283.

14. J. Shallit, 'Simple continued fractions for some irrational numbers', *J. Number Theory* **11** (1979), 209–217.

15. H. M. Stark, *An Introduction to Number Theory* (MIT Press, 1978).

16. Jun-Ichi Tamura, 'Symmetric continued fractions related to certain series', *J. Number Theory* **38** (1991), 251–264.

Alf van der Poorten
School of Mathematics, Physics, Computing and Electronics
Macquarie University
NSW 2109
Australia
e-mail alf@macadam.mpce.mq.edu.au

Chapter V

Special Session in Honour of
Paulo Ribenboim

A Plenary Address

Invited Addresses

and

Speeches at Banquet

Paulo Ribenboim, at the time of his retirement

Andrew Granville

University of Georgia

At this, the third meeting of the Canadian Number Theory Association, we have had the opportunity to enjoy a lot of good mathematics, while celebrating the career of one of Canada's most distinguished academics. Paulo Ribenboim, author of more than a hundred journal articles and of thirteen books, a Fellow of the Royal Society of Canada, and colleague and advisor to so many mathematicians, is retiring after a career spanning over forty years and three continents. In this article we present a biography of Paulo, together with a sampling of a few of his many interesting results.

Picture 1. Paulo Ribenboim

Paulo was born in 1928, into a middle class family in Recife, Brazil. In Picture 1 you can see him at a very young age with his two brothers. After high school, he did his military service (see Picture 2, p. 471), where his mathematical talents were recognized and he found himself teaching calculus. It amuses Paulo to think that one or two of his students were to become military rulers of Brazil – what did they learn from him in that calculus class?

After finishing in the army, Paulo entered the university in Rio de Ja-

Picture 2. Paulo and brothers

neiro, graduating with a B.Sc. in 1948. Following one more year in Rio, at the Instituto do Rio de Janeiro (the forerunner of today's IMPA), Paulo received a fellowship to study in France. So, in 1950, he went to Nancy, to work under Dieudonné's supervision.

In Nancy, he was close to Laurent Schwarz and Roger Godement; and had many colleagues and friends who went on to become important mathematicians, such as Grothendieck, Malliavin, Malgrange, Lions, P. M. Cohn and H. Bauer. This time in Nancy was important for Paulo in other ways too, for it was there that he met his lovely wife Huguette. The photographs in Picture 4 were both taken in Nancy, soon after they met.

Paulo found himself fascinated by the Bourbaki program, and returned to Rio in 1952 determined to pursue these studies. Soon after, he helped bring his friend Grothendieck over to Brazil for a couple of years, during which time Grothendieck wrote many of his important papers on topological vector spaces and had a lasting impact on the development of Brazilian mathematics.

Under Dieudonné's influence, Paulo had found himself particularly interested in the theory of valuations and, in 1954, headed back to Europe to study at the Mathematisches Institut in Bonn, under the direction of Krull.

As you can see from the photograph in Picture 4, Paulo soon found himself in the full swing of things in Bonn, and started proving noteworthy results in this rapidly developing area.

So what are valuations? The simplest example is the 'p-adic valuation', that function which gives the exact power of prime p dividing a given rational number; thus $v_p(a/b)$ is the power of p dividing a minus the power of p dividing b (so $v_2(16) = 4$, $v_2(1/2) = -1$, etc.). There are two obvious properties of such functions: For all a and b, we have

Picture 3. Paulo in military

$$v(ab) = v(a) + v(b) \quad \text{and} \quad v(a + b) \geq \min\{v(a), v(b)\}.$$

So a valuation is a homomorphism $v: K \rightarrow \Gamma \cup \{\infty\}$, from a field K to an additive, abelian group Γ which is totally ordered. We call $R = \{r \in K : v(r) \geq 0\}$ the *valuation ring* of v, so that $P = \{r \in K : v(r) > 0\}$ is the unique maximal ideal in R, and define $\bar{K} = R/P$ to be the *residue field* of K. In 1932, Krull wrote a seminal paper in which he developed a general theory of such 'valuations', based on the properties that we have just described.

We shall start by examining how Γ differs from being a subset of the real numbers, **R**. Suppose that $0 < a < b$ belong to Γ: is it true that there exists a positive integer n such that $b < an$? (Of course this would be the case for any subset of **R**). It turns out that not all totally ordered, additive

Picture 4. Paulo and Huguette in Nancy

abelian groups Γ satisfy this condition; for example, if $\Gamma = \mathbf{Z} \times \mathbf{Z}$, with
lexicographic ordering (that is $(x_1, y_1) < (x_2, y_2)$ if and only if $x_1 < x_2$, or
$x_1 = x_2$ and $y_1 < y_2$), then $n(0, 1) < (1, 0)$ for *all* positive integers n. We
thus introduce the notion of *archimedean equivalence*, so that for $a, b \in \Gamma$,

$$a \sim b \text{ if and only if } a < mb \text{ and } b < na$$

for some positive integers m and n. The *height* of a valuation is the totally
ordered set of such archimedean equivalence classes, or simply their number
if this set is finite. Notice that any height 1 valuation can be embedded
into the reals.

Now, Krull noted that if R is a height 1 valuation ring then (i) all ideals are primary and there is just one non-zero prime ideal, (ii) R is an integral domain, and (iii) R is completely integrally closed, and he asked whether the converse holds, which would lead to a nice classification of height 1 valuation rings. However, in 1955, Ribenboim exhibited a counterexample[1] as well providing various (iv)th conditions under which the converse would indeed hold (for instance, if R is Noetherian). This was Paulo's first major result.

Suppose that one is given a set of valuations $v_i : K \longrightarrow \Gamma_i$, for $i = 1, 2, \ldots, n$. In analogy with the Chinese Remainder Theorem (CRT) one might ask whether, for given $k_i \in K$ and $\gamma_i \in \Gamma_i$, there exists $x \in K$ for which $v_i(x - k_i) = \gamma_i$ for $i = 1, 2, \ldots, n$. To stress the importance of such a question we quote Schilling who wrote, 'Some of the more important tools in the theory of algebraic numbers and the arithmetic theory of covering varieties are furnished by appropriate generalizations of the CRT'. Krull showed that such an x exists if no two of the associated valuation rings have a common non-trivial prime ideal (corresponding to coprime moduli in CRT), leaving the more general question open; indeed Krull felt that a general solution to this problem was one of the most challenging open questions in the area. How-

Picture 5. Paulo in Bonn

ever, in 1957 Ribenboim solved this, giving an elegant 'if and only if' condition for whether such an x exists, depending only on the images of $P_i = \{r \in K : v_i(r) > 0\}$ under v_i. This condition and its proof are considered to be very clever, and stand as an important and attractive result in the subject of valuation rings.

Another nice result in this area, due to Ribenboim, is the generalization of Hasse's Theorem to valuation rings: In 1925, Hasse showed that given

---

[1] This appears in Bourbaki's *Commutative Algebra*, Chapter 6.

any algebraic number field K and 'partition' of a given integer n as $\sum_i e_i f_i$, and given any prime ideal $\mathbf{p}$ of K, there exists a field extension L of degree n over K such that $\mathbf{p}$ factors into prime ideals of L as $\prod_i p_i^{e_i}$, where p_i has relative degree f_i. In 1959, Paulo showed that this result can be extended to fields K that admit suitable valuations of finite height.

Ribenboim proved many other results on valuations and wrote a monograph [4] that quickly became a standard reference in the area. He is proud of the part that this monograph plays in Ax and Kochen's 1965 proof of Artin's Conjecture (that, for any given d, there exists a prime p_d such that any homogeneous form f of degree d in n ($> d^2$) variables, has a non-trivial p-adic zero for all primes $p \geq p_d$), which employs a non-standard model of the integers via valuation theory.

Getting back to Paulo's biography, he left Bonn in 1956, returning to Brazil until 1959. In 1957 he received the first Ph.D. ever granted by a Brazilian university (in São Paulo). From 1959 to 1962 Paulo was a Fulbright Fellow at Urbana-Champaign, and in 1962, moved to Queen's University, here in Canada, where he still works today.

During Paulo's first decade at Queen's, the mathematics department was small, yet there were students and faculty from all over the world. The atmosphere was friendly and there was an active departmental colloquium, as well as a steady stream of visitors from across Canada, the United States and Europe. Paulo invited a variety of visiting faculty members, including J. Neukirch, W. Scharlau and W.D. Geyer. Paulo was an enthusiastic organizer and, with his wife Huguette, provided a congenial atmosphere in their home for regular informal gatherings that brought together mathematicians with very different specialties.

In the early 1970s, under such an influence, Queen's became a center for research in Algebra and a 'place to be' for anyone in the area. In particular, Paulo was then working on generalizations and variants of Hilbert's 17th Problem. To remind the reader, in 1927 Artin solved Hilbert's 17th problem, which was to show that every positive definite rational function over the reals, equals the sum of squares of rational functions. Paulo and Gondard studied under what special conditions the same could be said of polynomial functions (this is not true in general). Also, in 1974, they proved the following delightful result, using methods from model theory: Let K be a real closed field, and m and n positive integers. A symmetric n-by-n matrix $M(X_1, \ldots, X_m)$, with entries in $K(X_1, \ldots, X_m)$, is said to be *positive definite* if the quadratic form associated to $M(x_1, \ldots, x_m)$ is positive definite whenever $M(x_1, \ldots, x_m)$ is well-defined for $x_1, \ldots, x_m \in K$. They proved that if M is positive definite then it is the sum of squares of symmetric matrices with entries from $K(X_1, \ldots, X_m)$.

In recent years, Paulo's research has mostly been in number theory, though with excursions into questions involving equations in groups, generalized power series, and ascending chain conditions.

In a series of interesting papers, Paulo has shown how to apply Baker's and Faltings' Theorems to a wide variety of Diophantine and related questions. For instance, in 1985, he and Powell used Faltings' Theorem to show that there are only finitely many coprime integer solutions to any one of the equations, $x^{2n} \pm y^{2n} = z^2$, $x^{2n} + y^{2n} = z^3$ or $x^4 - y^4 = z^n$, for any given $n > 3$ (while preparing this article, the author noted that their proof may be modified to prove that any equation $x^p \pm y^p = z^q$ has finitely many coprime integer solutions provided $2/p + 1/q < 1$).

An old problem of elementary number theory is to prove that no value occurs infinitely often in a (non-degenerate) linear recurrence (for second order recurrences, Beukers has proved that no value occurs more than three times, except in a few simple cases). A substantial generalization of this is to show that there are only finitely many pairs of distinct elements of any (greater than first order) linear recurrence whose product is a square. In 1989 Ribenboim proved this true of the Fibonacci and Lucas sequences, listing all such products. They are

$$F_1 F_2 = 1^2 \qquad F_1 F_{12} = F_2 F_{12} = 12^2 \qquad F_3 F_6 = 4^2$$

for the Fibonacci sequence, and

$$L_1 L_3 = 2^2 \quad \text{and} \quad L_0 L_6 = 6^2$$

for the Lucas sequence. Recently, with McDaniel, he has generalized the qualitative result to all Lucas-type sequences.

Paulo is admired for many of the qualities that he has brought to his career as a mathematician. Besides his research, some of which we have discussed above, he is renowned for being a marvelous author who enriches and enlivens our subject, and for his gentle yet exacting dealings with students and colleagues. His kind, gentlemanly manner makes him an excellent advisor of students, and he has succeeded in getting the most out of them.

In his educational writings, to wit, thirteen books and over thirty surveys and 'technical reports,' he has always kept in mind the varied backgrounds of his reading public. He succeeds admirably in reaching out and explaining difficult concepts to non-specialists as well as specialists, novices as well as experts. His book on valuations is the standard text in that area. *13 Lectures on Fermat's Last Theorem* is written in a very distinctive style, with charm and lightness of touch, yet with precision. It is guaranteed to excite any reader. His recent *Book of Prime Number Records*, recommended by Choice Magazine as one of the five best science books of 1990, is being printed in Japanese and French, as well as English, and allows any interested reader to obtain an excellent insight into many aspects of what is known about prime numbers. An abridged 'pocket-sized' version [10] was published to coincide with the start of the CNTA conference.

Picture 6. Cheers!

In 1965 Ribenboim graduated his first doctoral student, Malcolm Griffin, and has now had a total of thirteen doctoral students (including the author of this article). Add to this list his seven other students with varying degrees, and we find academics now working from Canada to Brazil, from Norway to Morocco, from India to France, and from Singapore to the United States. This is typical of Paulo, a truly international mathematician, who has spread his pleasure in mathematics, his infectious enthusiasm, to every corner of the globe. He is widely sought after as a lecturer, and has traveled extensively inspiring people everywhere. He has held academic positions in eleven countries, and given lectures in five South American countries, fifteen Western and six Eastern European countries, half a dozen countries in Africa and the Middle East, as well as a couple in the Far East, including such far-flung locations as Colombia, Curaçao, Egypt, Finland, Iran, Sudan, Taiwan and Tunisia. His invited addresses include a UNESCO conference in Buenos Aires in 1959, a 'Mathematics in the Third World' meeting in Khartoum in 1978, the 'Gauss Symposium' in Brazil in 1989, as well as a televised address in Québec in 1987.

In conclusion let us wish Paulo well in his retirement. It is our hope that the forthcoming years might remain as productive and worthwhile as his career has been so far.

Remark: This article is derived from a talk given at the CNTA conference on August 23rd, 1991. I would like to thank Ernst Kani and Jan Mináč, as well as Huguette and Paulo Ribenboim, for their help in the presentation of the mathematics, and also of the personal photographs and biography.

References:

1. [1] *Sur une conjecture de Krull en théorie des valuations,* Nagoya Math. J. **9** (1955), 87–97.

2. [2] *Le théorème d'approximation pour les valuations de Krull,* Math. Zeit. **68** (1957), 1–18.

3. [3] *Remarques sur le prolongement des valuations de Krull,* Rend. Circ. Mat. Palermo **(2) 8** (1959), 152–159.

4. [4] *Théorie des valuations,* Univ. de Montréal, Montréal, 1964.

5. [5] (with D. Gondard) *Le 17ᵉ problème de Hilbert pour les matrices,* Bull. Sci. Math. **98** (1974), 49–56.

6. [6] *13 Lectures on Fermat's Last Theorem,* (Springer-Verlag, New York, 1979).

7. [7] (with B. Powell) *Note on a paper of M. Filaseta regarding Fermat's Last Theorem,* Ann. Univ. Turku. Ser. A.1 **187** (1985), 22pp.

8. [8] *Square classes of Fibonacci and Lucas numbers,* Port. Math. **46** (1989), 159–175.

9. [9] *The Book of Prime Number Records,* (Springer-Verlag, New York, 1989).

10. [10] *The Little Book of Big Primes,* (Springer-Verlag, New York, 1991).

Andrew Granville
(Doctoral Student under Paulo Ribenboim, 1984-1987)
Department of Mathematics
University of Georgia
Athens, Georgia 30602
USA
e-mail andrew@sophie.uga.edu

The Kummer-Wieferich-Skula Approach to the First Case of Fermat's Last Theorem

Andrew Granville

University of Georgia

Dedicated to Paulo Ribenboim.

1. Introduction.

The First Case of Fermat's Last Theorem is said to be true for prime exponent p if there do not exist integers x, y, z such that

$$(1.1) \qquad x^p + y^p + z^p = 0 \quad \text{where} \quad p \text{ does not divide } xyz.$$

Instead we shall assume throughout this paper that (1.1) does have solutions and then deduce a variety of implausible consequences.

For example, in 1857 Kummer showed that if (1.1) has a solution then, for all n in the range $2 \le n \le p - 1$,

$$(1.2) \qquad \text{either} \quad B_{p-n} \equiv 0 \pmod{p} \quad \text{or} \quad \sum_{j=0}^{p-1} j^{n-1} t^j \equiv 0 \pmod{p}$$

for each $t \in \{-x/y, -y/z, -z/x\}$, where B_n, the nth Bernoulli number, is given by

$$\frac{X}{e^X - 1} = \sum_{n \ge 0} B_n \frac{X^n}{n!}.$$

As a consequence one can show that p must divide B_{p-3}, B_{p-5} and many other Bernoulli numbers too.

In 1909 Wieferich surprisingly deduced, from Kummer's criteria, that p^2 must divide $2^p - 2$. Soon others deduced that p^2 must also divide $3^p - 3$, $5^p - 5$ and so on; and Frobenius outlined a method to continue proving such criteria. However it requires a great deal of work to verify each successive

---

The author is supported, in part, by the National Science Foundation.

criteria, and so, even now, has only been computed up to p^2 divides $q^p - q$, for all $q \leq 89$.

In 1914 Vandiver proved that p divides $B_{p-1}\left(\frac{a}{5}\right) - B_{p-1}$ for $1 \leq a \leq 4$, as well as p^2 divides $5^p - 5$, where $B_{p-1}(t) = \sum_{j=0}^{p-1} \binom{p-1}{j} B_j t^{p-1-j}$ is the $(p-1)$st Bernoulli polynomial. In 1938, Emma Lehmer generalized this to p divides $B_{p-1}\left(\frac{a}{q}\right) - B_{p-1}$ for $1 \leq a \leq q$, for $q = 2, 3, 4$ and 6. Recently Skula showed that one could successively prove that p divides $B_{p-1}\left(\frac{a}{q}\right) - B_{p-1}$ for $1 \leq a \leq q$, for each q, at the same time as proving p^2 divides $q^p - q$.

In this paper we shall present (hopefully easier) proofs of these results, and of some consequences, and give some new results and possible directions.

2.   The structure of the p-th cyclotomic field and Kummer's Theorem.

Let $\xi = \xi_p$ be a primitive pth root of unity, $K = \mathbf{Q}(\xi)$, and C be the ideal class group of K. Let G be the Galois group of $K \mid \mathbf{Q}$, and χ be a generator of the character group of G (in the multiplicative group of the field of p elements, $\mathbf{F}_p^*$), so that

$$\chi(\sigma_a) = a \quad \text{where} \quad \sigma_a : \xi \mapsto \xi^a.$$

It is well-known (see [Wa], §6.3) that

$$C/C^p = \bigoplus_{i=1}^{p-1} (C/C^p)(\chi^i)$$

where

$$(C/C^p)(\chi^i) = \{I \in C/C^p : I^{\sigma - \chi'(\sigma)} \in C^p \text{ for all } \sigma \in G\};$$

moreover if $i \geq 3$ is odd and $(C/C^p)(\chi^i)$ is non-trivial (contains an element other than C^p) then p divides B_{p-i} (this result is known as Herbrand's Theorem). If we consider the natural homomorphism $\tau : C \mapsto C^p$, defined by $\tau(I) = I^p$, we see that $\ker \tau = \{I : I^p \text{ is principal}\}$. Moreover $C/\ker \tau \cong C^p$ (by the first isomorphism theorem), and so $C/C^p \cong \ker \tau$ (as C is an abelian group). Thus we may re-write Herbrand's Theorem to read:

If $i \geq 3$ is odd and

$$(\ker \tau)(\chi^i) := \{I \in C : I^p \text{ and } I^{\sigma - \chi'(\sigma)} \text{ are principal for all } \sigma \in G\}$$

is non-trivial then p divides B_{p-i}.

One of Kummer's more remarkable ideas was to understand A_p/A_p^p (where A is the ring of integers of K and $A_p = A/pA$) by using logarithmic derivatives (which was later generalized by Coates and Wiles - see §13.7 in [Wa]): For

$$\gamma = \gamma(\xi) = a_0 + a_1\xi + \ldots + a_{p-1}\xi^{p-1},$$

not divisible by $(1 - \xi)$, define

$$\gamma(e^X) = a_0 + a_1 e^X + \ldots + a_{p-1} e^{(p-1)X}$$

and

$$\ell_n(\gamma) = \left(\frac{\partial}{\partial X}\right)^n \{\log \gamma(e^X))\} \mid_{X=0},$$

for $1 \leq n \leq p - 2$, in $\mathbf{F}_p$ (note that this value is independent of the way that we write γ). Observe that $\ell_n(\gamma(\xi^j)) \equiv j^n \ell_n(\gamma(\xi))$ (mod p). Kummer noted that $\ell_n(\alpha\beta) \equiv \ell_n(\alpha) + \ell_n(\beta)$ (mod p) and so $\ell_n(\gamma^p) \equiv 0$ (mod p). Also $\ell_n(\xi^a) = a$ (for $n = 1$), 0 (otherwise), and if $\gamma \in \mathbf{Z}[\xi + \xi^{-1}]$ then $\ell_n(\gamma) \equiv 0$ (mod p) for odd n in the range $3 \leq n \leq p - 2$: therefore, as every unit u of A is a power of ξ times an element of $\mathbf{Z}[\xi + \xi^{-1}]$ thus $\ell_n(u) = 0$ for odd n in the range $3 \leq n \leq p - 2$. If $(\gamma) = (z)^p$ then $\gamma = uz^p$ for some unit u, and so

(2.1) $\quad \ell_n(\gamma) \equiv 0$ (mod p) for each odd n, $3 \leq n \leq p - 2$.

Let $\eta_i = \sum_{\sigma \in G} \chi^{-i}(\sigma)\sigma$. Observe that if $J = I^{\eta_i}$, for any ideal I, then $J^{\sigma - \chi'(\sigma)}$ is a power of I^p for every $\sigma \in G$; and so $J \in (\ker \tau)(\chi^i)$ if I^p is principal. Moreover, since $\ell_n(z^{\sigma_a}) \equiv a^n \ell_n(z)$ (mod p) we have

$$\ell_n(z^{\eta_i}) \equiv \sum_{a=1}^{p-1} \chi^{-i}(\sigma_a) a^n \ell_n(z)$$

(2.2) $\qquad \equiv \begin{cases} -\ell_n(z) \pmod{p} & \text{if } n \equiv i \pmod{p-1}; \\ 0 \pmod{p} & \text{otherwise.} \end{cases}$

Now suppose that we have a solution to (1.1). Factoring $x^p + y^p$ we get the ideal equation

$$(x + y)(x + \xi y) \ldots (x + \xi^{p-1}y) = (z)^p.$$

Evidently if $i \not\equiv j$ (mod p) then $(x + \xi^i y, x + \xi^j y)$ divides $(1 - \xi)$ (which divides (p)) as $(x, y) = 1$, and so $(x + \xi^i y, x + \xi^j y) = 1$ since p does not divide z. Therefore each $(x + \xi^j y) = I_j^p$ for some ideal I_j, by the unique factorization theorem for ideals. Thus the ideal

$$\theta_i := I^{\eta_i} \in (\ker \tau)(\chi^i) \quad \text{for every } 2 \le i \le p - 2 \quad (\text{where } I = I_1).$$

If θ_i is non-principal then p divides B_{p-i} by Herbrand's Theorem. On the other hand, if θ_i is principal, say $\theta_i = (z_i)$, then $(z_i)^p = \theta_i^p = I^{p\eta_i} = (x + \xi y)^{\eta_i}$, so that

$$(2.3) \qquad\qquad (x + \xi y)^{\eta_i} = u_i \quad \text{in } K^*/(K^*)^p,$$

for some unit u_i. Applying $\ell_i(.)$ to both sides of this equation, we deduce from (2.1) and (2.2) that $\ell_i(x + \xi y) \equiv 0 \pmod{p}$.

This is a 'new' proof of Kummer's result (stated slightly differently), though all the steps presented here are implicit in the literature. However we feel that this proof is more enlightening than the two standard ones, namely that of Kummer (see §7 of [Ri]) which is a mire of complicated details, and that which may be deduced from explicit reciprocity laws (see §9.5 of [Ri]) which is elegant but unilluminating.

Although the statement above is what Kummer actually proved, Mirimanoff simplified it to the statement given in the introduction by observing that, for $t = -y/x$,

$$\ell_n(x + \xi y) = \left(\frac{\partial}{\partial v}\right)^n \{\log(x + e^v y)\}\big|_{v=0} = \left(\frac{\partial}{\partial v}\right)^{n-1} \left\{1 - \frac{1}{1 - te^v}\right\}\big|_{v=0}$$

and, as $t \not\equiv 1 \pmod{p}$ (else p would divide z),

$$\frac{1}{1 - te^v} = \frac{1}{1 - t^p e^{pv}} \sum_{j=0}^{p-1} t^j e^{jv} \equiv \frac{1}{1 - t^p} \sum_{m \ge 0} \left\{\sum_{j=0}^{p-1} j^m t^j\right\} \frac{v^m}{m!} \pmod{p}$$

and so

$$(2.4) \qquad \ell_n(x + \xi y) \equiv -\frac{1}{1 - t^p} \sum_{j=0}^{p-1} j^{n-1} t^j \pmod{p}, \quad \text{for all } n \ge 2.$$

Pollaczek provided an explicit formula for the polynomials $\ell_n(x + \xi y)$: By noting that

$$\frac{\partial}{\partial v}\left(\frac{1}{1 - te^v}\right) = t \frac{\partial}{\partial t}\left(\frac{1}{1 - te^v}\right),$$

we deduce that, for $n \ge 2$,

$$\ell_n(x + \xi y) = - \left(\frac{\partial}{\partial v}\right)^{n-1} \frac{1}{1 - te^v}\Big|_{v=0}$$

$$= - \left(\frac{t\partial}{\partial t}\right)^{n-1} \frac{1}{1 - te^v}\Big|_{v=0} = - \left(\frac{t\partial}{\partial t}\right)^{n-1} \frac{1}{1 - t}.$$

Then, by a straightforward induction hypothesis, we get for $n \geq 2$,

$$(2.5) \qquad \ell_n(x + \xi y) = (-1)^{n+1} \sum_{j=1}^{n} \frac{s_{n,j}}{j} \frac{1}{(t-1)^j},$$

where $s_{n,j}$, the Stirling numbers of the second kind, are defined by $s_{1,j} = 1$ (for $j = 1$), 0 (otherwise), and then $s_{n+1,j} = j(s_{n,j} + s_{n,j-1})$ (so that, for instance, $X^n = \sum_{j=1}^{n} s_{n,j} \binom{X}{j}$).

It is possible to say a little more above in the case that θ_i is principal: Multiplying (2.3) by its complex conjugate we find that $u_i \bar{u}_i$ is a pth power in K and so, writing $u_i = \xi^a v_i$ for $v_i \in \mathbf{Z}[\xi + \xi^{-1}]$ we find that v_i^2 is a pth power and thus so is v_i. Therefore we can take $u_i = \xi^a$ in (2.3) and comparing $\ell_1(\cdot)$ of both sides gives $a \equiv 0 \pmod{p}$, and we deduce that $(x + \xi y)^{\eta_i} = 1$ in $K^*/(K^*)^p$. Therefore

$$(2.6) \qquad \prod_{j=1}^{p-1}(x + \xi^j y)^{j^{-i}} = z_i^p$$

for some $z_i \in K$. Thaine made this observation some time ago and has since worked to derive a similar criteria for even i, $2 \leq i \leq p - 3$:

3. Thaine's ideas.

Define U_+ to be the group of units of $\mathbf{Z}[\xi + \xi^{-1}]$, which contains the group of 'circular' units, V, generated by $(\xi^a - \xi^{-a})/(\xi - \xi^{-1})$ for $a = 2, 3, \ldots, \frac{p-1}{2}$. It is well known that U, the group of units of $\mathbf{Z}[\xi]$, is generated by U_+ together with ξ. It is of great interest to understand the structure of $W = U_+/V$, the non-circular units of $\mathbf{Z}[\xi + \xi^{-1}]$, and particularly its p-part. Again we can decompose

$$W/W^p = \bigoplus_{\substack{i=2 \\ i \text{ even}}}^{p-3} (W/W^p)(\chi^i),$$

and a remarkable result of Vandiver tells us that $(W/W^p)(\chi^i)$ is non-trivial if and only if $v_i := \prod_{a=1}^{p-1} \left(\frac{\xi^a - \xi^{-a}}{\xi - \xi^{-1}}\right)^{a^{-i}}$ is a pth power in K (see Theorem

8.14 in [Wa]), say α^p. (Indeed if we define $\theta : W \rightarrow W^p$ as $\theta(w) = w^p$ then, proceeding as in the previous section, $W/W^p \cong \ker \theta$, and $\alpha \in (\ker \theta)(\chi^i) = \{w \in W : w^p \text{ and } w^{\sigma - \chi'(\sigma)} \in V \text{ for all } \sigma \text{ in } G\}$.)

Since

$$(3.1) \qquad \left(\frac{\xi^a - \xi^{-a}}{\xi - \xi^{-1}} \right)^{\eta_i} = v_i^{a^i - 1} \text{ in } V/V^p,$$

we can deduce that

$$(3.2) \qquad v^{\eta_i} \text{ is a power of } v_i, \text{ for any } v \in V.$$

Moreover, applying (2.2) to (3.1) when a is a primitive root $\pmod p$ (so that $a^i - 1 \not\equiv 0 \pmod p$), we find that $\ell_n(v_i) \equiv 0 \pmod p$ except perhaps when $n = i$, in which case

$$\ell_i(v_i) \equiv \frac{-1}{a^i - 1} \ell_i \left(\frac{\xi^a - \xi^{-a}}{\xi - \xi^{-1}} \right) \qquad \pmod p$$

$$= \frac{-1}{a^i - 1} \left(\frac{\partial}{\partial v} \right)^i \left(\log(e^{2av} - 1) - \log(e^{2v} - 1) - (a-1)v \right) \big|_{v=0}$$

$$(3.3) \qquad = \frac{-1}{a^i - 1} \left(\frac{\partial}{\partial v} \right)^{i-1} \frac{1}{v} \left(\frac{2av}{e^{2av} - 1} - \frac{2v}{e^{2v} - 1} \right) \big|_{v=0} = -\frac{2^i}{i} B_i.$$

(Note that if v_i is a pth power then (3.3) implies that p divides B_i.)

Now, assume that $(W/W^p)(\chi^i)$ is trivial. In [Th1], Thaine proved that if $(W/W^p)(\chi^i)$ is trivial then $(C/C^p)(\chi^i)$ is trivial (this may be deduced directly from the 'Main Conjecture', first proved by Mazur and Wiles, but Thaine's proof is much easier, and its generalizations have led to the many recent important advances of Kolyvagin, Rubin and others). This implies that θ_i, from the previous section, must be principal.

Since $(x + \xi y)^{\eta_i} \in \mathbb{Z}[\xi + \xi^{-1}]$ we may assume that $u_i \in U_+$ in (2.3). Therefore $u_i^{\eta_i} \in (W/W^p)(\chi^i)$ (which we have assumed to be trivial), and so $u_i^{\eta_i} \in V$ in U/U^p. But then $u_i^{\eta_i^2} = (u_i^{\eta_i})^{\eta_i}$ is a power of v_i (say, $v_i^{\kappa_i}$) in U/U^p by (3.2). On the other hand, $\eta_i^2 = -\eta_i$ in $\mathbb{F}_p^*$, so that $\eta_i^3 = \eta_i$, and thus raising (2.3) to the power η_i^2 gives us

$$(3.4) \qquad (x + \xi y)^{\eta_i} = v_i^{\kappa_i} \text{ in } K^*/(K^*)^p$$

(which is Theorem 2 of [Th2]). Applying $\ell_i(\cdot)$ to both sides gives, by (3.3) and (2.2),

$$(3.5) \qquad \ell_i(x + \xi y) \equiv \kappa_i \frac{2^i}{i} B_i \pmod p$$

Vandiver conjectured that W/W^p is always trivial and this has recently been verified for all $p < 10^6$ in [B]. If this is true but p nonetheless divides B_i then $\ell_i(x + \xi y) \equiv 0 \pmod{p}$ by (3.3).

Similar arguments also apply in the 'second case' of Fermat's Last Theorem (see [Th2]).

4. p-Divisibility of Bernoulli numbers.

If $t \not\equiv -1, 0$ or $1 \pmod{p}$ then $\displaystyle\sum_{j=0}^{p-1} j^{3-1} t^j \equiv \frac{t(1+t)}{(1-t)^2}(1 - t^p) \not\equiv 0$

$\pmod{p}$. Since either $t = -x/y$ or $-y/z$ is $\not\equiv -1, 0$ or $1 \pmod{p}$, we deduce that p divides B_{p-3} by Kummer's Theorem (1.2). This argument generalizes for if p does not divide B_{p-n} (n odd) then

$$\sum_{j=0}^{p-1} j^{n-1} t^j \equiv \sum_{j=0}^{p-1} j^{n-1}(1 - t)^j \equiv 0 \pmod{p}$$

and so p must divide the resultant of these two polynomials (after dividing them both by appropriate powers of t and $(1-t)$). Krasner showed that this resultant is $< (n-1)!^{2(n-2)}$ (which is $< n^{2n^2}$) and thus $< p$ for $n < (\log p / \log\log p)^{1/2}$. Thus if (1.1) has a solution then p divides B_{p-n} for all $n \leq (\log p / \log\log p)^{1/2}$. Explicit computations of this resultant have shown that p must divide B_{p-n} for all $n \leq 45$.

Let $\Omega = \{i \text{ odd} : 3 \leq i \leq p-2 \text{ and } \theta_i \text{ is non-principal}\}$. Then

$$I_k/I_{-k} = \prod_{\substack{i=1 \\ i \text{ odd}}}^{p-1} \theta_i^{-2k^i} = \prod_{\substack{i=1 \\ i \text{ odd} \in \Omega}}^{p-1} \theta_i^{-2k^i} \quad \text{in} \quad C$$

(as $\theta_1 \in (\ker \tau)(\chi) \cong (C/C^p)(\chi)$ which is well-known to be trivial–see [Wa], §6.3). Let $r = |\Omega| + 1$. Evidently $\{I_k/I_{-k} : 1 \leq k \leq r\}$ must be multiplicatively dependent in C, so there exist integers $a_1, \ldots, a_r$ such that $\displaystyle\prod_{k=1}^{r} (I_k/I_{-k})^{a_k}$ is principal, and raising this to the pth power we get

$$\prod_{j=1}^{r} \left(\frac{x + \xi^j y}{x + \xi^{-j} y}\right)^{a_j} = uw^p \quad \text{for some } w \in K \text{ and unit } u.$$

Multiplying this equation by its conjugate we find that $u\bar{u}$ is a pth power, and so, as in section 2, $u = \xi^b$ for some integer b. So we may write

(4.1)
$$\prod_{j=1}^{r} \left(\frac{1-\xi^j t}{1-\xi^{-j}t} \right)^{a_j} = \xi^b w^p$$

where $t = -x/y$. Using (2.4), it is easily verified that

$$\sum_{n=1}^{p-2} \ell_n \left(\prod_{j=1}^{r} \left(\frac{1-\xi^j t}{1-\xi^{-j}t} \right)^{a_j} \right) \left(\sum_{k=1}^{p-1} k^{p-n} \xi^k \right) \equiv \frac{2 \sum_{j=1}^{r} j a_j}{1-t^p}$$
$$+ \sum_{j=1}^{r} j a_j \left(\frac{1}{1-\xi^j t} - \frac{1}{1-\xi^{-j}t} \right) \quad (\bmod\ p),$$

and from (4.1) this is $\equiv \sum_{k=1}^{p-1} \xi^k b \equiv -b \pmod{p}$. Therefore

$$\sum_{j=1}^{r} j a_j \left(\frac{1}{1-\xi^j t} - \frac{1}{1-\xi^{-j}t} \right) \equiv c \pmod{p}$$

for some integer c. Multiplying through by $\prod_{j=1}^{r}(1-\xi^j t)(1-\xi^{-j}t)$ we see that the coefficient of $\xi^{\frac{p-1}{2}}$ is $0 \pmod{p}$, and that the coefficient of ξ^i, for some smaller i, is $\not\equiv 0 \pmod{p}$ (provided $r(r+1) < p-1$) which is impossible. Thus $ii(p) = \#\{2n, 2 \le 2n \le p-3 \text{ and } p \mid B_{2n}\} \ge |\Omega| \ge \sqrt{p}-2$. (It seems that the proofs in the literature ([Wa], Thm 6.23), usually avoid explicitly using the Kummer homomorphism, but nonetheless take some equivalent logarithmic derivative).

5.   p-Divisibility of Fermat quotients.

There does not seem to be any particularly illuminating method of deducing the p-divisibility of Fermat quotients. This is perhaps because they seem to always be expressed as linear combinations of our polynomials $\sum_{j=0}^{p-1} j^{n-1} t^j$, and such combinations only seem to arise naturally in reference to explicit reciprocity laws.

Here we will develop a much shorter version of [GM]: Define

$$F_t(x) = \frac{1}{1-te^x} = \sum_{n \ge 0} f_n(t) \frac{x^n}{n!}$$

so that

$$f_n(t) = -\ell_{n+1}(1 - t\xi) \equiv \frac{1}{(1 - t^p)} \sum_{j=0}^{p-1} j^n t^j \quad (\text{mod } p)$$

by (2.2). From this we see that, for $t \not\equiv 0$ or 1 (mod p), $f_{p-1}(t) \equiv 0$ (mod p). By noting that

$$(-1)^{j-1} \cdot \frac{1}{p}\binom{p}{j} = \frac{1-p}{1} \cdot \frac{2-p}{2} \cdots \frac{(j-1)-p}{j-1} \cdot \frac{1}{j} \equiv \frac{1}{j} \quad (\text{mod } p)$$

we see that

$$f_{p-2}(t) \equiv \frac{1}{1-t^p} \cdot \left\{ \frac{(t-1)^p - t^p + 1}{p} \right\} \equiv \frac{1}{pz^p}\{x^p + y^p - (x+y)^p\} \quad (\text{mod } p)$$

for $t = -x/y$. Easy elementary arguments (see §4.3 of [Ri]) give $x^{p-1} \equiv y^{p-1} \equiv 1$ (mod p^2) and $x+y$ is a pth power, so $(x+y)^{p-1} \equiv 1$ (mod p^2); thus $f_{p-2}(t) \equiv 0$ (mod p). From this and Kummer's congruences (as proved in section 2), we have

$$(5.1) \qquad\qquad B_{p-1-n} f_n(t) \equiv 0 \quad (\text{mod } p)$$

for $n = 1, 2, \ldots, p - 1$.

In this section we will develop a theory of such congruences by considering the power series of which these functions are coefficients. Thus we rewrite (5.1) as:

For any integers q, r, s, we have

$(5.1)'$
$$\text{The coefficient of } X^{p-1} \text{ in } B(qX)\{F_t(rX) - F_t(sX)\}$$
$$\text{is } \equiv 0 \quad (\text{mod } p) \text{ if } p \nmid q$$

where $B(X) = X/(e^X - 1)$. Incidentally, from the identity

$$B((r - s)X)(F_t(rX) - F_t(sX)) = (r - s)XF_t(rX)(F_t(sX) - 1)$$

we know that

(5.2)
$$\text{The coefficient of } X^{p-2} \text{ in } F_t(rX)F_t(sX)$$
$$\text{is } \equiv 0 \quad (\text{mod } p) \text{ if } p \nmid r - s$$

(as $f_{p-2}(t) \equiv 0$ (mod p)) and so

$(5.2)'$ $\qquad f_n(t)f_{p-2-n}(t) \equiv 0$ (mod p) for $n = 0, 1, 2, \ldots, p - 2$.

There are two other functions, related to $f_{p-2}(t)$ and B_n, which will be of special interest: First, $W_u(t)$, the coefficient of $X^{p-2}/(p - 2)!$ in $F_{t,u}(X) = e^{uX}/(1 - te^X)$; proceeding as in the proof of (2.4), we have

$$W_u(t) \equiv \frac{1}{1-t^p} \sum_{j=0}^{p-1} (j+u)^{p-2} t^j \pmod{p}.$$ Second, the $(p-1)$st Bernoulli polynomial $B_{p-1}(u)$, the coefficient of $X^{p-1}/(p-1)!$ in $Xe^{uX}/(e^X-1)$. We also define

$$C_{m,n}(t) = \frac{1}{m} \sum_{j=0}^{m-1} \left\{ B_{p-1}\left(\frac{j}{m}\right) - B_{p-1} \right\} t^{\beta(j,n,m)},$$

where $\alpha(a,b,c)$, $\beta(a,b,c)$ are the least positive, non-negative residues of $a/b \pmod{c}$, respectively. Finally let

$$A_{m,n}(t) := \frac{1}{n} \sum_{j=1}^{n-1} t^{\alpha(j,n,m)} W_{j/n}(t).$$

Our starting point is the identity, for $(m,n)=1$,

$$X \sum_{j=1}^{n-1} t^{\alpha(j,n,m)} F_{t,j/n}(nX) + \sum_{j=0}^{m-1} \frac{X(e^{jX}-1)}{e^{mX}-1} t^{\beta(j,n,m)} + Xt^m F_t(nX) +$$
$$+ \frac{t^m-1}{m} B(mX)\{F_t(nX) - F_t(0)\} = 0.$$

Considering the coefficient of $X^{p-1}/(p-1)!$ here, and using (5.1)', we have

(5.3) $$C_{m,n}(t) \equiv A_{m,n}(t) \pmod{p}.$$

If $\ell \equiv n \pmod{m}$ then each $\beta(j,\ell,m) = \beta(j,n,m)$ so that $C_{m,\ell}(t) = C_{m,n}(t)$. Moreover if $\ell \equiv -n \pmod{m}$ then $\beta(j,\ell,m) = \beta(m-j,n,m)$ for $1 \le j \le m-1$ and $B_{p-1}\left(\frac{m-j}{m}\right) = B_{p-1}\left(\frac{j}{m}\right)$ (as $B_{p-1}(1-u) = B_{p-1}(u)$ for all u), so that $C_{m,\ell}(t) = C_{m,n}(t)$. We therefore deduce, from (5.3), that

(5.3)' If m divides $\ell \pm n$ then $A_{m,n}(t) \equiv A_{m,\ell}(t) \pmod{p}$.

Also, if m is an odd prime, then

$$2 \sum_{n=1}^{(m-1)/2} C_{m,n}(t) = \sum_{n=1}^{m-1} C_{m,n}(t),$$

which is the coefficient of $X^{p-1}/(p-1)!$ in

$$\frac{1}{m} \sum_{n=1}^{m-1} \sum_{j=0}^{m-1} X\left(\frac{e^{(j/m)X}-1}{e^X-1}\right) \cdot t^{\beta(j,n,m)} = \frac{t^m-t}{t-1}\left(\frac{X/m}{e^{X/m}-1} - \frac{X}{e^X-1}\right)$$

which equals $\frac{t^m - t}{t-1} B_{p-1}(m^{-(p-1)} - 1) \equiv \frac{t^m - t}{t-1} \cdot \frac{m^{p-1} - 1}{p}$ (mod p) by the Von Staudt-Clausen Theorem. Thus

$$(5.4) \qquad \sum_{\ell=1}^{(m-1)/2} C_{m,\ell}(t) \equiv \frac{t}{2(t-1)} \cdot (t^{m-1} - 1) \cdot \frac{m^{p-1} - 1}{p} \quad (\text{mod } p).$$

So we can now state our induction hypothesis:

$[W_{n,t}]$: $\qquad W_{j/\ell}(t) \equiv 0 \pmod{p}$ for $1 \leq j < \ell \leq n-1$.

Suppose that this holds:

By definition, $A_{m,\ell}(t) \equiv 0 \pmod{p}$ whenever $1 \leq \ell \leq n-1$ and $(m,\ell) = 1$. For any $m, 1 \leq m \leq 2n-1$ with $(m,n) = 1$, let $\ell = |n - m|$ so that $(m,\ell) = 1$. Then, by (5.3)', $A_{m,n}(t) \equiv A_{m,\ell}(t) \equiv 0 \pmod{p}$, and so

$$(5.5) \qquad \sum_{\substack{j=1 \\ (j,n)=1}}^{n-1} t^{\alpha(j,n,m)} W_{j/n}(t) \equiv 0 \quad (\text{mod } p).$$

If m is an odd prime then $C_{m,\ell}(t) \equiv 0 \pmod{p}$ for $1 \leq \ell \leq n-1$ by (5.3), and so

$$(5.6) \qquad (t^{m-1} - 1) \cdot \frac{m^{p-1} - 1}{p} \equiv 0 \quad (\text{mod } p)$$

by (5.4). Thus if there exists t satisfying $[W_{n,t}]$ of order $\geq 2n-1$ modulo p, then $m^{p-1} \equiv 1 \pmod{p^2}$ for all primes $m \leq 2n-1$.

We would like to be able to deduce $[W_{n+1,t}]$ from $[W_{n,t}]$. This follows from showing that the matrix

$$A_n(t) = \{t^{\alpha(j,n,m)}\}_{\substack{1 \leq m \leq 2n \\ 1 \leq j \leq n \\ (mj,n)=1}}$$

has full rank (i.e. $\phi(n)$) in $\mathbf{F}_p^*$, for then the only solutions to (5.5) come from each $W_{j/n}(t) \equiv 0 \pmod{p}$. Currently the only known method for doing this is to explicitly compute a few $\phi(n) \times \phi(n)$ subdeterminants and show that they cannot all be simultaneously 0 (mod p) (the record is for all $n \leq 46$, [GM], where this is all studied in minute detail).

We observe here that if $A_n(t)$ doesn't have full rank over the complex numbers then either $t = 0$ or t is an algebraic integer and unit: in particular t cannot be any rational integer other than $-1, 0$ or 1. To see this look at the square matrix formed by the top $\phi(n)$ rows of $A_n(t)$. The term of lowest (respectively, highest) degree in the mth row is t (t^m), and the furthest such term to the right (left) occurs on the reverse (main) diagonal, that is in column $j = n - m$ ($j = m$). Thus the term of lowest (highest) degree in the determinant of this matrix is $t^{\phi(n)}$ $(t^{n\phi(n)/2})$. So, if $A_n(t)$ does not have full rank then this determinant is 0; by expanding minors we observe that the determinant belongs to $\mathbf{Z}[t]$, and dividing through by $t^{\phi(n)}$ if $t \neq 0$, we see that t is an algebraic integer and a unit (since the polynomial is monic and has last term 1).

6. p-divisibility of Bernoulli polynomials.

(5.3) implies that $C_{m,\ell}(t) \equiv 0 \pmod{p}$ for $1 \leq \ell \leq n-1$ and any $(m, \ell) = 1$, if $[W_{n,t}]$ holds. We wish to deduce that

$$(6.1) \quad B_{p-1}\left(\frac{j}{m}\right) - B_{p-1} \equiv 0 \pmod{p} \quad \text{for all} \ \ 1 \leq j \leq m \leq 2n-1,$$

which we shall do by induction on m: So suppose (6.1) holds for all $m' < m$. Then, as $B_{p-1}\left(\frac{m-j}{m}\right) = B_{p-1}\left(\frac{j}{m}\right)$, we have

$$\sum_{\substack{1 \leq j < m/2 \\ (j,m)=1}} \left(B_{p-1}\left(\frac{j}{m}\right) - B_{p-1}\right)(t^{\beta(j,\ell,m)} + t^{m-\beta(j,\ell,m)}) \equiv 0 \pmod{p}$$

for each $\ell, 1 \leq \ell < m/2$ with $(\ell, m) = 1$. Thus (6.1) follows for m provided the matrix

$$S_m(t) = \{t^{\beta(j,\ell,m)} + t^{m-\beta(j,\ell,m)}\}_{\substack{1 \leq j, \ell < m/2 \\ (j\ell,m)=1}}$$

has full rank in $\mathbf{F}_p^*$; in other words, non-zero determinant. (Remark: If $[W_{n,1-t}]$ also holds then we can combine $S_m(t)$ and $S_m(1-t)$ to get double as many rows). However, it is easy to show that

$$(6.2) \qquad \det S_m(t) = \prod_{\substack{\chi \ an \ even \\ character \ \pmod{m}}} \left(\sum_{j=1}^{m} \chi(j)t^j\right).$$

Recently, Dilcher and Skula used this to establish (6.1) for all $m \leq 46$, and Cikánek for $m \leq 94$ with p sufficiently large.

Define the generalized Bernoulli number $B_{n,\chi} = m^{n-1} \sum_{j=1}^{m} \chi(j)B_n\left(\frac{j}{m}\right)$

for characters $\chi \pmod{m}$. Thus, (6.1) implies that

$$(6.3) \qquad\qquad B_{p-1,\chi} \equiv 0 \pmod{p}$$

for all non-principal characters χ (mod m). Let $L_p(s, \chi)$ be Leopoldt's p-adic L-function. By Theorem 5.11 of [Wa], (6.3) implies that $L_p(2-p, \chi) \equiv 0$ (mod p), and so, by Corollary 5.13 of [Wa],

(6.4)
$$L_p(n, \chi) \equiv 0 \ ^{\displaystyle\cdot} \ (\text{mod } p) \quad \text{for any integer } n \text{ and even,}$$
$$\text{non} - \text{principal character } \chi \quad (\text{mod } m).$$

Let u be the least positive residue of j/m (mod p). Then

$$\frac{X(e^{uX} - 1)}{e^X - 1} = \sum_{n \geq 1} \left(n \sum_{i=0}^{u-1} i^{n-1} \right) \frac{X^n}{n!}, \quad \text{so that}$$

$$B_{p-1}\left(\frac{j}{m}\right) - B_{p-1} \equiv (p-1) \sum_{1 \leq i \leq u-1} \frac{1}{i} \quad (\text{mod } p).$$

$u-1$ runs through $\left[\frac{p}{m}\right], \left[\frac{2p}{m}\right], \ldots, \left[\frac{(m-1)p}{m}\right]$ as j runs through $1, 2, \ldots, m-1$, so that

(6.5)
$$\sum_{1 \leq i \leq \left[\frac{jp}{m}\right]} \frac{1}{i} \equiv 0 \quad (\text{mod } p) \qquad \text{for } 1 \leq j \leq m.$$

For each i, $\frac{jp}{m} < i \leq \frac{(j+1)p}{m}$, we consider $k = mi - jp$ in (6.5); thus, for $h \equiv -jp$ (mod m),

(6.6)
$$\sum_{\substack{1 \leq k \leq p-1 \\ k \equiv h \ (\text{mod } m)}} \frac{1}{k} \equiv 0 \quad (\text{mod } p) \qquad \text{for } 0 \leq h \leq m - 1.$$

Therefore if g is any p-integral valued function, of period m, then

(6.7)
$$\sum_{i=1}^{p-1} \frac{g(i)}{i} \equiv 0 \quad (\text{mod } p).$$

A particular instance of this is for $g(i) = \xi^i$ for some mth root of unity ξ; giving $f_{p-2}(\xi) \equiv 0$ (mod p), and so

(6.8)
$$(1 - \xi)^p \equiv 1 - \xi^p \quad (\text{mod } p^2) \quad \text{for each } \xi^m = 1.$$

Suppose now that m is a prime $\equiv 1$ (mod 4) and let $\varepsilon = u + v\sqrt{m}$ and $h = h(m)$ be the fundamental unit and class number of $\mathbf{Q}(\sqrt{m})$, respectively. Define $u_n + v_n\sqrt{m} = \varepsilon^n$ and note that

$$\varepsilon^p = (u + v\sqrt{m})^p \equiv u^p + v^p m^{(p-1)/2}\sqrt{m} \equiv u + \left(\frac{m}{p}\right) v\sqrt{m} \bmod p,$$

where $\left(\frac{\cdot}{p}\right)$ is the Legendre symbol, so that $\varepsilon^{p-\left(\frac{m}{p}\right)} \equiv \pm 1 \pmod{p}$. Thus, as is well known, p divides $v_{p-\left(\frac{m}{p}\right)}$, analogous to Fermat's Little Theorem.

Now suppose that $\varepsilon^{p-\left(\frac{m}{p}\right)} = \pm 1 + p\omega$ for some $\omega \in \mathbf{Z}[1, \frac{1+\sqrt{m}}{2}]$, so that $\varepsilon^{2h(p-\left(\frac{m}{p}\right))} \equiv 1 \pm 2hp\omega \pmod{p^2}$. On the other hand it is well known that $\varepsilon^{2h} = \prod_{a=1}^{m-1}(1-\xi^a)^{-\left(\frac{a}{m}\right)}$ where ξ is a primitive mth root of unity, and so, by (6.8),

$$\varepsilon^{2hp} = \prod_{a=1}^{m-1}(1-\xi^a)^{-p\left(\frac{a}{m}\right)} \equiv \prod_{a=1}^{m-1}(1-\xi^{ap})^{-\left(\frac{a}{m}\right)}$$

$$\equiv \prod_{b=1}^{m-1}(1-\xi^b)^{-\left(\frac{a\frac{p-1}{m}}{}\right)} \equiv (\varepsilon^{2h})^{\left(\frac{p}{m}\right)} \pmod{p^2}.$$

Therefore

$$\varepsilon^{2h(p-\left(\frac{m}{p}\right))} \equiv 1 \pmod{p^2},$$

and, by comparing congruences, we find that p divides $2h\omega$. However $h < m$ so that if $m < p$ then p must divide ω and thus

(6.9) p^2 divides $v_{p-\left(\frac{m}{p}\right)}$.

One expects that this happens very rarely, and for $m = 5$, in which case the v_n are the Fibonacci numbers (see [SS]), it is known that (6.9) fails for all $p < 2^{32}$. In one final observation we note that, by the elementary theory of binary quadratic forms, (6.9) is equivalent to the assertion that

(6.10) $h(mp^4) = ph(mp^2)$

where, here, $h(d)$ is the number of equivalence classes of quadratic forms of discriminant d (usually $h(mp^4) = h(mp^2)$ rather than (6.10)).

7.   A special case.

In this section we focus on the special case where $x \equiv y \pmod{p}$.

If the elements of $\{-x/y, -y/z, -z/x, -y/x, -z/y, -x/z\}$ are not distinct $\pmod{p}$, then evidently they are either all sixth roots of unity $\pmod{p}$, or they are the set $\{-1, 2, 1/2\}$. The first case here was disposed of by Pollaczek [P] (though his argument needed correcting by Gunderson), while the second remains open. In this section we apply the criteria to this second case. So let $t = 2$.

By (2.5) we find that

$$(-1)^{n+1} f_n(2) = \sum_{j=1}^{n+1} \frac{s_{n+1,j}}{j} = 2 \sum_{j=1}^{n} s_{n,j}.$$

By definition each $s_{n,j}$ is a non-negative integer and $s_{n,1} = 1$ so that $(-1)^{n+1} f_n(2)$ is a positive integer. However since

$$s_{n,j} = j(s_{n-1,j} + s_{n-1,j-1}) \le n(s_{n-1,j} + s_{n-1,j-1}),$$

we deduce that

$$|f_n(2)| = \sum_{j=1}^{n} s_{n,j} \le 2n \sum_{j=1}^{n-1} s_{n-1,j} \le 2n \cdot 2(n-1) \sum_{j=1}^{n-2} s_{n-2,j} \le \ldots \le 2^{n-1} n!$$

Thus p does not divide $f_n(2)$ for any $n \le \log p / \log \log p$, and so, by Kummer's theorem,

$$(7.1) \qquad B_{p-n} \equiv 0 \pmod{p}, \text{ for all } n \le \log p / \log \log p.$$

By the final remarks of section 5 we know that the matrix formed by the top $\phi(n)$ rows of $A_n(2)$ has non-zero determinant. Since any entry in the mth row is smaller than 2^m, Hadamard's inequality tells us that this determinant is $< (\phi(n) 2^n)^{\phi(n)/2}$, and since the determinant is an integer, it cannot be divisible by p if $n < \sqrt{\log p}$. Thus $[W_{n,2}]$ holds in this range, and since 2 must have order $> \log p$ modulo p, we deduce from (5.6) that

$$(7.2) \qquad p^2 \text{ divides } m^p - m \text{ for all } m < 2\sqrt{\log p}.$$

Working in the field of the rationals extended by the primitive $\phi(m)$th roots of unity, we see that $(\det S_m(2))/2^{\phi(m)/2} \equiv 1 \pmod{2}$ (by (6.2)), and so is non-zero. Moreover by the definition of $S_m(2)$ it is clear that its determinant is an integer, and by Hadamard's inequality has absolute value

$$\le \left(\sum_{a=1, (a,m)=1}^{m/2} (2^a + 2^{m-a})^2 \right)^{\phi(m)/4} \le 2^{m\phi(m)/2}.$$

Thus this cannot be divisible by p if $m < 2\sqrt{\log p}$ and so, by the results of section 6, we know that

$$(7.3) \quad B_{p-1}\left(\frac{j}{m}\right) - B_{p-1} \equiv 0 \pmod{p} \text{ for all } 1 \le j \le m < 2\sqrt{\log p}.$$

Of course all of the equivalent formulations given in section 6, also follow for this range of values of m. (Remark: With a little care, the constant '2' in the '$2\sqrt{\log p}$' above can be improved.)

Collecting these criteria like this, it seems extremely unlikely that such implausible statements can all be simultaneously true, yet we are unable to show that this is the case (even for an infinite sequence of primes).

8. Bernoulli polynomials revisited.

Define

$$C^*_{m,n}(t) = \frac{1}{m} \sum_{\substack{j=0 \\ (j,m)=1}}^{m-1} \left(B_{p-1}\left(\frac{j}{m}\right) - B_{p-1}\right)t^{\beta(j,n,m)}.$$

Then

$$\sum_{n=1}^{m-1} \overline{\chi}(n)C^*_{m,n}(t) =$$

$$= \frac{1}{m} \sum_{\substack{j=1 \\ (j,m)=1}}^{m-1} \left(B_{p-1}\left(\frac{j}{m}\right) - B_{p-1}\right) \sum_{\substack{n=1 \\ (n,m)=1}}^{m-1} \overline{\chi}(j)\chi(j/n)t^{\beta(j,n,m)}$$

$$\equiv B_{p-1,\overline{\chi}}\left(\sum_{\substack{r=1 \\ (r,m)=1}}^{m} \chi(r)t^r\right) \quad (\text{mod } p)$$

if χ is non-principal. (Note that $C^*_{m,n}(t) = \sum_{d|m} \frac{\mu(d)}{d}C_{m/d,n}(t^d)$.)

Now suppose that m is prime so that $C^*_{m,n}(t) = C_{m,n}(t)$. From $[W_{n,t}]$ we obtain that, for even character χ (mod m),

$$(8.1) \qquad B_{p-1,\chi} \cdot \left(\sum_{r=0}^{m-1} \overline{\chi}(r)t^r\right) \equiv 0 \quad (\text{mod } p),$$

where $B_{p-1,\chi_0} = \frac{m^{p-1}-1}{p}$ by our (new) definition.

Evidently there exists a prime ideal $\mathbf{p}$ dividing p in $\mathbf{Q}(\xi_{m-1})$, and a character χ (mod m), such that

$$(8.2) \qquad \mathbf{p} \text{ divides } B_{p-1,\chi}$$

for, if not, each $\sum_{r=1}^{m-1} \overline{\chi}(r)t^r \equiv 0 \pmod{p}$ and so

$$0 \equiv \sum_{\substack{\chi \text{ even}}} \chi(s) \sum_{r=0}^{m-1} \overline{\chi}(r)t^r \equiv \frac{m-1}{2} \cdot (t^s + t^{m-s}) \pmod{p}$$

which is impossible for $s = \frac{m-1}{2}$, $t \not\equiv 0$ or $-1 \pmod{p}$.

By Iwasawa's *Main Conjecture*, proved by Coates and Wiles, we can deduce from (8.2) that

$$(8.3) \qquad (C/C^p)^{\frac{1}{\phi(n)(p-1)} \sum_{a=1}^{mp} \left(\sum_{\tau \in \text{Aut}(Q(\chi)/Q)} (\tau\chi)(a) a\sigma_a^{-1} \right)}$$

is non-trivial, where C is now the ideal class group of $Q(\xi_{np})$. It seems to me to be extremely likely that one should be able to explicitly identify some element of this component (like in section 2), though I have thus far been unable to do so. If one did, it would surely lead to a new approach to proving results like those in the last four sections, and perhaps a much better understanding as to why they hold.

9. A curve with many F_p^2-points.

Let g be a primitive root modulo p, and define $\log_p j$ to be that integer (modulo $p-1$) for which $g^{\log_p j} \equiv j \pmod{p}$. Define

$$G_p(X, Y) = \sum_{j=1}^{p-1} X^{\log_p j} Y^j,$$

so that

$$G_p(g^n, t) \equiv \sum_{j=1}^{p-1} j^n t^j \equiv (1 - t^p) f_n(t) \pmod{p},$$

and thus if (1.1) has solutions then, by (5.2)',

$$(9.1) \quad G_p(g^n, t) G_p(g^{p-2-n}, t) \equiv 0 \pmod{p} \quad \text{for } n = 0, 1, 2, \ldots, p-2.$$

This will usually lead to at least $3(p-1)$ non-trivial zeros of the curve $G_p(x, y) = 0$ in F_p^2. Generally we expect a curve to have around p such zeroes, and very rarely as many as $3(p-1)$. This is true of $G_p(x, y) = 0$ for the primes $p < 1000$. Unfortunately we are unable to use Weil's Theorem to *prove* something of this sort since there one needs a curve of low genus ($< \sqrt{p}$), while $G_p(x, y) = 0$ seems to have high genus (order p^2).

10. Our matrices revisited.

It seems to be difficult to prove that $A_n(t)$ has full rank directly. (If we could do so then one could deduce that p^2 divides $m^p - m$ for all $m < (\log p)^{1/4}$ — see [GM].) However the following generalization seems worth pursuing:

Let $A(X)$ be an m-by-n matrix (with $m \geq n$), in which each entry is a power of X, and suppose that every n-by-n submatrix of $A(X)$ has non-zero determinant. We conjecture that there is some function $m(n)$ such that if $n \geq m(n)$ and $A(t)$ does not have full rank then either t is 0 or a root of unity. Perhaps we can even take $m(n) = 2n$.

Van der Poorten and I observed that if $m \geq n(n^2 - 2n + 3)/2$ and $A(t)$ has rank $r < n$ then either t is 0 or an algebraic unit. We will prove this assuming also that t is an algebraic integer – the necessary modifications for arbitrary algebraic t are minor: If t is not a unit then select an r-by-$(r + 1)$ submatrix $B(X)$ of $A(X)$, such that $B(t)$ has rank r (which exists since $A(t)$ has rank r). Let $p_j(X)$ be the determinant of the r-by-r minor formed by deleting the jth column of $B(X)$ – at least one $p_j(X)$ must be non-zero else $B(t)$ would have rank $< r$. Now if $(X^{a_1}, \dots, X^{a_{r+1}})$ are the entries (in the same columns as in $B(X)$) of any other row of $A(X)$ then the determinant of the matrix formed by adjoining this new row to $B(X)$ is

(10.1) $X^{a_1} p_1(X) - X^{a_2} p_2(X) \dots \pm X^{a_{r+1}} p_{r+1}(X).$

Let $(t)^{\pi_j}$ be the largest power of the ideal (t) that divides $p_j(t)$, and define Q to be the smallest $a_j + \pi_j$. Since the sum in (10.1) with $X = t$ equals 0, we examine this sum modulo $(t)^{Q+1}$ and observe that there must be integers $i \neq j$ such that $a_i + \pi_i = Q = a_j + \pi_j$. Since there are $\binom{n}{2}$ possible pairs (i, j), some such pair must occur here at least n times since we have $m - (n-1) > \binom{n}{2}(n - 1)$ possibilities for the row that we adjoined to $B(X)$. But in the matrix M formed by n such rows we see that X^{π_i} times the ith column equals X^{π_j} times the jth column, and thus M does not have full rank, contradicting the hypothesis.

Acknowledgments: Much of my interest and understanding in this topic evolved during the enjoyable period in which I was a doctoral student under Paulo Ribenboim — thank you. This paper also benefitted from discussions and/or correspondence with Enrico Bombieri, Karl Dilcher, Ladja Skula, Zhi-Wei Sun and Alf Van der Poorten, as well as the excellent referee — thanks to you all.

References

[B] Buhler, J.P., Crandall, R.E. and Sompolski, R.W., *Irregular primes to one million*, (preprint).

[Fr] Frobenius, G., *Über den Fermatschen Satz* III, Sitzungsber. Adad. Wiss. Berlin (1914), 653-681.

[GM] Granville, A. and Monagan, M. B., *The First Case of Fermat's Last Theorem is True for all Prime Exponents up to 714,591,416,091,389*, Trans. A.M.S., **306** (1988), 329-359.

[Ku] Kummer, E. E., *Einige Sätze über die aus den Wurzeln der Gleichung $\alpha^\lambda = 1$ gebildeten complexen Zahlen, für den Fall dass die Klassenzahl durch λ theilbar ist, nebst Anwendungen derselben auf einen weiteren Beweis des letztes Fermat'schen Lehrsatzes*, Math. Abh. Akad. Wiss., Berlin, 1857, 41-74.

[Le] Lehmer, E., *On congruences involving Bernoulli numbers and the quotients of Fermat and Wilson*, Annals of Math., **39** (1938), 350-359.

[P] Pollaczek, F., *Über den grossen Fermat'schen Satz*, Sitzungsber. Akad. d. Wiss. Wien IIa, **126** (1917), 45-59.

[Ri] Ribenboim, P., *13 Lectures on Fermat's last theorem*, (Springer-Verlag, New York, 1979).

[Sk] Skula, L., *Fermat's Last Theorem and the Fermat quotients*, Comm. Math. Univ. Sancti Pauli, **41** (1992), 35-54.

[SS] Sun, Z.-H. and Sun, Z.-W., *Fibonacci numbers and Fermat's Last Theorem*, Acta Arith., **60** (1992), 371-388.

[Th1] Thaine, F., *On the ideal class groups of real abelian number fields*, Annals of Math., **128** (1988), 1-18.

[Th2] Thaine, F., *On Fermat's Last Theorem and the Arithmetic of $\mathbf{Z}[\xi_p + \xi_p^{-1}]$*, J. of Number Theory, **29** (1988), 297-299.

[Wa] Washington, L.C., *Introduction to Cyclotomic Fields*, Grad. Text in Math. 83, (Springer-Verlag, New York, 1982).

[Wi] Wieferich, A., *Zum letzten Fermat'schen Satz*, J. Reine Angew. Math. **136** (1909), 293-302.

Andrew Granville
Department of Mathematics
University of Georgia
Athens, GA 30602
USA
e-mail andrew@sophie.math.uga.edu

Class groups of exponent two
in real quadratic fields

S. Louboutin, R.A. Mollin and H.C. Williams

University of Caen, France
University of Calgary
and
University of Manitoba

Abstract

The purpose of this paper is to list all real quadratic fields $\mathbb{Q}(\sqrt{D})$ of Extended Richaud–Degert type (i.e, those square–free $D = m^2 + r$ with $4m \equiv 0 \pmod{r}$), (simply, ERD–types) with class group of exponent 2, and to prove that the list is complete with one possible exceptional value remaining, the existence of which would be a counter–example to the Generalized Riemann Hypothesis (GRH).

1 Introduction

In [4] and [5] Louboutin determined all real quadratic fields $K = \mathbb{Q}(\sqrt{D})$ where D is square–free of the form $D = m^2 \pm 1$ and the exponent of its class group C_K is 2, (with one possible exception). This extended much weaker results by Leu in [3]. The forms studied by Leu and Louboutin are very special cases of ERD–types. It is the purpose of this paper to complete the task by determining (with one possible exception) all real quadratic fields $K = \mathbb{Q}(\sqrt{D})$ having C_K of exponent 2 where D is of ERD–type. In [10] the last two authors determined all ERD–types having class number $h(D) = 1$ (with one possible exception), and all ERD–types having $h(D) = 2$ in [11]. Moreover the solution in [10] provided applications to conjectures in the literature; viz., one of Chowla [1], one of Yokoi [15], one of Mollin [7], and three of Mollin–Williams [8,9]. Therefore this paper completes the task of determining all real quadratic fields of ERD–type having class group C_K of exponent $e_K \leq 2$, (with one possible exceptional value ruled out by the GRH). The next section provides notation and preliminary results needed to provide the proof of the main result. A few remarks on our tables are given in Section 4.

2   Notations and Preliminaries

Throughout $K = \mathbb{Q}(\sqrt{D})$ where D is a positive square–free integer. The class group C_K has order h, (the class number), and exponent e_K. Equivalence of ideals I and J in C_K is denoted $I \sim J$. $N(\mathfrak{P})$ denotes the norm of an ideal $\mathfrak{P}$ in the ring of integers $\mathcal{O}_K$ of K. The discriminant of K is $\Delta = 4D/\sigma^2$ where

$$\sigma = \begin{cases} 2 & \text{if } D \equiv 1 \pmod 4 \\ 1 & \text{if } D \equiv 2, 3 \pmod 4. \end{cases}$$

Let $\omega = \frac{\sigma - 1 + \sqrt{D}}{\sigma}$, then the maximal order $\mathcal{O}_K$ in K is $\mathbb{Z} + \omega\mathbb{Z}$. The continued fraction expansion of ω is denoted by $\omega = < q, q_1, q_2, \cdots, q_\ell >$ of period length ℓ where $q = q_0 = \lfloor \omega \rfloor$ $q_i = \lfloor (P_i + \sqrt{D})/Q_i \rfloor$ for $i \geq 1$ (here $\lfloor \ \rfloor$ denotes the greatest integer function), $(P_0, Q_0) = (\sigma - 1, \sigma)$ and recursively for $i \geq 0$ we have $P_{i+1} = q_i Q_i - P_i$ and $Q_{i+1}Q_i = D - P_{i+1}^2$.

The above will be used extensively throughout the paper as well as the following general well known facts, (e.g., see [14]).

Theorem 2.1. C_K *is generated by the non–inert prime ideals $\mathfrak{P}$ with* $N(\mathfrak{P}) < \sqrt{\Delta}/2$.

Thus immediate from Theorem 2.1 is a class number one criterion.

Theorem 2.2. $h = 1$ *if and only if $p = Q_i/\sigma$ for some i with $0 < i < \ell$ whenever p is a non–inert prime less than $\sqrt{\Delta}/2$.*

Also we get

Theorem 2.3. $e_K \leq 2$ *if and only if for all primes $q < \sqrt{\Delta}/2$ with q a split prime in K we have $\mathfrak{q} \sim 1$ where $\mathfrak{q}$ is an $\mathcal{O}_K$–prime above q.*

The following will be useful in determining when $e_K \leq 2$.

Algorithm 2.1
   (1) For each prime $q < /\sqrt{\Delta}/2$ such that $(\Delta/q) = 1$, find P_0 such that $P_0^2 \equiv D \pmod{\sigma q}$ and put $Q_0 = \sigma q$.
   (2) Determine the continued fraction expansion of $(P_0 + \sqrt{\Delta})/Q_0$ until we find the least k such that $Q_k = Q_0$.
   (3) If for each such continued fraction we get $P_k \equiv -P_0 \pmod{Q_0}$, then the class group C_K has exponent $e_K \leq 2$.

Proof. Let $\mathfrak{q}$ be the ideal $(Q_0/\sigma)\mathbb{Z} + ((P_0 + \sqrt{D})/\sigma)\mathbb{Z}$. By Theorem 2.3 we must have $\mathfrak{q}^2 \sim 1$ or $\mathfrak{q} \sim \bar{\mathfrak{q}}$. Since $\mathfrak{q}$ is reduced, so is $\bar{\mathfrak{q}}$, and since they are in the same ideal class we must find some k such that

$$\mathfrak{q} = (Q_k/\sigma)\mathbb{Z} + ((P_k + \sqrt{D})/\sigma)\mathbb{Z}.$$

Since $N(\mathfrak{q}) = N(\bar{\mathfrak{q}}) = Q_0/\sigma$ and $\mathfrak{q} \neq \bar{\mathfrak{q}}$ (since $\mathfrak{q}$ is not ambiguous), we must have $P_k \equiv -P_0 \pmod{Q_0}$.

3 Exponent 2 and ERD–types

All of the notation in Section 2 is in force.

Remark 3.1 In Theorem 3.1. below we will prove:

(I) If D is of ERD–type, then any primitive principal ideal with norm N, not dividing D, must satisfy $N > \sqrt{D}/\gamma$ where

$$\gamma = \begin{cases} 2 & \text{if } D \not\equiv 1 \pmod 4 \\ 3 & \text{if } D \equiv 1 \pmod 4. \end{cases}$$

(II) If D is of ERD–type and C_K is of exponent 2, then $(D/p) \neq 1$ whenever $p^2 < \sqrt{D}/\gamma$.

Thus (II) provides us with an efficient test in order to sieve discriminants up to large values, and the Siegel–Tatsuzawa version of the Brauer–Siegel theorem provides us with an upper bound on the discriminants of the ERD fields with exponent 2.

Theorem 3.1. If D is of ERD–type, then $e_K = 2$ if and only if D is one of the 228 values in Tables 1 to 3, with possibly one exceptional value remaining, whose existence would be a counter–example to the Generalized Riemann Hypothesis (GRH).

Proof. Let $D = m^2 + k$ where $4m \equiv 0 \pmod k$. We consider the continued fraction expansion of ω in various cases.

<u>Case 1.</u> $D \equiv 1 \pmod 4$) $k > 0$ and $m \equiv 0 \pmod k$; (whence $m = \lfloor \sqrt{D} \rfloor$ is even and k is odd). Here $\ell = 6$, where the continued fraction expansion of ω is

n	0	1	2	3	4
P_n	1	$m\text{-}1$	$(k+1)/2$	$m\text{-}k$	$m\text{-}k$
Q_n	2	$m+(k\text{-}1)/2$	$m-(k\text{-}1)/2$	$2k$	$\vdots$
q_n	$m/2$	1	1	$m/k\text{-}1$	$\vdots$

Case 2. $D \not\equiv 1 \pmod 4$, $k > 0$ and $m \equiv 0 \pmod k$; (whence $m = \lfloor \sqrt{D} \rfloor$). Here $\ell = 2$.

n	0	1	2
P_n	0	m	m
Q_n	1	k	1
q_n	m	$2m/k$	$2m$

Case 3. $D \equiv 1 \pmod 4$), $k < 0$ and $m \equiv 0 \pmod k$; (whence $m - 1 = \lfloor \sqrt{D} \rfloor$ is odd and k is odd, $|k| \leq m/2$). Here $\ell = 4$.

n	0	1	2	3	4
P_n	1	m-1	$m+k$	$m+k$	m-1
Q_n	2	$m+(k$-$1)/2$	$2| k |$	$m+(k$-$1)/2$	2
q_n	$m/2$	2	$| m/k | -1$	2	m-1

Case 4. $D \not\equiv 1 \pmod 4$, $k < 0$ and $m \equiv 0 \pmod 4$; (whence $m - 1 = \lfloor \sqrt{D} \rfloor$ and $|k| \leq m/2$). Here $\ell = 4$.

n	0	1	2	3	4
P_n	0	m-1	$m+k$	$m+k$	m-1
Q_n	1	$2m+k$-1	$| k |$	$2m+k$-1	1
q_n	m-1	1	$2	m/k	-2$	1	$2m$-2

Case 5. $D \not\equiv 1 \pmod 4$, $k > 0$, $2m \equiv 0 \pmod k$ and $m \not\equiv 0 \pmod 4$; (whence $k \not\equiv 0 \pmod 4$, $k \equiv 0 \pmod 2$ and $m = \lfloor \sqrt{D} \rfloor$). Here $\ell = 2$.

n	0	1	2
P_n	0	m	m
Q_n	1	k	1
q_n	m	$2m/k$	$2m$

Case 6. $D \not\equiv 1 \pmod 4$, $-2m < k < 0$, $2m \equiv 0 \pmod k$ and $m \not\equiv 0 \pmod k$; (whence $m - 1 = \lfloor \sqrt{D} \rfloor$, and $|k| \leq 2m/3$). Here $\ell = 4$.

n	0	1	2	3	4
P_n	0	m-1	$m+k$	$m+k$	m-1
Q_n	1	$2m+k$-1	$| k |$	$2m+k$-1	1
q_n	m-1	1	$| 2m/k | -2$	1	$2m$-2

<u>Case 7.</u> $D \equiv 1 \pmod 4$, $k > 0$, $4m \equiv 0 \pmod 4$ and $2m \not\equiv 0 \pmod k$; (whence $m = \lfloor \sqrt{D} \rfloor$). Here $\ell = 2$.

n	0	1	2
P_n	1	m	m
Q_n	2	$k/2$	2
q_n	$(m+1)/2$	$4m/k$	m

<u>Case 8.</u> If $D \equiv 1 \pmod 4$, $k < 0$, $4m \equiv 0 \pmod k$ and $2m \not\equiv 0 \pmod k$; (whence $\lfloor \sqrt{D} \rfloor = m - 1$, and $|k| \leq 4m/3$). Here $\ell = 2$.
If $k = -4$ then the continued fraction expansion of ω is

n	0	1	2
P_n	1	m-2	m-2
Q_n	2	$2m$-4	2
q_n	$(m$-$1)/2$	1	m-2

If $|k| > 4$ then the continued fraction expansion of ω is

n	0	1	2	3	4
P_n	1	m-1	$m+k/2$	$m+k/2$	m-2
Q_n	2	$2m+k/2$-2	$\lvert k \rvert /2$	$2m+k/2$-2	2
q_n	$(m$-$1)/2$	1	$\lvert 4m/k \rvert -2$	1	m-2

This completes all possible cases for ERD–types $D = m^2 + k$. Examining these eight cases we find that if $D \not\equiv 1 \pmod 4$ or $D \equiv 1 \pmod 4$ and $m \not\equiv 0 \pmod k$, then $Q_i/\sigma > \sqrt{D}/2$ for $i \neq 0, \ell/2, \ell$. If $D \equiv 1 \pmod 4$ and $m \equiv 0 \pmod k$, then for $i \neq 0, \ell/2, \ell$ we have

$$Q_i/\sigma \geq \frac{1}{2}(m - (k-1)/2) \quad \text{when } k > 0 \qquad (3.1)$$

and

$$Q_i/\sigma > \frac{1}{2}(m - (|k|+1)/2) \quad \text{when } k < 0 \qquad (3.2).$$

Also m even and k odd together imply that k divides $m/2$, forcing $|k| \leq m/2$.

If (3.1) holds, then

$$Q_i/\sigma \geq \frac{1}{2}(m - k/2 + \frac{1}{2}) \geq \frac{1}{2}(m - m/4 + \frac{1}{2})$$
$$= 3m/8 + \frac{1}{4} > 3(\sqrt{D} - 1)/8 + \frac{1}{4}$$
$$= 3\sqrt{D}/8 - \frac{1}{8} > \sqrt{D}/3 \qquad \text{(for } D > 9\text{)}.$$

If (3.2) holds, then

$$Q_i/\sigma \geq \frac{1}{2}(m - |k/2| - \frac{1}{2}) \geq 3m/8 - \frac{1}{4}$$
$$> 3\sqrt{D}/8 - \frac{1}{4} > \sqrt{D}/3 \qquad \text{(for } D > 36\text{)}.$$

It follows that if $D > 36$ is of ERD–type and $i \neq 0, \ell/2, \ell$, then

$$Q_i/\sigma > \sqrt{D}/\gamma \quad \text{where } \gamma \text{ is as in (I)}.$$

Also, since $D \geq m^2 - |k|$, we get $m^2 - 4m \leq D$; whence $m - 2 \leq \sqrt{D + 4}$. If ε is the fundamental unit of $\mathbb{Q}(\sqrt{D})$, then $\varepsilon \leq (m + \sqrt{D})^2/|k|$, (the upper bound being a unit). Therefore,

$$\varepsilon \leq (\sqrt{D} + \sqrt{D + 4} + 2)^2/2 < 2(\sqrt{D} + 2)^2;$$

whence,

$$R = \log \varepsilon < \log 2(\sqrt{D} + 2)^2, \quad \text{(for } |k| \geq 2\text{)}.$$

If $|k| = 1$, then $\varepsilon \leq m + \sqrt{D}$ and $R < \log 2 (\sqrt{D} + 2)$.

Case A. $D \neq x^2 + y^2$. In this case, if C_K is to have exponent 2, we must have

$$h = |C_K| = \begin{cases} 2^{t-1} & \text{if } D \not\equiv 1 \pmod 4 \\ 2^{t-2} & \text{if } D \equiv 1 \pmod 4, \end{cases}$$

where t is the number of distinct odd prime divisors of D. (See Theorem 5 in [2], p. 225.)

By Tatsuzawa [13], we know

$$L(1, \chi_\Delta) \geq .655 \, \eta \, \Delta^{-\eta} \quad \text{for} \quad \Delta \geq \max\{ e^{11.2}, e^{1/\eta}\}$$

with $1/2 > \eta > 0$ where χ_Δ is a real, non–principal primitive character modulo Δ with one possible exception. Since

$$2 h R/\sqrt{\Delta} = L(1, \chi_\Delta) > .655\eta\Delta^{-\eta},$$

we get

$$2 h R/\eta\Delta^{1/2-\eta} > .655.$$

Moreover, $\Delta \geq e^{1/\eta}$ means $1/\eta < \log\Delta$; thus

$$.655 < 2 h R/(\eta\Delta^{1/2-\eta}) < 2 h R\log\Delta/\Delta^{1/2-\eta}.$$

If $d_t = p_1 p_2 \cdots p_t$ where $p_1 = 3$, $p_2 = 5$, $\cdots$ (the ordered product of the first t odd primes) then $D \geq d_t$. Put

$$F(D,\eta) = \log(2\sqrt{D}+2)^2)\log(4D)/(4D)^{1/2-\eta}.$$

If $t \geq 12$, then $d_t > 1.522 \times 10^{14}$. If $\eta = .02938$ and $D \geq d_t$, then we get $2^t F(D,\eta) < .51253 < .655$. We must have

$$2 h R\log(\Delta/\Delta^{1/2-\eta}) < (2/\sigma)\, 2^{t-1}\log(2(\sqrt{\Delta}+2)^2)\log(\Delta/\Delta^{1/2-\eta}).$$

Therefore, (since, for a fixed η, $F(D,\eta)$ is strictly decreasing),

$$2^t F(D,\eta) > 2 h R\log(\Delta/\Delta^{1/2-\eta}) > .655.$$

This is a contradiction. Thus, if $t \geq 12$ and $D > 1.522 \times 10^{14}$, then C_K cannot have exponent 2. On the other hand, if $D > 1.522 \times 10^{14}$ and $t < 12$, then we still have $2^t F(D,\eta) < .655$. Hence if C_K is to have exponent 2 we must have $D \leq 1.522 \times 10^{14}$ (with one possible exception).

Case B. $D = x^2 + y^2$. In this case if C_K is to have exponent 2 we must have ([2], op. cit) that

$$h = |C_K| = \begin{cases} 2^{t-1} & \text{if } D \equiv 1 \pmod 4 \\ 2^t & \text{if } D \equiv 2 \pmod 4. \end{cases}$$

In this case we can use the same argument as above with $t = 10$. We use the product $d_t = q_1 q_2 \cdots q_t$ where this represents the first t primes congruent to 1 modulo 4, ($d_{10} > 1.021 \times 10^{15}$). In this case, if q is an odd prime divisor of D, then $q \equiv 1 \pmod 4$. We find that we cannot have $t \geq 10$. If $D > 2 \times 10^{13}$, then

$$2^t F(D,\eta) < .2985988 < .655 \quad \text{where } \eta = .03266.$$

Thus, in this case, if C_K is to have exponent 2 we must have $D < 2 \times 10^{13}$, (again with one possible exception).

We now must find those values of D which are of ERD–type with $D < B = 1.53 \times 10^{14}$ such that C_K has exponent 2.

For a given prime p let $M_{p,i}$ be the least positive integer such that $M_{p,i} \equiv i \pmod 4$ and $(M_{p,i}/q) \neq 1$ for all primes q with $3 \leq q \leq p$, (where $(\,/\,)$ denotes the Legendre symbol).

D	r	k	D	r	k	D	r	k
10	3	1	903	10	3	3990	3	21
15	4	-1	915	2	15	4290	1	65
26	5	1	930	1	-31	4389	2	33
30	1	5	962	31	1	4758	23	-3
35	6	-1	1155	34	-1	4893	10	-7
39	2	3	1157	34	1	4935	2	35
42	1	-7	1173	2	17	5330	73	1
65	8	1	1190	1	-35	5610	5	-15
78	3	-3	1218	5	-7	5757	4	-19
95	2	-5	1230	7	5	6045	2	-39
105	2	5	1295	36	-1	6405	16	5
110	1	-11	1370	37	1	6765	2	41
122	11	1	1463	2	19	7035	4	-21
143	12	-1	1482	1	-39	7733	8	-11
170	13	1	1518	13	-3	7755	8	11
182	1	13	1595	8	-5	9030	19	5
195	14	-1	1605	8	5	9605	98	1
203	2	7	1722	1	41	12045	2	-55
210	1	-15	2030	9	5	12155	2	55
222	5	-3	2093	2	-23	13185	4	29
230	3	5	2117	46	1	14405	24	5
255	16	-1	2210	47	1	14630	11	-11
290	17	1	2301	16	-3	14885	122	1
327	19	3	2618	3	17	19565	4	-35
362	19	1	2717	4	13	19635	4	35
395	4	-5	3135	56	-1	20165	142	1
462	1	21	3230	3	19	33117	26	-7
483	22	-1	3335	2	-29	58565	242	1
485	22	1	3365	58	1	77285	278	1
530	23	1	3597	20	-3	108885	22	-15
663	2	-13	3605	12	5	451605	32	21
885	2	-15	3813	2	-31			

Table 1. $D = r^2 k^2 + k$

D	r	k	D	r	k	D	r	k
34	6	-1	623	25	-1	2607	17	3
51	7	1	627	25	1	2730	4	13
87	3	3	635	5	5	2910	18	-3
102	10	1	770	4	-7	3003	5	-11
138	4	-3	798	4	7	3255	19	3
194	14	-1	890	6	-5	3570	4	-15
215	3	-5	1022	32	-1	3723	61	1
231	5	3	1095	11	3	3927	3	-21
258	16	1	1298	36	1	5655	5	15
287	17	-1	1302	12	3	7215	17	-5
318	6	-3	1515	13	-3	10010	20	5
330	6	3	1547	3	13	10370	6	-17
390	4	-5	1610	8	5	12558	16	7
410	4	5	1770	14	3	13695	39	3
435	7	-3	1995	3	-15	19610	28	5
447	7	3	2015	9	-5	22490	30	-5
455	3	7	2387	7	-7	25935	23	7
570	8	-3	2415	7	7	81770	22	-13
615	5	-5	2595	17	-3			

Table 2. $D = r^2 k^2 + 2k$

<u>Claim</u> If D is of ERD-type with $\gamma^2 p^4 < D < M_{p,i}$ and $D \equiv i \pmod 4$, then C_K cannot have exponent 2.

Proof. Since $D < M_{p,i}$, there must exist some prime q such that $3 \le q \le p$ and $(D/q) = 1$. Furthermore since $D > \gamma^2 p^4 \ge \gamma^2 q^4$ we have $q^2 < \sqrt{D}/\gamma$, hence, there is some prime ideal $\mathfrak{q}$ over q such that $q^2 = N(\mathfrak{q}^2) < \sqrt{\Delta}/\gamma$. If C_K has exponent 2 then we must have $\mathfrak{q}^2$ principal. Also $D \not\equiv 0 \pmod q$ and $N(\mathfrak{q}^2) < \sqrt{D}/2$; whence $q^2 = Q_i/\sigma$ for some $i \ne 0, \pi/2, \pi$. However, we have seen that $Q_i/\sigma > \sqrt{\Delta}/\gamma$; hence, we get the claim.

In Tables 4, 5, and 6 we give opposite $q(i)$ the least positive integer N (congruent to 3, 2, or 1 (mod 4), respectively) such that $(N/q) \ne 1$ for all q such that $3 \le q \le p$ where $q(i-1) < p \le q(i)$. These tables were produced by using OASIS (see [12]). If for example we examine Table 5 we note that $M_{223,2} > B$. If $4 \times 223^4 < D < M_{223,2}$ (where $D \equiv 2 \pmod 4$)), then C_K cannot have exponent 2. However, $4 \times 223^4 < 9.9 \times 10^9 < M_{151,2}$. Thus if $D > 4 \times 151^4$, then C_K cannot have exponent 2. Furthermore,

$$4 \times 151^4 < 2.08 \times 10^9 < M_{127,2}$$

and

$$4 \times 127^4 < 1.04 \times 10^9 < M_{113,2}.$$

Thus if C_K is to have exponent 2 in the case $D \equiv 2 \pmod 4$, then we must have $D < 4 \times 113^4$, i.e., $\sqrt{D} < 25538$. By a similar analysis of these tables we see that if $D \equiv 3 \pmod 4$ we need $\sqrt{D} < 25538$, and if $D \equiv 1 \pmod 4$ we need $\sqrt{D} < 57963$ in order for C_K to be of exponent 2.

If D is of ERD–type, then

$$D = r^2 k^2 + k, \; r^2 k^2 + 2k \quad \text{or} \quad r^2 k^2 + 4k \quad \text{where } k \text{ is odd.}$$

We searched for all such D values with $r|k| < 30{,}000$ in the middle case and $r|k| < 60{,}000$ in the other two cases. We first eliminated any D values whenever we found a prime q such that $q^2 < \sqrt{D}/\gamma$ and $(D/q) = 1$, (since C_K cannot have exponent 2 in this case by the claim). This succeeded in eliminating almost all the possible D values. To test the remaining numbers we used Algorithm 2.1. After using this on the remaining D values we are left with only the values in Tables 1, 2, and 3. By the result of Tatsuzawa [op. cit] we now know that with one possible exception these are the only possible D values of ERD–type such that C_K has exponent 2. (See also Theorem 18 of [4]). $\qquad\square$

The following result is immediate from Theorem 3.1, and it solves a problem left in [6] wherein a classification was given of those D values such that there are nor *split* primes less than $\sqrt{\Delta}/2$. The authors in [6] determined that in order for this to happen the D's must be of ERD–type and $e_K \leq 2$. The following completes the classification by using Theorem 3.1 to list all such D values, with one possible exception remaining.

Corollary 3.1 $(D/p) \neq 1$ *for all primes* $p < \sqrt{\Delta}/2$ *if and only if D is a value in Table 6, with one possible exceptional value whose existence would be a counter–example to the Generalized Riemann Hypothesis (GRH).*

4 Tables

In Tables 1 to 6 we have *not* listed any D's with $h(D) = 1$ since these are known from [10]. Hence these represent the case where $e_K = 2$. However, in view of Corollary 3.1 we *have* listed D's with $h(D) = 1$ in Table 7.

Remark 4.1 In Table 2 we have eliminated 20 values of D, which would otherwise appear, but which are already in Table 1. For example, $15 = 4^2 - 1 = 5^2 - 10$, the first repetition, and $4290 = 65^2 + 65 = 2^2 \times$

D	r	k		D	r	k		D	r	k
85	9	1		2085	3	15		7917	1	-91
165	1	-15		2373	7	-7		8333	7	13
205	3	-5		2397	1	-51		8645	1	-95
221	1	-17		2405	49	1		9933	3	33
285	1	-19		2613	17	3		10205	101	1
357	1	-21		2805	1	-55		10965	7	-15
365	19	1		2813	53	1		11165	3	35
429	7	-3		3005	11	-5		13845	3	39
533	23	1		3045	11	5		14685	11	11
629	25	1		3237	19	-3		15645	25	5
645	5	5		3485	59	1		16133	127	1
741	9	3		3885	3	-21		17765	7	19
957	1	-33		3965	1	-65		20405	13	-11
965	31	1		4245	13	5		23205	3	-51
1085	1	-35		4277	5	13		24045	31	5
1205	7	-5		4485	1	-69		25493	3	53
1245	7	5		4773	23	3		26565	1	-165
1365	1	-39		5565	5	-15		30597	25	-7
1469	3	-13		5645	15	5		31317	59	-3
1517	1	-41		5885	7	-11		32045	179	1
1533	13	3		5957	11	7		35805	9	21
1685	41	1		6573	27	3		41093	7	-29
1853	43	1		7157	5	-17		55205	47	-5
1965	3	-15		7293	5	17		74613	13	21
2037	15	3		7565	1	-89				
2045	9	5		7685	3	29				

Table 3. $D = r^2 k^2 + 4k$

$(-33)^2 - 2 \cdot 33$, the last repetition. Table 3 contains no possible repetition from previous tables. Table 1 has 95 values, Table 2 has 56 values, and Table 3 has 76 values for a total of 227 distinct values of D of ERD–type for which $e_K = 2$. Moreover, it is interesting to note that for all these values $h \leq 16$ with only $D = 25935$ and $D = 81770$ from Table 2 having $h = 16$ and 19635 and 451605 from Table 1 having $h = 16$. All other values have $h \leq 8$.

Acknowledgments The last two authors' research is supported by NSERC Canada grants #A8484 and #A7649 respectively.

N	i	q(i)	N	i	q(i)
3	4	7	28203507	25	97
35	5	11	305875475	26	101
63	7	17	400744203	27	103
143	10	29	545559467	29	109
447	11	31	2342182895	33	137
635	12	37	7012246247	38	163
1295	17	59	2720673629663	39	167
101283	19	67	3166431076347	45	197
730847	21	73	37082770710347	47	211
18183483	22	79	163402643806023	49	227
26598843	23	83			

Table 4. $N \equiv 3 \pmod 4$

N	i	q(i)	N	i	q(i)
38	4	7	138614418	29	109
62	5	11	1164652502	30	113
98	6	13	2815100438	35	149
150	7	17	24453805998	36	151
318	9	23	348190770402	38	163
398	11	31	2295693610958	39	167
930	14	43	3017984461778	40	173
1722	17	59	4645014218918	41	179
314522	18	61	7767578138858	43	191
341502	20	71	12760252192170	46	199
845382	22	79	19285608519470	47	211
5505290	23	83	334664789952030	48	223
53772102	25	97			

Table 5. $N \equiv 2 \pmod 4$

References

[1] S. Chowla and J. Friedlander, *Class numbers and quadratic residues*, Glasgow Math. J., **17** (19761), pp. 47–52.

[2] H. Cohn, A Second Course in Number Theory, John Wiley & Sons, New York 1962.

[3] M.G. Leu, *On determination of certain real quadratic fields of class number two*, J. Number Theory **33** (1989), pp. 101–106.

N	i	$q(i)$		N	i	$q(i)$
5	4	7		46305413	30	113
17	5	11		1586592293	32	131
33	6	13		5702566397	34	139
437	11	31		15933687413	35	149
5297	12	37		25777678685	36	151
6905	13	41		181315486677	38	163
8393	14	43		413900743497	39	167
13593	15	47		1241083406505	42	181
64917	16	53		1519873554353	44	193
149597	17	59		31209910533489	45	197
524457	20	71		84542528922717	47	211
1878245	21	73		347478177155753	49	227
2628093	22	79		442815116635157	51	233
3009173	25	97				

Table 6. $D \equiv 1 \pmod 4$

[4] S. Louboutin, *Groupes des classes d'idéaux triviaux*, Acta Arith. **LIV** (1989), pp. 61–74.

[5] S. Louboutin, *Ideal class groups with exponent two in the real quadratic case*, Séminaire de Théorie des Nombres de Caen, VII, 1987–88.

[6] S. Louboutin, R.A. Mollin and H. C. Williams, *Class numbers of real quadratic fields, continued fractions, reduced ideals, prime producing polynomials and quadratic residue covers*, Canadian J. Math. (to appear).

[7] R.A. Mollin, *Class number one criteria for real quadratic fields I*, Proc. Japan Acad., Ser. A., **63** (1987), pp. 121–125.

[8] R.A. Mollin and H.C. Williams, *Prime-producing polynomials and real quadratic fields of class number one*, in Number Theory (J. M. DeKonick and C. Levesque (ed.)), Walter de Gruyter, Berlin (1989), pp. 654–663.

[9] R.A. Mollin and H.C. Williams, *On prime valued polynomials and class numbers of real quadratic fields*, Nagoya J. Math. **112** (1988), pp. 143–151.

[10] R.A. Mollin and H.C. Williams, *Solution of the class number one problem for real quadratic fields of Richaud–Degert type (with one possible exception)*, in Number Theory (R.A. Mollin (ed.)), Walter de Gruyter, Berlin (1990), pp. 417–425.

[11] R.A. Mollin and H.C. Williams, *On a solution of a class number two problem for a family of real quadratic fields*, in Computational Number Theory (A. Petho, M. Pohst, H.C. Williams and H. Zimmer (ed.)), Walter de Gruyter, Berlin (1991), pp. 95–101.

[12] A.J. Stephens and H.C. Williams, *An open architecture number*

Define $\mathcal{T} = \{\text{ramified primes} < \sqrt{\Delta}/2\}$

D	$\mathcal{T}$	D	$\mathcal{T}$	D	$\mathcal{T}$
2		77	7	357	3,7
3		83	2	398	2
5		87	2,3	413	7
6	2	93	3	437	19
7	2	110	2,5	447	2,3
11	2	138	2,3	453	3
13		143	2,11	483	2,3,7
14	2	165	3,5	635	2,5
15	3	167	2	717	3
21	3	173		930	2,3 5
23	2	182	2,7 13	957	3,11
29		195	2,3,5 13	1077	3
30	2,3,5	213	3	1085	5,7
35	2,5	215	2,5	1133	11
38	2	227	2	1253	7
42	2,3	237	3	1295	2,5,7
47	2	255	2,3,5	1722	2,3,7 41
53		285	3,5	1965	3,5
62	2	293		2085	3,5
69	3	318	2,3	2397	3,17

Table 7. *D*'s with no split prime $< \sqrt{\Delta}/2$

sieve, in *Number Theory and Cryptography* (J.J. Loxton (ed.)), London Math. Soc. Lecture Notes Series **154** (1990), pp. 38–75.

[13] T. Tatsuzawa, *On a theorem of Siegel*, Japan J. Math. **21** (1951), pp. 163–178.

[14] H.C. Williams and M.C. Wunderlich, *On the parallel generation of the residues for the continued fraction factoring algorithm*, Math. Comp. **177** (1987), pp. 405–423.

[15] H. Yokoi, *Class number one problem for certain kinds of real quadratic fields*, in Prod. Internat. Conf. on Class Numbers and Fundamental Units, Katata Japan (1986).

S. Louboutin
Department of Mathematics
University of Caen
France

Richard A. Mollin
Department of Mathematics and Statistics
University of Calgary
Calgary, Alberta T2N 1N4
Canada
e-mail ramollin@acs.ucalgary.ca

and

Hugh C. Williams
Department of Computer Science
University of Manitoba
Winnipeg, Manitoba R3T 2N2
Canada
e-mail Hugh_Williams@csmail.cs.umanitoba.ca

Poincaré Polynomials, Stability Indices and Number of Orderings I

Ján Mináč

University of Western Ontario

Dedicated to Paulo Ribenboim

1 Introduction

Suppose that F is a formally real pythagorean field with $|\dot{F}/\dot{F}^2| = 2^n$, $n \in \mathbb{N}$. An important question is: how many orderings can F possess? This question was completely answered by L. Bröcker in 1977 [Brö 1]. Define the sets $O(n) = \{k \mid k$ is the number of orderings of some pythagorean field F, such that $|\dot{F}/\dot{F}^2| = 2^n\}$, $n \in \mathbb{N}$.

A remarkable recurrence formula for $O(n)$, $n \in \mathbb{N}$, was discovered by Bröcker [Brö 1] and later simplified by Berman [Ber] in his thesis. We have

$$
\begin{aligned}
O(1) &= \{1\} \\
O(n) &= (O(n-1)+1) \cup 2 \cdot O(n), n \geq 2.
\end{aligned}
$$

Thus $O(1) = \{1\}$, $O(2) = \{2\}$, $O(3) = \{3,4\}$, $O(4) = \{4,5,6,8\}$, etc.

In 1985, while working on my thesis under the supervision of Paulo Ribenboim, I observed that one can obtain the sets $O(n)$ by defining the sets $B(n)$ consisting of Poincaré polynomials attached to fields F above with $|\dot{F}/\dot{F}^2| = 2^n$ (see below for details). Namely, it turns out that $B(n)$ satisfies the following recurrence relations:

$$
\begin{aligned}
B(1) &= \{1\} \\
B(n) &= (B(n-1)+t) \cup ((1+t)B(n-1)), n \geq 2.
\end{aligned}
$$

Thus

Supported in part by the Natural Sciences and Engineering Research Council of Canada.

$$B(1) = \{1\}$$
$$B(2) = \{1 + t\}$$
$$B(3) = \{1 + 2t, 1 + 2t + t^2\}$$
$$B(4) = \{1 + 3t, 1 + 3t + t^2, 1 + 3t + 2t^2, 1 + 3t + 3t^2 + t^3\}, \text{ etc.}$$

Of course, by putting $t = 1$, we get Bröcker's sets $O(n)$.

Poincaré polynomials capture not only the number of orderings, but also the stability indices, the number of square classes and other invariants of fields. They can be attached to any preordering T of a field F such that $|\dot{F}/T|$ is finite. (For the notion of preordering, stability indices and related topics, see [Lam 2].) They have remarkable combinatorial properties and it is possible that they can be successfully applied in other areas, besides ordered fields. (They can be assigned to certain graphs. See Theorem 1 below.) They have already proved quite useful in ordered fields. For example, the question about the precise relationship between the number of orderings, the stability indices and the number of square classes considered in [Mer] was completely solved using Poincaré polynomials. (See [Min 1] and [Min 4].)

The announcement of some results appeared in [Min 3] and the full proofs of them were circulated via copies of my thesis [Min 1]; however they have not yet been published.

The goal of this note is to define Poincaré polynomials, derive the recurrence formula for sets $B(n)$ and point out a few simple properties. Much more can be said about Poincaré polynomials and their connections with ordered fields. Some of these facts can be found in [Min 1] and [MST]. It would be also interesting to compare these polynomials with Poincaré series defined and investigated in [Sch] and [Kul]. All this is beyond the scope of this paper. However, now at least the stage is set and the play can begin.

2   Notation and Basic Notions

We define F to be a formally real pythagorean field. (This means that the sum of any two squares is a square and -1 isn't square in F.) $\dot{F}$ is the multiplicative group of F. We shall always assume that $|\dot{F}/\dot{F}^2| < \infty$. Let $F(2)$ be the maximal 2-extension of F and H the Galois group of $F(2)/F(\sqrt{-1})$.

Put $h_i = \dim_{\mathbb{Z}/2\mathbb{Z}} H^i(H, 2), 0 \leq i < \infty$, where $H^i(H, 2)$ is the i-th cohomological group of the group H with coefficients in the two element field. Then

$$P_H(t) = \sum_{i=0}^{\infty} h_i t^i$$

is the *Poincaré series* of the group H. Since the cohomological dimension $d(H) = d = st(F) = $ stability index of F, $P_H(t)$ is a polynomial (see

[Min 2].) We shall use freely notation, definitions and theorems in [Jac], [Lam 1], [Lam 2], [Mar 1], [Mar 2]. We shall recall here just the most essential notation and definitions. (However no extensive knowledge of the literature above is really necessary. Occasional recall of some basic definitions or theorems should be sufficient for reading this paper. Moreover, most of the invariants considered in this paper can be attached to the graphs fully described below. We shall recall some of these possible definitions; others will be clear from the proofs. For the reader not familiar with the notations above, Figure 1 should be useful.) $N(F)$ – the number of orderings of the field F; F is of type $(k, 2^n)$ if $N(F) = k$ and $|\dot{F}/\dot{F}^2| = 2^n$; (X, T) the abstract Marshall space of orderings; (we shall identify isomorphic order spaces); $(X_F, \dot{F}/\dot{F}^2)$, or simply X) space of orderings of the field F; $c\ell(X)$ chain length of X; $|S|$ the number of elements of a finite set S, in particular $|X|$ is the number of orderings of the order space $X = (X, T)$; $G = Gal(F(2)/F)$; $S_G(t)$ the Poincaré series of the group G.

For extensions, sums of order spaces and their valuation - theoretic interpretation we refer the reader to the papers [Jac], [Mar 1] and [Mar 2]. (*We shall always assume that in the decomposition of order spaces into sums all summands are indecomposable order spaces and each space with two orderings will be considered as a sum of 2 nonempty spaces.*)

Following [Brö 2], [Cra] we shall attach to any finite order space (X, T) the graph $Gr(X)$:

(1) *The set of vertices, $V(X)$, of the graph $Gr(X)$ is defined inductively by:*

    (1a) $X \in V(X)$.

    (1b) If $Y \in V(X)$, $Y = Z \times H$, where H is an elementary 2-group and Z is a decomposable space, then $Z \in V(X)$.

    (1c) If $W \in V(X)$ and $W = W_1 \oplus \cdots \oplus W_d$, then $W_i \in V(X)$, $i = 1, \ldots d$.

(2) *Edges*: Two vertices Y, Z are connected with oriented edge (Y, Z) if

    (2a) $Y = Z \times H$, for some elementary 2-group H, $H \neq \{1\}$, and Z is a decomposable order space.

    (2b) $Y = Z \oplus Z_2 \oplus \cdots \oplus Z_d$, $2 \leq d$.

We shall define functions n, m, s : $V(X) \to \mathbb{N} \cup \{0\}$ as follows.
Let $Y = Y_1 \oplus \cdots \oplus Y_d \in V(X)$. Then we put $n(Y) = d$. If $Y = Z \times H \in V(X)$, and Z is not an extension of any proper quotient space for Z, or $|Z| = 2$, we put $m(Y) = \log_2 |H|$.

For every vertex $Y \in V(X)$ there exists a unique shortest path $Y = Y_0, Y_1, \ldots, Y_k = X$, where (Y_i, Y_{i+1}), $i = 0, 1, \ldots, k-1$, belong to the set of edges of $Gr(X_F)$. Then we put

$$s(Y) = \sum_{i=1}^{k} m(Y_i).$$

We shall also define the function $L : V(X) \to \mathbb{Z}[t]$, $L(Y) = (1-b) + t(d-1)$, where $Y = Y_1 \oplus \cdots \oplus Y_a \oplus Z_1 \oplus \cdots \oplus Z_b$, $|Y_1| = \cdots = |Y_a| = 1$, $1 < |Z_1|, \ldots, 1 < |Z_b|$, $d = a + b$ and t is indeterminate.

Example 1. Let F be a field of type $(19, 2^7)$. Then one can prove that $(X_F, \dot{F}/\dot{F}^2)$ is uniquely determined and the graph $Gr(X_F)$ is:

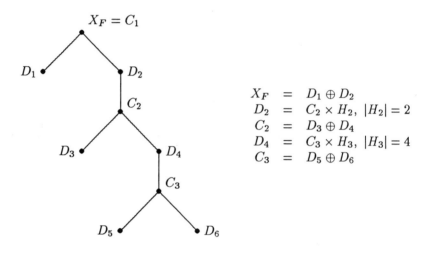

$$
\begin{aligned}
X_F &= D_1 \oplus D_2 \\
D_2 &= C_2 \times H_2, \ |H_2| = 2 \\
C_2 &= D_3 \oplus D_4 \\
D_4 &= C_3 \times H_3, \ |H_3| = 4 \\
C_3 &= D_5 \oplus D_6
\end{aligned}
$$

Figure 1.

We also have $0 = s(X_F) = s(D_1) = s(D_2)$, $1 = s(C_2) = s(D_3) = s(D_4)$, $3 = s(C_3) = s(D_5) = s(D_6)$ and $L(X_F) = t$, $L(C_2) = t$, $L(C_3) = 1 + t$.

We recall also that

$$
\begin{aligned}
c\ell(X_F) &= c\ell(D_1) + c\ell(D_2) = 1 + c\ell(C_2) \\
&= 1 + c\ell(D_3) + c\ell(D_4) = 2 + c\ell(C_3) \\
&= 2 + c\ell(D_5) + c\ell(D_6) = 4
\end{aligned}
$$

Finally,

$$
\begin{aligned}
st(X_F) &= \max\{st(D_1), st(D_2)\} = st(D_2) = \log_2 |H_2| + st(C_2) \\
&= 1 + \max\{st(D_3), st(D_4)\} = 1 + st(D_4) \\
&= 1 + \log_2 |H_3| + st(C_3) = 3 + st(C_3) = 4.
\end{aligned}
$$

(In general, $st(X) \in \mathbb{N} \cup \{0\}$, $c\ell(X) \in \mathbb{N}$, where $st(X) = 0$ if and only if $|X| = 1$, and if $|X_i| = 1, i = 1, \ldots, a$ and $2 \le a$, then $1 = st(X_1 \oplus \cdots \oplus X_a)$ and $a = c\ell(X_1 \oplus \cdots \oplus X_a)$.)

Moreover, if $|X_1 \oplus X_2| > 2$ then $st(X_1 \oplus X_2) = \max\{st(X_1), st(X_2)\}$ and $cl(X_1 \oplus X_2) = cl(X_1) + cl(X_2)$.

Finally, if $|Y| \neq \{1\}$ then $st(Y \times H) = st(Y) + \log_2 |H|$ and $cl(Y \times H) = cl(Y)$.

These properties define inductively both $cl(X)$ and $st(X)$ for any finite order space X.)

3 Theorems

We summarize the notation introduced above:

- F is a formally real pythagorean field.

- $G = Gal(F(2)/F)$.

- $H = Gal(F(2)/F(\sqrt{-1}))$.

- $S_G(t) =$ Poincaré series of the group G.

- $P_H(t) =$ Poincaré polynomial of the group H.

Theorem 2. *If F is of the type $(1, 2)$ then $P_H(t) = 1$, $S_G(t) = \dfrac{1}{1-t}$. If F is not of the type $(1, 2)$ then*

(1) $\quad S_G(t) = \dfrac{P_H(t)}{1-t}$,

(2) $\quad P_H(t) = \displaystyle\sum_{\substack{C \in V(X_F) \\ n(C) \geq 2}} (1 + t)^{s(C)} L(C)$.

Proof. Suppose that F is of the type $(1, 2)$. Then F is a Euclidean field. Hence from [Bec 1] we find $G \cong \mathbb{Z}/2\mathbb{Z}$ and $H = \{1\}$. Therefore $H^i(G, 2) \cong \mathbb{Z}/2\mathbb{Z}$ for $i \in \mathbb{N} \cup \{0\}$ and $H^i(H, 2) = \{0\}$ for $i \geq 1$. Therefore $S_G(t) = \frac{1}{1-t}$ and $P_H(t) = 1$ as we claimed.

Suppose that a field F is not of the type $(1, 2)$. We first show the formula (2) by induction on $cl(F)$ the chain length of X_F. We distinguish two cases:

<u>Case 1.</u> $X_F = X_1 \oplus \cdots \oplus X_a \oplus Y_1 \oplus \cdots \oplus Y_b$, where

$$|X_1| = \ldots = |X_a| = 1, \quad |Y_1| > 1, \ldots, |Y_b| > 1, \quad 2 \leq a + b.$$

From [Jac] we find that

$$F = \Big(\bigcap_{i=1}^{a} F_i\Big) \cap \Big(\bigcap_{j=1}^{b} F_j\Big),$$

$X_i \cong X_{F_i}$, $Y_j = X_{F_j}$, $i = 1, \ldots, a$; $j = 1, \ldots, b$ and

$$G = G_F \cong \left(\mathop{*}_{i=1}^{a} G_{F_i} \right) * \left(\mathop{*}_{j=1}^{b} G_{F_j} \right),$$

where $*$ means the free product in the category of pro-2-groups. Since H is the subgroup of index 2 of G we may apply the generalization of Kurosh's subgroup theorem to profinite groups (see [BNW]). We find that $H \cong J * \left(\mathop{*}_{\alpha, \sigma_\alpha} (G_\alpha^{\sigma_\alpha} \cap H) \right)$ where G_α runs through all G_{F_i} and G_{F_j} and σ_α runs through a system S_α of representatives of the double coset decomposition $\bigcup_\alpha G_\alpha \cdot \sigma_\alpha \cdot H$ of G_F, $G_\alpha^{\sigma_\alpha} = \sigma_\alpha G_\alpha \sigma_\alpha^{-1}$ and K is a free pro-2-group of rank $r = \sum_\alpha (2 - |S_\alpha|) - 1$. From [Bec 1] we see that each G_α contains a nontrivial involution which doesn't belong to H. Therefore G_α and H generate G for each α. Thus we can set $\sigma_\alpha = 1$ for each α, and we find $H \cong J * \left(\mathop{*}_\alpha (G_\alpha \cap H) \right)$, where J is a free pro-2-group of rank $a + b - 1$. On the other hand, since $H = G_{F(\sqrt{-1})}$, we see that $G_\alpha \cap H = G_{F_i(\sqrt{-1})}$ if $i = 1, \ldots, a$ and $G_\alpha \cap H = G_{F_j(\sqrt{-1})}$ if $j = 1, \ldots, b$. Therefore we find $h_0(H) = 1$ and

$$h_1(H) = a + b - 1 + \sum_{j=1}^{b} rank \, G_{F_j(\sqrt{-1})}$$

since $G_{F_i(\sqrt{-1})} = \{1\}$ for all $i = 1, \ldots, a$. Moreover,

$$h_t(H) = \sum_{j=1}^{b} h_t(G_{F_j(\sqrt{-1})}) \quad \text{for} \quad t \geq 2.$$

Since $c\ell(Y_j) < c\ell(X)$, $j = 1, \ldots, b$ we may use the induction hypothesis as follows:

$$\begin{aligned} P_H(t) &= \sum_{r=0}^{\infty} h_r t^r = \sum_{j=1}^{b} P_{F_j(\sqrt{-1})}(t) + (1 - b) + (a + b - 1)t \\ &= \sum_{\substack{C \in V(X_F) \\ n(C) \geq 2}} (1 + t)^{s(C)} L(C), \end{aligned}$$

which completes the proof in case 1.

Case 2. Assume that $X_F = Y \times H$, where H is an elementary abelian 2-group and Y is a direct sum of at least two order spaces. From [Jac] we know that there exists a 2-Henselian valuation v on F such that $X_{F_v} \cong Y$ and $\dot{F}/U_v \dot{F}^2 \cong H$, where F_v is the residue field of v and U_v is the group of units of the valuation ring of v. On the other hand from [War 2] we find $G_{F(\sqrt{-1})} \cong G_{F_v(\sqrt{-1})} \times \mathbb{Z}_2^I$, where $\mathbb{Z}_2$ is the additive group of 2-adic integers and $|I| = \log_2 |H|$. In order to compute the cohomology group of $G_{F(\sqrt{-1})}$, recall that by a standard application of the Hochschild-Serre spectral sequence we find for any pro-2-group M,

$$H^n(M \times \mathbb{Z}_2, 2) \cong H^{n-1}(M, 2) \oplus H^n(M, 2),$$

for $1 \leq n$. Therefore

$$P_{G_{F(\sqrt{-1})}}(t) = (1+t)^{|I|} P_{G_{F_v(\sqrt{-1})}}(t).$$

Since F_v is again a pythagorean field and X_{F_v} is a direct sum of at least two order spaces, from Case 1 above we find

$$P_{G_{F_v(\sqrt{-1})}}(t) = \sum_{\substack{C \in V(X) \\ n(C) \geq 2}} (1+t)^{s(C)-|I|} L(C).$$

This together with the relationship between $P_{G_{F(\sqrt{-1})}}(t)$ and $P_{G_{F_v(\sqrt{-1})}}(t)$ above proves our formula for $P_{G_{F(\sqrt{-1})}}(t)$ also in Case 2. Thus the formula (2) is established.

In order to show (1) we shall use Arason's long exact sequence applied to the field extensions $F \subset F(\sqrt{-1}) \subset F(2)$ (see [Ara]). Then we have

$$0 \longrightarrow H^0(G, 2) \xrightarrow{Res} H^0(H, 2) \xrightarrow{Cor} H^0(G, 2) \xrightarrow{\cup(-1)}$$
$$H^1(G, 2) \xrightarrow{Res} H^1(H, 2) \xrightarrow{Cor} H^1(G, 2) \xrightarrow{\cup(-1)} \cdots$$

where Res, Cor and $\cup(-1)$ mean restriction, corestriction and cup product by (-1), respectively. Here (-1) is the element of $H^1(G, 2)$ corresponding to the field element -1. Let I be the fundamental ideal of the Witt ring $W(F)$ (see [Lam 1]). Then from [AEJ 2] we find that there exist canonical isomorphisms $\psi_n : I^n/I^{n+1} \to H^n(G_F, 2), n \in \mathbb{N}$ such that the following diagrams commute:

$$
\begin{array}{ccc}
H^n(G, 2) & \xrightarrow{\cup(-1)} & H^{n+1}(G, 2) \\
\psi_n \uparrow & & \uparrow \psi_{n+1} \\
I^n/I^{n+1} & \xrightarrow{<1,1>_L} & I^{n+1}/I^{n+2}.
\end{array}
$$

Here $< 1,1 >_L$ means the map $I^n/I^{n+1} \to I^{n+1}/I^{n+2}$ obtained from multiplication by the class $[< 1,1 >] \in I$. Since F is a pythagorean field we see that $< 1,1 >_L$ is injective (see [Lam 3]). Consequently the map $\cup(-1)$ is injective as well.

Therefore we get by induction on n that

$$\sum_{i=0}^{n} h_i(H) = h_n(G)$$

which proves that $S_G(t) = \frac{P_G(t)}{1-t}$, as we claimed. ∎

Example 3. Let F be a pythagorean field of type $(19, 2^7)$. With the notation of Figure 1,

$$
\begin{aligned}
P_F(t) &= L(C_1) + (1+t)L(C_2) + (1+t)^3 L(C_3) \\
&= t + (1+t)t + (1+t)^4 \\
&= t^4 + 4t^3 + 7t^2 + 6t + 1.
\end{aligned}
$$

Theorem 1 above suggests the following definition:

Definition 4. *Let X be a finite order space, $Gr(X)$ its graph. Then we define the Poincaré polynomial $P_X(t)$ of X by the formula*

$$
P_X(t) = \begin{cases}
\displaystyle\sum_{\substack{C \in V(X) \\ n(C) \ge 2}} (1+t)^{s(C)} L(C) & \text{if } |X| \ne 1 \\
1 & \text{if } |X| = 1.
\end{cases}
$$

Remark 5. It is known that each finite order space X is realisable as X_F for some formally real pythagorean field. (See [Cra], [Jac].)

The order space (X, D) captures the "number of square classes", the number of orderings $|X|$ and the stability index $st(X)$. More precisely, we have:

Theorem 6. *Let (X, T) be a finite order space and $P_X(t)$ its Poincaré polynomial. Then we have*

(1) $P_X(1) = |X|$, $P_X(0) = 1$, $\deg P_X(t) = st(X)$, $1 + P_X'(0) = \log_2 |T|$. *($P_X'(t)$ is the derivative of $P_X(t)$).*

(2) *Suppose that $X = X_1 \oplus \ldots \oplus X_a \oplus Y_1 \oplus \ldots \oplus Y_b$, where $|X_i| = 1, i = 1, \ldots, a$ and $|Y_j| > 1$, $j = 1, \ldots, b$. Then $P_X(-1) = 2 - a - 2b$.*

Proof. Since all assertions are true for the case $|X| = 1$, we shall assume that $1 < |X|$. We shall prove our theorem by induction on $cl(X)$.

Because the proofs of all the statements are similar to each other we shall restrict ourselves to proving the claim: $1 + P_X'(0) = \log_2 |T|$. We assume that $P_X(0) = 1$ has already been proved.

If $cl(X) = 2$, then $X = Y \times H$, $|Y| = 2$ and $P_X(t) = (1+t)^{n+1}$, where $n = \log_2 |H|$. Thus $P_X'(0) = n + 1$ and $1 + P_X'(0) = \log_2 |T|$.

Suppose that $cl(X) > 1$ and write X as sum of order spaces as in (2).

Case 1. $a + b \ge 2$. Then $P_X(t) = (1 - b) + (a + b - 1)t + \displaystyle\sum_{j=1}^{b} P_{Y_j}(t)$.

Hence:

$$
\begin{aligned}
1 + P_X'(0) &= 1 + (a + b - 1) + \textstyle\sum_{j=1}^{b} P_{Y_j}'(0) \\
&= a + \textstyle\sum_{j=1}^{b} (P_{Y_j}'(0) + 1) \\
&= a + \textstyle\sum_{j=1}^{b} \log_2 |T_j| \\
&= \log_2 |T|
\end{aligned}
$$

<u>Case 2.</u> $X_F = Y \times H, Y = W \oplus V$ for some nonempty order spaces W and V. Then $P_X(t) = (1 + t)^{\log_2 |H|} P_Y(t)$. Therefore $1 + P'_X(0) = (\log_2 |H|) \cdot P_Y(0) + P'_Y(0) + 1$. Since $P_Y(0) = 1$ and $P'_Y(0) + 1 = \log_2 |T_Y|$ by case 1, we see that $1 + P'_X(0) = \log_2 |H| + \log_2 |T_Y| = \log_2 |H \times T_Y| = \log_2 |T_X|$, which completes the proof of our claim. ∎

Remark 7A. One can check the relation $P_X(1) = |X|$ also in a different way as follows:

Let I be the fundamental ideal of the Witt ring $W\left(F(\sqrt{-1})\right)$. Then from [AEJ 2] we find $h_n(G_{F(\sqrt{-1})}) = \dim_{\mathbb{Z}/2\mathbb{Z}} I^n / I^{n+1}$, $n = 0, 1 \ldots$ (Here by I^0 we mean $W\left(F(\sqrt{-1})\right)$. Then

$$
\begin{aligned}
\sum_{n=0}^{st(F)} h_n\left(G_{F(\sqrt{-1})}\right) &= \sum_{n=0}^{\infty} h_n\left(G_{F(\sqrt{-1})}\right) = \sum_{n=0}^{\infty} \dim_{\mathbb{Z}/2\mathbb{Z}} I^n / I^{n+1} \\
&= \sum_{n=0}^{\infty} (\dim_{\mathbb{Z}/2\mathbb{Z}} I^n - \dim_{\mathbb{Z}/2\mathbb{Z}} I^{n+1}) \\
&= \dim_{\mathbb{Z}/2\mathbb{Z}} W\left(F(\sqrt{-1})\right) \\
&= \dim_{\mathbb{Z}/2\mathbb{Z}} W(F)/2W(F) = |X_F| \text{ (see [Lam 3])}.
\end{aligned}
$$

Here it is useful to observe that $I^k = \{0\}$ for $k > |\dot{F}(\sqrt{-1})/\dot{F}(\sqrt{-1})^2|$. Indeed I^k is generated by k-fold Pfister forms over $F(\sqrt{-1})$. However, since $\sqrt{-1} \in F(\sqrt{-1})$, each Pfister form $\ll a_1, \ldots, a_k \gg$, $a_i \in F(\sqrt{-1})$; $i = 1, \ldots, k$, with two elements a_i, a_j, $i \neq j$ which differ by square in $F(\sqrt{-1})$ is hyperbolic (see [Lam 1]; Chapter X).

Finally, observe that $\dim_{\mathbb{Z}/2\mathbb{Z}} \dot{F}(\sqrt{-1})/\dot{F}(\sqrt{-1})^2 = \dim_{\mathbb{Z}/2\mathbb{Z}} \dot{F}/\dot{F}^2 - 1$ (see [Lam 1], page 202).

Remark 7B. Observe that $P_{X_{F(-1)}}$ is the Euler-Poincaré characteristic of $G_{F(\sqrt{-1})}$. It is interesting to clarify when $P_X(-1) = 0$. From our theorem we see that $P_X(-1) = 0 = 2 - a - 2b$ if $a = 2$ and $b = 0$ or $a = 0$ and $b = 1$. Thus if $|X| > 2$ we see that $P_X(-1) = 0$ if $X = Y \times H$, $|H| \geq 2$, Y is some order space. In other words $(1 + t)/P_X(t), |X| > 2$, implies $X = Y \times H$, $|H| \geq 2$. This fact will be used in the proof of the lemma below. It also has interesting consequence for $G_{F(\sqrt{-1})}$ (see [Min 1]). Finally, we shall define our sets $B(n)$.

Definition 8. *For each $n \in \mathbb{N}$ define $B(n) = \{P_X(t) \mid X$ is some order space (X, T) with $|T| = 2^n\}$.*

In order to prove the recurrence formula for sets of polynomials $B(n)$, $n = 1, 2, \ldots$, we shall prove the following lemma analogous to Proposition 2.6 in [Mer].

Lemma 9. *Let $X = (X, T)$ be a nontrivial sum of finite order spaces X_1 and X_2. Then there exists an order space Y such that $P_X(t) = P_Y(t) + t$.*

Proof. Write $P_{X_i}(t) \in B(n_i)$, $i = 1, 2$; $P_X(t) \in B(n)$. Hence $n = n_1 + n_2$. We shall prove our assertion by induction on n.

If $n = 2$, $|X_1| = |X_2| = 1$; $P_X(t) = 1 + t = t + P_Y(t)$; where $|Y| = 1$.

Suppose that $2 < n$ and that our lemma is true for each $P_X(t) \in B(m)$ for all $m < n$. Let d_i be the highest power of $(1 + t)$ which divides $P_{X_i}(t)$, $i = 1, 2$. Set $d = min\{d_1, d_2\}$. Without loss of generality we assume that $d = d_1$. Set $f = d$ if $P_{X_1}(t) \neq (1 + t)^{d_1}$ and $f = d - 1$ if $P_{X_1}(t) = (1 + t)^{d_1}$. Then

$$\frac{P_{X_1}(t)}{(1 + t)^f} \in B(n_1 - f),$$

$n_1 - f > 1$, $n_1 - f \leq n_1 < n$. Hence there exists an order space (Y, C) such that

$$P_Y(t) = \frac{P_{X_1}(t)}{(1 + t)^f} \in B(n_1 - f).$$

Since either $|Y| = 2$ or $(1+t) \nmid P_Y(t)$ we see that Y is a non-trivial direct sum of order spaces. Since $n_1 - f < n$, from the induction hypothesis we deduce that $\frac{P_{X_1}(t)}{(1 + t)^f} - t \in B(n_1 - f - 1)$. We also have $\frac{P_{X_2}(t)}{(1 + t)^f} \in B(n_2 - f)$ and therefore there exists an order spaces (Z_1, T_1) and (Z_2, T_2) such that $P_{Z_1}(t) = \frac{P_{X_1}(t)}{(1 + t)^f} - t$ and $P_{Z_2}(t) = \frac{P_{X_2}(t)}{(1 + t)^f}$. Set $Z = Z_1 \oplus Z_2$. Then $P_Z(t) = P_{Z_1}(t) + P_{Z_2}(t) + t - 1$. (This can be easily seen from the formula for $P_Z(t)$, see Theorem 1.) Therefore

$$P_Z(t) = \frac{P_{X_1}(t) + P_{X_2}(t)}{(1 + t)^f} - 1 \in B(n_1 + n_2 - 2f - 1) = B(n - 2f - 1).$$

If $f = 0$, then $P_{X_1}(t) + P_{X_2}(t) - 1 = P_{X_1 \oplus X_2}(t) - t \in B(n - 1)$ as we were to prove.

If $f > 0$ we shall show below that

$$P_Z(t) = \frac{P_{X_1}(t) + P_{X_2}(t)}{(1 + t)^f} - 1 \in B(n - 2f - 1)$$

implies that

$$\frac{P_{X_1}(t) + P_{X_2}(t)}{(1 + t)^{(f-1)}} - 1 \in B(n - 2(f - 1) - 1).$$

Indeed, multiplying $P_Z(t)$ by $(1 + t)$ we find

$$\frac{P_{X_1}(t) + P_{X_2}(t)}{(1 + t)^{f-1}} - 1 - t \in B(n - 2f).$$

Adding t we get

$$\frac{P_{X_1}(t) + P_{X_2}(t)}{(1+t)^{f-1}} - 1 = B(n - 2(f-1) - 1).$$

Repeat this until we reach $f = 0$ in which case we know that $P_X(t) = P_{X_1}(t) + P_{X_2}(t) - 1 = P_{X_1 \oplus X_2}(t) - t \in B(n-1)$ as were to prove. ∎

Now we are prepared to prove the recurrence formula for the sets $B(n)$.

Theorem 10. $B(1) = \{1\}$ *and for all* $n \geq 2$ *we have*

$$B(n) = \Big(B(n-1) + t\Big) \cup \Big((1+t)B(n-1)\Big).$$

Proof. Since clearly $B(1) = \{1\}$, we shall assume $2 \leq n$. First we shall show that $\Big(B(n-1) + t\Big) \cup \Big((1+t)B(n-1)\Big) \subset B(n)$. Assume that X is an order space such that $P_X(t) \in B(n-1)$. Then $P_{X \times H}(t) = (1+t)P_X(t)$, where $H \cong \mathbb{Z}/2\mathbb{Z}$. Hence $(1+t)P_X(t) \in B(n)$. Let Y be an order space consisting of the single ordering. Then $P_{X \oplus Y}(t) = P_X(t) + t \in B(n)$.

Suppose that X is an order space such that $P_X(t) \in B(n)$. If $n = 2$, then $P_X(t) = 1 + t \in B(1) + t = (1+t)B(1)$. Assume, therefore that $3 \leq n$. If $(1+t) | P_X(t)$ then $X = Y \times H$, where $|H| = 2$ (see Remark 7B). Hence $P_X(t) = P_Y(t)(1+t)$, $P_Y(t) \in B(n-1)$. If $(1+t) \nmid P_X(t)$ then from Remark 7B we conclude that X is a nontrivial sum of two order spaces. Therefore from the lemma above we conclude that $P_X(t) - t \in B(n-1)$.

Thus in any case $P_X(t) \in \Big(B(n-1) + t\Big) \cup (1+t)B(n-1)$. Our equality is proved. ∎

We shall finish our paper with determining our sets $B(n)$ internally independent of our recurrence formula as well as showing that $| B(n) | = 2^{n-2}$.

Theorem 11. *Suppose that* $2 \leq n$. *Then* $|B(n)| = 2^{n-2}$. *The polynomial* $g(t) \epsilon \mathbb{Z}[t]$ *belongs to* $B(n)$ *if and only if*

$$g(t) = (1+t)^{s-1} + t\Big((1+t)^{s-1}a_{s-1} + \ldots + a_0\Big), \qquad (*)$$

where $0 \leq a_0, \ldots a_{s-2}$, $1 \leq a_{s-1}$, s *and* $a_0 + a_1 + \ldots + a_{s-1} + s = n$.

Proof. First we shall show by induction on n that for $n \geq 2$, $|B(n)| = 2^{n-2}$. For $n = 2$ we have $|B(2)| = 1$ which provides the basis for the induction.

We shall now assume $3 \leq n$ and $|B(n-1)| = 2^{n-3}$. From the recurrence formula for $B(n)$ we see that it is enough to show: $\Big(B(n-1) + t\Big) \cap$

$(1 + t)B(n - 1) = \emptyset$. Suppose it is false. Then there exists a polynomial $h(t) \in B(n - 1)$ such that $(1 + t)h(t) - t \in B(n - 1)$.

Let $a \in \mathbb{N}$ be the largest integer such that $(1 + t)h(t) - at \in B(n - a)$. Then $n - a \neq 1$. Indeed, otherwise $(1 + t)h(t) - at = 1$ and $h(t) = a = 1$. Thus $n = 2$, a contradiction!

Therefore $n - a > 1$. From the recurrence formula and the choice of a we deduce that there exists a polynomial $g(t) \in B(n - a - 1)$ such that $(1 + t)h(t) - at = (1 + t)g(t)$. Therefore $a = 0$ a contradiction! This completes the proof of the equality $|B(n)| = 2^{n-2}$ for all $n \geq 2$.

We shall now prove the relation $(*)$ by induction on n. If $2 = n$, $B(2) = \{(1 + t)\}$ and our assertion is true.

Suppose $3 \leq n$ and that our assertion is true for $(n - 1)$. Then

$$g(t) \in B(n) \text{ if } g(t) = (1 + t)\Big[(1 + t)^{s-1} + t\big((1 + t)^{s-1}a_{s-1} + \ldots + a_0\big)\Big],$$

or

$$g(t) = (1 + t)^{s-1} + t\big((1 + t)^{s-1}a_{s-1} + \ldots + (a_0 + 1)\big),$$

where $0 \leq a_0, \ldots + a_{s-2}, 1 \leq a_{s-1}, s$, and $a_0 + a_1 + \ldots + a_{s-1} + s = n - 1$. This proves our theorem. ∎

Acknowledgment

Paulo Ribenboim helped me to escape from the madhouse and made it possible for me to devote my time to things like Poincaré polynomials. To him I owe my utmost respect, gratitude and admiration.

T.Y. Lam read carefully parts of my thesis and made a number of invaluable comments and suggestions for all my further writing. Also a number of useful comments were made by Tony Geramita and Charles Small.

Clive Reis is trying, in vain, to save me from errors in English. If some article in this paper is correct it is entirely due to him.

References

[Ara] Arason, J. (1975). Cohomologische Invarianten Quadratischer Formen. *Journal of Algebra*, **36(3)**, 448–491.

[AEJ 1] Arason, J., Elman, R. and Jacob, B. (1984). The graded Witt ring and Galois cohomology I. *Canad. Math. Soc. Conference Proceedings (edited by Riehm, C. and Hambleton, I.)*, **4**, Quadratic and Hermitian forms, 17–50.

[AEJ 2] Arason, J., Elman, R. and Jacob, B. (1985). Graded Witt rings of elementary type. *Math. Ann.*, **272**, 267–280.

[Bec 1] Becker, E. (1974). Euklidische Körper and Euklidische Hüllen von Körpers. *J. Reine and Angew. Math.*, **268/269**, 41–52.

[Bec 2] Becker, E. (1978). *Hereditarily Pythagorean fields and orderings of higher level*, IMPA, Rio de Janeiro, **29**.

[Ber] Berman, L. (1978). *The Kaplansky radical and values of binary quadratic forms over fields*, Thesis, University of California, Berkeley.

[BNW] Binz, E., Neukirch, J. and Wenzel, G. H. (1971). A subgroup theorem for free products of pro-finite groups. *Journal of Algebra*, **19**, 104–109.

[Brö 1] Bröcker, L. (1977). Über die Anzahl der Anordnungen eines kommutativen Körpers. *Arch. Math.*, **29**, 458–464.

[Brö 2] Bröcker, L. (1984). Spaces of orderings and semialgebraic sets. *Canad. Math. Soc. Conference Proc.*, **4**, *Quadratic and Hermitian forms (edited by Riehm, C. and Hambleton, I.)*, 231–248.

[Cra] Craven, T. (1978). Characterizing reduced Witt rings of fields. *Journal of Algebra*, **53(1)**, 68–77.

[Jac] Jacob, B. (1981). On the structure of Pythagorean fields. *Journal of Algebra*, **68(2)**, 247–267.

[Kul] Kula, M. (1991). Crystal growth and Witt rings. *Journal of Algebra*, **136**, 190–196.

[Lam 1] Lam, T. Y. (1980). *Algebraic theory of quadratic forms*, W.A. Benjamin, Reading, Mass., 2nd printing with revisions.

[Lam 2] Lam, T. Y. (1983). Orderings, valuations and quadratic forms. *CBMS*, **52**, A.M.S., Providence, R.I.

[Lam 3] Lam, T. Y. (1977). Ten lectures on quadratic forms over fields. *Proceedings of Quadratic Form Conference (edited by Orzech, G.)*, **46**, Queen's papers in Pure and Applied Mathematics, Kingston, Ontario, 1–102.

[Mar 1] Marshall, M. (1979). Classification of finite spaces of orderings. *Canad. J. Math.*, **31**, 320–330.

[Mar 2] Marshall, M. (1980). Spaces of orderings IV. *Canad. J. Math.*, **XXXII(3)**, 603–627.

[Mer] Merzel, J. (1982). Quadratic forms over fields with finitely many orderings. *Contemporary Math.*, **8**, 185–229.

[Min 1] Mináč, J. (1986). *Galois groups, order spaces and valuations.* Thesis, Queen's University, Kingston.

[Min 2] Mináč, J. (1986). Stability and cohomological dimension. *C.R. Math. Rep. Acad. Sci. Canada,* **VIII(1)**, 13–18.

[Min 3] Mináč, J. (1986). Poincaré polynomials and ordered fields. *C.R. Math. Acad. Sci. Canada,* **VIII(4)**, 411–416.

[Min 4] Mináč, J. (1987). Stability indices and Poincaré polynomials. *C.R. Math. Acad. Sci. Canada,* **IX(5)**, 253–257.

[MST] Mináč, J. and Sterry, C. (in preparation). Poincaré polynomials which determine their order spaces.

[Sch] Scharlau, W. (1989). Generating functions of finitely generated Witt rings. *Acta Arithmetica,* **LIV**, 51–59.

[War 1] Ware, R. (1979). Quadratic forms and profinite 2-groups. *Journal of Algebra,* **58(1)**, 227–237.

[War 2] Ware, R. (1985). Quadratic forms and pro-2-groups III. *Comm. in Algebra,* **8(13)**, 1713–1736.

Jan Mináč
Department of Mathematics
University of Western Ontario
London, Ontario N6A 5B7
Canada
e-mail 7116_440@uwovax.uwo.ca

Paulo Ribenboim: three aspects of his career

T. M. Viswanathan

University of North Carolina at Wilmington

Distinguished Guests, Ladies and Gentlemen:

This is a unique occasion bringing together a gathering of educators, scientists, professionals and leaders of the larger community. Before this august gathering, it is my privilege and honor to say a few words about the remarkable career of Paulo Ribenboim.

His is a career dedicated to mathematics in its diverse manifestations. If I am asked to summarize his career, I would like to use a trio of phrases: Paulo's global view of mathematics, his goal-oriented methodology, and his international commitment which took him to other centers of learning and research. I would like to say a few words about each one of these three aspects.

His mathematical work is characterized by a global view of our task as educators relating mathematics to teaching and the education of the individual, a global view of our task as researchers, and a global reach out aimed at the mathematical literacy of the non-specialist and the specialist, the the novice and the expert. This view of the place of mathematics in the larger context of education has formed the philosophical basis of his diverse activities.

Secondly, his activities have always been goal-oriented. This penchant for well-defined goals has defined the methodology of his teaching and research. He has always set himself definite goals be it in research, be it in the supervision of the research of his graduate students, be it in teaching. He chose to enlist a very powerful medium to accomplish his goals: The medium of writing, and of print.

Paulo preferred to lecture from handwritten notes, which he prepared with great care. Likewise he trained his students to have their seminars and talks prepared in writing. In research he would write down his results systematically, and demanded the same of his research students. He has been prolific of ideas and his powerful urge for writing translated his ideas

---

Revised version of a speech delivered at the banquet in Honor of Paulo Ribenboim on Thursday August 22, 1991 at Leonard Hall.

into finished products. Eventually writing would make of him a messenger
of mathematics, a messenger of mathematical research, and of scholarship.

Queen's has always emphasized a global view of mathematics. During
the first decade of Paulo's career at Queen's in the sixties, the department
of mathematics was small and a personal touch was brought to bear upon
the department. It was a department teeming with global fellowship, as
students and faculty from almost every part of the world were to be found
in the department. There were students and faculty from Newzealand,
from China, Malaysia and Singapore, from India and Pakistan, from Hun-
gary, Poland, and Romania, from England, France, and Germany, and from
Brazil. A cordial sense of fellowship was actively cultivated through the departmental colloquium. There was a steady stream of weekly visitors from
other Canadian universities, from U.S. and other foreign universities. A
systematic effort was made to cut across the diverse areas of mathematics.
Paulo and Huguette Ribenboim provided a congenial atmosphere in their
home for regular informal gatherings bringing together the non-specialist
and the specialist.

Paulo added yet another dimension to his teaching and research by his
international commitment which took him far and wide to great centers of
learning, to national and international gatherings and centers promoting
research. He has been an enthusiastic expositor of research and curricula
conducive to research. He has a gift for by-passing large chunks of material
and his well-prepared survey talks were accessible to beginners. As a result
he has been avidly sought after by developing institutions. Peter Taylor
once wrote to me about Paulo: "... he [Ribenboim] seems to be one person
who does not seem to change much around here. ..". Actually, the changes
in Ribenboim are seen through his writings, which, in their turn, were
influenced by his travels. I would like to illustrate my remark with an
example.

During the academic year 1963-64 he gave a course on rank-1 valuations with applications to number fields and function fields. As the course
progressed, his handwritten notes were being typed and by the end of
the academic year he had more than 300 pages of typed notes from the
course. During the summer of 1964 he was invited to give a course on
general valuation theory in the Séminaire de Mathématiques Supérieures
at the Univesité de Montréal. His course notes later appeared in that famous series [1]. At Montreal that summer he was profoundly influenced by
Jean Dieudonné who was also giving a course on algebraic geometry [2].
When Paulo came back to Queen's in the fall of 1964, he abandoned his
notes on rank-1 valuations and started instead a seminar on homological
algebra. The following year he went on sabbatical to Boston and regularly
attended David Mumford's classes in algebraic geometry. From then on, he
began to travel regularly. For a while he was commuting to the Univesité
de Montréal and in the seventies and eighties he was well into his annual

semester visits to Paris. While he delivered lectures, he was also regularly sitting in courses and seminars. These travels changed him profoundly. If he was specializing and doing research on the one hand, he also became a messenger of mathematics to large segments of the international scientific community. His book on algebraic numbers [3] and the French book on the arithmetic of fields [4] are masterpieces as textbooks. His book on Fermat's Last Theorem [5] caters to the specialist and the research student. On the other hand, his latest Book of Prime Number Records [6] reaches out to the non-specialist as well as the specialist. The success of this latest book can be measured by the fact that the book is already on its way to the third edition, and has been translated into other languages, French, Japanese, and Spanish among them. A pocket-book edition [7] has just appeared. A second edition of his book on Fermat's Last Theorem has already appeared in Japanese.

Summing up one would say that Paulo Ribenboim adapted himself very well to a situation reminiscent of the Red Queen's Race described by Lewis Carroll [8] in the following words: "Now, here, you see, it takes all the running you can do, to keep in the same place." Given the global character of mathematics and of the modern world, the writings of Paulo had to change so much for him to remain "unchanged".

Thank you.

Acknowledgment: The author is thankful to the referee for many positive remarks, and for many corrections in style and language.

T.M. Viswanathan

REFERENCES

[1]   *Théorie des valuations.* Séminaire de Mathématiques Supérieures, 9 (été 1971). Montreal: Les presses de l'université de Montréal, deuxième édition, 1968.

[2]   Jean Dieudonné. *Fondements de la géométrie algébrique moderne.* Montreal: Les presses de l'université de Montréal, 1968.

[3]   *Algebraic Numbers. Pure and Applied Mathematics*, 27. New York: Wiley-Interscience, 1972. Ann Arbor, MI.: Bks Demand, Univ. Microfilms International, 300 N. Zeeb Rd.

[4]   *L'Arithmétique des Corps.* Paris: Hermann, 1972.

[5]   *Thirteen Lectures on Fermat's Last Theorem.* New York: Springer, 1979.

[6]   *The Book of Prime Number Records.* New York: Springer, 2nd edition, 1989.

[7]   *The Little Book of Big Primes.* New York: Springer, paper ed., 1991

[8]   Lewis Carroll. *Through the Looking Glass* in *Alice in Wonderland.* Ed. Donald J. Gray. New York: W.W. Norton & Co., 500 Fifth Avenue, N.Y. 10110, 1971.

T. M. Viswanathan
Department of Mathematics
University of North Carolina at Wilmington
Wilmington, NC 28403
USA
e-mail viswanathant@uncwil.edu

Response

Paulo Ribenboim

Queen's University at Kingston

I am very happy with all my friends who are here, and who have come to Kingston to hear many outstanding speakers, and perhaps also a little bit to comfort me as I approach my retirement. So I will begin with a warning: I am not yet retiring! In fact, I will only do so two years from now. I hope not everyone here is saddened by that news!

Looking back to all my years of mathematical activity, I see little merit in what I have done. I will explain why.

I was indeed blessed—having started in Brazil—with three of the main conditions which are required in Mathematics.

The first condition is **appetite**. You see—now that we all ate and drank to our contentment—that we feel no urge to do mathematics. We would have to whet our appetite.

In Brazil, I was always tempted by the very flavourful foods, the exotic and tasty fruits. I think of *moqueca de siri, vatapá,* and *churrasco,* and of fruits such as *fruta do conde, avocado, banana ouro, manga, sapoti* .... All of these excite the appetite.

So, in another plane, mathematicians must have the appetite for tasteful and spicy problems—not dull ones. But perhaps I should now drop this topic, since we have all just finished eating.

The second condition is **imagination**. In the 1960's, Smale frequented the Copacabana and Ipanema beaches while working on solving Poincaré's problem. And, being a courageous person, he bravely reported it to the NSF—and got into big trouble. But I understand that all that he was looking for on the beaches was to excite his imagination.

In front of the voluptuous beauties, scantily dressed, your imagination goes by leaps and bounds. We mathematicians also need, in the face of a problem full of secrets, to imagine what is beneath the surface—I might say, to mentally undress our problem.

I have often been to the beaches in Brazil, and there I developed my imagination.

The final condition is **ignorance**. Ignorance is almost synonymous with youth; perhaps I am no longer young—though I am not quite senile yet—but I still have plenty of ignorance.

This is how I began—a B. Sc. in Rio de Janeiro, at age 20, in 1948 (please don't subtract!). It was an excellent diploma in ignorance. I remember that when I first came to Nancy, to work with Dieudonné in 1950, he said right away: "You know nothing," and gave me Bourbaki's *Algèbre I* to study. Of course, he did not know that I had already studied carefully his own Brazilian notes on commutative fields...But I agree that this is not much.

It is thanks to my ignorance that I undertook to write books—my goal was to become ignorant at a higher level.

In conclusion, it has been my good luck to receive in large amounts these three main ingredients for doing mathematics: **appetite**, **imagination**, and **ignorance**, and so my only merit is to be the friend and admirer of all excellent practitioners of Mathematics.

Thank you all for being here!

J